29 Springer Series in Chemical Physics

Edited by J. P. Toennies

Springer Series in Chemical Physics

Editors: Vitalii I. Goldanskii Fritz P. Schäfer J. Peter Toennies

Volume 40 **High-Resolution Spectroscopy of Transient Molecules**
Editor: E. Hirota

Volume 41 **High Resolution Spectral Atlas of Nitrogen Dioxide 559–597 nm**
By K. Uehara and H. Sasada

Volume 42 **Antennas and Reaction Centers of Photosynthetic Bacteria**
Structure, Interactions, and Dynamics
Editor: M. E. Michel-Beyerle

Volume 43 **The Atom-Atom Potential Method for Organic Molecular Solids**
By A. J. Pertsin and A. I. Kitaigorodsky

Volume 44 **Secondary Ion Mass Spectrometry SIMS V**
Editors: A. Benninghoven, R. J. Colton, D. S. Simons, and H. W. Werner

Volumes 1–39 are listed on the back inside cover

U. G. Pirumov G. S. Roslyakov

Gas Flow in Nozzles

With 149 Figures

Springer-Verlag
Berlin Heidelberg New York Tokyo

Professor Ul'yan G. Pirumov
Moscow Aviation Institute, SU-125871 Moscow, USSR

Professor Gennadi S. Roslyakov, Ph.D.
Moscow University, SU-119899 Moscow, USSR

Series Editors

Professor Dr. Fritz Peter Schäfer
Max-Planck-Institut für
Biophysikalische Chemie
D-3400 Göttingen-Nikolausberg
Fed. Rep. of Germany

Professor Vitalii I. Goldanskii
Institute of Chemical Physics
Academy of Sciences
Vorobyevskoye Chaussee 2-b
Moscow V-334, USSR

Professor Dr. J. Peter Toennies
Max-Planck-Institut für Strömungsforschung
Böttingerstraße 6–8
D-3400 Göttingen
Fed. Rep. of Germany

Title of the original Russian edition: *Techeniya gaza v soplakh*

ISBN 978-3-642-86792-7 ISBN 978-3-642-86790-3 (eBook)
DOI 10.1007/978-3-642-86790-3

Library of Congress Cataloging-in-Publication Data. Pirumov. Ul'îan Gaĭkovich. Gas flow in nozzles. (Springer series in chemical physics ; 29) Translation of: Techeniya gaza v soplakh. Bibliography: p. Includes index. 1. Nozzles–Fluid dynamics. I. Roslîakov, Gennadiĭ Stepanovich. II. Title. III. Series: Springer series in chemical physics ; v. 29. TL567.N6P5713 1986 620.1'064 85-24966

Softcover reprint of the hardcover 1st edition 1986

2153/3130-543210

Preface

This monograph treats, for the first time, major aspects of gas dynamics of nozzles from a general point of view. Its outstanding feature is the presentation of the modern theory of gas flows and modern analytical and numerical methods, together with numerous examples of practical applications. At the same time, quite diverse physico-chemical processes, such as dissociation and recombination, relaxation of vibrational degrees of freedom, two-phase flows with phase transformations and electromagnetic influences, are considered.

The material is presented in such a way as to help the reader to use numerous methods and approaches, not only for the study of gas flows in nozzles, but also for the investigation of a wide variety of problems of physical gas dynamics in different areas of application. The number of applications which may benefit from the use of the methods and results presented in this book is constantly growing. Theoretical, numerical and analytical methods of physical gas dynamics of internal flows may be, and are nowadays, applied to solving the problem of preventing pollution of the air basin with toxic substances. These methods make it possible to describe the formation and transformation of toxic components in the vapour generators of thermal power plants, internal combustion engines and various metallurgical installations. The methods of physical gas dynamics may be used in meteorology and powder metallurgy to create ultradispersed media and predict their properties.

Some chapters of the monograph are based upon lecture courses delivered by the authors at Moscow University, Moscow Physico-Technical Institute and Moscow Aviation Institute. The first chapters may therefore be used in universities and higher technical educational establishments for students and postgraduates studying gas dynamics and applied mathematics.

As a whole, the monograph is adressed to those who are interested in the problems of gas dynamics of nozzles, as well as to engineers and research scientists working in the fields of gas dynamics and applied mathematics.

The authors are glad to use this opportunity to express their profound gratitude to Acad. G.I. Petrov for his constant attention and his simulation of many studies included in the book. They wish to thank A.M. Ovsyannikov, who has taken part in writing Sects. 3.1 and 6.3; the authors are also much indebted to D.A. Mel'nikov, G.F. Telenin, A.A. Sergienko, E.A. Ashratov, Yu.D. Shmyglevskii, O.S. Ryzhov, L.E. Sternin, A.N. Kraiko and A.P. Tishin for numerous useful discussions. They thank also V.A. Khokhryakov for translating the book.

U.G. Pirumov
G.S. Roslyakov

Moscow, October 1985

Contents

1. Introduction

Nozzles are extensively used for solving various scientific and technological problems. They were first utilized by the Swedish engineer Laval in stem turbines. Progress in highspeed aerodynamics made it necessary to develop laboratory means of studying hyper-velocity flows which cannot be obtained without nozzles. Since then, the latter have been used for generating uniform velocity fields. A nozzle is a necessary component of any jet engine whose performance depends to a large degree on the efficiency of the nozzle.

The last decade witnessed another important application of nozzles – their use for generating the working media of MGD (magnetogasdynamic) installations and gasdynamic lasers. In the latter case nozzles proved to be a convenient means of obtaining large pressure gradients allowing the inverse population of vibrational levels of molecules to be attained. Still another sphere in which nozzles have recently been applied is their use as "gasdynamic" lenses. Large longitudinal and zero lateral density gradients in plane nozzles allow the deflection of a light beam without causing substantial divergence. Thus, the range of practical applications of nozzles is quite extensive.

Physical processes accompanying nozzle gas flows display a wide variety of phenomena, including nonequilibrium excitation of vibrational degrees of freedom of molecules, processes of mass, momentum and energy exchange between gas and liquid/solid phases and interaction of moving paticles with magnetic and electric fields.

No less diverse are the methods of solving equations describing nozzle flows. The problem is complicated not only by the large number of equations necessary to describe nonequilibrium processes but also because the latter differ for different parts of a nozzle. In the case of a stationary flow the system of partial differential equations is elliptic in the subsonic region, parabolic in the transonic region and hyperbolic in the supersonic region. Thus, in studying nozzle flows one deals with many areas of modern physics, and in solving the governing equations one must cope with major types of equations of mathematical physics. This situation has resulted in the formation of a special division of gas dynamics called "physical gas dynamics of internal flows".

Despite a large number of periodicals and monographs dedicated to special aspects of nozzle gas flows, there has been, up to now, no book in the world literature presenting from a single point of view the general theory of nozzle gas flows, methods of solving equations and major particularities of such flows associated with the above physical processes. The authors hope that this monograph will help to fill the gap. Its purpose is to acquaint the reader not only with fundamentals of gas dynamics and modern analytical and numerical methods, but also with some applications and their results. The authors have tried, as far as possible, to supply physical constants and tables necessary for the study of certain problems and data which enable some engineering estimates to be made. Besides original works by the authors, the monograph is based upon recent results obtained by other researchers. Soviet and Western publications have been used equally; the list of references includes work up to 1984.

Let us briefly consider the general contents of the monograph. The Chaps. 2-4 and some sections of Chap.6 are of rather a general character and can be used in the study of diverse problems of physical gas dynamics, such as external flows and jet flows. They contain fundamental notions of gas dynamics, analytical methods, modern numerical methods and some fragments of physical gas dynamics incorporating kinetics of nonequilibrium processes. This first part of the monograph can be used as a textbook in universities and higher educational establishments, by students and postgraduates studying gas dynamics and applied mathematics.

The other part of the book, including Chaps.5 and 7 and the larger part of Chap.6, deals with applications. It mostly demonstrates applications of numerical methods to solving practical problems and illustrates major qualitative and quantitative dependences. Here, a lot of experimental data, many physical constants and tables and graphs, which allow some calculations to be carried out, are presented.

We now consider the book in detail.

Chapter 2 gives an account of basic equations for the general case of three-dimensional nonstationary gas flow with nonequilibrium physico-chemical transformations. Some special cases of the general system of equations describing ideal gas flows without physico-chemical transformations, gas flows with nonequilibrium chemical reactions and two-phase flows are considered. Special attention is paid to characteristics of equations of gas dynamics, since these are extremely important for the understanding of gas-dynamic phenomena; the direct and the inverse problems of the theory of nozzles are formulated. Some elementary theories (one-dimensional theory, source flow, Prandtl-Meyer flow) which allow one to study major qualitative properties of nozzle gas flows are discussed in Sect.2.2. The last section of Chap.2 is dedicated to variational problems of the theory of nozzles. It has been included into Chap.2 because, despite the obvious complexity of solving the general variational problem, some quite refined results may be obtained in a closed form for a certain class of problems, thus exposing some important qualitative features of nozzle gas flows. Such results have been obtained mostly for two-dimensional (plane and axisymmetric) flows.

In Chap.3 numerical methods for studying nozzle gas flows are discussed. The numerical methods included in the chapter are quite versatile and widely used to explore nozzle gas flows using digital computers. The reader is supposed to be acquainted with fundamentals of the method of grids. Section 3.1 deals with a method of solving relaxation equations which belong to the so-called "rigid" equations describing nonequilibrium processes. A parametric implicit difference scheme is presented. Some widely used methods of calculating two-dimensional stationary supersonic nozzle gas flows are discussed in Sect.3.2. Numerical schemes of the method of characteristics are considered including the cases of flows with equilibrium and nonequilibrium reactions. Two finite-difference methods are discussed, adjusted to calculating discontinuous gas flows and belonging to the class of the so-called shock-smearing methods. Section 3.3 describes methods of calculating three-dimensional flows in nozzles, among them being the method of characteristics of bicharacteristic type and a three-dimensional analogy of the shock-smearing method presented in Sect.3.2. Section 3.4 contains a description of an original numerical method of solving the inverse problem of the theory of nozzles which is quite versatile and ensures high precision. It allows flows in sub-, trans- and supersonic regions of a nozzle to be simultaneously calculated. The last section of this chapter deals with methods of solving the direct problem using the method of stabilization. Two variations of the method are presented, both based upon the divergent form of equations of gas dynamics.

Chapter 4 is dedicated to a discussion of analytical methods of the theory of nozzles. Here, among others, the source-sink method for plane, axisymmetric and three-dimensional flows is presented. By choosing a system of sources and sinks appropriately one can construct a flow in a channel of a given shape. Analytical solutions of the inverse problem of the theory of nozzles for incompressible fluid are considered for plane and axisymmetric flows. An analytical solution obtained by expansion in the stream function series is also considered. It has been obtained for a general case of three-dimensional flow and is a generalization of classic results in the field. The plane and axisymmetric solutions represent special cases of this general solution. An original solution for the case of two-phase flows is also presented. The same chapter contains descriptions of asymptotic methods for the transonic part of a nozzle. Gas flow in this part of a nozzle may be studied by means of the small perturbation method, which allows one to obtain a solution for the velocity potential, and by means of a method based upon series expansion in the vicinity of a rectilinear sonic line. Section 4.4 discusses solution in the neighbourhood of the infinite point in the subsonic region. This solution allows one to formulate boundary conditions for the direct and inverse problems of the theory of nozzles. The last section of this chapter deals with the small perturbation method for the investigation of flows similar to those generated by a source. The theory has been developed for plane, axisymmetric and three-dimensional flows. Much attention is paid to comparison of analytic and numerical methods.

Chapter 5, which deals mostly with applications, considers peculiarities of gas flows in nozzles of jet engines, methods of profiling sub-, trans- and supersonic parts of nozzles, methods of calculating impulse losses and methods of simulating parameters at a nozzle exit section. Section 5.1 describes flow in sub- and transonic parts of a nozzle; this depends on geometric characteristics and the adiabatic exponent. Patterns of streamlines, isotaches, limiting characteristics and sonic lines are presented. Special attention is paid to local stagnation zones. Values of the flow rate coefficient and critical pressure drop are presented, as well as distributions of Mach number for transonic regions of various configurations. Modern methods of profiling supersonic parts of jet engine nozzles are discussed in Sect.5.2. Here some families of nozzles are considered, including nozzles with uniform, variational and symmetric characteristics at the exit section. Graphic and tabular data are presented allowing impulse losses due to divergence and friction to be calculated. Methods of calculating the thrust of a nozzle at off-design regimes are described. The last section of the chapter is dedicated to the major principles underlying the choice of an optimum jet engine nozzle, which realizes maximum specific thrust at given isoperimetric conditions.

Nozzle gas flows with physico-chemical transformations are discussed in Chap.6. It is one of the most important chapters of the monograph because it deals with modern and rather complicated theoretical and numerical methods of physical gas dynamics and presents results of solving a number of practical problems. Equilibrium and frozen isentropic flows are considered in Sect.6.1. Here a method is presented taking into account real gas properties which become important at high pressures. Sections 6.2 and 6.3 deal with flows with nonequilibrium chemical reactions and with nonequilibrium excitation of vibrational degrees of freedom. Here, together with a presentation of the general theory, we give methods of calculation and data for systems of reactions, examples of calculating flows in plane and axisymmetric nozzles, as well as data concerning specific impulse losses in jet engine nozzles due to the nonequilibrium flow of chemical reactions and relaxation of vibrational degrees of freedom. Results of studying two-phase flows in nozzles are presented in Sect.6.4, which contains general theory, modern methods of calculations and values of physical constants both for flows with and without phase transformations. Results of calculating one- and two-dimensional flows are presented. For the case of two-dimensional flows the peculiarities of the trajectories of particles in nozzles are described. The interactions of particles in nozzles, their coagulation and breaking are considered. Considerable attention is paid to flows with nonequilibrium homogeneous condensation and the determining parameters are introduced. Flows of conductive gases in nozzles in the presence of electromagnetic fields are considered in Sect.6.5. It contains basic equations, methods of calculation and results obtained at certain values of the interaction parameters. Section 6.6 has been prepared especially for the English edition. Here, general theory and some results of studying multi-layered flows with motion of gases which have different physical properties are presented. The flows have been studied in the framework of one- and two-dimensional formulations, ignoring viscosity, diffusion and mixing of layers.

Chapter 7 deals with nozzles of special configurations, such as various annular nozzles, aerodynamic and conical nozzles. Some results of up-to-date rare investigations of three-dimensional nozzle gas flows are presented. In particular, swirling flows are considered. Much attention is paid to slightly perturbed flows and the corresponding perturbations of the side forces and moments. Section 7.5 presents calculated and experimental results of studying viscous nozzle flows at low Reynolds numbers when the influence of viscosity is not restricted by the thin boundary layer near a nozzle's wall but embraces the whole of the flow.

2. General Theory of Nozzle Gas Flows

The basic equations for the general case of a three-dimensional nonstationary gas flow with nonequilibrium physico-chemical transformations are presented in the opening part of the chapter. Some special cases of the general system of equations, describing ideal gas flows without physico-chemical transformations, gas flows with nonequilibrium chemical reactions and two-phase flows are considered. Special attention is paid to characteristics of gasdynamic equations, since these are extremely important for understanding gasdynamic phenomena, and the direct and inverse problems of the theory of nozzles are formulated. Some elementary theories (the one-dimensional theory, the source flow, the Prandtl-Meyer flow) which allow the study of major qualitative particularities of nozzle gas flows are discussed in Sect.2.2. The last section is devoted to variational problems in the theory of nozzles. It has been included in Chap.2 because, despite the obvious complexity of solving the general variational problem, some quite refined results may be obtained in closed form for a certain class of problems, thus exposing some important qualitative features of nozzle gas flows. Such results have been obtained mostly for two-dimensional (plane and axisymmetric) flows.

2.1 Basic Equations and Problem Formulation

The state of a gas with known thermodynamic properties flowing in a nozzle is defined by means of its velocity, density and pressure as functions of coordinates and time. The functions can be obtained from a system of gasdynamic equations representing general laws of mass, momentum and energy conservation expressed in a differential form. The system is closed by means of thermal and caloric equations of state. However, nozzle gas flows are usually accompanied by various nonequilibrium processes which can be described if the corresponding kinetic and relaxation relations are added to the gasdynamic equations. Additionally, terms accounting for the influence of nonequilibrium processes upon the gasdynamic parameters should be included in the equations. The nonequilibrium processes taking place in nozzle gas flows are quite diverse. Usually investigators deal with nonequilibrium excitation of vibrational degrees of freedom, nonequilibrium dissociation and recombination and nonequilibrium motions of liquid and solid particles, while nonequilibrium condensation and evaporation take place.

2.1.1 Basic Equations

Let us write down differential equations for mass, momentum and energy conservation for the general case of unsteady flow in multiphase media with relaxation processes and interaction of phases taking place. The equations will be written down in Euler coordinates which are commonly used to describe nozzle gas flows. The interested reader will find derivations of the equations in, for example, [2.1-3].

Homogeneous and heterogeneous mixtures will be considered as multivelocity continua with interpenetrating motion of components and exchange of mass, momentum and energy. A multivelocity continuum will be represented as a sum of N continua each referring to its component and occupying the same volume which is occupied by the mixture. Each continuum will be characterized by its average density ρ_i (the mass of the ith component divided by the volume occupied by the mixture), velocity $\mathbf{W}_i$, internal energy e_i and stress tensor $\mathbf{\Pi}_i$. Besides average density, true density ρ_i^0 will be introduced equal to the ith phase mass divided by the volume it occupies. By components we shall mean gas components corresponding to different molecules and their various quantum states (for instance, vibrational degrees of freedom of molecules) and particles of different sizes possessing various physical properties. Thus N densities, N velocities, etc. will be defined at every point of the volume occupied by the mixture.

The equation of mass conservation is written down for every component in the form

$$\frac{d\rho_i}{dt} + \rho_i \nabla \mathbf{W}_i = \sum_{k=1}^{N} R_{ik} \quad , \qquad (i = 1,2,\ldots,N) \quad , \tag{2.1}$$

where ∇ is a gradient or divergence operator and R_{ik} is the variation of the ith component mass due to the kth component (for instance, due to chemical reactions, condensation or transition of molecules from one quantum state to another) referred to unit volume and time. It follows from the law of mass conservation for various physico-chemical processes that $R_{ik} = -R_{ki}$. Obviously $R_{ii} = 0$.

The equation of motion for a component is

$$\rho_i \frac{d\mathbf{W}_i}{dt} = \nabla^n \Pi_i^n + \rho_i \mathbf{F}_i + \sum_{k=1}^{N} (\mathbf{P}_{ik} - R_{ik}\mathbf{W}_i) \quad , \quad (i = 1,2,\ldots,N) \quad , \tag{2.2}$$

d/dt being the substantial derivative associated with the motion of the ith component,

$$\frac{d}{dt} = \frac{\partial}{\partial t} + \mathbf{W}_i \nabla = \frac{\partial}{\partial t} + W_i^n \frac{\partial}{\partial x^n} = \frac{\partial}{\partial t} + W_i^n \nabla^n \quad .$$

In these formulas summation is performed only for superscripts denoting the coordinate axes. The first term on the right side of (2.2) corresponds to the external surface forces and the second one corresponds to external mass forces characterized by the vector $\mathbf{F}_i$ and referring to the ith component. $\mathbf{P}_{ik}$ is the change of the ith component momentum due to the kth component, referred to unit volume and time. The law of momentum conservation in the case of various physico-chemical transformations and interactions between components of a mixture results in $\mathbf{P}_{ik} = -\mathbf{P}_{ki}$. Obviously $\mathbf{P}_{ii} = 0$.

The energy equation for a component of a mixture is

$$\begin{aligned} \rho_i \frac{d}{dt} (e_i + W_i^2/2) &= \nabla(\mathbf{L}_i - \mathbf{Q}_i) + \rho_i \mathbf{F}_i \mathbf{W}_i \\ &+ \sum_{k=1}^{N} \left[E_{ik} - R_{ik}(e_i + W_i^2/2)\right] \quad , \quad (i = 1,2,\ldots,N) \quad . \end{aligned} \tag{2.3}$$

Here the vector $\mathbf{L}_i$ characterizes the work done by external surface forces, $\mathbf{Q}_i$ is the heat flow vector and E_{ik} is the change of energy of the ith component due to the kth component, referred to unit volume and time. It follows from the law of energy conservation that $E_{ik} = -E_{ki}$ and $E_{ii} = 0$.

The system of equations (2.1-3) is closed by means of thermal and caloric equations of state which allow (if local thermodynamic equilibrium is assumed when the temperature T_i is defined at all points) the stress tensor Π_i and the internal energy e_i to be expressed through the other parameters of the mixture and some physico-chemical variables. To solve particular problems one also needs some phenomenological equations to determine the parameters of mass, R_{ik}, force, $\mathbf{P}_{ik}$, and energy, E_{ik}, interactions between phases. If the departure from equilibrium is insignificant, the necessary relations for R_{ik}, $\mathbf{P}_{ik}$ and E_{ik} can be obtained by means of Onsager's principle and the postulation of linear relations for thermodynamic fluxes (for instance, heat exchange, friction between phases, phase transition intensity, etc.) [2.1]. If chemical reactions take place, then the necessary relations for R_{ik} are defined by chemical kinetics.

The partial differential equations (2.1-3) are quite general and their solution without some simplifying assumptions is an extremely difficult task. At the same time, in the description of nozzle gas flows, it is often possible to neglect mass forces, external heat sources, viscosity, heat conduction and diffusion. Viscosity and heat conduction, which play an important role near nozzle walls, can, as a rule, be considered in the framework of the boundary layer approximation. However, viscosity, heat conduction and diffusion must be taken into account when interaction of phases is considered.

Let us formulate some further assumptions usually made while describing nozzle gas flows. At present, when describing multiphase flows in nozzles, investigators widely use models of two-fluid and multifluid media in which, instead of a real flow of a dusty gas, an interpenetrating flow of two or more media—gas proper and "gas" of particles—is considered. In this case all flow parameters represent values obtained by means of averaging over a small volume containing a large number of foreign particles. The following assumptions are made for the gas proper:

a) The gas is a mixture of perfect gases each governed by its own thermal and caloric equations of state. As a rule, departures from perfect gas properties (such as triple collisions, van der Waals' forces, etc.) are ignored.

b) Translational and rotational degrees of freedom of molecules are supposed to refer to a state of thermodynamic equilibrium. Nonequilibrium excitation of vibrational degrees of freedom, chemical reactions in the gaseous phase, condensation, etc. are allowed. Usually the assumption of the equilibrium of translational and rotational degrees of freedom of molecules is fulfilled quite precisely because to acquire equilibrium states, molecules have to collide only 10 to 100 times.

c) All gas components of a mixture possess equal velocities of directed macroscopic motion because mutual diffusion of components is not considered.

As for the "gas" of particles, the following assumptions are made:

a) Generally, a component of the "gas" of particles with known thermal and calorical equations of state may correspond to a component of the gaseous phase. However, there are cases when it is not related to any component of the gas proper.

b) All the particles are spherical and can be sorted into a finite number of groups, each containing particles of the same size.

c) The distances over which flow-characteristics change significantly are much larger than the distances between particles.

d) Pressure developed by the "gas" of particles due to their chaotic motion is negligible compared to the pressure of the gas proper.

Now we shall consider some special cases of Eqs. (2.1-3) keeping in mind the assumptions formulated above.

If a mixture consists of N gaseous components only, then neglecting viscosity, heat conduction and diffusion, one gets

$$\mathbf{W} = \mathbf{W}_i , \quad \rho = \sum_{i=1}^{N} \rho_i , \quad p = \sum_{i=1}^{N} p_i \quad ,$$

where $\mathbf{W}$, ρ and p are velocity, density and static pressure of the mixture respectively. Summing over all i equations in (2.1-3) we obtain the usual equations of continuity, motion and energy for a single-velocity medium:

$$\frac{d\rho}{dt} + \rho\nabla\mathbf{W} = 0 \quad , \tag{2.4}$$

$$\rho \frac{d\mathbf{W}}{dt} + \nabla p = 0 \quad , \tag{2.5}$$

$$\rho \frac{d}{dt} (h + W^2/2) = \frac{\partial p}{\partial t} \quad , \tag{2.6}$$

where $h = \Sigma_{i=1}^{N} h_i\rho_i/\rho$ is the enthalpy referred to unit mass of mixture and the h_i are the unit mass enthalpies of individual components.

The system (2.1-3) is closed by means of equations of state which, in the case of a mixture of perfect gases are

$$h_i = h_i(T) \ , \quad p_i = R\rho_i T/\mu_i \ , \quad p = R\rho T/\mu \quad .$$

Here R is the universal gas constant, μ_i is the molecular weight of the ith component and $\mu = \Sigma_{i=1}^{N} \mu_i p_i/p$ is the molecular weight of the mixture.

If reactions take place, then equations similar to (2.1) which allow ρ_i to be found must be added, where the R_{ik} are now the rates of reactions. In this case there is no need to use (2.2,3) for each component because values of e_i and p_i can be found from corresponding equations of state using the known common pressure, temperature and densities of individual components of the mixture. Note that upon the introduction of mass concentrations

$$\alpha_i = \rho_i/\rho \ , \quad \sum_{i=1}^{N} \alpha_i = 1 \ ,$$

(2.1), with the help of (2.4), may be transformed to

$$\rho \frac{d\alpha_i}{dt} = \sum_{k=1}^{N} R_{ik} \ , \quad (i = 1,2,\ldots,N) \ . \tag{2.7}$$

If no reactions take place in the mixture, $R_{ik} = 0$, $\alpha_i = \text{const}$ and the system (2.4-6) describes an ideal gas with a constant specific heat ratio.

Later we shall find it useful to have an equation for entropy generation which can be derived from (2.4-7) and the first law of thermodynamics.

Let the system contain small elements of mass in the state of local equilibrium, for which the local entropy is the same function of T, ρ and α_i as in the state of total equilibrium, though the system itself may depart from the state of equilibrium [2.1]. In particular, the formula of equilibrium thermodynamics for the entropy of an element of mass along its path,

$$T \frac{dS}{dt} = \frac{de}{dt} + p \frac{d}{dt}\left(\frac{1}{\rho}\right) - \sum_{k=1}^{N} \chi_k \frac{d\alpha_k}{dt} \ , \tag{2.8}$$

remains valid, where S is the entropy of a mixture taken as the sum of the entropies of individual components and χ_k is the thermodynamic or chemical potential of the kth component (Gibbs' partial specific function). Using (2.5,6), we get the equation for entropy generation,

$$T \frac{dS}{dt} = - \sum_{k=1}^{N} \chi_k \frac{d\alpha_k}{dt} \ . \tag{2.9}$$

In the general case of a viscous, heat-conducting, multivelocity medium with external forces and heat sources, the right side of (2.9) is a sum of products of thermodynamic forces with corresponding fluxes of various tensor dimensions. Later we shall write down specific relations for a number of special cases. If the dependence of entropy on other thermodynamic parameters is known [e.g., for an ideal gas $S = \ln(p\rho^{-\gamma})$] then (2.9) does not provide additional information compared to (2.5,6) because it has been derived with the aid of the latter equations. Therefore, considering specific problems, one can use either (2.5,6) or, instead of one of them, (2.9).

Two limiting cases of isentropic flows — an equilibrium flow and a frozen flow — are of principal significance. For the former, characteristic times of relaxation processes are much smaller than the characteristic gasdynamic time; for a frozen flow they are much longer. Parameters characterizing a relaxation process in the case of an equilibrium flow can be found from thermodynamic relations (the latter are obtained by equating the right sides of the corresbonding relaxation equations to zero); in the case of a frozen flow they are supposed to be constant and equal to their values at a characteristic point in the flow.

If the mixture consists of a nonreactive gas ($R_{ik}=0$) and uniform particles, if condensation and evaporation are absent, if the volume occupied by particles is ignored and the above-mentioned assumptions, referring to a "gas" of particles, are taken into consideration, then (2.1-3) can be reduced to

$$\frac{d\rho}{dt} + \rho\nabla\mathbf{W} = 0 \quad , \quad \rho\,\frac{d\mathbf{W}}{dt} + \nabla p + \rho_s\mathbf{f} = 0 \quad , \tag{2.10}$$

$$\rho\,\frac{d}{dt}\,(h + W^2/2) + \rho_s\,\frac{d}{dt}\,(e_s + W_s^2/2) = \frac{\partial p}{\partial t} \quad , \tag{2.11}$$

$$\frac{d\rho_s}{dt} + \rho_s\nabla\mathbf{W}_s = 0 \quad , \tag{2.12}$$

$$\frac{d\mathbf{W}_s}{dt} = \mathbf{f} \ , \quad \frac{de_s}{dt} = q \quad . \tag{2.13}$$

Here $\mathbf{f}$ is the force of interaction of the gas proper and the "gas" of particles, referred to unit mass of the "gas" of particles, and q is the heat flow between gas and particles (parameters of particles are denoted by the subscript s). The system of equations is closed by means of equations of state of the gas proper and the "gas" of particles which in this case are

$$h = h(p,T) \ , \quad p = R\rho T/\mu \ , \quad e_s = e_s(T) \quad .$$

Here μ is the molecular weight of the gas proper. We shall also need additional relationships describing the dependence of the force of interaction and the heat flow on parameters of the gas and particles.

2.1.2 Characteristics

Characteristic surfaces play an important role in gasdynamic problems in general and in the study of nozzle gas flows in particular. The general theory of characteristics is treated in many works (e.g., [2.4-6]). Therefore, we shall present here only major ideas and definitions nedded for further discussion. Following the pattern of the references above, let us consider a system of quasi-linear equations,

$$\sum_{j=1}^{m} \sum_{k=1}^{n} a_{ijk} \frac{\partial u_j}{\partial x_k} = b_i \quad , \quad (i = 1,2,\dots,m) \quad , \tag{2.14}$$

where a_{ijk} and b_i are functions of x_k and u_j. We rewrite (2.14) in the form

$$\sum_{j=1}^{m} (\mathbf{a}_{ij}\nabla)u_j = b_i \quad , \quad (i = 1,2,\dots,m) \quad , \tag{2.15}$$

where $\mathbf{a}_{ij}$ are vectors with components $a_{ij1}, a_{ij2},\dots, a_{ijn}$.

Now we shall consider a linear combination of equations (2.15) with coefficients $\omega_1,\omega_2,\dots,\omega_m(\Sigma_{i=1}^{m} \omega^2_i \neq 0)$:

$$\sum_{i=1}^{m} \omega_i \sum_{j=1}^{m} (\mathbf{a}_{ij}\nabla)u_j = \sum_{i=1}^{m} \omega_i b_i \quad \text{or} \tag{2.16}$$

$$\sum_{j=1}^{m} (\mathbf{A}_j\nabla)u_j = B \quad , \quad \text{where} \tag{2.17}$$

$$\mathbf{A}_j = \sum_{i=1}^{m} \omega_i \mathbf{a}_{ij} \quad , \quad B = \sum_{i=1}^{m} \omega_i b_i \quad . \tag{2.18}$$

Let us try to choose ω_i in such a way that all vectors $\mathbf{A}_j$ lie in the same plane (or hyperplane, if $n > 3$). If such ω_i do exist, then the plane containing vectors $\mathbf{A}_j$ is called a characteristic plane, its normal $\mathbf{n}$ is a characteristic normal and (2.17) is a characteristic relation corresponding to the fixed normal. Relation (2.17) contains differentiation only in the direction of vectors $\mathbf{A}_j$ which lie in the same plane. Thus, if $\mathbf{n}$ is a characteristic normal, then the following conditions must be satisfied:

$$\mathbf{n}\mathbf{A}_j = 0 \quad , \quad (j = 1,2,\dots,m) \quad , \quad \text{or} \tag{2.19}$$

$$\mathbf{n} \sum_{i=1}^{m} \omega_i \mathbf{a}_{ij} = \sum_{i=1}^{m} \omega_i (\mathbf{n}\mathbf{a}_{ij}) = 0 \quad . \tag{2.20}$$

The system (2.20) is a linear homogeneous system of equations for ω_i. It has a nontrivial solution if the determinant of its matrix equals zero:

$$\begin{vmatrix} na_{11} & na_{21} & \cdots & na_{m1} \\ na_{12} & na_{22} & \cdots & na_{m2} \\ \cdots & \cdots & \cdots & \cdots \\ na_{1m} & na_{2m} & \cdots & na_{mm} \end{vmatrix} = 0 \; . \tag{2.21}$$

This equation is called a characteristic equation of the system (2.14). It is an algebraic homogeneous m-order equation for the vector **n** components $n_1, n_2, \ldots, n_n$ which, hence, can be determined to an approximation including a constant multiplier. Thus, (2.21) at every point P defines a cone (hypercone) of order m, which can degenerate or decompose into a number of separate cones. Every plane passing through the point P and perpendicular to one of the normals which form the cone, is a characteristic plane. The envelope of the family of characteristic planes is called a characteristic surface. Thus through every point P, infinitely many characteristic surfaces pass, all of them tangent to the characteristic cone at the point. The envelope of these surfaces is called a characteristic conoid. The lines along which characteristic surfaces contact the conoid are called bicharacteristics.

If a characteristic normal defined by (2.21) is given, then the system (2.20) has $m_1 = m - m_2$ independent solutions, where m_2 is the rank of its matrix. Hence, there are m_1 independent characteristic relations (2.17) corresponding to a given normal. In every particular situation it is necessary to find out which of the characteristic relations are linearly independent and how many of them belong to this category. For the case of three-dimensional gas flows (stationary and nonstationary) this question has been discussed in detail in [2.6]. Often characteristic surfaces are defined as the surfaces upon which it is impossible to put initial conditions. Evidently, this definition is equivalent to the one introduced above. In fact, if the surface of initial data is a characteristic one, then on it, m_1 ($m_1 < m$) independent compatibility conditions (2.17) following from the main system of differential equations are satisfied. The remaining $m - m_1$ main equations, which in terms of [2.5] are called additional relations and contain at least one derivative of the functions sought with respect to the direction normal to the characteristic surface, are insufficient for finding a solution outside the surface of initial data.

Consider a three-dimensional stationary gas flow with equilibrium physicochemical processes in a Cartesian coordinate system. Using the general theory, we shall derive characteristic equations in the form proposed in [2.7,8] leaving for Sect.3.2 the discussion of the numerical method of characteristics developed in these references.

Equations (2.4,5,9) in the case of stationary equilibrium flow have the form

$$\nabla(\rho\mathbf{W}) = 0 \quad , \tag{2.22}$$

$$\rho(\mathbf{W}\nabla)\mathbf{W} + \nabla p = 0 \quad , \tag{2.23}$$

$$(\mathbf{W}\nabla)S = 0 \quad . \tag{2.24}$$

The equation for the square of the sonic speed, $a^2 = (\partial p/\partial\rho)_S$, helps to reduce (2.24) to

$$(\mathbf{W}\nabla)\rho = a^{-2}(\mathbf{W}\nabla)p$$

which enables derivatives of density to be eliminated from the continuity equation (2.22):

$$\rho a^2\nabla\mathbf{W} + (\mathbf{W}\nabla)p = 0 \quad . \tag{2.25}$$

In the case of an equilibrium flow the stagnation enthalpy H_0 and stagnation pressure p_0 are constant along streamlines; thus,

$$(\mathbf{W}\nabla)H_0 = 0 \quad , \quad (\mathbf{W}\nabla)p_0 = 0 \quad , \tag{2.26}$$

where $H_0 = h + W^2/2$.

Let us further consider the system (2.23,25,26). The density and sonic speed factors in these expressions are functions of p, p_0, H_0:

$$\rho = \rho(p,p_0,H_0) \quad , \quad a = a(p,p_0,H_0) \quad . \tag{2.27}$$

For a perfect gas, ρ and a can be expressed explicitly. In the case of equilibrium flows they can be obtained from the equations for chemical equilibrium.

Following the general theory, we shall multiply every i^{th} equation of the system (2.23,25) by ω_i ($i = 1,2,\ldots,6$) and sum the products. The result is (2.17) which, in detailed form, is

$$\omega_1\rho(\mathbf{W}\nabla)u + \omega_2\rho(\mathbf{W}\nabla)v + \omega_3\rho(\mathbf{W}\nabla)w$$

$$+ \omega_4\rho a^2\nabla\mathbf{W} + \omega_1\frac{\partial p}{\partial x} + \omega_2\frac{\partial p}{\partial y} + \omega_3\frac{\partial p}{\partial z}$$

$$+ \omega_4(\mathbf{W}\nabla)p + \omega_5(\mathbf{W}\nabla)p_0 + \omega_6(\mathbf{W}\nabla)H_0 = 0 \quad , \tag{2.28}$$

where u, v and w are projections of the velocity vector **W** upon the axes of a Cartesian coordinate system.

The characteristic equation (2.21) which enables the vectors **n** to be found has the following form:

$$\begin{vmatrix} \rho(\mathbf{Wn}) & 0 & 0 & \rho a^2 n_x & 0 & 0 \\ 0 & \rho(\mathbf{Wn}) & 0 & \rho a^2 n_y & 0 & 0 \\ 0 & 0 & \rho(\mathbf{Wn}) & \rho a^2 n_z & 0 & 0 \\ n_x & n_y & n_z & (\mathbf{Wn}) & 0 & 0 \\ 0 & 0 & 0 & 0 & (\mathbf{Wn}) & 0 \\ 0 & 0 & 0 & 0 & 0 & (\mathbf{Wn}) \end{vmatrix} = 0 \quad ,$$

where n_x, n_y, n_z are components of the vector **n**. Hence,

$$\rho^3(\mathbf{Wn})^4[(\mathbf{Wn})^2 - a^2(\mathbf{nn})] = 0 \quad . \tag{2.29}$$

The cone of normals decomposes into two cones having no intersection. One of them is described by

$$\mathbf{Wn} = 0 \quad . \tag{2.30}$$

This cone has degenerated into a plane perpendicular to the velocity vector **W**. The corresponding characteristic cone degenerates into a straight line parallel to the vector **W**. Thus stream surfaces are characteristic surfaces. The equation of the second cone is

$$\mathbf{Wn} = a \quad , \quad \mathbf{nn} = 1 \quad , \tag{2.31}$$

which is a circular cone. Its axis is parallel to the velocity vector and its vertex angle $\beta = \arccos M^{-1}$, where $M = W/a$ is the Mach number. The axis of the corresponding characteristic cone is also parallel to **W** and its vertex angle α is equal to $\arcsin M^{-1}$. This cone is also called the Mach cone while the corresponding characteristic surfaces are known as wave surfaces. Figure 2.1 shows the relative positions of the characteristic surfaces, cones and conoid. Obviously, the Mach cone is real only if $M > 1$, that is, if the flow is supersonic and stationary. It is not difficult to obtain the equation of the bicharacteristics. Let $\boldsymbol{\ell}$ be a vector of infinitesimal displacement along a bicharacteristic; then $\boldsymbol{\ell}$ lies on one of the generatrices of the Mach cone tangent to the bicharacteristic at the given point. Evidently, $\mathbf{W}\boldsymbol{\ell} = \ell(W^2 - a^2)^{\frac{1}{2}}$, i.e.

$$(u\,dx + v\,dy + w\,dz)^2 = (W^2 - a^2)[(dx)^2 + (dy)^2 + (dz)^2] \quad . \tag{2.32}$$

These equations define characteristic surfaces. Now we shall consider equations which are valid on these surfaces and are called the equations of com-

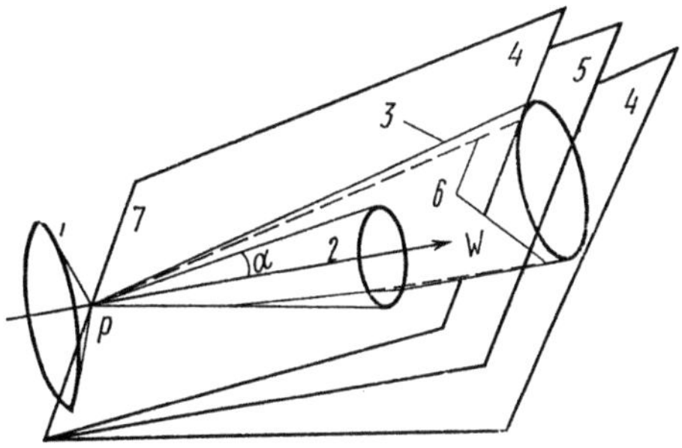

Fig.2.1. Cone of normals (*1*), characteristic cone (*2*), characteristic conoid (*3*) drawn at the point P. The stream surface (*5*) and two wave surfaces (*4*) contacting the conoid along bicharacteristics (*6*) are drawn through the curve (*7*)

patibility. Equations for $\omega_1,\omega_2,\ldots,\omega_6$ form the following system:

$$\begin{aligned}
&\rho(\mathbf{Wn})\omega_1 + \rho a^2 n_x\omega_4 = 0 \quad ,\\
&\rho(\mathbf{Wn})\omega_2 + \rho a^2 n_y\omega_4 = 0 \quad ,\\
&\rho(\mathbf{Wn})\omega_3 + \rho a^2 n_z\omega_4 = 0 \quad ,\\
&n_x\omega_1 + n_y\omega_2 + n_z\omega_3 + (\mathbf{Wn})\omega_4 = 0 \quad ,\\
&(\mathbf{Wn})\omega_5 = 0 \quad ,\\
&(\mathbf{Wn})\omega_6 = 0 \quad .
\end{aligned} \tag{2.33}$$

Let us consider stream surfaces. Every normal to a stream surface at a given point is perpendicular to a streamline, that is,

$$\mathbf{Wn} = 0 \quad . \tag{2.34}$$

In this case the rank of the matrix of (2.33) equals 2; hence, there are four independent solutions corresponding to one and the same normal. By (2.34) it follows from (2.33) that $\omega_4 = 0$. In the original system, equations (2.26) already have the characteristic form because they contain derivatives with respect to the streamlines. Thus, it is convenient to choose the following two solutions of the system (2.33): $\omega_1 = \omega_2 = \omega_3 = \omega_4 = \omega_5 = 0$, $\omega_6 \neq 0$ and $\omega_1 = \omega_2 = \omega_3 = \omega_4 = \omega_6 = 0$, $\omega_5 \neq 0$. The other two solutions can be obtained by putting $\omega_4 = \omega_5 = \omega_6 = 0$, $\omega_1 = u$, $\omega_2 = v$, $\omega_3 = w$ or $\omega_1 = m_x$, $\omega_2 = m_y$, $\omega_3 = m_z$, where $\mathbf{mn} = 0$ and vector $\mathbf{m}$ is independent of $\mathbf{W}$.

Subsituting independent solutions $\omega_1,\omega_2,\ldots,\omega_6$ into (2.28) one obtains equations of compatibility corresponding to the given stream surface normal:

$$(\mathbf{W}\nabla)p_0 = 0 \quad , \tag{2.35}$$

$$(\mathbf{W}\nabla)H_0 = 0 \quad , \tag{2.36}$$

$$\rho u(\mathbf{W}\nabla)u + \rho v(\mathbf{W}\nabla)v + \rho w(\mathbf{W}\nabla)w + (\mathbf{W}\nabla)p = 0 \quad , \tag{2.37}$$

$$\rho m_x(\mathbf{W}\nabla)u + \rho m_y(\mathbf{W}\nabla)v + \rho m_z(\mathbf{W}\nabla)w + (\mathbf{W}\nabla)p = 0 \quad . \tag{2.38}$$

Equation (2.35) is evidently equivalent to the condition of conservation of entropy along a streamline, while (2.37) is the Bernoulli equation.

Next we shall consider wave surfaces. The scalar product of the unit normal to a wave surface and the velocity vector is equal to the local sound velocity,

$$\mathbf{Wn} = a \quad . \tag{2.39}$$

Substituting (2.39) into (2.33), we find that the matrix of (2.33) has rank 5; it has only one independent solution. By (2.39) it follows from (2.33) that $\omega_5 = \omega_6 = 0$, while ω_4 can be chosen arbitrarily. Note that $\omega_1 = -an_x\omega_4$, $\omega_2 = -an_y\omega_4$, $\omega_3 = -an_z\omega_4$. Putting $\omega_4 = -1$, one gets from (2.28)

$$\rho an_x(\mathbf{W}\nabla)u + \rho an_y(\mathbf{W}\nabla)v + \rho an_z(\mathbf{W}\nabla)w$$
$$- \rho a^2\nabla \mathbf{W} + [(a\mathbf{n} - \mathbf{W})\nabla]p = 0 \quad . \tag{2.40}$$

Among the infinity of characteristic conditions (2.35-38,40) corresponding to different normals, the number of independent conditions cannot exceed the number of the initial differential equations. Equations (2.35-37) along streamlines and (2.40) along the three characteristic surfaces can be chosen for the role of such independent characteristic relations.

The general theory developed earlier in this section allows characteristic equations to be obtained for systems describing even more complicated types of gas flow, e.g., three-dimensional flows with nonequilibrium physico-chemical processes or multiphase flows [2.9]. In the following chapters we shall consider similar flows only for the case of two independent variables. Therefore we shall consider this case in greater detail. Doing so, we shall use an alternative approach based on the definition of characteristics as the lines at which the initial data cannot be specified while solving an initial-value problem. This approach facilitates the reduction of equations with two independent variables to the characteristic form.

Consider a quasi-linear system of order m:

$$A \frac{\partial \mathbf{u}}{\partial x_1} + B \frac{\partial \mathbf{u}}{\partial x_2} = \mathbf{f} \quad , \tag{2.41}$$

where a_{ij}, b_{ij}, f_i are elements of matrices A, B and vectors $\mathbf{f}$, (i,j = (1, 2, ..., m), are functions of x_1, x_2, and are the required vector function $\mathbf{u}$ has components $u_1, u_2, \ldots, u_m$.

Let a smooth solution **u** of the system (2.41) be known in the region G of the plane x_1,x_2 and a curve Γ be given. We shall try to find out whether it is possible, knowing u along Γ, to reconstruct solutions in a certain vicinity of Γ or, in other words, to define uniquely all partial derivatives $\partial u/\partial x_1$, $\partial \mathbf{u}/\partial x_2$ along Γ using differential equations.

Having written down the following expression for the total differential,

$$du = \frac{\partial \mathbf{u}}{\partial x_1} dx_1 + \frac{\partial \mathbf{u}}{\partial x_2} dx_2 \quad , \tag{2.42}$$

where (dx_1,dx_2) is a vector of infinitesimal displacement along Γ, one gets an algebraic system (2.41,42) of order 2n which can be used to find derivatives $\partial u/\partial x_1$, $\partial \mathbf{u}/\partial x_2$. Solving (2.42) for $\partial \mathbf{u}/\partial x_1$ and substituting the result into the original system of equations, one obtains

$$(B\,dx_1 - A\,dx_2)\frac{\partial \mathbf{u}}{\partial x_2} = \mathbf{f}\,dx_1 - A\,d\mathbf{u} \quad . \tag{2.43}$$

If the determinant of the matrix

$$B\,dx_1 - A\,dx_2 \tag{2.44}$$

is nonzero at every point along Γ, then all the derivatives are defined uniquely. If this determinant is identically zero along Γ, then it follows, from the assumption that the system (2.41) possesses a solution, that the derivatives $\partial \mathbf{u}/\partial x_1$ and $\partial \mathbf{u}/\partial x_2$ are defined ambiguously. In this case curve Γ is called a characteristic. Equating the determinant of the system (2.43) to zero, one obtains

$$\left| B - A\frac{dx_2}{dx_1} \right| = 0 \quad , \tag{2.45}$$

which allows one to find the values of dx_2/dx_1 which define the characteristic directions. If the equation has m different real roots $dx_2/dx_1 = \lambda_i$, $(i = 1,2,\ldots,m)$, then the original system of differential equations is called hyperbolic. Note the following circumstance. The slope of the characteristic λ_i depends not only on the coordinates but also on the solution **u**. Thus, characteristic curves can be drawn on the plane x_1,x_2 only for a specific solution of the system of differential equations. In a linear case, λ_i are functions of x_1 and x_2 only and the characteristic curves can be drawn even if the solution **u** is unknown.

Let us show that at the characteristics a solution of (2.41) must satisfy certain conditions. Suppose the system (2.41) is a hyperbolic one. This means that the rank of its matrix (2.44) is equal to $m-1$. On the other hand, from the assumption of solution-existence, it follows that the rank of the extended matrix,

$$(B\,dx_1 - A\,dx_2 \quad , \quad \mathbf{f}\,dx_1 - A\,d\mathbf{u}) \quad , \tag{2.46}$$

is also equal to $m-1$. Equating determinants of order m to zero, one finds relations between the components of the vector of displacement and the solution differentials. Due to the linear relation between the columns of matrix (2.44), the minors of the extended matrix differ from each other by a constant multiplier only; thus it makes no difference which of them is equated to zero, for only the cases of trivial identity are to be avoided.

Equation (2.45) is called an equation of characteristic directions while the relation obtained by equating to zero a determinant composed of any $m-1$ columns of matrix (2.44) and the column $\mathbf{f}\,dx_1 - A\,d\mathbf{u}$ is called a differential relation at a characteristic or a condition of compatibility (similar to the general case of three-dimensional flows). We shall proceed by considering some particular systems which will be used later on.

Let us begin with equations describing one-dimensional nonstationary flow of a perfect gas:

$$\begin{aligned} &\frac{\partial u}{\partial t} + u\,\frac{\partial u}{\partial x} + \frac{1}{\rho}\,\frac{\partial p}{\partial x} = 0 \quad , \\ &\frac{\partial \rho}{\partial t} + u\,\frac{\partial \rho}{\partial x} + \rho\,\frac{\partial u}{\partial x} = 0 \quad , \\ &\frac{\partial p}{\partial t} + u\,\frac{\partial p}{\partial x} + \rho a^2\,\frac{\partial u}{\partial x} = 0 \quad . \end{aligned} \tag{2.47}$$

In this case (2.45) transforms to $(u - dx/dt)[(u - dx/dt)^2 - a^2] = 0$, which yields three equations for the characteristic directions,

$$dx/dt = u \quad , \quad dx/dt = u + a \quad , \quad dx/dt = u - a \quad ,$$

which are called C_0, C_+ and C_- characteristics, respectively. Matrix (2.44) attains the form

$$\begin{pmatrix} u\,dt - dx & 0 & dt/\rho \\ \rho\,dt & u\,dt - dx & 0 \\ a^2\rho\,dt & 0 & u\,dt - dx \end{pmatrix} \quad .$$

Let us compose the determinant of this matrix using vector **f** dt - A**du** instead of the second column, and equate it to zero:

$$dp - a^2 d\rho + (u - dx/dt)\,[(u - dx/dt)d\rho - \rho\, du] = 0 \quad .$$

Substituting u, u + a and u - a instead of dx/dt, one obtains conditions of compatibility at the characteristics:

$$\begin{aligned} dp - a^2 d\rho &= 0 \quad \text{along} \quad C_0: dx/dt = u \ , \\ dp + a\rho\, du &= 0 \quad \text{along} \quad C_+: dx/dt = u + a \ , \\ dp - a\rho\, du &= 0 \quad \text{along} \quad C_-: dx/dt = u - a \ . \end{aligned} \tag{2.48}$$

The relation $dp - a^2 d\rho = 0$ is equivalent to the condition of constant entropy along the path of particles.

If the flow is isentropic then expressions $du + dp/\rho a$ and $du - dp/\rho a$ are total differentials of I_+, I_- (Riemann's invariants), where

$$I_\pm = u \pm \int \frac{dp}{\rho a} = u \pm \frac{2a}{\gamma - 1} \quad .$$

It follows from (2.48) that

$$I_\pm = \text{const along } C_\pm: dx/dt = u \pm a \quad . \tag{2.49}$$

We note that if the term $\rho u\, d(\ln F)/dx$ is added to the continuity equations (2.47), then this system of equations will describe one-dimensional gas flow in a nozzle whose contour is defined by the equation $y = F(x)$. If the flow is isentropic, then the direction equations and compatibility conditions attain the following form:

$$\begin{aligned} dx/dt &= u \pm a \ , \\ dI_\pm \pm au[d(\ln F)\, dx]\, dt &= 0 \ . \end{aligned} \tag{2.50}$$

Now let us consider the equations of characteristics for nonequilibrium stationary gas flows. Equations (2.7) describing the change of parameters α_i, which characterize nonequilibrium states of the medium, will be written down as

$$d\alpha_i/dt = F_i(p, T, \alpha_1, \ldots \alpha_N) \quad , \quad (i = 1, 2, \ldots, N) \quad . \tag{2.51}$$

We have already pointed out that mass concentrations of gaseous components, energies of internal degrees of freedom, etc., may be looked upon as such parameters. We have also at our disposal the equations of state, which may be presented in the form

$$h = h(p,T,\alpha_1,\ldots,\alpha_N) \quad , \tag{2.52}$$

$$\rho = \rho(p,T,\alpha_1,\ldots,\alpha_N) \quad . \tag{2.53}$$

Let us write down equations of motion, continuity and energy in the form

$$u\,\frac{\partial u}{\partial x} + v\,\frac{\partial u}{\partial y} + \frac{1}{\rho}\,\frac{\partial p}{\partial x} = 0 \quad ,$$

$$u\,\frac{\partial v}{\partial x} + v\,\frac{\partial v}{\partial y} + \frac{1}{\rho}\,\frac{\partial p}{\partial y} = 0 \quad , \tag{2.54}$$

$$\frac{d\rho}{dt} + \rho\,\frac{\partial u}{\partial x} + \rho\,\frac{\partial v}{\partial y} + \nu\,\frac{\rho v}{y} = 0 \quad , \tag{2.55}$$

$$\frac{dh}{dt} - \frac{1}{\rho}\,\frac{dp}{dt} = 0 \quad . \tag{2.56}$$

Here and below for plane flows $\nu = 0$, while for axisymmetric flows $\nu = 1$. Using (2.51-53), we shall represent the total derivatives $d\rho/dt$, dh/dt in the form

$$\frac{d\rho}{dt} = \frac{\partial\rho}{\partial p}\,\frac{dp}{dt} + \frac{\partial\rho}{\partial T}\,\frac{dT}{dt} + \sum_{i=1}^{N} \frac{\partial\rho}{\partial\alpha_i}\,F_i \quad , \tag{2.57}$$

$$\frac{dh}{dt} = \frac{\partial h}{\partial p}\,\frac{dp}{dt} + \frac{\partial h}{\partial T}\,\frac{dT}{dt} + \sum_{i=1}^{N} \frac{\partial h}{\partial\alpha_i}\,F_i \quad . \tag{2.58}$$

Equation (2.56), upon substitution of (2.58), becomes

$$\left(\frac{\partial h}{\partial p} - \frac{1}{\rho}\right)\frac{dp}{dt} + \frac{\partial h}{\partial T}\,\frac{dT}{dt} + \sum_{i=1}^{N} \frac{\partial h}{\partial\alpha_i}\,F_i = 0 \quad . \tag{2.59}$$

Following [2.10], let us introduce the frozen sound velocity,

$$a = \left(\frac{\partial p}{\partial\rho}\right)^{\frac{1}{2}}_{s,\alpha_i} = \left[\frac{\partial\rho}{\partial p} + \frac{\partial\rho/\partial T}{\partial h/\partial T}\left(\frac{1}{\rho} - \frac{\partial h}{\partial p}\right)\right]^{-\frac{1}{2}} \quad . \tag{2.60}$$

With the aid of (2.57) and (2.59), (2.55) can be reduced to

$$\rho\,\frac{\partial u}{\partial x} + \rho\,\frac{\partial v}{\partial y} + \frac{u}{a^2}\,\frac{\partial p}{\partial x} + \frac{v}{a^2}\,\frac{\partial p}{\partial y} = -\,\nu\,\frac{\rho v}{y} + \left(\frac{\partial h}{\partial T}\right)^{-1} \sum_{i=1}^{N} \left(\frac{\partial\rho}{\partial T}\,\frac{\partial h}{\partial\alpha_i} - \frac{\partial h}{\partial T}\,\frac{\partial\rho}{\partial\alpha_i}\right)F_i \quad . \tag{2.61}$$

Using a general approach, one can obtain the equations of characteristics for the system (2.51,54,55,61). There exist three families of characteristics in supersonic flow — two families of Mach lines (characteristics C_+ and C_-) and the streamlines (characteristic C_0). The direction equations and compatibility conditions for Mach lines are

$$dy = \tan(\theta \pm \alpha)dx \quad , \tag{2.62}$$

$$d\theta \pm \frac{\cot\text{an}\alpha}{\rho W^2}\,dp \pm \left(\frac{\nu}{y}\sin\theta\sin\alpha - \Phi_1\right)\frac{dx}{\cos(\theta \pm \alpha)} = 0 \quad , \tag{2.63}$$

where

$$\Phi_1 = \frac{a}{\rho W^2}\left(\frac{\partial h}{\partial T}\right)^{-1}\sum_{i=1}^{N}\left(\frac{\partial\rho}{\partial T}\frac{\partial h}{\partial\alpha_i} - \frac{\partial h}{\partial T}\frac{\partial\rho}{\partial\alpha_i}\right)F_i \quad ,$$

and for streamlines,

$$dy = \tan\theta\, dx \quad , \tag{2.64}$$

$$W\,dW + dp/\rho = 0 \quad , \tag{2.65}$$

$$d\alpha_i = (F_i/W\cos\theta)dx \quad , \quad (i = 1,2,\ldots,N) \quad , \tag{2.66}$$

$$\left(\frac{\partial h}{\partial p} - \frac{1}{\rho}\right)dp + \frac{\partial h}{\partial T}\,dT + \sum_{i=1}^{N}\frac{\partial h}{\partial\alpha_i}\,d\alpha_i = 0 \quad . \tag{2.67}$$

Here $\alpha = \arcsin M^{-1}$ and $M = W/a$ are the frozen Mach angle and Mach number respectively, and θ is the angle between the velocity and the x direction. It should be emphasized that in nonequilibrium flows the direction of the characteristics is defined not by the sound velocity $a = (dp/d\rho)^{\frac{1}{2}}$ or the equilibrium sound velocity $a_\ell = (dp/d\rho)_s^{\frac{1}{2}}$ but by the frozen sound velocity $a_f = (dp/d\rho)_{s,\alpha_i}^{\frac{1}{2}}$. In nonequilibrium flows small disturbances also propagate with the frozen sound velocity [2.10].

Let us introduce the stream function ψ by means of the following equation:

$$d\psi = \rho W y^\nu(\cos\theta\, dy - \sin\theta\, dx) \quad .$$

Then

$$d\psi = \pm\frac{\rho W y^\nu \sin\alpha}{\sin(\theta \pm \alpha)}\,dy, \quad d\psi = 0 \tag{2.68}$$

along characteristics $C_\pm$ and C_0.

It follows from (2.65,67) that along a streamline

$$\frac{1}{2} W^2 + h = H_0 \quad . \tag{2.69}$$

Equations (2.62-69) are valid for equilibrium and frozen flows if we formally put $F_i \equiv 0$. Then, the sound velocity is defined by (2.60) while ρ and h can be found from the corresponding state equations,

$$h = h(p,T) \quad , \quad \rho = \rho(p,T) \quad . \tag{2.70}$$

From (2.9) it follows that in such flows (unlike nonequilibrium flows) entropy remains constant along streamlines:

$$dS = 0 \quad . \tag{2.71}$$

If the relations

$$a = \left(\frac{\gamma p}{\rho}\right)^{\frac{1}{2}} \quad , \quad h = \frac{\gamma}{\gamma - 1} \frac{p}{\rho} \quad , \quad S = \ln \frac{p}{\rho^{\nu}} \tag{2.72}$$

are used instead of (2.60,70), then (2.62-64,71) at $F_i = 0$ become characteristic relations for equations describing the flow of a perfect gas with constant specific heats.

Let us consider another important case — the flow of a two-phase medium. The stationary flow of such a medium under the assumptions made in Sect.2.1.1 is described by (2.10-13) if all partial derivatives with respect to time are equated to zero. The system (2.10-13) in the cases of plane and axisymmetric flows can be used to determine u, u_s, v, v_s, ρ, ρ_s, T, T_s, h, e_s. The frozen sound velocity is defined by (2.60) and if one neglects the volume occupied by particles, then it is equal to the sound velocity in the gas. Just as in the case of the flow of a one-phase medium, the Mach lines (2.62) and the streamlines (2.64) represent characteristics of the system. In the case we are considering, there is also another family of characteristics — the streamlines of particles.

Let us write down the characteristic relations without derivation (see [2.11]). At the characteristics $C_{\pm}$ (Mach lines), $dy = \tan(\theta \pm \alpha)dx$, the following relations are valid:

$$d\theta \pm \frac{\operatorname{cotan}\alpha}{\rho W^2} dp \pm \left(\frac{\nu}{y} \sin\theta \sin\alpha - \Phi_2\right) \frac{dx}{\cos(\theta \pm \alpha)} = 0 \quad , \tag{2.73}$$

where

$$\Phi_2 = \frac{\rho_s}{\rho W^2} \left\{ - \frac{a}{\rho} \frac{\partial \rho / \partial T}{\partial h / \partial T} [\mathbf{W}_s - \mathbf{W})\mathbf{f} + q] \mp [f_x \sin(\theta \pm \alpha) - f_y \cos(\theta \pm \alpha)] \right\} \quad ,$$

and f_x and f_y are projections of vector $\mathbf{f}$ on the axes of a Cartesian coordinate system.

At the streamlines (characteristics C_0), $dy = \tan\theta\, dx$, one has

$$\frac{\partial h}{\partial T}\, dT + \left(\frac{\partial h}{\partial p} - \frac{1}{\rho}\right)dp + \frac{\rho_s}{\rho W \cos\theta}\,[(\mathbf{W}_s - \mathbf{W})\mathbf{f} + q]dx = 0 \quad , \tag{2.74}$$

$$dH_0 + \frac{\rho_s}{\rho W \cos\theta}\,(\mathbf{f}\mathbf{W}_s + q)dx = 0 \quad . \tag{2.75}$$

It follows from the latter equation that if there is no interaction between the gas and particles, that is if $\mathbf{f} = 0$ and $q = 0$, then the stagnation enthalpy remains constant along the gas streamlines.

Equations (2.13), which are in characteristic form already, are valid along the streamlines of particles. Thus, we have the following direction and compatibility equations:

$$v_s dx - u_s dy = 0 \quad ,$$

$$u_s du_s - f_x dx = 0 \;, \quad u_s dv_s - f_y dx = 0 \;, \quad u_s de_s - q\, dx = 0 \quad . \tag{2.76}$$

Using the continuity equation for the "gas" of particles, we obtain another relation for the determination of ρ_s which is valid along the streamlines of particles:

$$d\rho_s + \rho_s(\partial\xi/\partial y + \nu\xi/y + f_x/u_s^2)dx = 0 \quad , \tag{2.77}$$

where $\xi = v_s/u_s$.

The theory of characteristics is extremely important for the formulation of boundary conditions in gasdynamic problems. Also, properties of characteristics are widely used for numerical solutions. Below we shall often deal with these questions while discussing various problems of nozzle gas flows. Now we present only a brief explanation of the main ideas of numerical methods of characteristics.

For the sake of simplicity we shall consider the hyperbolic system (2.41) for the case of two independent variables and two unknown functions designated x, y and u, v, respectively. We have

$$a_{11}\frac{\partial u}{\partial x} + a_{12}\frac{\partial v}{\partial x} + b_{11}\frac{\partial u}{\partial y} + b_{12}\frac{\partial v}{\partial y} = f_1 \quad .$$

$$a_{21}\frac{\partial u}{\partial x} + a_{22}\frac{\partial v}{\partial x} + b_{21}\frac{\partial u}{\partial y} + b_{22}\frac{\partial v}{\partial y} = f_2 \quad , \tag{2.78}$$

where a_{ij}, b_{ij} are functions of x, y, u, v. This system has the following equations for the characteristics:

$$dy/dx = \lambda_i \quad , \quad (i = 1,2) \quad , \tag{2.79}$$

$$(a\lambda_i + b)du + c\,dv + h\,dx + e\,dy = 0 \quad , \quad \text{where} \tag{2.80}$$

$$a = a_{11}a_{22} - a_{12}a_{21} \; , \quad b = b_{11}b_{22} - b_{12}b_{21} \; , \quad c = a_{22}b_{12} - a_{12}b_{22} \quad ,$$

$$h = b_{22}f_1 - b_{12}f_2 \; , \quad e = a_{12}f_2 - a_{22}f_1 \quad .$$

Let us consider two points 1 and 2 lying close together in the plane x, y on the same characteristic and try to find u and v at point 3 where the characteristics passing through the two initial points intersect. The coordinates of a point i will be denoted by x_i, y_i, and the value of an arbitrary function g at this point by g_i. Substituting differentials, we get the following system of equations for the determination of x_3, y_3:

$$y_3 - y_1 = (\lambda_1)_1(x_3 - x_1) \; , \quad y_3 - y_2 = (\lambda_2)_2(x_3 - x_2) \quad . \tag{2.81}$$

Similarly, from (2.80) we obtain

$$\begin{aligned} &(\lambda_1 a + b)_1(u_3 - u_1) + c_1(v_3 - v_1) + h_1(x_3 - x_1) \\ &\qquad + e_1(y_3 - y_1) = 0 \quad , \\ &(\lambda_2 a + b)_2(u_3 - u_2) + c_2(v_3 - v_2) + h_2(x_3 - x_2) \\ &\qquad + e_2(y_3 - y_2) = 0 \quad , \end{aligned} \tag{2.82}$$

which allows us to find u_3, v_3. The error in the values of x_3, y_3, u_3, v_3 obtained in this way is of the order of the square of the distances from point 3 to points 1 and 2. The accuracy of the solution could be raised by an order of magnitude if, in the process of calulating the coefficients before the differentials in (2.79,80), the values of the required functions u and v at point 3 were used. This can be done by means of iteration:

$$y_3^{(s)} - y_1 = (\lambda_1)_{13}^{(s)}(x_3^{(s)} - x_1) \; , \quad y_3^{(s)} - y_2 = (\lambda_2)_{23}^{(s)}(x_3^{(s)} - x_2) \quad ,$$

$$\begin{aligned} &(\lambda_1 a + b)_{13}^{(s)}(u_3^{(s)} - u_1) + c_{13}^{(s)}(v_3^{(s)} - v_1) \\ &\qquad + h_{13}^{(s)}(x_3^{(s)} - x_1) + e_{13}^{(s)}(y_3^{(s)} - y_1) = 0 \quad , \\ &(\lambda_2 a + b)_{23}^{(s)}(u_3^{(s)} - u_2) + c_{23}^{(s)}(v_3^{(s)} - v_2) \end{aligned}$$

$$+ h_{23}^{(s)}(x_3^{(s)} - x_2) + e_{23}^{(s)}(y_3^{(s)} - y_2) = 0 \quad ,$$

$$(s = 1,2,\dots) \quad ,$$

where

$$g_{13}^{(s)} = \frac{1}{2}\left[g_3^{(s-1)} + g_1\right] \quad , \quad g_{23}^{(s)} = \frac{1}{2}\left[g_3^{(s-1)} + g_2\right] \quad .$$

In the process of calculating $g_{13}^{(0)}$, $g_{23}^{(0)}$, a solution of (2.81,82) is used.

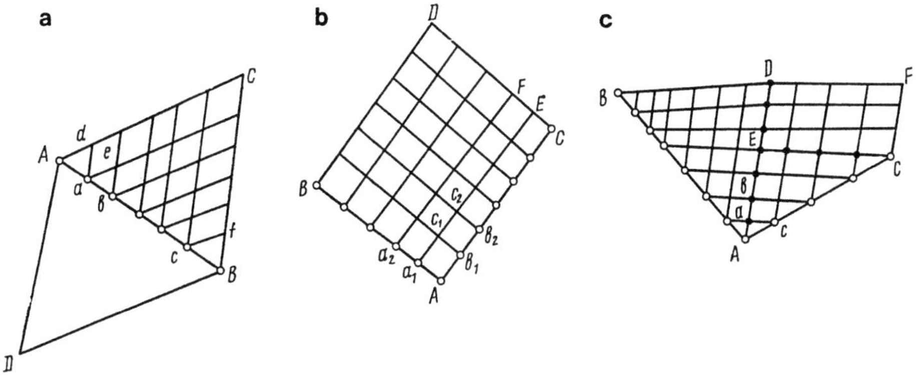

Fig.2.2a-c. Application of the method of characteristics to solving (**a**) the initial-value problem, (**b**) the Goursat problem and (**c**) the boundary-value problem

Let u and v be given on the arc AB (Fig.2.2) by a curve which has no characteristic direction at any point on AB, and let a solution be sought in the vicinity of AB (the initial-value problem). We choose on AB a sequence of points A, a, b, ..., c, B (Fig.2.2a) and draw characteristics of both families through each of them. Using the method which has been described above, one can calculate the unknown functions at the points of intersection d, e, ..., f. Knowing the solution at these points, it is possible to move to the next "layer", and so on until solution at point C is obtained. Thus the solution is gradually found simultaneously with the formation of the characteristic grid. The solution in the characteristic triangle ABD is sought in the same way. Such a procedure is possible only if a continuous solution exists in the region ACBD. It is known that the existence of a continuous solution of a quasi-linear system can be guaranteed only in the vicinity of the line of initial data. Even in the case of quite smooth initial data, discontinuities may appear in the domain of influence of arc AB (region ACBD). In this case calculation by the method of characteristics becomes much more complicated. However, this question is beyond the scope of our present elementary consideration of the numerical method of characteristics.

Let us next consider Goursat's problem: u and v are given at arcs AB and AC of characteristics belonging to different families. In addition, u and v satisfy the compatibility condition. We choose two sequences of points A, $a_1, a_2, \dots$, B and A, $b_1, b_2, \dots$, C at the arcs AB and AC, Fig.2.2b. Using points a_1 and b_1 as references, let us find the values of the required functions at point c_1. Then the numerical solution can be found at point c_2 and all other points of the characteristic a_1E. After that, the process is repeated for the next characteristic a_2F, and so on, until the unknown values are found at the boundary characteristic BD of the characteristic quadrangle ABDC.

To conclude the section, let us consider one of the possible boundary problems. Let boundary conditions be put at arcs AB and AC of two noncharacteristic curves, in such a way that u and v are given at AB, a linear combination $\alpha u + \beta v$ at AC, and arc AC lies inside the angle formed by characteristics of different families passing through point A, Fig.2.2c. The data given at AB allow us to calculate u, v within the triangle ABD including the points belonging to characteristic AD (points a, b, etc.). To define u, v at point c one may use the characteristic condition along arc ac and the combination $\alpha u + \beta v$ given at point c. After calculating the values sought in the triangle AEC, Goursat's problem is solved with the data on ED and EC.

2.1.3 The Direct Problem

While solving specific applied problems, it is necessary to supplement the system (2.1-3) with corresponding initial and boundary conditions. Two problems are distinguished in the theory of Laval nozzles—the direct problem and the inverse problem. The direct problem is a problem of defining the flow field when the nozzle's contour and some conditions in its initial and exit cross-sections are given. This problem can be reduced to a boundary-value problem for gasdynamic equations. The inverse problem is a problem of defining the flow field under conditions specified at a known surface and in the initial cross-section of a nozzle. This problem can be reduced to an initial-value problem. Evidently, the character of the initial and boundary conditions depends upon the type of flow and is different for subsonic and supersonic flows.

Let us start with consideration of the direct problem for the general case of a nonstationary three-dimensional flow of a nonreactive mixture of gases. The impermeability condition,

$$(W\nabla)F = 0 \quad , \tag{2.83}$$

is specified at the nozzle's contour described by $F(x,y,z) = 0$. To define the necessary initial conditions at $t = 0$ we shall fix values of all gasdynamic parameters of the flow in the entire region it occupies (surfaces of discontinuity are allowed).

At the initial cross-section of a nozzle where flow will be supposed always to be subsonic, either all components of velocity and pressure (or density) or two components of velocity, pressure and density may be specified as functions of time. If all components of a solution (the velocity vector, pressure and density) are given in the initial plane then the problem is overdetermined. (This statement may be proved on the basis of the theory of characteristics.) The exit section data are specified in the same manner if the exhaust velocity is subsonic. However, if the velocity at the exit cross-section of a nozzle exceeds the sound speed, then no boundary conditions can be given for the section because here the solution is fully defined by specifying the initial data and boundary conditions in the initial section of the nozzle.

The existence of only one solution is due to the fact that it can be found uniquely by means of the method of characteristics. Let us demonstrate this for the example of a one-dimensional nonstationary isentropic gas flow in a nozzle. There exist two families of characteristics in this case. Characteristic relations in the form of (2.50) connect the differentials of velocity u and sound speed a.

As initial data at $t = 0$, it is necessary to specify functions u and a for all x from $0 \leqslant x \leqslant x_k$, where $x = 0$ and $x = x_k$ are the coordinates of the initial and exit cross-sections of a nozzle. Let us begin with the case when flow is subsonic in all parts of a nozzle. The initial conditions uniquely define this flow inside the characteristic triangle OAB, Fig.2.3a. It is impossible to define both functions u and a at the left boundary ($x = 0$) because a characteristic C_- enters every point of the left boundary from the side of smaller values of t where the solution is already known. The characteristic relation referring to this characteristic will generally contradict values of u and a arbitrarily defined at $x = 0$. In order to find a unique solution in the region OBD, Fig.2.3a, by the method of characteristics, it is necessary to define at $x = 0$ either one of the functions u and a or a

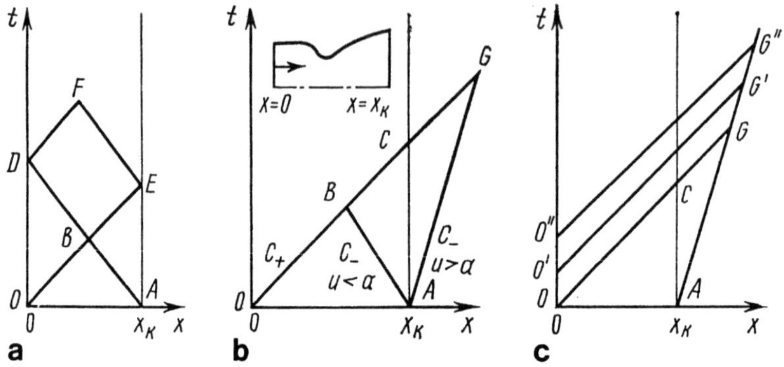

Fig.2.3a-c. Formulation of boundary conditions for a one-dimensional non-stationary gas flow in a nozzle

linear combination of them. The same may be said about the right boundary. Thus, specification of one of the functions (or of their linear combinations) at the initial and exit sections of a nozzle enables one to find a unique solution successively in regions OAB, OBD, ABE, BDFE, etc.

If a gas exits with a supersonic velocity, then the initial data allow the solution inside the characteristic triangle OAG, Fig.2.3b, to be found, in particular, at the segment AC of the straight line $x = x_k$. Hence, boundary conditions cannot be put at AC. They cannot be put at $x = x_k$ either, or above point C because here the solution is fully defined by boundary conditions at $x = 0$ and the known solution at OC. The calculation can be done, for instance, successively along characteristics O'G', O"G" and so on, Fig.2.3c. Thus the correctness of the above formulation of initial and boundary conditions is confirmed.

If relaxation processes take place in the flow, then the parameters characterizing them (concentrations of mixture components, velocities and temperatures of particles, etc.) must be specified at $t = 0$ and at the initial cross-section of the nozzle.

When a direct problem is being solved for a stationary flow the impermeability condition at the nozzle walls must also be satisfied. If the flow is fully supersonic in some part of a nozzle, then to find a solution in this region it is necessary to specify all the required functions at a certain surface AB, Fig.2.4. This surface can be arbitrarily oriented in space, the only condition being that the velocity at every point must exceed the sound velocity. The uniqueness of the solution follows because its construction by the method of characteristics is only possible in a single way. In this

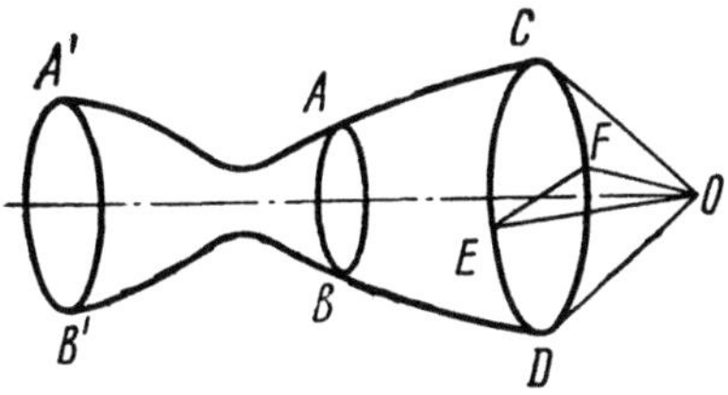

Fig.2.4. The formulation of boundary conditions for a supersonic stationary gas flow in a nozzle

case the solution is defined in ABDOC, where OCEDF is a characteristic conoid.

If the flow is subsonic, then it is necessary to specify at the initial plane either all n velocity components and pressure (or density), or n-1 velocity components, pressure and density (here n is the dimension of the velocity vector).

If a special case of isentropic isoenergetic flow is being considered, then only n-1 velocity components need to be specified at the initial plane. The correctness of such a specification of boundary conditions can easily be proved if one constructs, for example, a solution of the equation for velocity potential in an incompressible plane flow. The solution of this equation can always be presented as a real part of an analytical function of a complex argument $x + iy$. If an analytical function and all its derivatives are known at a point, then by the principle of analytical prolongation the function is known everywhere in the plane. Hence, if a solution of the equation for potential is known along an infinitesimal arc of a curve (as in the case when both velocity components are given), then it is known along the remaining part of the curve. Thus, if velocity components at points of the initial plane are given arbitrarily, then, in general, a contradiction may result and the impermeability condition may not be satisfied. Thus, contrary to the case of supersonic flow, one cannot specify the velocity vector in the subsonic part of a nozzle in the initial plane A'B' [2.12].

At the same time, if a direct problem is being solved for the region A'B'AB, there is no need to put any boundary conditions at AB located in the supersonic region, Fig.2.4. Up to now, the uniqueness of the solution of a boundary problem for nonlinear gasdynamic equations in the region A'B'AB has not been proved for a general case, though a number of numerical solutions has been obtained. It has been proved only for the case of supersonic outflow from a plane orifice, when the problem is reduced to Trikomi's problem. For this case an analytical solution has been obtained in the form of a series [2.13]. A solution of the direct problem in the region A'B'AB exists only for dimensionless flow rates $\psi \leqslant \psi^*$, and the critical value ψ^* diminishes with the contour curvature radius at the nozzle throat.

In [2.14] an attempt has been made to prove the ambiguity of ψ^* for a nozzle of fixed configuration. In this case the flow was supposed to change its velocity from subsonic to supersonic in the vicinity of the throat. The character of the flow is defined by its prehistory and depends on the way in which the critical value of the flow rate has been achieved. The statement was not proved strictly. At the same time uniqueness of the critical flow rate for the case when a nozzle works in the design regime has been proved (by solving variations of the direct problem) [2.14,15]. The idea of ambiguity of the critical flow rate seems to be most likely, especially in the case of a gas flow with a nonequilibrium physico-chemical transformation.

2.1.4 The Inverse Problem

It is convenient to solve the inverse problem and to formulate boundary conditions employing equations in which the length of an arc along a curve and the stream function are used as independent variables. The use of the stream function is extremely convenient because initial conditions in the inverse problem are usually put at a stream surface (a solid wall or a symmetry axis). Let us write down the corresponding equations, supposing for the sake of simplicity that the mixture does not contain foreign particles, but that there can take place nonequilibrium chemical reactions and excitation of vibrational degrees of freedom of molecules. In addition, the flow will be assumed stationary.

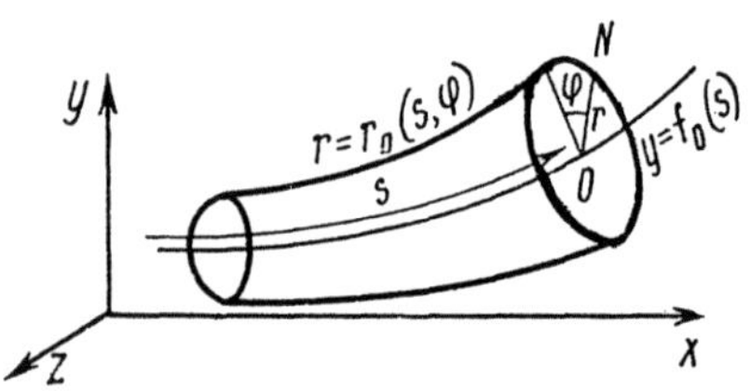

Fig.2.5. Curvilinear coordinate system s, r, φ

We introduce a curvilinear orthogonal coordinate system associated with the curve $y = f_0(s)$ lying in the plane x, y, Fig.2.5. Coordinates of a point in this system are defined by the length of the arc s, the distance along the normal to this curve r and the angle φ. Elementary geometrical considerations show that the Cartesian coordinates x, y, z are related to the curvilinear coordinates s, r, φ by the following formulas:

$$x = \int_{s_0}^{s} \cos\delta(s)\,ds - r\cos\varphi\sin\delta(s) \quad ,$$

$$y = f_0(s) + r\cos\varphi\cos\delta(s) \quad , \tag{2.84}$$

$$z = r\sin\varphi \quad ,$$

where $\delta(s)$ is an angle between a tangent to the curve $f_0(s)$ and the x axis. In the curvilinear system of coordinates s, r, φ, (2.4-7) acquire the following form [2.16]:

$$\frac{\partial}{\partial s}(\rho u H_2 H_3) + \frac{\partial}{\partial r}(\rho v H_1 H_3) + \frac{\partial}{\partial \varphi}(\rho w H_1 H_2) = 0 \quad , \tag{2.85}$$

$$\frac{u}{H_1}\frac{\partial u}{\partial s} + \frac{v}{H_2}\frac{\partial u}{\partial r} + \frac{w}{H_3}\frac{\partial u}{\partial \varphi} \pm \nu\frac{wu\sin\varphi}{RH_1}$$

$$\mp \frac{vu\cos^{\nu}\varphi}{RH_1} + \frac{1}{\gamma\rho H_1}\frac{\partial p}{\partial s} = 0 \quad , \tag{2.86}$$

$$\frac{u}{H_1}\frac{\partial v}{\partial s} + \frac{v}{H_2}\frac{\partial v}{\partial r} + \frac{w}{H_3}\frac{\partial v}{\partial \varphi} \pm \frac{u^2\cos^{\nu}\varphi}{RH_1} - \frac{w^2}{H_3} + \frac{1}{\gamma\rho H_2}\frac{\partial p}{\partial r} = 0 \quad , \tag{2.87}$$

$$\frac{u}{H_1}\frac{\partial w}{\partial s} + \frac{v}{H_2}\frac{\partial w}{\partial r} + \frac{w}{H_3}\frac{\partial w}{\partial \varphi} \mp \nu\frac{u^2\sin\varphi}{RH_1} + \frac{wv}{H_3} + \frac{1}{\gamma\rho H_3}\frac{\partial p}{\partial \varphi} = 0 \quad , \tag{2.88}$$

$$\frac{u}{H_1}\frac{\partial}{\partial s}\left(h + \frac{W^2}{2}\right) + \frac{v}{H_2}\frac{\partial}{\partial r}\left(h + \frac{W^2}{2}\right) + \frac{w}{H_3}\frac{\partial}{\partial \varphi}\left(h + \frac{W^2}{2}\right) = 0 \quad , \tag{2.89}$$

$$\frac{u}{H_1}\frac{\partial \alpha_i}{\partial s} + \frac{v}{H_2}\frac{\partial \alpha_i}{\partial r} + \frac{w}{H_3}\frac{\partial \alpha_i}{\partial \varphi} - \frac{1}{\rho}\sum_{k=1}^{N} R_{ik} = 0 \quad , \qquad (i = 1,2,\ldots,N). \tag{2.90}$$

Here u, v, w are projections of the velocity vector **W** on the axes of the curvilinear system of coordinates s, r, φ referred to the critical sound speed a_*, p and ρ are the pressure and density referred to their values p_* and ρ_* in the critical section. Parameters having dimension of length are referred to the characteristic length r_*, R(s) is the radius of curvature of the curve $f_0(s)$, $\gamma = a_*^2\rho_*/p_*$; $H_1 = 1 \mp r\cos^{\nu}\varphi/R$, $H_2 = 1$, $H_3 = 1$ are Lamé's coefficients.

The minus-sign in the expression for H_1 corresponds to concave curves $f_0(s)$ while the plus-sign corresponds to convex ones; for plane flows $\nu = 0$, while for axisymmetric and three-dimensional flows $\nu = 1$.

The system of partial differential equations (2.86-90) together with the corresponding thermal and caloric equations of state is a rather general one and describes the nonisentropic, rotational flow of a mixture of perfect gases in which relaxation processes take place and stagnation enthalpy is different for different streamlines.

It is known that the continuity equation written in a divergence form defines a solenoidal vector which may be presented as a vector product of the gradients of two functions; the continuity equation is satisifed identically in this case. Two stream functions for stationary three-dimensional flows can then be introduced [2.17]. Let us introduce a new independent variable ψ by means of an equation, the left side of which is a scalar product of the velocity vector and the vector $\nabla\psi$:

$$\frac{u}{H_1}\frac{\partial\psi}{\partial s}+\frac{v}{H_2}\frac{\partial\psi}{\partial r}+\frac{w}{H_3}\frac{\partial\psi}{\partial\varphi}=0 \quad . \tag{2.91}$$

Evidently, the surfaces $\psi\ (s,r,\varphi)=\text{const}$ are stream surfaces. Using the relations

$$\frac{\partial\psi}{\partial s}=-\frac{\partial r/\partial s}{\partial r/\partial\psi}\ ,\quad \frac{\partial\psi}{\partial r}=\left(\frac{\partial r}{\partial\psi}\right)^{-1}\ ,\quad \frac{\partial\psi}{\partial\varphi}=-\frac{\partial r/\partial\varphi}{\partial r/\partial\psi}\ , \tag{2.92}$$

we shall obtain the following formulas for transformation to new variables:

$$\left.\frac{\partial}{\partial s}\right|_{r,\varphi}=\left.\frac{\partial}{\partial s}\right|_{\psi,\varphi}-\left.\frac{\partial r/\partial s}{\partial r/\partial\psi}\frac{\partial}{\partial\psi}\right|_{s,\varphi}\ ,\qquad \left.\frac{\partial}{\partial r}\right|_{s,\varphi}=\left.\frac{1}{\partial r/\partial\psi}\frac{\partial}{\partial\psi}\right|_{s,\varphi}\ ,$$

$$\left.\frac{\partial}{\partial\varphi}\right|_{s,r}=\left.\frac{\partial}{\partial\varphi}\right|_{s,\psi}-\left.\frac{\partial r/\partial\varphi}{\partial r/\partial\psi}\frac{\partial}{\partial\psi}\right|_{s,\varphi}\quad . \tag{2.93}$$

We introduce another independent variable θ, again by means of an equation, the left side of which is a scalar product of the velocity vector and the vector $\nabla\theta$:

$$\frac{u}{H_1}\frac{\partial\theta}{\partial s}+\frac{w}{H_2}\frac{\partial\theta}{\partial\varphi}=0 \quad . \tag{2.94}$$

It is obvious that the surfaces $\theta(s,\psi)=\text{const}$ are also stream surfaces and that both vectors $\nabla\theta$ and $\nabla\varphi$ are orthogonal to the velocity vector. The lines of intersection of the surfaces $\psi=\text{const}$ and $\theta=\text{const}$ are streamlines. Thus, it follows from (2.94), that the differential equation of a streamline at the surface $\psi=\text{const}$ is

$$\frac{d\varphi}{ds} = \frac{H_1 w}{H_3 u} \quad . \tag{2.95}$$

Replacing the variables of the system (2.85-90) by a new set of variables s, ψ, θ and using transformation formulas similar to (2.93), we obtain the following system of equations which can be used to find the functions u, v, w, p, ρ, α_i, r and φ:

$$\frac{\partial p}{\partial \psi} = - \gamma \frac{G(s,\theta,\psi)}{r^\nu u(\partial\varphi/\partial\theta)^\nu} + \nu \frac{\partial p/\partial\theta}{\partial\varphi/\partial\theta}\frac{\partial\varphi}{\partial\psi} \quad , \tag{2.96}$$

$$\frac{\partial r^{1+\nu}}{\partial \psi} = \frac{1}{(\partial\varphi/\partial\theta)^\nu}\left(\frac{1+\nu}{\rho u} + \nu \frac{\partial\varphi}{\partial\psi}\frac{\partial r^{1+\nu}}{\partial\theta}\right) \quad , \tag{2.97}$$

$$\frac{\partial w r}{\partial s} = \pm \frac{ur \sin\varphi}{R} - \frac{H_1}{u\partial\varphi/\partial\theta}\left[\frac{1}{\gamma\rho}\frac{\partial p}{\partial\theta} + G(s,\theta,\psi)\frac{\partial r}{\partial\theta}\right] \quad , \tag{2.98}$$

$$\gamma \frac{\partial}{\partial s}\left(\frac{1}{2} W^2\right) + \frac{1}{\rho}\frac{\partial p}{\partial s} = 0 \quad , \tag{2.99}$$

$$\frac{\partial r}{\partial s} = \frac{v}{u} H_1 \quad , \tag{2.100}$$

$$\frac{\partial\varphi}{\partial s} = \frac{w}{ru} H_1 \quad , \tag{2.101}$$

$$h + \frac{1}{2} W^2 = H_0(\psi,\theta) \quad , \tag{2.102}$$

$$\frac{u}{H_1}\frac{\partial\alpha_i}{\partial s} = \frac{1}{\rho}\sum_{k=1}^{N} R_{ik} \quad , \quad (i = 1,2,\ldots,N) \quad , \tag{2.103}$$

$$h = \sum_{i=1}^{N} h_i\alpha_i \quad , \quad h_i = h_i(T) \quad , \quad p = \frac{R\rho T}{\mu} \quad , \tag{2.104}$$

$$G(s,\theta,\psi) = \frac{u}{H_1}\frac{\partial v}{\partial s} \pm \frac{u^2\cos^\nu\varphi}{RH_1} - \nu\frac{w^2}{H_3} \quad .$$

Here ψ and θ are referred to $\rho_* a_* r_*^2$ and $H_0(\psi,\theta)$ is the stagnation enthalpy.

The system (2.96-103) without (2.98) and (2.101) describes a plane gas flow, when $\nu = 0$.

If the mixture is nonreactive, then it follows from (2.103) that $\alpha_i = \Phi_1(\psi,\theta)$ and from (2.9) it follows that the entropy S stays constant along streamlines, that is,

$$\ln(p\rho^{-\gamma}) = \Phi_2(\psi,\theta) \quad . \tag{2.105}$$

From (2.99) the Bernoulli integral is obtained:

$$\frac{1}{2} W^2 + \frac{1}{\gamma - 1}\frac{p}{\rho} = \frac{1}{2}\frac{\gamma + 1}{\gamma - 1} \quad , \tag{2.106}$$

where γ is the specific heat ratio which is generally also a function of ψ and θ, because the critical parameters may change in the course of transfer from one streamline to another. Also, if the radius of curvature $R = \infty$ and $w = 0$, then the above system of equations coincides with the system obtained in [2.18,19] for plane and axisymmetric flows. It is obvious that in this case stream surfaces $\theta = \text{const}$ are planes and $\theta = \varphi$.

For plane and axisymmetric isentropic flows with a constant specific heat ratio the system (2.96-103) attains quite a simple form:

$$\frac{\partial p}{\partial \psi} = -\frac{\gamma}{r^\nu}\frac{\partial v}{\partial s} \quad , \qquad \frac{\partial r^{1+\nu}}{\partial \psi} = \frac{1 + \nu}{\rho u} \quad , \tag{2.107}$$

$$\frac{\partial r}{\partial s} = \frac{v}{u} \quad , \tag{2.108}$$

$$p = \rho^\gamma \quad , \tag{2.109}$$

$$u = \left[\frac{\gamma + 1}{\gamma - 1} - \frac{2}{\gamma - 1}\, p^{(\gamma-1)/\gamma} - v^2\right]^{\frac{1}{2}} \quad . \tag{2.110}$$

In the case of two-phase flows a general system of equations may be written, with s,ψ,θ or s,ψ_s,θ_s where ψ_s,θ_s are stream functions (trajectories) of particles, being used as independent variables. If a polydisperse mixture is considered, then the stream function corresponding to a certain particle may be chosen and used as ψ_s and θ_s. Let us write down a system of equations describing plane and axisymmetric flows of a two-phase mixture with uniform particles in s,ψ coordinates. We have

$$\frac{\partial p}{\partial \psi} = -\frac{\gamma}{r^\nu}\left(\frac{\rho_s}{\rho u} f_r + \frac{\partial v}{\partial s}\right) \quad , \tag{2.111}$$

$$\frac{\partial r^{1+\nu}}{\partial \psi} = \frac{1 + \nu}{\rho u} \quad , \qquad \frac{\partial r}{\partial s} = \frac{v}{u} \quad ,$$

$$\frac{\partial}{\partial s}\left(\frac{1}{2} W^2\right) + \frac{1}{\gamma\rho}\frac{\partial p}{\partial s} + \frac{\rho_s}{\rho u}\, \mathbf{fW} = 0 \quad ,$$

$$\frac{\partial}{\partial s}\left(h + \frac{1}{2} W^2\right) + \frac{\rho_s}{\rho u} (\mathbf{W}_s f + q) = 0 \quad ,$$

$$\frac{\partial(u_s \rho_s)}{\partial s} + r^{\nu}\rho\left[\frac{\partial \rho_s}{\partial \psi} (v_s u - u_s v) + \rho_s\left(u \frac{\partial v_s}{\partial \psi} - v \frac{\partial u_s}{\partial \psi}\right)\right] + \frac{\nu \rho_s v_s}{r} = 0 \quad , \qquad (2.112)$$

$$u_s \frac{\partial u_s}{\partial s} + r\rho(uv_s - vu_s) \frac{\partial u_s}{\partial \psi} = f_x \quad ,$$

$$u_s \frac{\partial v_s}{\partial s} + r\rho(uv_s - vu_s) \frac{\partial v_s}{\partial \psi} = f_r \quad ,$$

$$u_s \frac{\partial e_s}{\partial s} + r\rho(uv_s - vu_s) \frac{\partial e_s}{\partial \psi} = q \quad ,$$

$$h = h(p,T) \quad , \quad p = \rho T \quad , \quad e_s = e_s(T_s) \quad , \qquad (2.113)$$

where f_x, f_r are projections of the vector **f** upon the axes x and r and all parameters have been made dimensionless by referring them to the corresponding critical values.

The system of coordinates s,ψ,φ introduced above, is not the only possible one, but it has an advantage in comparison with Cartesian, cylindrical and spherical coordinate systems, because the boundaries of the investigated regions in the s,ψ,θ coordinates are usually planes or straight lines. This system of coordinates is not orthogonal; its analogues are called normal systems of coordinates. Coordinates which allow the boundaries of the regions being considered to be transformed into planes or straight lines by means of a suitable normalization of variables y and z are also used while solving the direct problem.

If the inverse problem is being solved for nozzles or channels of complicated curvilinear geometry with contours that are multivalued functions of Cartesian coordinates, it is convenient, instead of normal coordinate systems such as s,ψ,φ, to use orthogonal systems associated with streamlines. A well-known example of orthogonal coordinates for potential flows is the ψ,Φ coordinate system, where Φ is the velocity potential.

Let us consider another orthogonal system of coordinates used for the calculation of gas flows in complicated curvilinear channels [2.20]. This system will be introduced for flows whose parameters do not depend on the angular coordinate φ but which may have swirl (or, in a plane case, downwash parallel to the axis Oz). Surfaces $\psi = \text{const}$, defined by (2.91), represent the first family of coordinate surfaces in this system. Planes $\varphi = \text{const}$ are chosen as the second family. Projections of streamlines on these planes coincide with the projections of surfaces $\psi = \text{const}$.

Now we shall consider a method of introducing the third family of coordinate surfaces. Let s be a distance along the projection of a streamline on a plane $\varphi = \text{const}$. Let us further write down an expression for the differential of an arc:

$$\delta s = -V^{-1}(u\,dx + v\,dr) \quad , \tag{2.114}$$

where u,v,V are projections of the velocity vector **W** on axes x,r and a plane $\psi = \text{const}$, respectively, that is $V^2 = W^2 - w^2$. Generally, δs is not a total differential, and hence there exists an integrating factor, such that

$$d\sigma = g\delta s = (-guV^{-1})dx + (-gvV^{-1})dr \tag{2.115}$$

is a total differential. Therefore,

$$\frac{\partial\sigma}{\partial x} = -guV^{-1} \quad , \quad \frac{\partial\sigma}{\partial r} = -gvV^{-1} \quad , \quad \frac{\partial\sigma}{\partial\varphi} = 0 \quad . \tag{2.116}$$

It can easily be shown that surfaces $\psi = \text{const}$, $\varphi = \text{const}$, $\sigma = \text{const}$ are orthogonal, which is why surfaces $\sigma = \text{const}$ may be chosen as coordinate surfaces of the third family.

In the case of two-dimensional potential flows, $\sigma = \text{const}$ at the lines of constant potential $\Phi = \text{const}$, because in this case velocity serves as an integrating factor for the length of arc δs. The length of an arc along the projection of a characteristic streamline on a plane $\psi = \text{const}$ may be considered as σ. Using variables ψ,σ,φ under the condition that the functions being sought do not depend on φ, although the circumferential component of velocity w may exist (if swirl takes place), one gets the following system of equations:

$$\frac{\partial r^{1+\nu}}{\partial\psi} = \frac{(1+\nu)u}{\rho V^2} \quad , \tag{2.117}$$

$$\frac{\partial x}{\partial\psi} = -\frac{v}{r^{\nu}\rho V^2} \quad , \tag{2.118}$$

$$\frac{\partial p}{\partial\psi} = \frac{\gamma g}{r^{\nu}}\left[\frac{u}{V}\left(\frac{\partial v}{\partial\sigma} + \frac{\nu w^2}{r^{\nu} g V^2}\right) - \frac{v}{V}\frac{\partial u}{\partial\sigma}\right] \quad , \tag{2.119}$$

$$\ln\frac{p}{\rho^{\nu}} = \Phi_2(\psi) \quad , \quad \frac{1}{2}W^2 + \frac{1}{\gamma-1}\frac{p}{\rho} = \frac{1}{2}\frac{\gamma+1}{\gamma-1} \quad ,$$

$$r^{\nu} w = C(\psi) \quad , \quad V = (W^2 - w^2)^{\frac{1}{2}} \quad , \quad g = \left[\left(\frac{\partial x}{\partial \sigma}\right)^2 + \left(\frac{\partial r}{\partial \sigma}\right)^2\right]^{-\frac{1}{2}} \quad , \tag{2.120}$$

$$u = -gV \frac{\partial x}{\partial \sigma} \quad , \quad v = -gV \frac{\partial r}{\partial \sigma} \quad . \tag{2.121}$$

Let us now consider the formulation of initial and boundary conditions for the inverse problem of nozzle gas flows. We shall use again the theory of characteristic surfaces for a quasi-linear system of first-order partial differential equations. According to the general theory stated in Sect.2.1.2, it is possible, having found characteristic surfaces and the normals to them, to replace the initial system of m equations by an equivalent system composed of a linear combination of these equations, which can be divided into two parts. One of the parts consists of m_1 compatibility equations and the other consists of $m-m_1$ additional equations [2.5]. In particular, at the stream surfaces, which in the case of steady inviscid subsonic or supersonic flows coincide with the characteristic surfaces, two equations of motion (2.98,99), the energy equation (2.102) and N relaxation equations (2.103) containing no derivatives with respect to normals to streamlines, play the role of compatibility equations. One of the equations of motion (2.96) and the continuity equation [or (2.97) which follows from it] containing derivatives with respect to normals to stream surfaces serve as the additional equations. Equations (2.100,101) are the equations of direction. Note that the functions, derivatives of which enter the additional equations, cannot be discontinuous in the direction normal to a stream surface.

For the general case of a three-dimensional stationary flow of a reacting gas the inverse problem of the theory of nozzles can be formulated as follows. A velocity component $u = u_0(s,\theta)$ (or pressure/density) and a function $r = r_0(s,\theta)$ are specified at a surface $\psi = \psi_0 = \text{const}$. Functions $w = w_0(\theta,\psi)$, $\rho = \rho_0(\theta,\psi)$, $\alpha_i = \alpha_{i0}(\theta,\psi)$, $\varphi = \varphi_0(\theta,\psi)$ are given at the plane $s = s_0$. It is desired to find a family of stream surfaces and the values of flow parameters in the vicinity of the initial stream surface $\psi = \psi_0$. Obviously, the function $H_0(\psi,\theta)$ entering (2.102) must be given. It should be emphasized that specification of velocity components v and w at a stream surface $\psi = \psi_0$ will only overdetermine the problem, because having equations of compatibility and direction, one can find these components using the given velocity component and the known equation of the stream surface. If the flow of a nonreacting mixture is considered, then instead of density, entropy $S = \ln(p\rho^{-\gamma}) = \Phi_2(\psi,\theta)$ may be given at the plane $s = s_0$. Additionally, instead of (2.99,102), one can use Bernoulli's equation (2.106) with a fixed function $\gamma(\psi,\theta)$.

In the case of plane or axisymmetric isentropic ($S = \text{const}$) or isoenergetic ($H_0 = \text{const}$) flows without swirl the two additional equations (2.107) allow boundary conditions to be put only at a streamline. Thus, the inverse problem is reduced to an initial-value problem for the system (2.107-110) with data given at a streamline, that is, at a characteristic. The possibility of formulating an initial-value problem is explained by the fact that a streamline is at the same time a degenerate characteristic.

If the flow is not isentropic and isoenergetic or if it is swirled, then a streamline is no longer a degenerate characteristic and two additional equations are insufficient to find the unknown values in its vicinity. In such a case it is necessary to specify in addition the entropy and stagnation enthalpy distribution or the circumferential velocity component at the initial plane.

A similar situation exists in the case of a three-dimensional flow even in the case when it is isentropic and isoenergetic. Also, specifying data at a stream surface which is at the same time a characteristic surface, it is necessary to put boundary conditions at a surface which is not characteristic, for instance, at the plane $s = s_0$.

2.2 Some Elementary Theories

To get an analytical or even numerical solution of the system (2.3-7) is a very difficult task, not only due to the complexity of the physico-chemical processes taking place in nozzle gas flows but also due to the presence of subsonic, transonic and supersonic zones in the same nozzle flow, these being described by different types of partial differential equations. To calculate a nozzle flow one has to solve simultaneously elliptic, parabolic and hyperbolic equations. Luckily, the construction of certain analytical solutions based on approximate assumptions allows the methods of solution to be simplified and a number of significant qualitative results to be obtained. Therefore, we shall consider some elementary theories which reveal the most important laws of nozzle gas flows. These are: (i) the one-dimensional theory of nozzle flows, (ii) the theory of source-sink flows and (iii) the theory of a simple wave or the Prandtl-Meyer flow.

2.2.1 One-Dimensional Theory

Let us begin with the one-dimensional theory of nozzle flows. We shall consider a steady flow of a perfect gas without relaxation processes. (In Chap.6 the reader will find a detailed discussion of the application of the one-dimensional theory to the calculation of flows with nonequilibrium physico-chemical processes.) By the main hypothesis of the one-dimensional theory the flow is uniform over any cross-section perpendicular to the nozzle axis and the velocity vector is directed practically along the axis which may be curvilinear. The assumption holds either in the case where the nozzle is relatively slender or in the case where the cross-section of the streamtube considered is small compared to the characteristic transverse dimensions of the flow field and, hence, the transverse velocity components may be neglected as a first approximation. Then the gas parameters depend on the longitudinal coordinate only and (2.99) and (2.102), valid for a streamline, can be used to calculate them. It follows from (2.97) that

$$r^{1+\nu}\rho W = (1+\nu)\psi \quad . \tag{2.122}$$

Denoting the area of a nozzle's cross-section by F, the total mass flow by Q and the mass flow by j, one gets the following continuity equation:

$$\rho WF = jF = Q \quad . \tag{2.123}$$

In the case of one-dimensional flow it follows from (2.99,102,123) that

$$\frac{d\bar{j}}{dW} = \rho(1-M^2) \qquad \text{or} \qquad \frac{d\bar{j}}{dp} = \frac{M^2-1}{\gamma W} \quad , \tag{2.124}$$

where $\bar{j} = j/\rho_* a_*$ is the dimensionless mass flow, $M = W/a$ is the Mach number and the pressure, velocity and density are referred to their critical values (which means that the dimensionless velocity W coincides with the velocity coefficient λ). Also,

$$\frac{dF}{F} = (M^2-1)\frac{dW}{W} = (1-M^2)\frac{dp}{\gamma\rho W^2} \quad . \tag{2.125}$$

Relations (2.124) show that in a subsonic flow j grows with the growth of velocity and the fall of pressure; on the contrary, in a supersonic flow it becomes smaller.

The mass flow attains its maximum value $j_* = \rho_* a_*$ in those parts of a flow field where the velocity and density of the gas acquire their critical values. It becomes zero for $W = 0$ and $p = p_0$ (stagnation pressure), or for $p = 0$ and $W = W_{max}$ — the velocity attained by the gas discharged into vacuum. The dimensionless mass flow j depends on the Mach number (or λ) and specific heat ratio γ. For a perfect gas the relation is as follows:

$$\bar{j} = q(\lambda) = \left(\frac{\gamma + 1}{2}\right)^{1/(\gamma-1)} \lambda \left(1 - \frac{\gamma - 1}{\gamma + 1}\lambda^2\right)^{1/(\gamma-1)} \quad . \tag{2.126}$$

Evidently $j = F_*/F$ where F_* is the area of the nozzle's throat. Figure 2.6 shows the function $q(\lambda)$ for various γ. It can be seen that two values of the velocity coefficient correspond to each value of $\bar{F} = F/F_*$ —a subsonic one ($\lambda < 1$) and a supersonic one ($\lambda > 1$). It follows from (2.125) that if a flow is subsonic, then its velocity grows and pressure falls as the nozzle narrows and vice versa for a supersonic flow.

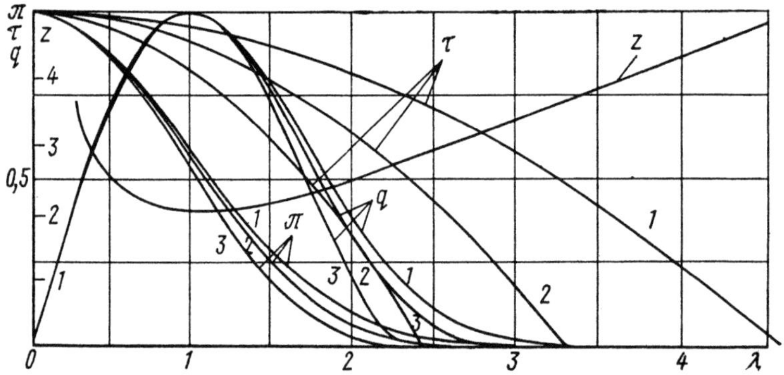

Fig.2.6. Dependence of gasdynamic functions on the velocity coefficient λ at various values of adiabatic exponent γ: (*1*) $\gamma = 1.1$, (*2*) $\gamma = 1.2$, (*3*) $\gamma = 1.4$

According to (2.125), the critical velocity can be attained only in the throat of a Laval nozzle, where $dF = 0$. Downstream of the throat the velocity may either continue growing or once again become smaller, depending on the ratio of the external pressure to the stagnation pressure. If the flow velocity is not critical in the throat, then it has an extreme value which is a maximum if the velocity at the initial section of a nozzle is subsonic, or a minimum if it is supersonic. Obviously, in the framework of the one-dimensional approximation, supersonic velocity cannot be attained in a convergent channel if the initial velocity is subsonic (in the case of a three-dimensional flow it can). The maximum velocity in such a channel is equal to the sound velocity. Supersonic velocity can be obtained only in a nozzle which has a narrowing part, which widens again beyond the throat.

The maximum value of j_* is attained when the flow velocity becomes critical. The maximum possible value of the total mass flow Q_m for a given nozzle

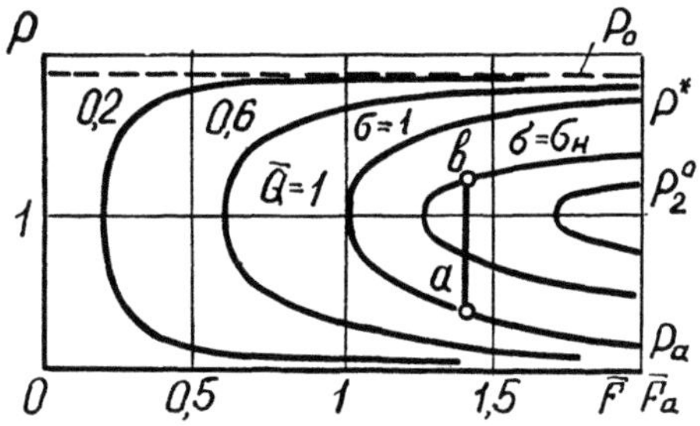

Fig.2.7. Dependence of pressure on the cross-sectional area at different regimes of nozzle flow

is obviously equal to j_*F_*. Using (2.126) and allowing the existence of a shock wave in the supersonic part of a nozzle, one obtains for the dimensionless total mass flow $\bar{Q}$:

$$\bar{Q} = \frac{Q}{Q_m} = p^{1/\gamma}\left[\frac{\gamma+1}{\gamma-1} - \frac{2}{\gamma-1}\,p^{(\gamma-1)/\gamma}\right]^{\frac{1}{2}}\bar{F}\sigma \quad , \qquad (2.127)$$

where $\sigma = p_{02}/p_{01}$ is the coefficient of the total pressure loss in a shock wave, p_{01} and p_{02} are the stagnation pressure values in front of and behind a shock wave and p is referred to its critical value as before. For different values of $\bar{Q}$ and σ one can obtain the dependence of p on the dimensionless area of a nozzle's cross-section $p = p(\bar{F})$. Figure 2.7 shows the corresponding curves for $\gamma = 1.4$. Here σ has been put equal to 1 for $\bar{Q} < 1$ and varied for $\bar{Q} = 1$. We see that two values of p—a subsonic one and a supersonic one—correspond to each value of F. The upper branches of the curves asymptotically approach the straight line $p = p_0 = [(\gamma+1)/2]^{\gamma/(\gamma-1)}$, while the lower branches approach in the same manner the straight line $p = 0$. Obviously, for a given nozzle, the physically meaningful parts of the curves are those which lie to the left of the straight line $\bar{F} = 1$. The branches of the curves corresponding to $\bar{Q} < 1$ and $p > 1$ describe the pressure change between the initial section of a nozzle and its throat if the flow is subsonic at the initial section, while the branches corresponding to $\bar{Q} < 1$ and $p < 1$ do the same if the flow velocity in the nozzle's initial section is supersonic.

Let us see what happens in a Laval nozzle when the external pressure p_H is varied. Let the gas enter the nozzle from a sufficiently large reservoir where it is stagnant and has uniform pressure p_0 which for all curves in Fig.2.7 is approached asymptotically as $\bar{F} \to \infty$. The dimensionless area of the exit section will be denoted by $\bar{F}_a$.

If the external pressure p_H is equal to p_0, the gas remains everywhere at rest. If p_H is only slightly below p_0, the entire flow is subsonic and attains its maximum speed at the throat, where $\bar{F} = 1$. Evidently, to determine the flow parameters, one should select the curve at the upper branch of which $p = p_H$ and $\bar{F} = \bar{F}_a$. The part of the curve from the asymptote $p = p_0$ to its intersection with the straight line $\bar{F} = 1$ corresponds to the flow in the convergent part of a nozzle. In its divergent part the flow velocity decreases while pressure increases. Here the flow is described by the same branch of the curve, but in the opposite direction — from $F = 1$ to $F = F_a$. The dimensionless total mass flow $\bar{Q}$ is uniquely determined by the value of p_H.

This method of flow description, when the upper branches of the curves in the plane $p\bar{F}$ are followed twice, once in each direction, is applicable only till they cross the straight line $\bar{F} = 1$. In this case the flow is everywhere subsonic and the total mass flow $\bar{Q} < 1$. While the external pressure decreases, the velocity at the throat and the total mass flow increase. At a certain $p_H = p^*$ the sound velocity is attained at the throat; however, elsewhere the flow remains subsonic. Now the total mass flow is a maximum ($\bar{Q} = 1$) and doesn't change with a further decrease in p_H. The magnitude of p^* depends on $\bar{F}_a$ and can be found from

$$(p^*)^{1/\gamma}\left[\frac{\gamma + 1}{\gamma - 1} - \frac{2}{\gamma - 1}(p^*)^{(\gamma-1)/\gamma}\right]^{\frac{1}{2}}\bar{F}_a = 1 \quad . \tag{2.128}$$

Hence, for $\bar{F}_a = 1$, $p^* = 1$. Thus, in a convergent nozzle the sound velocity is attained in the narrowest section — the nozzle's throat. The minimum external pressure at which this happens is equal to the critical pressure — in other words it is $[2/(\gamma + 1)]^{\gamma/(\gamma-1)}$ times smaller than the stagnation pressure.

When $\bar{F}_a$ increases, p^* grows and for $\bar{F}_a \to \infty$, $p^* \to p_0$. However, in practice the presence of the boundary layer prevents sonic velocity being attained in the throat when the difference between the stagnation pressure and the external pressure is negligible. Nevertheless, a Laval nozzle allows the sound velocity to be reached when the stagnation pressure is only 10 to 20% higher than the external one, while to do the same in a purely convergent channel for a diatomic gas, the stagnation pressure should exceed the external one by a factor of 1.89.

If the external pressure is below p^*, another type of flow takes place. In front of the throat the flow remains subsonic; it is represented by the upper branch of the curve $p = p(\bar{F})$ for $\bar{Q} = 1$ and doesn't depend on the external pressure. Beyond the throat the flow accelerates to supersonic speeds and corresponds to the lower branch of the curve which crosses the straight line $\bar{F} = \bar{F}_a$ at the point $p = p_a$. As earlier, the value of p_a is determined by (2.128). If $p_a = p_H$, then the nozzle is said to work in the design regime.

In this case the nozzle flow is isentropic; its velocity increases while the pressure, density and temperature decrease monotonically.

Let us consider the case when $p_a < p_H < p^*$. It is evident that this condition cannot be fulfilled in the course of isentropic expansion and a shock wave must appear in the supersonic part of the flow. In the framework of the one-dimensional approximation such a shock wave is normal and forms a right angle with the nozzle axis. The pressure necessary at the exit section can be obtained by introducing a shock wave with a given σ. The value of σ is found from the condition that the upper branch of the curve $p = p(\bar{F})$, obtained from (2.127) at $\bar{Q} = 1$, crosses the straight line $\bar{F} = \bar{F}_a$ at the point $p = p_H$. Point b on the curve (Fig.2.7) and point a on the lower branch of the curve $p = p(\bar{F})$ at $\bar{Q} = 1$ and $\sigma = 1$ correspond to the location where the shock wave is formed having the specified value $\sigma = \sigma_H$. The points can be found from (2.127) and Hugoniot's relation for a normal shock wave,

$$p_2/p_1 = \frac{2\gamma}{\gamma + 1} M^2 - \frac{\gamma - 1}{\gamma + 1} \quad . \tag{2.129}$$

Thus, the curve $p = p(\bar{F})$ at $\bar{Q} = 1$, $\sigma = 1$ as far as point a corresponds to isentropic flow in the subsonic and supersonic parts of a nozzle up to the location where the shock wave appears. Then the normal shock transfers the flow into state b at the curve corresponding to $\bar{Q} = 1$ and $\sigma = \sigma_H$. Behind the shock wave the flow continues as an isentropic subsonic decelerating flow with pressure increasing downstream until p_H is attained. Here the stagnation pressure is lower than in front of the shock wave; therefore the critical parameters p_{*2} and ρ_{*2} differ from those at the initial section of the nozzle while the critical velocity remains continuous in the shock wave.

However, such a flow exists only if the values of p_H satisfy the condition $p_2^0 \leqslant p_H < p^*$. The value of p_2^0 is calculated from (2.129) into which M_a, corresponding to p_a, is substituted. If p_H is only slightly below p^*, then a weak normal shock wave appears in the supersonic vicinity of the throat. As p_H decreases the shock wave moves downstream and reaches the exit section at $p_H = p_2^0$. Calculations give $p_2^0/p_{*2} \approx 1.7$ for $M_a > 2$.

As the external pressure decreases further, that is at $p_a < p_H \leqslant p_2^0$, oblique shock waves form in the jet flowing from the nozzle; here compression from p_a to p_H takes place. In the framework of the one-dimensional approximation it may be assumed that down to the exit section the gas flow is described at these values of p_H and at other $p_H < p_a$ by the curve $p = p(\bar{F})$ for $\bar{Q} = 1$ and $\sigma = 1$. At $p_H < p_a$ the Prandtl-Meyer flow develops in the vicinity of the nozzle

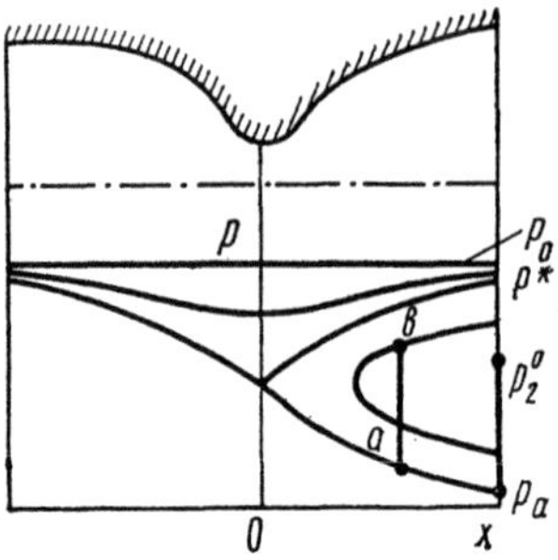

Fig.2.8. Pressure distribution along a nozzle at different values of pressure in the environment

edges. Evidently this flow and flows with oblique shock waves are essentially three-dimensional and cannot be described in the framework of the one-dimensional approximation.

With the help of these results and using the dependence of the area F of the nozzle's cross-section on the coordinate x, one is able to obtain pressure as a function of x. Figure 2.8 shows pressure distribution for the types of flow considered above. The curve corresponding to the maximum total mass flow $\bar{Q}=1$ bifurcates at the throat where the flow attains the sound velocity. In the exit section $p=p^*$ corresponds to the upper curve and $p=p_a$ to the lower one. The slopes of the branches of this curve at the throat are, according to (2.128),

$$\frac{dp}{dx} = \pm\gamma\left(\frac{1}{\gamma+1}\,\frac{d^2F}{dx^2}\right)^{\frac{1}{2}} \quad , \tag{2.130}$$

where the plus-sign refers to the upper curve. The corresponding derivatives of the velocity coefficient at the throat are equal to

$$\frac{d\lambda}{dx} = \mp\left(\frac{1}{\gamma+1}\,\frac{d^2F}{dx^2}\right)^{\frac{1}{2}} \quad . \tag{2.131}$$

Naturally, the real structure of flow in a nozzle is more complicated than that described by the one-dimensional theory. Nevertheless, the theory has made it possible to predict that a pressure p^* exists, such that at $p_H<p^*$ the total mass flow and the other flow parameters in the subsonic part of a nozzle do not depend on the external conditions, and that also a pressure p_2^0 exists, such that at $p_H<p_2^0$ the entire nozzle flow does not depend on the external conditions. Both these extremely important conclusions are drawn from the fact that small disturbances propagate in a gas with the sound velocity and cannot penetrate into a nozzle if they encounter a section where the flow velocity is sonic or supersonic.

The one-dimensional theory cannot generally be used to calculate local parameters of three-dimensional flows because of the large transverse gradients of gasdynamic parameters. At the same time the theory allows integral flow parameters, such as the total mass flow or the nozzle impulse, to be found much more precisely than the local ones. The total mass flow and momentum flux, which is usually called the nozzle impulse and is equal to the thrust developed by the nozzle in the direction x, can be written for an arbitrary section of a nozzle in the form

$$Q = \int_0^F \rho W \cos\theta \; dF \quad , \qquad I = \int_0^F (\rho W^2 \cos^2\theta + p) dF \quad , \tag{2.132}$$

where θ is the angle between the velocity vector and the x direction. Denoting parameters of the one-dimensional flow by the superscript 0 and using (2.126), one obtains from (2.132):

$$\bar{Q} = \int_0^{\bar{F}} \left[q(\lambda^0) + \left.\frac{dq}{d\lambda}\right|_{\lambda=\lambda^0} (\lambda - \lambda^0) + \frac{1}{2} \left.\frac{d^2q}{d\lambda^2}\right|_{\lambda=\lambda^0} (\lambda - \lambda^0)^2 + \ldots \right] \left(1 - \frac{\theta^2}{2} + \ldots\right) d\bar{F} \quad , \tag{2.133}$$

$$\bar{I} = \int_0^{\bar{F}} \left[f(\lambda^0) + \left.\frac{df}{d\lambda}\right|_{\lambda=\lambda^0} (\lambda - \lambda^0) + \frac{1}{2} \left.\frac{d^2f}{d\lambda^2}\right|_{\lambda=\lambda^0} (\lambda - \lambda^0)^2 + \ldots - q(\lambda^0)\lambda^0(\theta^2 + \ldots) \right] d\bar{F} \quad , \tag{2.134}$$

where

$$\bar{Q} = Q/\rho_* a_* F_* \quad , \qquad \bar{I} = I/\rho_* a_*^2 F_* \quad , \qquad f(\lambda) = \frac{\gamma + 1}{2\gamma} q(\lambda)(\lambda + \lambda^{-1}) \quad .$$

It can be easily shown that $dq/d\lambda|_{\lambda=1} = df/d\lambda|_{\lambda=1} = 0$. Hence, for integration at the throat section, (2.133) and (2.134) acquire the form

$$\bar{Q} - \bar{Q}^0 \approx - \int_0^1 \left[\frac{\theta^2}{2} - \frac{1}{2} \left.\frac{d^2q}{d\lambda^2}\right|_{\lambda=1} (\lambda - 1)^2 \right] d\bar{F} \quad , \tag{2.135}$$

$$\bar{I} - \bar{I}^0 \approx - \int_0^1 \left[\theta^2 - \frac{1}{2} \left.\frac{d^2f}{d\lambda^2}\right|_{\lambda=1} (\lambda - 1)^2 \right] d\bar{F} \quad , \tag{2.136}$$

$$\bar{Q}^0 = q(\lambda^0)\bar{F} \quad , \qquad \bar{I}^0 = f(\lambda^0)\bar{F} \quad .$$

It follows from the mean-value theorem that the total mass flow and momentum flux in the throat differ from the corresponding values given by the one-dimensional approximation by second-order quantities only because at the throat of real nozzles $\theta < 0.1$ and $\lambda - 1 < 0.1$, while $\bar{Q}^0 = 1$ and $\bar{I}^0 = 2$.

In addition, the values of $\bar{Q}$ and $\bar{I}$ at the throat are smaller than $\bar{Q}^0$ and $\bar{I}^0$ because for $\lambda = 1$, $d^2q/d\lambda^2 < 0$ and $d^2f/d\lambda^2 < 0$. The same result follows from physical considerations because the maximum total mass flow and momentum flux are attained at the uniform flow parameters.

Since the total mass flow through any section of a nozzle is equal to the mass flow through the throat, the difference between the true value of the total mass flow and its value given by the one-dimensional theory is of second order of magnitude.

Using (2.133,134) and the mean-value theorem, one can easily show that at an arbitrary section the difference $\bar{I} - \bar{I}^0$ is proportional to θ^2. In other words, in cases of practical interest, $\bar{I}$ differs from $\bar{I}^0$ by no more than 10%. It should be noted that at the same time differences in the values of local parameters may be quite large.

In accordance with the facts mentioned above, the one-dimensional theory is at present the basis for calculating various parameters of nozzle flows for engineering purposes. There exist widely-used tables of such flows, where ratios of the pressure, density and temperatures to their stagnation values and the dimensionless cross-section area $\bar{F}$ are presented as functions of λ (or M) for various values of the adiabatic exponent γ (see, for example, [2.21]).

It should also be noted than the one-dimensional theory in the case of a perfect gas without relaxation processes allows the state of flow in any cross-section of a nozzle to be found if $\bar{F}$ is given and the type of flow—subsonic or supersonic—is known. The absolute dimensions of the nozzle and the shape of the channel upstream and downstream of the section considered play no role because (2.123-125) do not contain any characteristic length. The one-dimensional theory gives an analogous result for equilibrium or frozen flows. On the other hand, in the case of the nonequilibrium flow of a reactive gas, flow parameters for a fixed $\bar{F}$ depend also on the nozzle configuration upstream and downstream of the section and on its absolute dimensions because there is a characteristic length in such flows, i.e., the size of the relaxation zone.

To conclude this section, let us write down formulas allowing flow parameters in the one-dimensional approximation to be calculated if the geometry of a nozzle, i.e.,, $\bar{F}(x) = F/F_*$, is known.

Consider a case when the flow in the divergent part of a nozzle is supersonic and, hence, parameters upstream of a cross-section do not depend on the external pressure. Standard gasdynamic functions of the velocity coefficient λ (or the Mach number M) will be used. Then the ratios of gasdynamic parameters to their stagnation values (the latter denoted by the subscript 0) are determined by the following formulas:

$$\frac{p}{p_0} = \pi(\lambda) = \left(1 - \frac{\gamma - 1}{\gamma + 1}\lambda^2\right)^{\gamma/(\gamma-1)} , \tag{2.137}$$

$$\frac{T}{T_0} = \tau(\lambda) = \left(1 - \frac{\gamma - 1}{\gamma + 1}\lambda^2\right) , \tag{2.138}$$

$$\frac{\rho}{\rho_0} = \varepsilon(\lambda) = \left(1 - \frac{\gamma - 1}{\gamma + 1}\lambda^2\right)^{1/(\gamma-1)} . \tag{2.139}$$

The velocity coefficient λ is calculated with the help of (2.126) if $\bar{F}$ and γ are given. If tables of gasdynamic functions are used, then it is convenient to take $q(\lambda) = F_*/F$ as an independent variable and to look for λ and other functions in the tables.

The Mach number, flow velocity, flow rate, impulse and specific impulse $I_s = I/Q$ of a nozzle are calculated from

$$M^2 = \frac{2}{\gamma + 1}\lambda^2\left(1 - \frac{\gamma - 1}{\gamma + 1}\lambda^2\right)^{-1} , \tag{2.140}$$

$$W = \left\{\frac{2\gamma}{\gamma - 1} RT_0\left[1 - [\pi(\lambda)]^{(\gamma-1)/\gamma}\right]\right\}^{\frac{1}{2}} , \tag{2.141}$$

$$Q = A(\gamma)\frac{p_0 F_*}{\sqrt{RT_0}} , \quad A(\gamma) = \left(\frac{2}{\gamma + 1}\right)^{(\gamma+1)/2(\gamma-1)} \sqrt{\gamma} , \tag{2.142}$$

$$I = \varepsilon(1) p_0 F_* z(\lambda) , \quad I_s = \left(\frac{\gamma + 1}{2\gamma} RT_0\right)^{\frac{1}{2}} z(\lambda) , \quad z(\lambda) = \lambda + \frac{1}{\lambda} , \tag{2.143}$$

where R is the gas constant. The functions $q(\lambda)$, $\pi(\lambda)$, $\tau(\lambda)$, $z(\lambda)$ are shown in Fig.2.6 for different values of γ. It follows from the figure that at $q(\lambda) = \text{const}$ (i.e., if a nozzle with a given area ratio $\bar{F}$ is considered, λ, $\pi(\lambda)$, $\tau(\lambda)$ and $z(\lambda)$ grow while γ decreases. If the stagnation parameters stay constant, the specific impulse, pressure, exit velocity and temperature increase. The result is physically evident, because a decrease in γ is due to an increase in the number of degrees of freedom of the molecules. The greater the number of excited degrees of freedom, the greater is their contribution of energy to translational degrees of freedom and the energy of directed motion. At $\gamma = 1$ the flow is isothermal; hence, $T = T_0$, $\rho/\rho_0 = p/p_0 = \exp(-\lambda^2/2)$ and $M = \lambda$.

If the flow in the convergent and divergent parts of a Laval nozzle is subsonic, then its parameters are calculated in another manner. Usually the external pressure is known. Since for a subsonic flow the external pressure is equal to the exit pressure, (2.137) may be used to determine λ at the exit section and the total mass flow Q,

$$Q = A(\gamma)p_0Fq(\lambda)(RT_0)^{-\frac{1}{2}} \quad . \tag{2.144}$$

Since the mass flow is the same for all cross-sections of a nozzle, this formula allows $q(\lambda)$ to be found for any cross-section F and then, using (2.126,137-139) or gasdynamic tables, also allows λ and the other functions to be found. The Mach number, velocity and specific impulse are calculated from (2.140,141,143). If the flow is subsonic, (2.142) cannot be used to calculate the total mass flow Q because in this case the flow velocity at the throat is below the sound velocity and $q(\lambda) \neq 1$.

Engineering calculations of flows in curvilinear channels are often done with the help of the method of inscribed circles which gives better results than the classical one-dimensional theory. The method is based on the one-dimensional theory, the difference being that in its framework, the velocity is considered to be constant along circular arcs orthogonal to the walls. However, it should be noted that this method and other known variations of the classical one-dimensional theory do not give the exact values of gasdynamic parameters at the walls of a curvilinear channel. Strictly speaking, they are valid only for channels with rectilinear axes.

Let us consider a modification of the classical one-dimensional theory. While doing so, we shall use the system (2.177-120). From (2.119), for plane and axisymmetric flows without swirl ($w = 0$),

$$r^{\nu} \frac{\partial p}{\partial \psi} = \gamma VK \quad , \tag{2.145}$$

where

$$K = \left(\frac{\partial x}{\partial \sigma}\frac{\partial^2 r}{\partial \sigma^2} - \frac{\partial r}{\partial \sigma}\frac{\partial^2 x}{\partial \sigma^2}\right)g^3$$

is the curvature of streamlines. Assuming as a first approximation that the sections $\sigma = \text{const}$ are plane, we obtain the following relation for determining the pressure along the normal to the channel's axis n:

$$\frac{\partial p}{\partial n} = \gamma \rho V^2 K \quad . \tag{2.146}$$

For rectilinear channels the curvature of streamlines is zero. By (2.146), the pressure and hence the other parameters remain constant across the channel in accordance with the classical one-dimensional theory. However, in the case of curvilinear channels, $\partial p/\partial n \neq 0$ for all points at which $VK \neq 0$. Thus the one-dimensional theory in its classical form is valid only for configurations with rectilinear axes. For curvilinear channels it gives only mean values of flow parameters.

With the help of (2.146) the distribution of parameters in the transverse direction can be obtained. From (2.146) one gets the following expression for pressure at the walls of a curvilinear channel:

$$p = p^0 \pm \frac{1}{2} \gamma h \rho^0 (V^0)^2 (R^0)^{-1} \quad , \tag{2.147}$$

where p^0, ρ^0, V^0 are the mean values of parameters given by the classical one-dimensional theory, h is the width of the channel and $R^0 = (K^0)^{-1}$ is the mean radius of curvature of a streamline. In this formula all gasdynamic parameters are related to their critical values. The plus-sign refers to the external, and the minus-sign to the internal wall of the channel. Consideration of the second term on the right side of (2.147) shows that this correction is due to the action of centrifugal forces. When the gas velocity is high or the channel has a sharp turn, these forces cannot be neglected.

Equation (2.147) can be generalized for the case of a three-dimensional swirling flow to give

$$p = p^0 \pm \frac{1}{2} \gamma h \rho^0 (V^2/R_1^0 + w^2/R_2^0) \quad , \tag{2.148}$$

where R_1^0 and R_2^0 are the mean radii of the streamlines' major curvatures.

Let us compare the values of pressure at the walls of a channel calculated on the basis of the classical one-dimensional theory and from (2.147) with the exact data. In [2.22] the following values of pressure at the upper and lower walls at $x = 0$ were found for the channel shown in Fig.2.9: $p_u = 0.8975$, $p_l = 1.1036$. According to the one-dimensional theory $p = 1$ at the throat. Equation (2.147) gives $p_u = 0.8963$, $p_l = 1.1037$. Thus, the errors are quite small. In the throat of a Laval nozzle with the radius of curvature of the contour $R = 1.6\ r_*$, the pressure at the wall is $p_u = 0.7808$ and the pressure at the axis is $p_H = 1.201$. Equation (2.147) gives for $p^0 = \rho^0 = V^0 = 1$ and $R^0 = 3.2\ r_*$ (here the mean value of the axis and the contour curvature is taken) $p_u = 0.782$, $p_H = 1.22$, i.e. the error is within 1.5%.

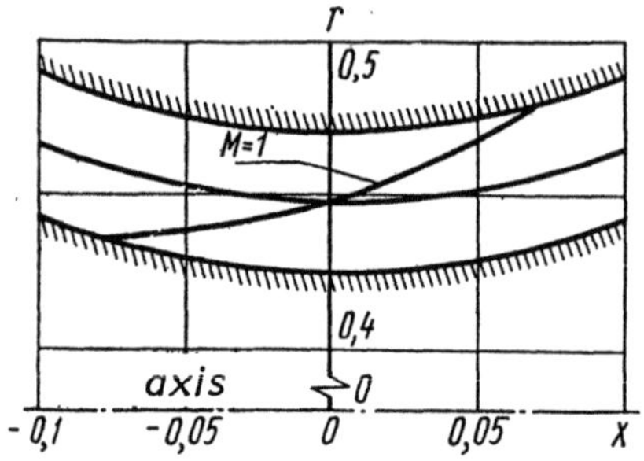

Fig.2.9. Geometry of a curvilinear annular channel. Position of the sonic line

While calculating pressure from (2.147) for nozzles with small radii of curvature ($R/r_* < 1$) the flow field should be divided into a number of zones by means of a fictitious orthogonal grid analogous to the ψ, σ coordinate grid; then by determination of the radii of curvature for a number of streamlines the average radius should be calculated.

The one-dimensional theory can also be developed for multilayer flows [2.23]. The static pressure in all cross-sections of a nozzle is supposed to be constant. The other parameters – temperature, density, velocity, adiabatic exponent, stagnation pressure and temperature – may be different for different layers. An analysis similar to the one above shows that choking of flow takes place when

$$\beta = \int \frac{M_i^2 - 1}{\gamma_i M_i^2} dF = 0$$

in the throat. Here the parameters of the ith layer are denoted by the subscript i. The choking condition for a one-layer flow is obtained from the formula as a special case: $M = 1$ at the throat. For $\beta > 0$ a multilayer flow is supersonic, while for $\beta < 0$ it is subsonic though in the throat the velocities of individual layers may be either subsonic or supersonic.

2.2.2 Radial Flows or Source/Sink Flows

Another example of an idealized one-dimensional flow is that in a channel formed by streamlines of a source or a sink, Fig.2.10a. In a spherical system of coordinates the dependence of the radius r on λ is determined by the expression

$$r^{1+\nu} = r_*^{1+\nu} q^{-1}(\lambda) \quad . \tag{2.149}$$

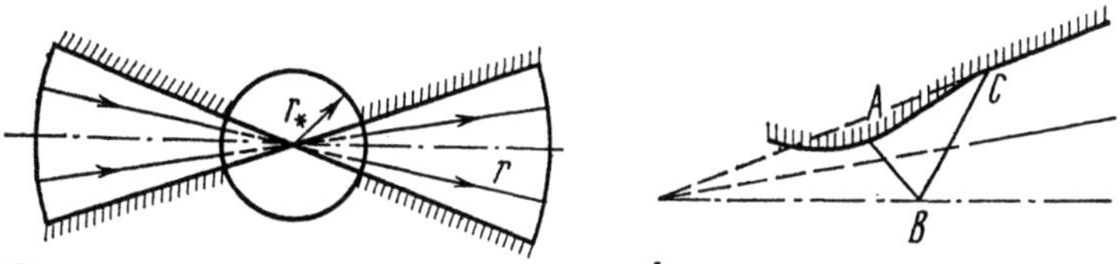

a **b**

Fig.2.10. Source-sink flow (**a**), profiling of a nozzle which realizes flow from a source upstream of the characteristic BC (**b**)

The minimum value of r is equal to r_* and the flow exists only outside a sphere/circle $r = r_*$ and cannot be continued inside the nucleus, at the surface of which the flow velocity equals the sound velocity. For $r > r_*$ two solutions are possible: a subsonic one with the velocity decreasing monotonically from $\lambda = 1$ to $\lambda = 0$ (a sink) and a supersonic one with the velocity increasing monotonically from $\lambda = 1$ to $\lambda = \lambda_{max}$ (a source). Neither flow can cross the limiting line $r = r_*$; therefore, near the line such flows are physically impossible. Also, the circular sound line cannot exist since it coincides with the limiting line [2.5]. In the source/sink flow the lines of constant λ are concentric circles with the centre in the nucleus of the source/sink. For incompressible fluids the nucleus shrinks to a point.

Flows similar to the source/sink flow exist, for instance, in the subsonic and supersonic parts of conical nozzles. By special profiling of the supersonic part of such a nozzle an exact source flow can be obtained.

Let AB be a second family characteristic (see Fig.2.10b) defined by the transonic zone of the nozzle. We shall draw a characteristic of the first family BC which will be a characteristic of the source, and the flow rate across it will be equal to the flow rate across AB. Solving Goursat's problem we shall find in the space between characteristics AB and BC, a contour AC which will make the flow radial with respect to the source. Then, drawing a straight generatrix we shall obtain downstream of the characteristic BC the flow in a supersonic source. Evidently the presence of only straight generatrices without special profiling of the upstream region does not guarantee an exactly radial flow.

It is impossible to construct an exact sink flow in the subsonic part of a nozzle because disturbances propagate upstream from the transonic region of the flow. The existence of a simple analytical relation between r and λ allows a theory of flows similar to a source/sink flow to be constructed.

2.2.3 Prandtl-Meyer Flow

If a gas flow is plane and irrotational in a certain region, and a family of Mach lines consists of straight lines with a constant velocity at each of them, then the flow in the region is called a simple wave or the Prandtl-Meyer flow. Riemann's wave is its analogue in the case of a one-dimensional unsteady flow. The following statement formulates the main property of a simple wave. A region of flow with constant parameters can only border either on another analogous region of flow with constant parameters or on a simple wave. A simple wave or the Prandtl-Meyer flow is an example of flow which allows a plane nozzle flow to be described by means of rather simple analytical dependences. (Sometimes an approximate description of axisymmetrical flows can also be given.) From (2.63) we get for a plane equilibrium or frozen irrotational flow ($\nu = 0$, $\Phi_1 = 0$, $S = \text{const}$) the following relations:

$$d\theta \pm \frac{\text{cotan}\alpha}{\rho W^2}\,dp = 0 \quad , \quad d\theta \mp \frac{\sqrt{M^2-1}}{M\left(1+\frac{\gamma-1}{2}M^2\right)}\,dM = 0 \quad . \qquad (2.150)$$

Then

$$\theta \mp f(\alpha) = \text{const} \quad , \quad f(\alpha) = -\int \frac{\text{cotan}\alpha}{\rho W^2}\,dp \quad ,$$

where $f(\alpha)$ is the Prandtl-Meyer angle. For an ideal gas with a constant specific heat ratio,

$$f(\alpha) = \sqrt{\frac{\gamma+1}{\gamma-1}}\,\arctan\left(\text{cotan}\alpha\,\sqrt{\frac{\gamma-1}{\gamma+1}}\right) + \alpha - \frac{\pi}{2} \quad , \qquad (2.151)$$

or

$$f(M) = \sqrt{\frac{\gamma+1}{\gamma-1}}\,\arctan\sqrt{\frac{\gamma-1}{\gamma+1}(M^2-1)} - \arctan\sqrt{M^2-1} \quad .$$

For a supersonic flow from a source, in the vicinity of the axis of symmetry of an axisymmetric flow where $\theta \to 0$ for $y \to 0$, one has

$$\theta \mp \frac{1}{2} f(\alpha) = \text{const} \quad .$$

Note that (2.150) is valid at the corner points of a supersonic flow (both plane and axisymmetric).

Let us consider flow in a plane, generally nonsymmetric, channel, Fig. 2.11. Let the velocity vector be known and constant in the exit cross-section BB'. Then characteristics OB and OB' are rectilinear. Due to the properties of a simple wave, characteristics of the first and second families will be rectilinear in regions AOB and A'OB' respectively (AO and A'O are

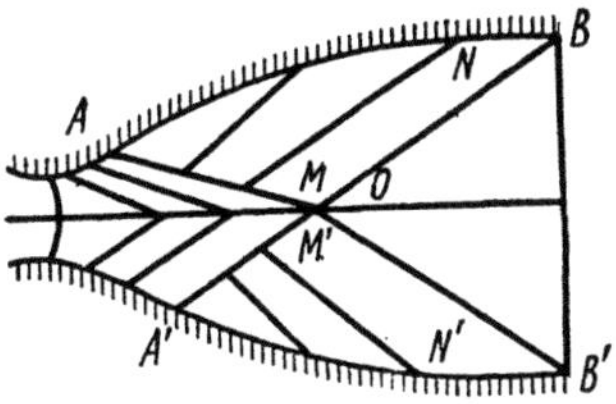

Fig.2.11. Prandtl-Meyer flow in a plane nozzle

characteristics of the first and the second families correspondingly). By (2.150) we have

$$\begin{aligned} \theta - f(\alpha) &= f(\alpha_0) \quad \text{in region AOB} \ , \\ \theta + f(\alpha) &= f(\alpha_0) \quad \text{in region A'OB'} \ , \end{aligned} \tag{2.152}$$

where α_0 is the Mach angle at point O. Then, knowing the configuration of the solid walls AB and A'B' and, hence, the angles at which the velocity vector is inclined to the walls, one can use (2.152) to determine α along AB and A'B'.

Evidently in the case of a smooth nozzle contour the Mach number can be found from (2.150), usable not only for plane but for axisymmetric flows too.

Equations (2.152) may be used for an approximate determination of flow parameters at solid walls even if the flow in the exit cross section is non-uniform. In this case α_0 in (2.152) must be substituted by α^0 calculated from the one-dimensional theory.

Sometimes it is necessary to design a nozzle generating a uniform flow with parameters defined at characteristics AO and A'O. In such cases rectilinear characteristics OB and OB' are drawn from point O, such that the mass flows across OB and OB' are equal to the mass flows across AO and A'O respectively. The data at characteristics AO and OB, A'O and OB' are used to solve Goursat's problem and to determine the coordinates of the streamlines AB and A'B'. Since in this case the Prandtl-Meyer flow exists in regions AOB and A'OB', the solution of Goursat's problem is essentially simplified. It is enough to draw rectilinear characteristics from some points of the characteristics AO and A'O and to terminate them when the condition of equality of the corresponding mass flows is fulfilled. The mass flow across a characteristic is calculated from (2.68). Then coordinates of point N (similarly for N') belonging to the solid wall are determined from the known parameters at point M and the mass flow ψ_M across AM with the help of

$$y_N^{\nu+1} = y_M^{\nu+1} + \frac{1}{\rho W}\left[(\nu + 1)\psi_M \frac{\sin(\theta + \alpha)}{\sin\alpha}\right] \quad ,$$

$$x_N = x_M + (y_M - y_N)\cot an(\theta + \alpha) \quad . \tag{2.153}$$

Here the first formula can be used to calculate contours of plane nozzles only. However, it may also help to construct an approximate contour of an axisymmetric nozzle if applied to parts of flow far from the axis of symmetry.

2.3 Variational Problems of Gas Dynamics of Internal Flows

Fundamental results in the theory of variational problems of gas dynamics have been obtained in [2.24-27]. In [2.28-31] they were generalized for the case of gas flows with equilibrium and nonequilibrium physico-chemical transformations and for two-phase flows. Below we summarize the results of these studies and consider in detail only some problems involving determination of the optimum contour. The most rigorous approach based on the general method of Lagrange multipliers is only outlined.

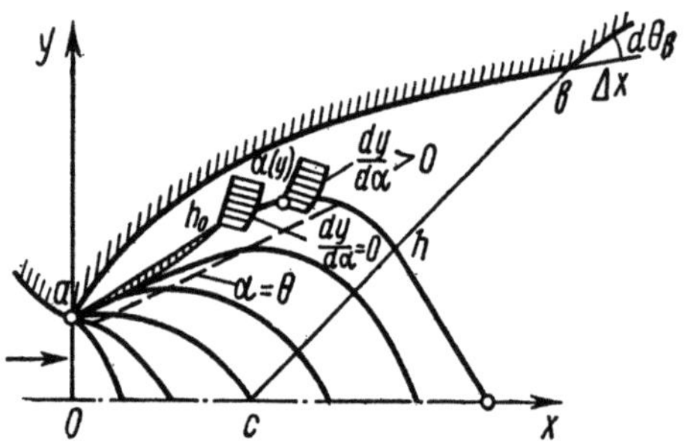

Fig.2.12. Parameters used in the solution of variational problems. The line $\alpha = \theta$ is drawn through extremal points. No shockless solutions of variational problems exist to the left of point h_0

Let us begin by constructing a contour of the supersonic part of a plane/axisymmetric nozzle developing maximum thrust, when its length and the flow parameters at the initial cross-section are specified, the flow being equilibrium and rotational, but without shock waves. Obviously, such a problem is equivalent to the problem of constructing the shortest contour for a fixed thrust. The contour sought must either pass through a given point b (Fig. 2.12) or, if the ordinate is not fixed, provide the maximum thrust at a given external pressure p_H. Below we consider the problem at $p_H = U$.

If the curvature of the optimum contour is unlimited, then it must have a corner point a. This principal result postulated in [2.25] is a consequence of the fact that the slope of the given contour to the left of point a and that of the required contour to the right of the same point, generally do not coincide. If, however, the radius of curvature has a finite value, then the corner point may be replaced by an arc of the maximum permissible curvature. In the framework of the variational problem formulated at the control contour, the necessity of a corner point or an arc of maximum curvature has been accounted for in [2.32], where some limiting properties of rarefaction flows have been found.

Thus, it is necessary to design a nozzle contour ab ensuring the maximum thrust J, which, to an insignificant addend (the momentum at the initial section of the supersonic part of a nozzle) and a positive multiplicative factor (2π for axisymmetric flows), is equal to

$$J = \int_a^b p y^\nu dy \quad , \tag{2.154}$$

while the fixed length of the contour is

$$X = \int_a^b \cot\mathrm{an}\theta \, dy \quad . \tag{2.155}$$

The dependence of pressure on the shape of the contour is determined by the system of partial differential equations (2.4-7) and, therefore, is unknown beforehand. Furthermore, the pressure at a point of the contour depends on its form upstream of the point, i.e., is a functional of an unknown structure. The formulated two-dimensional problem in which (2.4-7), general isoperimetric conditions similar to (2.155) and the impermeability condition at the contour are used, is the Boltz problem and may be solved only with the help of the general method of Lagrange multipliers for two independent variables [2.24,33,34]. However, sometimes the number of independent variables can be reduced if the integral along contour ab is replaced by the integrals along characteristics ab and cb, as proposed in [2.26]. It is convenient to solve the variational problem using the stream function ψ as an independent variable along a characteristic. If irreversible processes are prohibited in the region considered, then the pressure, the density, temperature, entropy, enthalpy and the Mach angle depend only on ψ and W. If the flow is irrotational and equilibrium, then the parameters are functions of W only. The equations of characteristics are

$$\frac{dx}{d\psi} \mp \frac{\cos(\beta \pm f \pm \alpha)}{y^\nu \rho W \sin\alpha} = 0 \quad , \tag{2.156}$$

$$\Phi_3^{\pm} \equiv \frac{dy}{d\psi} \mp \frac{\sin(\beta \pm f \pm \alpha)}{y^{\nu}\rho W \sin\alpha} = 0 \quad , \tag{2.157}$$

$$\Phi_4^{\pm} \equiv \frac{d\beta}{d\psi} \pm \frac{\partial f}{d\psi} \pm \frac{\text{cotan}\alpha}{\rho W^2}\frac{\partial p}{\partial\psi} + \nu\frac{\sin(\beta \pm f)}{y^{1+\nu}\rho W} = 0 \quad . \tag{2.158}$$

In the case of an irrotational flow f is the Prandtl-Meyer angle and $\partial f/\partial\psi = \partial p/\partial\psi \equiv 0$. If the flow is rotational, then $\partial f/\partial W = \text{cotan}\alpha/\rho W$, $\beta = \theta \mp f$. The dimensionless parameters are introduced just as in the preceding sections with the exception of pressure which is referred to $\rho_* W_*^2$ and temperature which is divided by $W_*^2 R^{-1}$.

For the variational problem with a fixed endpoint and nozzle length, the integral J may be easily replaced by the integrals along ac and cb if the law of momentum conservation is applied to the contour abc. Evidently, the length of the nozzle may also be expressed by integrals along these characteristics. The flow at the characteristic ac and at the arc hc of characteristic bc may be considered known and unchanging for small variations of the contour ab, while at the arc hb relations among the flow parameters are given with the help of ordinary differential equations (2.156-158). Thus this problem, which is generally a two-dimensional Boltz problem, is reduced to a one-dimensional Lagrange problem by a transition to the control contour.

Expressing J and X by integrals along characteristics ac and cb and using the stream function ψ as an independent variable along the characteristics, we have

$$J = \int_a^c F_1 d\psi - \int_b^c \Phi_1 d\psi = \int_a^h F_1 d\psi - \int_b^h \Phi_1 d\psi \quad , \tag{2.159}$$

$$X = \int_a^c F_2 d\psi + \int_b^c \Phi_2 d\psi = \int_a^h F_2 d\psi + \int_b^h \Phi_2 d\psi \quad , \tag{2.160}$$

where

$$F_1 = W\cos\theta - \frac{p\sin(\theta - \alpha)}{\rho W\sin\alpha} \quad , \quad F_2 = \frac{\cos(\theta - \alpha)}{y^{\nu}\rho W\sin\alpha} \quad ,$$

$$\Phi_1 = W\cos(\theta + f) + \frac{p\sin(\beta + f + \alpha)}{\rho W\sin\alpha} \quad , \tag{2.161}$$

$$\Phi_2 = \frac{\cos(\beta + f + \alpha)}{y^{\nu}\rho W\sin\alpha} \quad .$$

Now the variational problem can be formulated as follows: it is necessary to determine the magnitude of the corner in the contour at point a, the position of point h at the closing characteristic of the rarefaction wave and the function $W(\psi)$ at hb, so that a maximum value of (2.159) results satisfying, at the same time the isoperimetric condition (2.160) and the differential relations (2.157,158). The value of X, the coordinates of points a and b and the flow parameters at the characteristic ac are considered known. After the characteristic hb is found, the nozzle contour and the values of the parameters along it are determined from the solution of Goursat's problem in the space between the known characteristics ah and hb.

The problem formulated above is a degenerate one, because functions Φ_1, Φ_2, Φ_3, Φ_4 do not contain derivatives $dW/d\psi$, while $d\beta/d\psi$ and $dy/d\psi$ enter in linear combinations. Usually this results in the absence of a two-sided extremum. The solution of the problem may completely or partly differ from the integral of the system of Euler equations. A one-sided (boundary) extremum defined by regional inequalities may exist instead of a two-sided one. This means, in particular, that while solving the problem for fixed points h and b, one cannot generally draw an extremal through points h and b. The passage of the extremal through the fixed point b is possible due to the freedom of choice of the intensity of the rarefaction wave (i.e. due to the cusp in the contour at point a) and of the position of point h at the closing characteristic. Evidently, characteristics cb of the Euler equation are not satisfied at the arc ch.

Let us construct an auxiliary function, as is usually done while solving the variational problems, with isoperimetric conditions and differential relations:

$$\chi = \int_a^c F\,d\psi + \int_c^b \Phi\,d\psi = \int_a^h F\,d\psi + \int_h^b \Phi\,d\psi \quad . \tag{2.162}$$

Here $F = F_1 + \lambda_1 F_2$, $\Phi = \Phi_1 - \lambda_1\Phi_2 + \lambda_2\overset{+}{\Phi}_3 + \lambda_3\overset{+}{\Phi}_4$, λ_1 is a constant Lagrange multiplier and λ_2, λ_3 are variable Lagrange multipliers. We consider a small variation in the functional (2.162), assuming that all the parameters are continuous at hb. Allowing the position of point h to vary both along cb and ah, recalling that $\psi_a = \psi_b$ and $\delta\psi_b = 0$ and denoting by $\delta\psi_{h1}$ the variation of ψ along the characteristic ah, we have

$$\delta\chi = \delta J = [F - \Phi + \lambda_2(\dot{y}_+ - \dot{y}_-) + \lambda_3(\dot{\beta}_+ - \dot{\beta}_-)]\delta\psi_{h1}$$

$$+ \lambda_{3b}\delta\beta_b + \int_a^b [(\Phi_y - \dot{\lambda}_2)\delta y + \Phi_W\delta W + (\Phi_\beta - \dot{\lambda}_3)\delta\beta]d\psi \quad , \tag{2.163}$$

where the dot denotes total differentiation with respect to ψ, the + and - subscripts of the derivatives mean differentiation along characteristics of the first and the second families, respectively, and the other subscripts of functions designate the corresponding partial derivatives.

Let us choose the Lagrange multipliers λ_2 and λ_3 in such a way that the coefficients of variations δy and $\delta\beta$ at hb are zero and let us put the multiplier of $\delta\psi_{h1}$ equal to zero due to the arbitrariness of the variations $\delta\psi_{h1}$. Then,

$$F - \Phi + \lambda_2(\dot{y}_+ - \dot{y}_-) + \lambda_3(\dot{\beta}_+ - \dot{\beta}_-) = 0 \quad , \tag{2.164}$$

$$\lambda_{3b} = 0 \; , \; \dot{\lambda}_2 - \Phi_y = 0 \; , \; \dot{\lambda}_3 - \Phi_\beta = 0 \quad . \tag{2.165}$$

To obtain a two-sided extremum, it is necessary to equate the expression to zero before varying δW in (2.163). Together with (2.165) this gives the following system of Euler equations:

$$\Phi_W = 0 \quad , \quad \dot{\lambda}_2 = \Phi_y \quad , \quad \dot{\lambda}_3 = \Phi_\beta \quad . \tag{2.166}$$

If the contour ab is optimum, then due to the supersonic character of the flow, any of the part between a and b must be optimum [2.27]. Then (2.164) must be fulfilled at any point of characteristic hb. Doing the necessary calculations, we find that everywhere along hb,

$$\begin{aligned} py^\nu \sin(\beta + f) - \lambda_1 \cos(\beta + f) - \lambda_2 \sin(\beta + f) \\ - \frac{1}{2}\lambda_3(\beta_+ - \beta_-)y^\nu \rho W \tan\alpha = 0 \quad . \end{aligned} \tag{2.167}$$

This relation together with (2.166) results in a uniform first-order differential equation for the variable Lagrange multiplier λ_3. Then it follows from the boundary condition $\lambda_{3b} = 0$ that $\lambda_3 = 0$ everywhere upon hb. After some transformations we get

$$y^\nu \rho W^2 \tan\alpha \sin^2\theta = \lambda_1 \quad , \tag{2.168}$$

$$y^\nu (p - \rho W^2 \tan\alpha \sin\theta \cos\theta) = \lambda_2 \quad , \tag{2.169}$$

$$\lambda_2 = - \frac{\nu W}{y}\left[\tan\alpha \sin\theta - \frac{p \sin(\theta + \alpha)}{\rho W^2 \sin\alpha}\right] \quad . \tag{2.170}$$

The constant λ_1 and the initial value of λ_2 are determined from (2.168,169) and the known parameters at point h. Thus the necessity of defining λ_3 is avoided, the problem is significantly simplified and one can obtain some closed relationships. The compatibility relation (2.158) at the characteristic hb, until now unused, is identically satisfied upon substitution of (2.168-170).

Let us point out some particular features of the solution obtained above. It follows from (2.168) that the function θ preserves its sign at the characteristic hb, and if it is zero at a point belonging to hb, then it is zero at all other points of hb, while $p = 0$. In such a case point h belongs to the axis of symmetry and the maximum thrust is attained in a nozzle with a flow parallel to the axis x at the exit cross-section and with a constant pressure in the section.

In the case of a plane flow $\nu = 0$ and it follows from (2.170) that $\lambda_2 = \text{const}$. If the flow is at the same time irrotational, i.e. all its parameters are independent of ψ while p, ρ and α depend on W only, then it follows from (2.168,169) that the flow parameters are constant at the characteristic hb while the characteristic itself is rectilinear. Hence, the Prandtl-Meyer flow takes place in the region ahb. Therefore, the arc ab of optimum contour, in the case of a plane flow, must belong to a one-parametric family of nozzles producing at the exit cross-section uniform flow parallel to the axis of symmetry. If y is used as an independent variable instead of ψ at the characteristic hb, then

$$y^{\nu}\rho W^2 \tan\alpha \sin^2\theta = -\bar{\lambda}_1 \quad , \tag{2.171}$$

$$W \cos(\theta - \alpha)/\cos\alpha = \bar{\lambda}_2 \quad , \tag{2.172}$$

$$\frac{d\bar{\lambda}_2}{dy} = y^{\nu} \left[\frac{\partial p}{\partial\psi} - \frac{\partial\rho}{\partial\psi} \frac{W^2 \sin\theta \sin^2\alpha}{\cos\alpha \sin(\theta + \alpha)}\right] \quad . \tag{2.173}$$

If the flow is irrotational, then $\partial p/\partial\psi = \partial\rho/\partial\psi = 0$ and the solution may be written in the form of the algebraic formulas (2.171,172) with constant $\bar{\lambda}_1$ and $\bar{\lambda}_2$.

Thus, in the case of irrotational equilibrium flow, relations at the characteristic hb corresponding to the maximum thrust have rather a simple form. However, if the point b is fixed, then to find the corresponding extremal hb passing through the point, one has, due to the degenerate character of the problem, to vary point h. Generally speaking, this is a time-consuming procedure though the relations used are finite.

The inverse problem is much easier to solve. In this case point h is fixed and the system (2.171-173) is integrated along hb up to $y = y_b$. Sometimes it is more convenient to use the compatibility equation instead of (2.171) or (2.172). Solving this problem gives the coordinates of point b. The contour passing through the point which is obtained by solving Goursat's problem between the calculated characteristic hb and the characteristic ah corresponds to the maximum thrust.

If the ordinate of point b is not fixed and the external pressure p_H is nonzero, then to determine the ordinate of point b of a nozzle developing the maximum thrust, Busemann's condition [2.25],

$$\sin 2\theta_b = 2\cot\alpha_b(p_b - p_H)/\rho_b W_b^2 \quad , \tag{2.174}$$

is used, obtained from both the transversality condition at the point b and an elementary consideration of an infinitesimal edge element of the contour, which for a maximum thrust nozzle, must give the maximum contribution to the total thrust.

If, instead of the nozzle length X, the ordinate of point b is fixed, then the maximum thrust is developed by a nozzle which, at a fixed y_b, forms at the exit section a uniform flow parallel to the axis x.

In an axisymmetric case it is sometimes impossible to construct a characteristic hb satisfying the condition of continuity for the gasdynamic parameters [2.27,35]. The rarefaction wave proves to have a characteristic possessing the following property: the derivatives with respect to y tend to infinity at the extremum characteristics issuing from the points to the left of it. No shock-free solution is possible in this case. As a result of a most refined analysis of the ncessary conditions for the extremum and the maximum thrust, the so-called "discontinuous shock-free" solutions containing no shock waves in the domain of influence have been obtained [2.27]. Figure 2.13 shows such a flow schematically. Here ahk and lhm are compression and rarefaction waves, hn is a shock wave and ht is a streamline. To construct such a flow, the Prandtl-Meyer relations are used, together with the equation

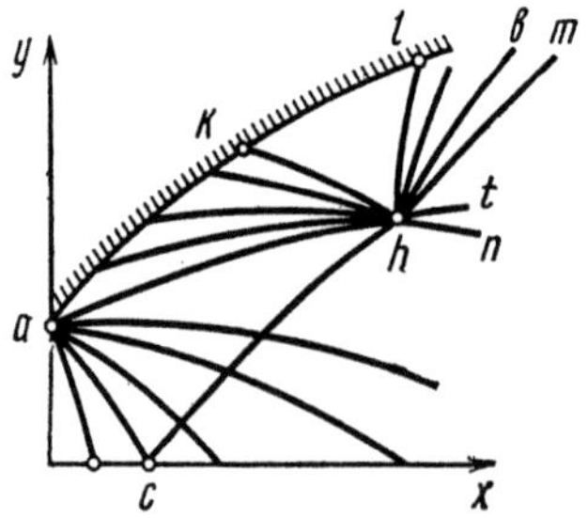

Fig.2.13. Diagram of flow in an optimum nozzle with a shock wave. lhm are the compression and rarefaction waves, hn is the shock wave

of the shock adiabatic and the condition of equality of pressure and velocity vector inclination angles to the left and right of ht. The relations used above, e.g. (2.171-174) are satisfied at hb while the arc ak of the contour ab is a zone of the boundary extremum corresponding to the boundary of flows which have no shock waves in the domain of influence of the contour being sought.

In order to find the region of continuous shock-free solutions, let us consider relations (2.171,172) for the case of irrotational flows of an ideal gas with a constant specific heat ratio [2.35]. Recalling that in this case $\bar{\lambda}_1$ and $\bar{\lambda}_2$ are constants, we have

$$y = -\frac{\bar{\lambda}_1(-\bar{\lambda}_2 \sin\alpha \pm \sqrt{W^2 - \bar{\lambda}_2^2 \cos^2\alpha})^2}{\rho \sin 2\alpha(\bar{\lambda}_2^2 - W^2)^2} \quad . \tag{2.175}$$

Here the plus-sign corresponds to $\alpha_h < \theta_h$ and the minus-sign to $\alpha_h > \theta_h$. It can be easily deduced that at $\alpha_h \geqslant \theta_h$ along an extremal the values of α and θ decrease while y increases. At $\alpha_h < \theta_h$, α increases and θ decreases. At a certain point of the extremal the equality $\alpha = \theta$ is attained and for further calculation of the extremal the sign of the radical in (2.175) should be changed. This results in a nonmonotonical velocity change along an extremal.

It follows from (2.175) that while point h moves along the characteristic ah to point a (in the region where $\alpha < \theta$), $dy/d\alpha$ decreases in the direction of the extremal; at a certain point $h = h_0$, $dy/d\alpha$ equals zero and later becomes negative. This means that construction of shock-free solutions for all points of the characteristic ah lying to the left of point h_0 is impossible (see Fig.2.12). However, if α decreases along the extremal while y increases, then the resulting nozzles will not possess the maximum thrust. From (2.175) a relation between α and θ at the curve ah_0, to the left of which there are no shock-free solutions of the variational problem, can be easily obtained. We have

$$\gamma \sin\theta \sin(\alpha + \theta) - \cos\alpha \sin^2 2\alpha + \sin\theta \sin(\theta - \alpha) - \sin\alpha \sin 2\theta \cos 2\alpha = 0 \quad . \tag{2.176}$$

If the isoperimetric conditions prohibit transfer to the control contour (e.g. under the condition that the area of the side surface or the volume remains constant) the solution can be obtained only by means of the general method of Lagrange multipliers [2.24,34]. The equations of motion containing variable Lagrange multipliers $h_i(x,y)$ are introduced into the function which contains a double integral over the region ahb while another function containing integrals along boundaries of ahb consists of isoperimetric conditions (with constant Lagrange multipliers), the impermeability condition, the compatibility equation with corresponding Lagrange multipliers $\beta_i(y)$ and an expression for the nozzle thrust, which is being optimized.

The equations for Lagrange multipliers are linear and have the same characteristics as the equations of motion. The best nozzle is the one for which the optimum condition containing flow parameters and multipliers h_i is satisfied everywhere on ab. The condition, naturally, contains both constant Lagrange multipliers and multipliers $\beta_i(y)$.

The method described in [2.24] is generalized in [2.28,30,31]. Flows were considered with nonequilibrium transformations, which make it necessary to introduce additional Lagrange multipliers, the number of which is equal to the number of nonequilibrium parameters [2.30], and two-phase flows [2.31]. The possible existence of discontinuous multipliers h_i in the case when flow parameters are continuous was demonstrated and it was shown that if the closing characteristic of the contour's domain of influence originates at the axis of symmetry outside the rarefaction wave, then the optimum contour has internal corner points [2.31].

Despite the broad possibilities of the general method of Lagrange multipliers as a tool for solving quite general variational problems, its practical realization involves considerable computational difficulties and nowadays examples of its successful application are encountered only rarely.

3. Numerical Methods of Studying Nozzle Gas Flows

In this chapter we discuss numerical methods of studying nozzle gas flows. One of the sections is dedicated to the solution of relaxation equations. The numerical methods included in this chapter are, from our point of view, quite versatile and are widely used to explore nozzle gas flows by means of digital computers. The reader is assumed to be acquainted with the fundamentals of the method of grids for which the basic theory is discussed in [3.1-5].

3.1 Methods of Solving Relaxation Equations

While calculating nonequilibrium flows, one has to perform numerical integration of equations describing the nonequilibrium relaxation process under consideration. Kinetic, or relaxation, equations describing the process in the vicinity of equilibrium belong, as a rule, to a class of equations with a small parameter before the highest derivative, which makes their numerical integration quite complicated. The equations of conservation of chemical components, equations for the determination of vibrational energy, velocities and temperatures of particles in two-phase flows, equations of radiation transfer and so on, are among these relaxation equations. Nonequilibrium nozzle flows differ from other kinds of flows because they start from a state of rest when the system is close to thermodynamic equilibrium. In similar cases the time of relaxation and, hence, the length of the relaxation zone, are small. Therefore it is very difficult to choose an appropriate step length for integration. It has been found that the step length in numerical integration using an explicit method, e.g. that of Euler or Runge-Kutta, must, in order to ensure stability of the procedure, be so small that it is practically impossible to obtain a solution even with the help of modern computers.

Recently a number of numerical methods of integrating relaxation equations have been proposed, among which methods based on implicit difference schemes are the most efficient and versatile [3.6,7]. The main requirement is to be able, using a general scheme, to calculate accurately enough both the regions where all or some of the nonequilibrium parameters are close to equilibrium, and the regions where they depart from that state considerably. Below we give recommendations for the choice of implicit difference schemes for numerical solution of relaxation equations.

A relaxation equation for the determination of a nonequilibrium parameter α_i may be presented in the following general form:

$$\frac{d\alpha_i}{dt} = \frac{1}{\tau_i}(\alpha_i^* - \alpha_i) \quad , \qquad (i = 1,2,\dots,N) \quad , \tag{3.1}$$

where τ_i is the relaxation time, generally a function of t, p, T and all α_i, α_i^* is the local equilibrium value of α_i (also a function of t,p,T and all α_i) which is equal to α_i in the case of equilibrium isentropic flow. From (3.1), making the natural assumption that the derivative $d\alpha_i/dt$ is limited, we deduce that when $\tau_i \to 0$ the flow approaches the state of equilibrium, i.e. $\alpha_i \to \alpha_i^*$.

Let us, instead of using (3.1), consider a model equation with constant τ_i and α_i^*. Its general solution has the form

$$\alpha = (\alpha_0 - \alpha^*)e^{-t/\tau} + \alpha^* \quad , \tag{3.2}$$

the subscripts i are omitted for the sake of simplicity and $\alpha_0 = \alpha(0)$. Let us further solve the difference equation

$$\frac{\alpha_{n+1} - \alpha_n}{h} = \frac{s}{\tau}(\alpha^* - \alpha_n) + \frac{1-s}{\tau}(\alpha^* - \alpha_{n+1}) \quad , \tag{3.3}$$

which approximates the differential equation (3.1). Here h is the step of integration, s is a parameter ($0 \leqslant s \leqslant 1$) and the subscripts n and n+1 denote parameters at the known and the desired points, respectively. Using the recursion relation

$$\alpha_{n+1} = \frac{1-\kappa s}{1+\kappa(1-s)}\alpha_n + \frac{\alpha^*\kappa}{1+\kappa(1-s)} \quad , \qquad \kappa = \frac{h}{\tau} \quad , \tag{3.4}$$

which follows directly from (3.3), we obtain the following solution of the difference equation (3.3):

$$\alpha_{n+1} = (\alpha_0 - \alpha^*)\left[\frac{1-\kappa s}{1+\kappa(1-s)}\right]^{n+1} + \alpha^* \quad . \tag{3.5}$$

The exact solution of the differential equation at the point $n+1$, according to (3.2) is

$$\alpha_{n+1} = (\alpha_0 - \alpha^*)e^{-\kappa(n+1)} + \alpha^* \quad . \tag{3.6}$$

It follows from (3.5,6) that the solution of the difference equation (3.5) approaches the exact solution at $\kappa \to 0$ for all s from $0 \leqslant s \leqslant 1$. However, the accuracy and stability of the solution of the difference equation depend on the magnitude of this parameter. At $s = 0.5$ the solution of the difference equation is of the second order of accuracy. At $s = 1$ (an explicit scheme like that of Euler or Runge-Kutta) and $\kappa \gg 1$ the solution of the difference equation deviates considerably from the exact one. The maximum step length which may be used for numerical integration of (3.1) is equal to 2τ. Therefore, near equilibrium where the time of relaxation is very small, the explicit schemes allow integration of relaxation equations to be performed only with a very small step length ($h \approx 2\tau$), for which even the fastest modern computers are inadequate.

At $s = 0.5$ and $\kappa \gg 1$, it follows from (3.5) that the solution of the difference equation oscillates about the exact one with an amplitude less than or equal to $|\alpha_0 - \alpha^*|$, and these oscillations decrease rather slowly as the step size increases. Nevertheless, the difference scheme gives, at $s = 0.5$, solutions close to the exact one in the regions where $|\alpha_0 - \alpha^*|$ is small compared to α^*. Such regions correspond to small values of τ and near-equilibrium flows. At the same time, in those regions where τ is large and the flow departs substantially from the equilibrium state (e.g. at a corner point or behind a shock wave), $|\alpha_0 - \alpha^*|$ may be comparable with α^* and the use of a difference scheme at $s = 0.5$ may result in large errors. An analysis of the model equation at $s = 0.5$ and large κ shows that errors in the initial data attenuate after a sufficiently large number of steps. This circumstance may prevent departure from the starting points near equilibrium, especially in the case of multicomponent mixtures.

Consequently, implicit schems with $0 \leqslant s \leqslant 0.5$ are preferable because at $\kappa \gg 1$ practically exact solutions can be obtained which ensure rapid attenuation of initial data errors. Such schemes allow the size of the integration step to be substantially increased in comparison with the explicit scheme and ensure stability of the procedure. In [3.7,8], $s = 0.4$ was used for the calculation of particular flows giving close to second-order accuracy. The above analysis of the model equation may be generalized to the case when α^* is not constant but depends on t.

Let us now consider a concrete algorithm for the solution of difference equations for complex nonequilibrium systems. We shall assume that there are N nonequilibrium parameters α_i and N relaxation equations similar to (3.1) for their determination. Without losing generality we shall assume that τ_i and α_i^* are functions of t and α_i only and do not depend on p and T. Then, according to (3.3), the system of differential equations (3.1) may be approximated by the following system of difference equations:

$$f_i(\alpha_{i(n+1)}) \equiv \alpha_{i(n+1)} - \frac{1}{1 + \kappa_{i(n+1)}(1 - s)} \cdot [\alpha_{in} + \kappa_{in}s(\alpha_{in}^* - \alpha_{in}) + \kappa_{i(n+1)}(1 - s)\alpha_{i(n+1)}^*] = 0 \quad , \tag{3.7}$$

where

$$\kappa_i = h/\tau_i \quad , \quad (i = 1,2,\ldots,N) \quad , \quad \text{or}$$

$$\mathbf{f}(\alpha_{n+1}) = 0 \quad , \tag{3.8}$$

where $\mathbf{f} = (f_1,f_2,\ldots,f_N)$, $\boldsymbol{\alpha} = (\alpha_1,\alpha_2,\ldots,\alpha_N)$. The system (3.8) is a transcendental system of equations in $\alpha_{i(n+1)}$ $(i = 1,2,\ldots,N)$. It should not be solved by the method of direct iteration because the condition of convergence results in the same limitation of the stepsize h $(h \approx \tau_i)$ as in the case of explicit schemes. Therefore, some other method should be used, e.g. Newton's method with a variable Jacobi matrix D, which can be readily obtained if the analytical expressions for the derivatives $\partial f_i/\partial\alpha_k$ are used. The unknown $\alpha_{i(n+1)}$ are found by iterations according to the formula

$$\boldsymbol{\alpha}_{n+1}^{(j+1)} = \boldsymbol{\alpha}_{n+1}^{(j)} - D_{n+1}^{-1}\mathbf{f}(\alpha_{n+1}^{(j)}) \quad . \tag{3.9}$$

The iteration is terminated upon fulfilment of the conditions

$$|\alpha_{i(n+1)}^{(j+1)} - \alpha_{i(n+1)}^{(j)}| \leqslant \varepsilon_i |\alpha_{i(n+1)}^{(j+1)}| \quad , \quad |f_i(\alpha_{n+1}^{(j+1)})| \leqslant \delta_i \quad ,$$

$$(i = 1,2,\ldots,N) \quad ,$$

where ε_i, δ_i are the allowable errors in the components of the solution and the errors of closure, respectively.

Many approaches have been proposed which ensure fast convergence of iterations to the solution [3.7,8]. When calculations are being done according to (3.9), about 90% of the time is spent in determining the components of matrix D and its inversion. Hence, a lot of time may be saved while calculating D^{-1} if it is calculated at the point $n+1$ only in the first two iterations; then it is fixed and further iterations are being performed with the same matrix D^{-1}. Furthermore, matrix D^{-1} calculated at the point $n+1$ is used to find the solution from (3.9) at the points $n+2$, $n+3$, etc. The "new" matrix D^{-1} is calculated at the point $n+k$ only after the number of iterations needed to ensure convergence to the solution with a specified accuracy exceeds a prescribed number (4-6). The fixing of the inverse matrix allows the time of calculation to be shortened by an order of magnitude.

It is known that convergence of Newton's method to the solution depends on the proximity of the initial approximation to the solution. Therefore, the initial approximation at the point $n+1$ should be chosen by extrapolating the desired functions using their values at the previous points. Finally, the time of calculation depends very much on the accuracy of the data specification at the initial point of integration, which, as has been mentioned, lies, in the case of nozzle flows, in the near-equilibrium region. Even slight errors in the initial data (in the fourth or fifth significant figure) may, due to small values of τ, result in a large time needed for calculation at the initial segment because of a slow convergence of iterations. Therefore, at the initial point, the system (3.7) should also be solved by Newton's method with a variable matrix putting the second term on the left side of (3.7) equal to α_i^*. The implicit difference scheme at $s \leqslant 0.5$ with the use of the modification described above has been successfully used for calculation of various non-equilibrium nozzle flows both in the one-dimensional aporoximation and in the cases of plane and axisymmetric flows [3.7-10].

3.2 Methods of Calculating Plane and Axisymmetric Supersonic Flows

This section presents some widely used methods of calculating two-dimensional stationary supersonic nozzle gas flows. Numerical schemes of the method of characteristics are considered including the cases of flows with equilibrium and nonequilibrium reactions. Two finite-difference methods are discussed, adjusted for calculating discontinuous gas flows and belonging to the class of the so-called shock-smearing methods.

3.2.1 Method of Characteristics

With the advent of computers the method of characteristics remained for a long time practically the only tool for the numerical investigation of nozzle gas flows. It is widely used nowadays and is the subject of numerous publications which, unfortunately, are not systematized up to date. It is impossible in the framework of this book to consider in detail the various numerical schemes of the method, even those applied to nozzle gas flows. Therefore, we shall briefly consider here flows of a perfect gas only. As for more complicated types of flows, we shall limit our consideration to some remarks and references to the most important publications. Among the works dedicated to the development and application of the method of characteristics to the calculation of flow for a perfect gas, we shall point out [3.11-14] and survey articels [3.15,16].

The equations of characteristics (2.62-64,67) at $F_i \equiv 0$ can be written in the form

$$dy = \tan(\theta \pm \alpha)dx \quad ,$$

$$d\theta \pm \frac{1 + \cos 2\alpha}{\gamma - \cos 2\alpha} d\alpha \pm \frac{\nu \sin\alpha \sin\theta}{y \cos(\theta \pm \alpha)} dx \pm \frac{\sin 2\alpha}{2\gamma(\gamma - 1)} dS = 0 \quad , \tag{3.10}$$

$$dy = \tan\theta \, dx \quad , \quad dS = 0 \quad .$$

Introduction of the variables

$$\zeta = \tan\theta \quad , \quad \beta = \mathrm{cotan}\alpha \tag{3.11}$$

allows one to reduce the volume of the necessary calculations. With these variables, (3.10) and the relations for the stream function (2.69) have the form

$$dy = \lambda^{\pm} dx \quad , \quad A\,d\zeta \mp B\,d\beta \pm C^{\pm} dx \mp D\,dS = 0 \quad , \tag{3.12}$$

$$dy = \lambda^{0} dx \quad , \quad dS = 0 \quad , \tag{3.13}$$

$$d\psi = \pm E^{\pm} dx \quad , \tag{3.14}$$

where

$$\lambda^{0} = \zeta \ , \quad \lambda^{\pm} = \frac{\beta\zeta \pm 1}{\beta \mp \zeta} \quad , \quad A = \frac{1}{1 + \zeta^2} \quad ,$$

$$B = \frac{2\beta^2}{(1 + \beta^2)[(\gamma + 1) + (\gamma - 1)\beta^2]} \quad , \quad C^{\pm} = \frac{\nu\zeta}{y(\beta \mp \zeta)} \quad ,$$

$$D = \frac{\beta}{\gamma(\gamma - 1)(1 + \beta^2)} \quad , \tag{3.15}$$

$$E^{\pm} = y^{\nu} \gamma^{-1/(\gamma-1)} e^{-S/(\gamma-1)} \frac{\sqrt{(1 + \beta^2)(1 + \zeta^2)}}{\beta \mp \zeta} \left(1 + \frac{\gamma - 1}{\gamma + 1} \beta^2\right)^{-(\gamma+1)/2(\gamma-1)} \quad .$$

In (3.10-15) the upper signs correspond to the characteristics of the first family and the lower ones to those of the second family. For irrotational flows of a perfect gas in nozzles the multiplier $\gamma^{-1/(\gamma-1)}e^{-S/(\gamma-1)}$ must be replaced by 1; then the term in (3.12) containing dS evidently vanishes.

While calculating stationary flows of a perfect gas in nozzles for the case of two independent variables, one has to deal with the solution of a relatively small number of elementary problems associated with the determination of unknown quantities at the internal and boundary nodes of the characteristic grid. Below we write down formulas for the calculation of gas-dynamic parameters at the internal nodes and the nodes belonging to the wall or to the nozzle axis.

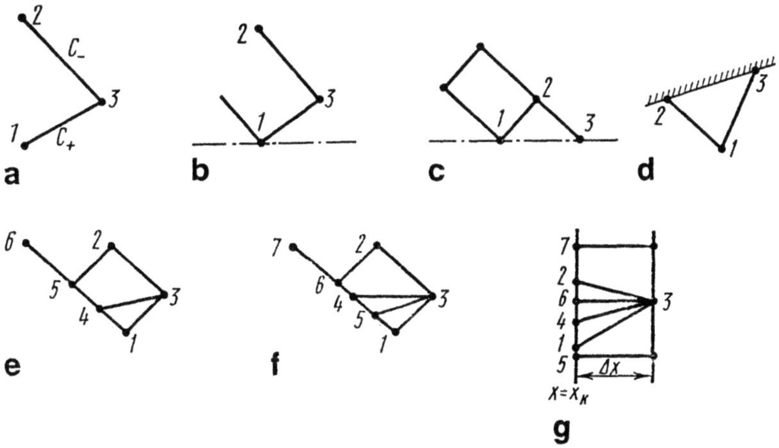

Fig.3.1a-g. Determination of flow parameters at the nodes of the characteristic net. (**a**) Point of the flow field; (**b**) points near the symmetry axis; (**c**) point at the symmetry axis; (**d**) point at a rigid wall; (**e**) point of the flow field (nonequilibrium flow); (**f**) point of the flow field (two-phase flow); (**g**) point of the flow field (layerwise net method)

We shall begin with the internal nodes (the idea of using the method of characteristics in this case has already been discussed in Sect.2.2). Let us determine the required functions at the point 3 (Fig.3.1a), lying at the intersection of the characteristics of the first (1-3) and the second (2-3) families, passing through points 1 and 2, where the solution is known. Doing so, we shall assume that the dependence of the entropy S on ψ is known: $S = S(\psi)$. Replacing differentials in (3.12,14) by finite differences at segments 1-3, 2-3, we have

$$x_3 = x_1 + \frac{y_2 - y_1 - \lambda_{23}^-(x_2 - x_1)}{\lambda_{13}^+ - \lambda_{23}^-} \quad ,$$

$$y_3 = y_1 + \lambda_{13}^+(x_3 - x_1) \quad , \qquad (3.16)$$

$$\psi_3 = \psi_1 + E_{13}^+(x_3 - x_1) = \psi_2 - E_{23}^-(x_3 - x_2) \quad ,$$

$$S_3 = S(\psi_3) \quad ,$$

$$\beta_3 = \beta_1 + \frac{1}{A_{13}B_{23} + A_{23}B_{13}} [A_{13}A_{23}(\zeta_2 - \zeta_1) + A_{13}B_{23}(\beta_2 - \beta_1)$$

$$+ A_{13}C_{23}^-(x_3 - x_2) - A_{13}D_{23}(S_3 - S_2) + A_{23}C_{13}^+(x_3 - x_1)$$

$$- A_{23}D_{13}(S_3 - S_1)] \quad ,$$

$$\zeta_3 = \zeta_1 + \frac{1}{A_{13}} [B_{13}(\beta_3 - \beta_1) - C_{13}^+(x_3 - x_1) + D_{13}(S_3 - S_1)] \quad .$$

A two-digit subscript means that the corresponding coefficient is an average of two neighbouring coefficients, e.g. $A_{13} = (A_1 + A_3)/2$, $A_{23} = (A_2 + A_3)/2$, etc.

The solution is sought by means of iterations (see Sect.2.2.2), and as a first approximation, instead of the semisums, values of the coefficients at the corresponding reference points (1 or 2) are taken. If the initial approximation is well chosen, then the required accuracy is usually attained after two or three iterations; however, for each concrete problem special attention must be paid to an adequate, from the point of view of errors of approximation, choice of the characteristic grid.

It should be pointed out that (3.16) cannot be used for calculations if, in the vicinity of the point in question, the inclinations of the characteristics of the first and second families are close to $\pi/2$, i.e., if $\beta + \zeta$ or $\beta - \zeta$ tend to zero, because in this case the errors are too large. If, for instance, $\beta + \zeta$ differs only slightly from zero, then the equations of the characteristics of the second family (3.12) and the corresponding equation for ψ (3.14) should be transformed to the form

$$dx = \tilde{\lambda}^- dy \quad , \qquad A\,d\zeta + B\,d\beta - \tilde{C}^- dy + D\,dS = 0 \quad ,$$

$$d\psi = -\tilde{E}^- dy \quad . \qquad (3.17)$$

where $\tilde{\lambda}^- = (\beta + \zeta)/(\beta\zeta - 1)$ and $\tilde{C}^-$, $\tilde{E}^-$ are obtained from C^- and E^- upon replacing the multiplier $\beta + \zeta$ by $\beta\zeta - 1$. For the same reason the transformed equations (3.17) cannot be used where the directions of the characteristics of the second family are close to the horizontal, i.e. where $\beta\zeta - 1$ is close to zero. The same is true for characteristics of the first family. The rules of transition from one modification of the calculation formulas to the other are quite evident. In every approximation the entropy S is sought on the basis of the calculated value of ψ by means of quadratic interpolation of the known tabulated values of $S = S(\psi)$. The table fo values of $S(\psi)$ is provided beforehand or is calculated in the course of calculations (e.g. when the isentropic character of the flow is broken due to the appearance of a shock wave in the flow field).

If the points 3 and 1 belong to the axis of symmetry (when $\nu = 1$), then the calculation formulas must be modified because equations (3.12) have a singularity at the axis of symmetry ($y = \zeta = 0$). Let the point 1 belong to the axis, Fig.3.1b,c. To exclude the singularity (just as in the case of an internal point) we shall write in the form of a finite difference only those equations (3.12-14) which are valid along the characteristics of the first family, having multiplied them, beforehand, by y. Then we once again get (3.16) for the quantities desired, in which we must, however, put

$$A_{13} = \frac{1}{2}\left(A_3 + \frac{x_3 - x_1}{y_3(\beta_3 - \zeta_3)}\right) ,$$

$$B_{13} = \frac{1}{2} B_3 , \quad C_{13}^{+} = 0 , \quad D_{13} = \frac{1}{2} D_3 .$$

If the point 3 lies on the axis of symmetry, then $y_3 = \zeta_3 = \psi_3 = 0$. In order to determine x_3, β_3 we shall use the equation of characteristics of the second family and the corresponding characteristic relation multiplied by y:

$$x_3 = x_2 - y_2/\lambda_{23} ,$$
$$\beta_3 = \beta_2 + \frac{1}{B_2^-} [A_2\zeta_2 + C_2^-(x_3 - x_2) - D_2(S_3 - S_2)] . \tag{3.18}$$

As far as the axis of symmetry is a streamline, $\psi = C = \mathrm{const}$ at all its points. The difference between ψ_3 calculated from the formula $\psi_3 = \psi_2 - E_{23}^-(x_3 - x_2)$ [see (3.16)], and C may be used to measure the calculation error. Note that there exist other means of calculating points in the neighbourhood of the axis of symmetry.

Let the point 3 now belong to the wall of a nozzle described by an equation $y = f(x)$, Fig.3.1d. We have

$$x_3 = \frac{y_2 - y_1 + \lambda^+_{13}x_1 - f'_{23}x_2}{\lambda^+_{13} - f'_{23}} ,$$

$$y_3 = f(x_3) \quad , \quad \zeta_3 = f'(x_3) \quad , \quad S_3 = S_2 \quad , \quad \psi_3 = \psi_2 \quad ,$$

$$\beta_3 = \beta_1 + \frac{1}{B_{13}} [A_{13}(\zeta_3 - \zeta_1) + C^+_{13}(x_3 - x_1) - D_{13}(S_3 - S_1)] \quad . \qquad (3.19)$$

Note that, as in the case of an internal point, the formulas for the determination of parameters at the points of the axis and the nozzle wall must be changed if the inclination of the characteristics tends to $\pi/2$. Here we shall not consider calculation points belonging to other families. The interested reader is directed to specific references, e.g. [3.13,14].

As an example, let us consider isentropic supersonic flow in a nozzle with a corner point, in whose section OA (Fig.3.2a) all gasdynamic parameters are constant, and where the velocity is equal to the sound speed. It is necessary to find the form of the supersonic part of the nozzle which ensures uniform flow parallel to the axis at the exit section and which has a fixed Mach number $M = M_0$.

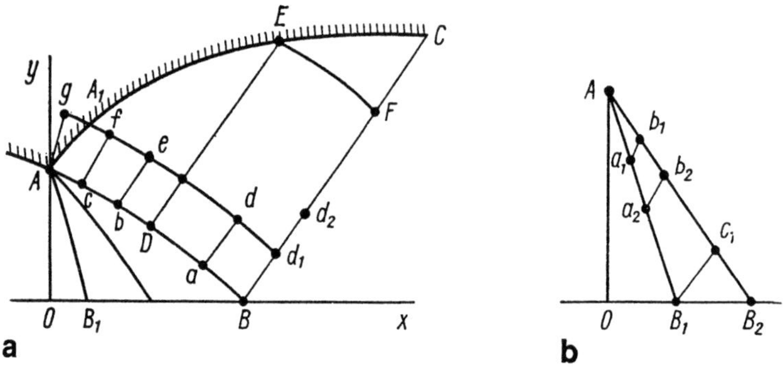

Fig.3.2a,b. Calculation of a flow in a nozzle with a corner point using the method of characteristics. (**a**) Solution of Goursat problem and determination of nozzle contour. (**b**) Calculation of rarefaction fan

The problem is divided into two parts: calculation of gasdynamics parameters in the rarefaction wave OAB (the accelerating part of the nozzle) and determination of the nozzle contour AC and flow parameters in the region ABC (the straightening part). The boundary characteristic AB of the rarefaction wave is such that at the point B, $\beta = \beta_0 = \mathrm{cotan}\,\alpha_0$, where $\alpha_0 = \arcsin M_0^{-1}$. The characteristic BC is, due to the requirement of flow uniformity at the exit cross-section of the nozzle, a straight line inclined to the axis of symmetry at an angle α_0.

Let us start with a calculation for the region of the rarefaction wave OAB. The method of characteristics cannot be used to depart from the sonic line OA because here characteristics of both families converge. However, the solution can be expanded into a series. The problem will be considered in detail in Sect.4.3.

To calculate the flow parameters at a characteristic close to the sonic line we shall use an expansion of the solution in characteristic coordinates [3.17]. The following simple approach is often used in practice. A characteristic close to the sonic line is substituted by a straight line, at every point of which the velocity vector is parallel to the axis while the Mach number is constant and equal to the Mach number at the corner point. Calculations show that if the characteristic corresponding to $M \approx 1.001$ is thus approximated at the corner point, then quite satisfactory accuracy is obtained.

Let all the flow parameters be known at the points A, a_1, a_2, ..., B_1 of the characteristic AB_1, Fig.3.2b. The Prandtl-Meyer flow takes place in the vicinity of the corner point A where $x = x_A$, $y = y_A$, $\psi = \psi_A$ are constant, while β and ζ are related by [see (2.151)]

$$f(\beta) - \arctan\zeta = 0 \quad . \tag{3.20}$$

Knowing the values of the variables at the point A corresponding to the characteristic AB_1, and incrementing, for instance, to $\Delta\beta$, one can find from (3.20) the values of β and ζ at the point A corresponding to the characteristic AB_2 close to AB_1. Then, using the parameters known at the points A and a_1 and (3.16) (for $S = \text{const}$) we can calculate all the required parameters at the point b_1. Using the points a_2 and b_1 as references, we can calculate the flow parameters at the point b_2, and so on. While calculating the parameters at the points C_1 and B_2 at $\nu = 1$, (3.16-18) are used. The parameters at the next characteristic are calculated in a similar way. The acceleration region is calculated until $\beta = \beta_0$ (which corresponds to the given Mach number M_0 at the nozzle's exit) is achieved at the axis of the nozzle. Then the straightening region of the nozzle is calculated.

Having the values of the parameters at the points B, a, ..., c, A of the boundary characteristic of the rarefaction wave and at the point d_1 of the straight closing characteristic (along which $\zeta = 0$, $\beta = \beta_0 = \sqrt{M_0^2 - 1}$), one can calculate parameters in succession, starting from the point d (Fig.3.2a), at the points of the second family characteristic issuing from the point d_1.

The calculation continues until ψ exceeds the given value ψ_A at a point of the characteristic. Then the coordinates of the point A_1 at the contour are determined together with the values of the gasdynamic parameters. This is done by means of quadratic interpolation for the given $\psi = \psi_A$ with the help of the parameters calculated at the last three points (the points e, f, g in Fig.3.2a). Further calculations are carried out in a similar manner.

Besides the main contour AC containing a corner point, it is possible, by means of interpolation for fixed intermediate values $0 < \psi_i < \psi_A$, to obtain, in the process of calculation, coordinates of the streamlines and the distribution of parameters along them. These streamlines may be chosen as contours of ordinary smooth nozzles with a plane surface of transition to supersonic flow.

If the flow at the section OA is supersonic and uniform, then the second family characteristic issuing from the corner point and separating the regions of the uniform and rarefaction flows will be at the same time the line of discontinuity of derivatives of the flow parameters. Let the point 1 in Fig.3.1b be the point where this characteristic intersects the axis of symmetry. It may be shown [3.18] that

$$\lim_{y \to 0} (\zeta/y) = (d\zeta/dy)_1 = \infty \quad ,$$

where the limit and the derivative are taken along the segment 1-3. In this case the calculation formulas may be obtained similarly to (3.16,17), the only difference being that the compatibility equation for the second family characteristics must be multiplied not by y but by y/ζ. To calculate the flow parameters at the point 3 we shall use again (3.16), putting this time

$$A_{13} = \frac{1}{2} A_3 + \frac{x_3 - x_1}{y_2(\beta_3 - \zeta_3)} \;, \quad B_{13} = \frac{1}{2} B_3 \;, \quad C^+_{13} = 0 \;, \quad D_{13} = \frac{1}{2} D_3 \;. \tag{3.21}$$

The accuracy of the calculations may be controlled by the values of the function ψ at the points belonging to the axis of symmetry ($\psi = 0$) obtained by integration of (3.14) along the second family characteristics while calculating the region of the rarefaction wave. Also, due to the fact that ψ is calculated at every point with the help of two different formulas [see (3.16)], the accuracy of calculation is defined by the difference of the values of ψ obtained by integrating along two different paths (e.g. ADE and BFE for the point E in Fig.3.2a).

Now let us consider the case of a real gas. We shall rewrite the equations of characteristics (2.62-69) in the form

$$dy = \frac{\beta\zeta \pm 1}{\beta \mp \zeta}\, dx \quad , \tag{3.22}$$

$$\frac{1}{1+\zeta^2}\, d\zeta \pm \frac{\beta}{\rho W^2}\, dp \pm \left(\frac{\nu\zeta}{y} - \sqrt{1+\zeta^2}\Phi_1\right) \frac{dx}{\beta \mp \zeta} = 0 \quad , \tag{3.23}$$

$$d\psi = \pm \frac{\rho W y^\nu \sqrt{1+\zeta^2}}{\beta \mp \zeta}\, dx \quad , \tag{3.24}$$

$$dy = \zeta\, dx \quad , \tag{3.25}$$

$$d\alpha_i = \frac{1}{W} \sqrt{1+\zeta^2} F_i\, dx \quad , \tag{3.26}$$

$$dT + \left(\frac{\partial h}{\partial T}\right)^{-1} \left(\frac{\partial h}{\partial p} - \frac{1}{\rho}\right) dp + \left(\frac{\partial h}{\partial T}\right)^{-1} \sum_{i=1}^{N} \frac{\partial h}{\partial \alpha_i}\, d\alpha_i = 0 \quad , \tag{3.27}$$

$$\frac{1}{2} W^2 + h = H_0 \quad . \tag{3.28}$$

Calculation schemes of the method of characteristics for equilibrium and frozen gas flows, when $F \equiv 0$ and the state equations are given in the form $h = h(p,T)$, $\rho = \rho(p,T)$, are only slightly more complicated than in the case of a perfect gas (see, e.g. [3.19-21]). Let the values of the flow parameters be known at the reference points 1 and 2 (Fig.3.1a). Then, writing (3.22-24) in the difference form and using the thermodynamic identity

$$dT = \left[T\, dS + \left(\frac{1}{\rho} - \frac{\partial h}{\partial p}\right) dp\right] \left(\frac{\partial h}{\partial T}\right)^{-1} \tag{3.29}$$

and considering entropy a known function of ψ, $S = S(\psi)$ from (2.71), we obtain a system of relations for determining x_3, y_3, p_3, ζ_3, T_3, ψ_3, S_3. To determine the values of the remaining parameters, the following formulas are used:

$$h_3 = h(p_3,T_3) \quad , \quad \rho_3 = \rho(p_3,T_3) \quad , \quad W_3 = \sqrt{2(H_0 - h_3)} \quad ,$$

$$a_3^{-2} = \left(\frac{\partial \rho}{\partial p}\right)_3 + \frac{(\partial \rho/\partial T)_3}{(\partial h/\partial T)_3} \left[\frac{1}{\rho_3} - \left(\frac{\partial h}{\partial p}\right)_3\right] \quad , \quad \beta_3 = \sqrt{W_3^2 a_3^{-2} - 1} \quad . \tag{3.30}$$

This system may be solved by means of iteration, similarly to the system (3.16).

The calculation of equilibrium flows is more complicated than that of a perfect gas because one needs to calculate the thermodynamic functions for a mixture of reactive gases. Usually the functions are given in the form of tables (e.g., [3.22]), the use of which is associated with certain inconveniences in the process of calculation. Therefore, a number of authors (e.g., [3.20], see bibliography in [3.1]) have developed, on the basis of existing tables, approximate formulas for the equations of state. The use of such approximations considerably simplifies the calculation of equilibrium flows. In this connection the reference [3.23] should be pointed out where, for the effective adiabatic exponent $\kappa = \rho a^2/p$, convenient approximating formulas have been proposed which allow flows of an equilibrium and perfect gas to be calculated using the same computer program.

Let us consider the case of nonequilibrium flow. Replacing the differential equations (3.22-24) by the difference equations, we obtain relations for the determination of x_3, y_3, ψ_3, p_3 and ζ_3 at the point 3 through their known values at the points 1 and 2 (Fig.3.1e). Then, we can calculate all the parameters at the point where the streamline entering the point 3 intersects the known segment of characteristic 1-6 (the point 4). As far as $\psi_4 = \psi_3$, the coordinate of the point 4 and the values of flow parameters at the point may be found by means of quadratic interpolation in ψ, using the known values at the nodes of the characteristic 1-5-6, Fig.3.1e.

Then, having written (3.26,27) as finite differences at the streamline segment 3-4, we obtain relations for

$$\alpha_{i3} = \alpha_{i4} + G_{i34}(x_3 - x_4) \quad , \qquad (i = 1,\ldots,N) \quad , \tag{3.31}$$

$$T_3 = T_4 + K_{34}(p_3 - p_4) - \sum_{i=1}^{N} M_{i34}(\alpha_{i3} - \alpha_{i4}) \quad ,$$

where

$$G_i = \frac{1}{W}\sqrt{1 + \zeta^2}F_i \quad , \qquad K = \left(\frac{\partial h}{\partial T}\right)^{-1}\left(\frac{1}{\rho} - \frac{\partial h}{\partial p}\right) \; , \qquad M_i = \left(\frac{\partial h}{\partial T}\right)^{-1} \frac{\partial h}{\partial \alpha_i} \quad .$$

Now h_3 and ρ_3 can be calculated from (2.52,53):

$$h_3 = h(p_3, T_3, \alpha_{13},\ldots, \alpha_{N3}) \quad , \qquad \rho_3 = \rho(p_3, T_3, \alpha_{13}, \ldots, \alpha_{N3}) \quad . \tag{3.32}$$

The values of W_3, a_3, β_3 may be calculated from (3.30). As usual, coefficients with two-figure indices in the first approximation are calculated on the basis of the known parameters at the corresponding reference points and their accuracy is improved in the course of subsequent iterations.

However, it has been found that the iteration may diverge if the values of some (or all) α_i are close to the equilibrium values. This is due to the fact that the relaxation equations (3.26) near equilibrium turn out to be equations with a small parameter before the highest derivative (see Sect.3.1).

Near equilibrium the function F_i is highly sensitive to variations in α_i and T. The iteration fails to converge because using (3.31) one has to calculate F_i at the point 3 from the values the parameters had in the previous iteration. To ensure convergence of iterations one has to use a grid with much smaller spacing than is needed for the solution of gasdynamic equations.

Many authors proposed different ways of circumventing the difficulty. Note, in particular, the work described in [3.24] where a technique has been developed ensuring convergence of iterations both near equilibrium and far from it. However, the method based on the use of implicit approximations of the equations of chemical kinetics described in Sect.3.1, is preferable.

We rewrite (3.26) in the form

$$d\alpha_i = [\bar{L}_i(p, T, \alpha_1, \ldots, \alpha_N) - \alpha_i \bar{N}_i(p, T, \alpha_1, \ldots, \alpha_N)]dx \tag{3.33}$$

[which is evidently equivalent to (3.1)]. Besides (3.33), we shall use (2.65) and (3.28) along the streamlines. Approximating these equations with the help of a difference scheme analogous to (3.7), we have

$$\alpha_{i4} - \alpha_{i3} + \frac{x_3 - x_4}{1 + (1 - s)(x_3 - x_4)N_{i3}} \{s\bar{L}_{i4} + (1 - s)\bar{L}_{i3} - \alpha_{i4} \times [s\bar{N}_{i4} + (1 - s)\bar{N}_{i3}]\} = 0 \quad , \qquad (i = 1,2,\ldots,N) \quad , \tag{3.34}$$

$$W_3^2 - W_4^2 + \left[\frac{1}{\rho(p_3, T_3, \alpha_{13}, \ldots, \alpha_{N3})} + \frac{1}{\rho(p_4, T_4, \alpha_{14}, \ldots, \alpha_{N4})}\right](p_3 - p_4) = 0 \quad , \tag{3.35}$$

$$\frac{1}{2} W_3^2 + h(p_3, T_3, \alpha_{13}, \ldots, \alpha_{N3}) - H_0 = 0 \quad . \tag{3.36}$$

Having determined in the current iteration the values of x_3, x_4, ψ_3, p_3 and ζ_3 and calculated the flow parameters at the point 4, one solves the system (3.34-36) by Newton's method with a variable Jacobi matrix, as described in Sect.3.1, and determines the values of α_{i3}, T_3, W_3. Then, h_3 and ρ_3 are calculated from (3.32) and a_3, β_3 from (3.30), and thus the iteration is concluded.

The method of characteristics is also successfully applied to the calculation of two-phase flows. In this case equations of characteristics of the first and the second families differ from (3.23) only in the coefficient before dx — the function Φ_1 should be replaced by Φ_2 defined by (2.73). The stream function is given by (3.24) as earlier. The following relation is valid for the stream function of the particles:

$$d\psi_s = \pm \frac{1 + \zeta\xi \pm \beta(\zeta - \xi)}{\beta \mp \zeta} y^\nu \rho_s u_s \, dx \ , \quad \xi = \frac{v_s}{u_s} \ . \tag{3.37}$$

The equations valid along the streamlines of the gas and of the particles have been given in Chap.2 (2.74-77).

Here we do not present the logic flow diagrams. In [3.25,26] the reader will find a detailed description of algorithms for calculation of gasdynamic functions at different types of points of the characteristic grid. Approximation of (3.22-24,37) allows x_3, y_3, ψ_3, p_3, ζ_3 to be calculated at an internal point (Fig.3.1f). The values of flow parameters at the reference points 4 and 5 of the gas and particle streamlines are determined by quadratic interpolation in ψ and ψ_s. Equations (2.74,75) are approximated along the gas streamlines in a standard way. In the case of (2.76) for the streamlines of particles, the standard approximation leads to the same difficulties as in the case of chemical nonequilibrium flows. Implicit schemes can be used in this case too.

Let us make some general remarks concerning the method of characteristics. This method is widely used in solving both stationary and nonstationary problems of gas dynamics. This is not accidental. By its very physical essence (small disturbances propagate along characteristics) this method allows the realization of a numerical solution at a grid which, to the maximum extent, takes into account the structure of flow. Thus, the method allows rarefaction waves to be conveniently calculated, the lines of weak discontinuities to be accounted for and the formation of suspended shocks to be seen [3.14].

However, the method of characteristics has a number of drawbacks. One of them is due to the fact that the required values are calculated at nodes of a characteristic grid which is not known beforehand. In practice it is often desirable to know distributions of parameters at fixed values of x. To find the values of x required, interpolation is necessary which makes the computer program more complicated. Sometimes calculations along characteristics result in a highly non uniform distribution of nodes or in an exponential growth of the number of points at the characteristics (for instance, while calculating the rarefaction wave in the problem of nozzle gas flow considered above). Evidently, in similar situations it is necessary to redistribute points at the characteristics from time to time and diminish their number if necessary. This procedure also involves interpolation. Finally, it should be pointed out that the process of numerical solution becomes much more complicated when the gas flow is accompanied by the formation of interacting discontinuities. Then the so-called shock-smearing methods (without an explicit definition of the discontinuity surfaces) should be used. This question is dealt with in Sect.3.2.2.

The need to eliminate the first problem resulted in the development of the so-called layerwise net method of characteristics (e.g. [3.1,12,27]). Let us explain the idea of this method using as examples (3.12,13). We shall consider a rectangular grid of coordinate lines parallel to the axes x, y. Let the solution be known at the nodes 5, 6, 7 of the layer $x = x_k$, Fig.3.1g. We shall denote by 1, 2, 4 the points of intersection of characteristics of the first and the second families and a streamline with the line $x = x_k$ and write down (3.12,13) in difference form. We have

$$y_3 - y_1 = \overset{+}{\lambda}_{13}\Delta x \quad , \quad y_3 - y_2 = \overset{-}{\lambda}_{23}\Delta x \quad , \quad y_3 - y_4 = \overset{0}{\lambda}_{34}\Delta x \quad , \tag{3.38}$$

$$A_{13}(\zeta_3 - \zeta_1) - B_{13}(\beta_3 - \beta_1) + \overset{+}{C}_{13}\Delta x - D_{13}(S_3 - S_1) = 0 \quad , \tag{3.39}$$

$$A_{23}(\zeta_3 - \zeta_2) + B_{23}(\beta_3 - \beta_2) - \overset{-}{C}_{23}\Delta x + D_{23}(S_3 - S_2) = 0 \quad , \tag{3.40}$$

$$S_3 = S_4 \quad . \tag{3.41}$$

Let the parameters at the point 3 be equal to their values at, for instance, the point 6. Then we shall find the coordinates of the points 1, 2, 4 from (3.38) and calculate at them the required functions by means of interpolation between the nodes 5, 6, 7. Our next step is to find ζ_3, β_3, S_3 in the first approximation from (3.39-41). Then the values of the coefficients are corrected and the process repeated. The scheme is quite simple and can be easily used for calculation of three-dimensional flows, though the determination of parameters at the boundary points is somewhat more complicated.

Evidently, due to the influence of the boundary lines (which cross the grid in arbitrary directions), the grid will be distorted in the process of calculation for the step size in the y direction will be nonuniform. This step nonuniformity is undesirable, for it is determined only by the form of the boundaries and has nothing to do with the character of variation of parameters in the y direction. Usually, after determination of the parameters at a new layer, the nodes are redistributed. Sometimes it is advantageous to transform the coordinates, including boundaries, into the coordinate grid; then the calculation can be carried out with a fixed grid.

3.2.2 Shock-Smearing Methods

If a nozzle gas flow is accompanied by the formation of shock waves the use of the method of characteristics (or any other method with an explicit determination of discontinuities) becomes difficult. The difficulties are especially pronounced when the flow field contains suspended shock waves or a complex system of interacting discontinuities. In these and similar cases shock-smearing methods should be used; they allow the entire flow field to be calculated using a general algorithm without an explicit determination of discontinuities. There exists a number of different approaches to the construction of such methods: introduction of artificial viscosity into the equations of gas dynamics, smoothing of difference solutions, construction of difference schemes possessing smoothing qualities [3.5,14,28]. The degree of perfection of such methods is determined by the width of discontinuity smearing and the accuracy of calculation outside the regions of large gradients. Two numerical methods of calculating nozzle flows will be considered here. One of them is based on the application of explicit difference schemes of second order of accuracy together with a smoothing procedure [3.29], and the other is based on the use of a scheme possessing smoothing qualities [3.28].

Let us consider the stationary supersonic flow of a perfect gas in a nozzle. We shall be looking for a solution in the region $x \geqslant 0$, $G(x) \leqslant y \leqslant F(x)$, where x, y are Cartesian or cylindrical coordinates, whose x axes coincide with the line/axis of symmetry and where the equations $y = F(x)$, $y = G(x)$ define the upper and lower boundaries of the region respectively, Fig.3.3. A boundary may be represented by a free surface, a solid wall or a line/axis of symmetry. We shall introduce new variables

$$x = x \quad , \quad \xi = \frac{y - G(x)}{F(x) - G(x)} \quad , \tag{3.42}$$

transforming the region into a semistrip $x \geqslant 0$, $0 \leqslant \xi \leqslant 1$.

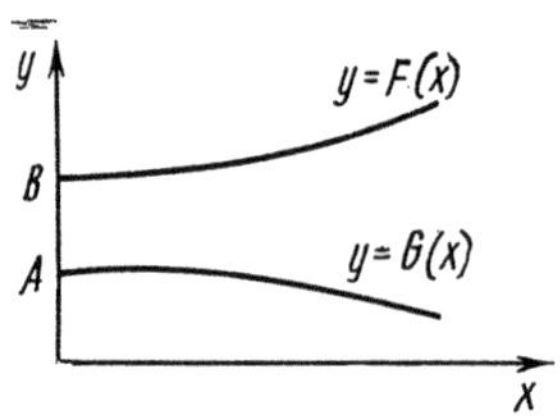

Fig.3.3. Region of calculation for shock-smearing methods. $y = F(x)$ and $y = G(x)$ define boundaries of a region (rigid walls, free surfaces, symmetry axis)

Let the equations of gas dynamics be written down in the divergent form:

$$\frac{\partial A_i}{\partial x} + \frac{\partial B_i}{\partial \xi} = T_i \quad , \qquad \text{where} \tag{3.43}$$

$$A_i = P_i\varepsilon \quad , \qquad B_i = Q_i - P_i(G' + \xi\varepsilon') \quad ,$$

$$P_1 = (p + \rho u^2)y^\nu \quad , \qquad Q_1 = \rho uvy^\nu \quad , \qquad T_1 = 0 \quad ,$$

$$P_2 = \rho uvy^\nu \quad , \qquad Q_2 = (p + \rho v^2)y^\nu \quad , \qquad T_2 = \nu p\varepsilon \quad ,$$

$$P_3 = \rho uy^\nu \quad , \qquad Q_3 = \rho vy^\nu \quad , \qquad T_3 = 0 \quad ,$$

$$P_4 = (\rho e + p)uy^\nu \quad , \qquad Q_4 = (\rho e + p)vy^\nu \quad , \qquad T_4 = 0 \quad ,$$

$$\varepsilon = F - G \quad , \qquad e = p/(\gamma - 1)\rho + W^2/2 \quad .$$

Here velocities are referred to the maximum velocity W_{max}, the density, to ρ_* and the pressure, to $\rho_* W^2_{max}$.

Next we shall introduce in the region $x \geqslant 0$, $0 \leqslant \xi \leqslant 1$ a grid $x^n = n\tau$, $(n = 0,1,\ldots)$, $\xi_m = mh$, $(m = 0,1,\ldots,M)$ and denote $f^n_m = f(n\tau, mh)$. Let the solution be known at the layer n. We shall start with calculation of internal points at the layer $(n+1)$. To find the solution at the point x^{n+1}, one has, using the known values at the nodes x^n, ξ_{m-1}; x^n, ξ_m; x^n, ξ_{m+1}, to determine the unknown values at the intermediate nodes $x^{n+\sigma}$, $\xi_{m-\frac{1}{2}}$ and $x^{n+\sigma}$, $\xi_{m+\frac{1}{2}}$, $(0 < \sigma \leqslant 1)$ applying the Lax scheme. Then, with the known values at the intermediate nodes and the node x^n, ξ_m the solution is sought at the node x^{n+1}, ξ_m by means of the "cross" scheme. Let us write down the difference equations. For the node $x^{n+\sigma}$, $\xi_{n-\frac{1}{2}}$ we have from (3.43):

$$A^{n+\sigma}_{i(m-\frac{1}{2})} = \frac{1}{2}\left(A^n_{im} + A^n_{i(m-1)}\right) - \frac{\tau_1}{h}\left(B^n_{im} - B^n_{i(m-1)}\right) + \frac{\tau_1}{2}\left(T^n_{im} + T^n_{i(m-1)}\right) \equiv f_i \quad , \tag{3.44}$$

where $\tau_1 = \sigma\tau$. Solving the system for the unknown values, we get

$$u^{n+\sigma}_{m-\frac{1}{2}} = \frac{1}{2a_1}\left(b_1 + \sqrt{b_1^2 - 4a_1c_1}\right) \quad , \qquad v^{n+\sigma}_{m-\frac{1}{2}} = \frac{f_2}{f_3} \quad ,$$
$$\rho^{n+\sigma}_{m-\frac{1}{2}} = \left(\frac{f_3}{\varepsilon uy^\nu}\right)^{n+\sigma}_{m-\frac{1}{2}} \quad , \qquad p^{n+\sigma}_{m-\frac{1}{2}} = \left(\frac{f_1 - f_3u}{\varepsilon y^\nu}\right)^{n+\sigma}_{m-\frac{1}{2}} \quad , \tag{3.45}$$

where

$$a_1 = \frac{\gamma + 1}{2(\gamma - 1)} f_3 \quad , \quad b_1 = \frac{\gamma}{\gamma - 1} f_1 \quad , \quad c_1 = f_4 - \frac{f_2^2}{2f_3} \quad .$$

Similarly, the solution at the node $x^{n+\sigma}$, $\xi_{m+\frac{1}{2}}$ is found. Using the "cross" scheme, we obtain

$$A_{im}^{n+1} = A_{im}^{n} - \frac{\tau}{h}\left(B_{i(m+\frac{1}{2})}^{n+\sigma} - B_{i(m-\frac{1}{2})}^{n+\sigma}\right)$$

$$+ \frac{\tau}{2}\left(T_{i(m+\frac{1}{2})}^{n+\sigma} + T_{i(m-\frac{1}{2})}^{n+\sigma}\right) \equiv g_i \quad . \tag{3.46}$$

Solving this system of equations, we get

$$u_m^{n+1} = \frac{1}{2a_2}\left(b_2 + \sqrt{b_2^2 - 4a_2c_2}\right) \quad , \quad v_m^{n+1} = \frac{g_2}{g_3} \quad ,$$

$$\rho_m^{n+1} = \left(\frac{g_3}{\varepsilon u y^{\nu}}\right)_m^{n+1} \quad , \quad p_m^{n+1} = \left(\frac{g_1 - g_3 u}{\varepsilon y^{\nu}}\right)_m^{n+1} \quad , \tag{3.47}$$

where

$$a_2 = \frac{\gamma + 1}{2(\gamma - 1)} g_3 \quad , \quad b_2 = \frac{\gamma}{\gamma - 1} g_1 \quad , \quad c_2 = g_4 - \frac{g_2^2}{2g_3} \quad .$$

This scheme at $\sigma = 1/2$ is accurate to second order. If $\sigma \neq 1/2$, then the approximation is of the first order of accuracy, but the remainder term contains the multiplier $(\sigma - \frac{1}{2})$ and hence, for values of σ close to 1/2, the accuracy of the scheme tends to second order. Note that the functions A_i, B_i, T_i contain F, G and their derivatives with respect to x at the layers n and $n + \sigma$.

Next we shall write down the algorithms of solution at the boundary nodes. Let $y = F(x)$ be the equation of the solid wall. To determine parameters at the node x^{n+1}, ξ_M we shall use the "rectangle" scheme with weights σ_1 and $1 - \sigma_1$ for the derivatives with respect to ξ at the layers $n+1$ and n. We have

$$A_{iM}^{n+1} + \frac{2\tau}{h}\sigma_1 B_{iM}^{n+1} = A_{i(M-1)}^{n} - A_{i(M-1)}^{n+1} + A_{iM}^{n} + \frac{2\tau}{h}\sigma_1 B_{i(M-1)}^{n+1}$$

$$- \frac{2\tau}{h}(1 - \sigma_1)\left(B_{iM}^{n} - B_{i(M-1)}^{n}\right) + 2\tau T_{i(M-\frac{1}{2})}^{n+\sigma} \equiv h_i \quad ,$$

$$(i = 1,2,3,4) \quad . \tag{3.48}$$

We shall use the first, the third and the fourth of these equations together with the boundary condition

$$v_M^{n+1} = (F')^{n+1} u_M^{n+1} \quad . \tag{3.49}$$

Solving the system and taking into consideration that $\xi_M = 1$, $y_M = F$, we get

$$u_M^{n+1} = \frac{1}{2a_3}\left(-b_3 - \sqrt{b_3^2 + 4a_3c_3}\right) \quad , \quad v_M^{n+1} = (F')^{n+1} u_M^{n+1} \quad ,$$

$$\rho_M^{n+1} = \left(\frac{h_3}{F^\nu u\varepsilon}\right)_M^{n+1} \quad , \quad p_M^{n+1} = \left[\frac{h_1 - h_3 u}{F^\nu\left(\varepsilon - \frac{2\tau}{h}\sigma_1 F'\right)}\right]_M^{n+1} \quad , \quad \text{where} \tag{3.50}$$

$$a_3 = h_3\left[\frac{1}{2}(1 + F'^2) - \frac{\gamma\varepsilon}{\gamma - 1}\left(\varepsilon - \frac{2\tau}{h}\sigma_1 F'\right)^{-1}\right]_M^{n+1} \quad ,$$

$$b_3 = \frac{\gamma h_1}{\gamma - 1}\left[\varepsilon\left(\varepsilon - \frac{2\tau}{h}\sigma_1 F'\right)^{-1}\right]_M^{n+1} \quad , \quad c_3 = h_4 \quad .$$

When $y = G(x)$ is the solid wall equation, a similar algorithm for determining the solution at the node x^{n+1}, ξ_0 is used.

One of the boundaries of the region may be formed by a free surface; let it be the upper boundary. We shall suppose that the pressure p is constant along an unknown free surface and that no shock waves terminate here (i.e. the entropy does not change). Then the boundary conditions may be written as

$$F' = v/u \quad , \quad p = p_M = \text{const} \quad , \quad \rho = \rho_M = \text{const} \quad ,$$
$$M = M_M = \text{const} \quad . \tag{3.51}$$

We shall supplement these relations with the second of the equations (3.43). Approximating, as earlier, using a "rectangle" scheme, we get for the node x^{n+1}, ξ_M:

$$p_M^{n+1} = p_M \quad , \quad \rho_M^{n+1} = \rho_M \quad , \quad u_M^{n+1} = \left[\frac{\gamma p_M}{\rho_M} M_M^2 - (v^2)_M^{n+1}\right]^{1/2} \quad ,$$

$$(\rho u v \varepsilon)_M^{n+1} + \frac{2\tau}{h}\sigma_1\left[(p + \rho v^2) - \rho u v F'\right]_M^{n+1}$$
$$= \frac{1}{(F^\nu)^{n+1}}\left[A_{2M}^n + A_{2(M-1)}^n - A_{2(M-1)}^{n+1} + \frac{2\tau}{h}\sigma_1 B_{2(M-1)}^{n+1}\right. \tag{3.52}$$
$$\left. - \frac{2\tau}{h}(1 - \sigma_1)\left(B_{2M}^n - B_{2(M-1)}^n\right) + 2\tau T_{2(M-\frac{1}{2})}^{n+\sigma}\right] \quad ,$$

$$(F')^{n+1} = (v/u)_M^{n+1} \quad .$$

To determine u_M^{n+1}, v_M^{n+1}, the third and fourth of the euqations must be solved simultaneously by Newton's method.

If a boundary point belongs to the axis of symmetry (for $\nu = 1$), then the system (3.43) cannot be used straightforwardly because then $y = v = 0$. We shall write the equations in the form

$$v = 0 \quad ,$$

$$\frac{\partial \bar{A}_i}{\partial x} - \frac{\partial \bar{B}_i}{\partial \xi} = 0 \quad , \qquad \text{where}$$

$$\bar{A}_i = \bar{P}_i \varepsilon \quad , \qquad \bar{B}_i = \bar{Q}_i - \bar{P}_i \xi \varepsilon' \quad ,$$

$$\bar{P}_1 = p + \rho u^2 \quad , \quad \bar{P}_2 = \rho u v \quad , \quad \bar{P}_3 = \rho u \quad , \quad \bar{P}_4 = (\rho e + p) u \quad ,$$

$$\bar{Q}_1 = 2\rho u v \quad , \quad \bar{Q}_2 = p + \rho v^2 \quad , \quad \bar{Q}_3 = 2\rho v \quad , \quad \bar{Q}_4 = 2(\rho e + p) v \quad .$$

The symmetry implies that

$$u_{-\frac{1}{2}}^{n+\sigma} = u_{\frac{1}{2}}^{n+\sigma} \quad , \quad \rho_{-\frac{1}{2}}^{n+\sigma} = \rho_{\frac{1}{2}}^{n+\sigma} \quad , \quad p_{-\frac{1}{2}}^{n+\sigma} = p_{\frac{1}{2}}^{n+\sigma} \quad ,$$

$$v_{-\frac{1}{2}}^{n+\sigma} = -v_{\frac{1}{2}}^{n+\sigma} \quad , \quad \varepsilon_{-\frac{1}{2}}^{n+\sigma} = \varepsilon_{\frac{1}{2}}^{n+\sigma} \quad , \quad (\varepsilon')_{-\frac{1}{2}}^{n+\sigma} = -(\varepsilon')_{\frac{1}{2}}^{n+\sigma} \quad .$$

The quantities at the node $x^{n+\sigma}$, $\xi_{\frac{1}{2}}$ are calculated from (3.44,45) at $m = 1$, where the capital letters should be replaced by barred quantities. Further, we have

$$\frac{1}{\tau}\left(\bar{A}_{i0}^{n+1} - \bar{A}_{i0}^{n}\right) + \frac{1}{h}\left(\bar{B}_{i\frac{1}{2}}^{n+\sigma} - \bar{B}_{i(-\frac{1}{2})}^{n+\sigma}\right) = 0 \quad ,$$

and hence

$$\bar{A}_{i0}^{n+1} = \bar{A}_{i0}^{n} - \frac{\tau}{h}\left(\bar{B}_{i\frac{1}{2}}^{n+\sigma} - \bar{B}_{i(-\frac{1}{2})}^{n+\sigma}\right) \equiv \bar{g}_i \quad .$$

Finally, we obtain

$$\begin{aligned} &v_0^{n+1} = 0 \quad , \quad u_0^{n+1} = \frac{1}{2\bar{a}_2}\left(\bar{b}_2 + \sqrt{\bar{b}_2^2 - 4\bar{a}_2\bar{c}_2}\right) \quad , \\ &p_0^{n+1} = \left[\frac{1}{\varepsilon}\left(\bar{g}_1 - \bar{g}_3 u\right)\right]_0^{n+1} \quad , \quad \rho_0^{n+1} = \left(\frac{\bar{g}_3}{\varepsilon u}\right)_0^{n+1} \quad , \end{aligned} \qquad (3.53)$$

where

$$\bar{a}_2 = \frac{\gamma + 1}{2(\gamma - 1)}\bar{g}_3 \quad , \quad \bar{b}_2 = \frac{\gamma}{\gamma - 1}\bar{g}_1 \quad , \quad \bar{c}_2 = \bar{g}_4 \quad .$$

A straightforward application of the difference technique described above results in the appearence of oscillations of the solution in the regions of large gradients as is typical for schemes of second order of accuracy. A difference solution may be perfected, if smoothing is done in separate layers with the use of five nodes,

$$f^{n+1}_{m\,sm} = \sum_{k=-2}^{2} \alpha_k f^{n+1}_{m-k} \quad , \quad \alpha_k = \alpha_{-k} \quad , \quad \sum_{k=-2}^{2} \alpha_k = 1 \quad ,$$

and if the values of α_0 and q are fixed, the latter determining the character of diminishing of coefficients α_k which satisfy the conditions $\alpha_2 = q\alpha_1$ and $\alpha_0 + 2\alpha_1(1+q) = 1$. At the boundary points the solution is not smoothed, while at the points adjacent to the boundary it is smoothed using three nodes. The smoothing is applied not to the functions themselves, but to functions A_i; then the final values of gasdynamic parameters in the layer $n+1$ are found from (3.47). The size of step needed to move to the next layer is determined from the condition,

$$\tau^{n+1} = K \min_{m} \left\{ \left| \frac{\varepsilon h}{\tan(\theta \pm \alpha)} \right| \right\}^{n}_{m} \quad , \quad K \leqslant \frac{1}{2\sigma} \quad , \quad \sigma \geqslant \frac{1}{2}$$

where K is the Courant number, usually put equal to 0.6-0.8. The method has been used, in particular, for calculation of gas flows in conical nozzles ([3.30]; see also Sect.7.2).

Let us proceed with a description of the method proposed in [3.28]. It is based on a difference scheme, analogous to the one of [3.31] developed for calculations of one-dimensional nonstationary gas flows. We shall again consider the motion of a perfect gas in the region shown in Fig.3.3. The equations of gas dynamics will be written in the form of the following integral laws of conservation, where the velocity, density and pressure are referred to a_*, ρ_* and $\rho_* a_*^2$ respectively:

$$\frac{d}{dx}\int_{y_-}^{y_+} \mathbf{a}\, dy = (\mathbf{a}y' - \mathbf{b})\Big|_{y_-}^{y_+} + \int_{y_-}^{y_+} \mathbf{f}\, dy \quad , \tag{3.54}$$

where

$$\mathbf{a} = \begin{pmatrix} \rho u \\ p + \rho u^2 \\ \rho uv \end{pmatrix}, \quad \mathbf{b} = \begin{pmatrix} \rho v \\ \rho uv \\ p + \rho v^2 \end{pmatrix}, \quad \mathbf{f} = -\nu \frac{\rho}{y} \begin{pmatrix} v \\ uv \\ v^2 \end{pmatrix}, \quad y'_{\pm} = \frac{dy_{\pm}(x)}{dx} \quad .$$

The system is closed by Bernoulli's equation,

$$\frac{1}{2} W^2 + \frac{\gamma}{\gamma - 1}\frac{p}{\rho} = \frac{\gamma + 1}{2(\gamma - 1)} \quad . \tag{3.55}$$

Let us introduce a grid with nodes placed at sections $x = \text{const}$. In each section the nodes are equally spaced by the step $h = (F - G)/N$, i.e. the step length is changed from section to section. The step between sections τ is nonuniform. Nodes at a section and segments are numbered from bottom to top: the nodes are ascribed numbers n, $(n = 0,1,\dots,N)$ and the segments have numbers $n - 1/2$, $(n = 1,2,\dots,N)$. The flow parameters are averaged for each segment. The mean values of the parameters at the segments of a fixed layer $x = x_0$ are denoted by a subscript corresponding to the number of a segment ($\rho_{n-\frac{1}{2}}$, $u_{n-\frac{1}{2}}$ and so on). The nodes with equal numbers at the sections $x = x_0$ and $x = x_0 + \tau$ are joined with straight-line segments. The mean values of parameters at all these segments are denoted by capital letters (e.g. R, U, V instead of the usual ρ, u, v) and are supplied with subcripts coinciding with the numbers of nodes joined with the given segment.

Let us write down (3.54) for each segment, taking the ordinates of the end points of the segments for y_-, y_+, and integrate with respect to x between x_0 and $x_0 + \tau$. Then the following difference scheme is obtained:

$$\begin{aligned}
a^{n-\frac{1}{2}} &= a_{n-\frac{1}{2}} + D_n - D_{n-1} - d_{n-\frac{1}{2}} - d^{n-\frac{1}{2}} \quad , \\
b^{n-\frac{1}{2}} &= b_{n-\frac{1}{2}} + (Ph)_n - (Ph)_{n-1} \\
&\quad + (UD)_n - (UD)_{n-1} - (ud)_{n-\frac{1}{2}} - (ud)^{n-\frac{1}{2}} \quad , \\
c^{n-\frac{1}{2}} &= c_{n-\frac{1}{2}} + (P_{n-1} - P_n)\tau + (VD)_n \\
&\quad - (VD)_{n-1} - (vd)_{n-\frac{1}{2}} - (vd)^{n-\frac{1}{2}} \quad ,
\end{aligned} \tag{3.56}$$

where $a = \rho uh$, $b = (p + \rho u^2)h$, $c = \rho uvh$, $d = \nu\rho vh\tau/2y$, $D = R(Uh - V\tau)$, $h_j = y^j - y_j$, $j = n - 1$, n; the indices of h and y have the same meaning as in the case of the functions. Knowing $a^{n-\frac{1}{2}}$, $b^{n-\frac{1}{2}}$, $c^{n-\frac{1}{2}}$, one can calculatee $u^{n-\frac{1}{2}}$, $v^{n-\frac{1}{2}}$, $\rho^{n-\frac{1}{2}}$, $p^{n-\frac{1}{2}}$:

$$u = \frac{1}{ka}\left\{\frac{\gamma b}{\gamma - 1} + \left[\left(\frac{\gamma b}{\gamma - 1}\right)^2 - k(ka^2 - c^2)\right]^{\frac{1}{2}}\right\} \quad , \tag{3.57}$$

$$v = c/a \quad , \quad \rho = a/uh \quad , \quad p = (b - au)/h \quad .$$

Here the superscripts are omitted and $k = (\gamma + 1)/(\gamma - 1)$.

The values of U, V, P, R are found from the solution of a plane problem of the interaction of two uniform supersonic flows having parameters $u_{n-\frac{1}{2}}$, $v_{n-\frac{1}{2}}$, ... and $u_{n+\frac{1}{2}}$, $v_{n+\frac{1}{2}}$, ... respectively. It allows the "large" quantities in (3.56) to be expressed through the known "small" quantities with sub-

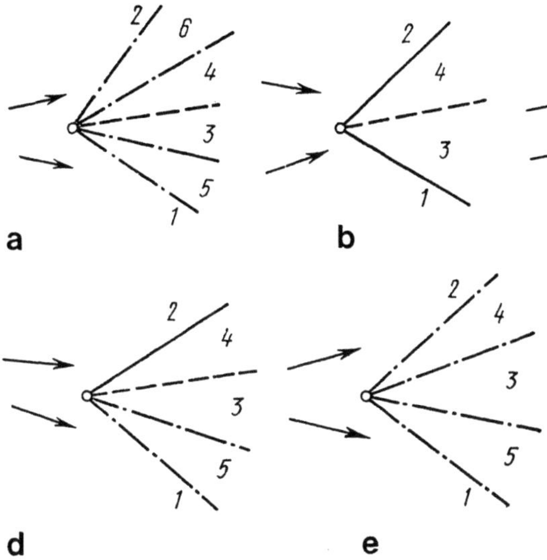

Fig.3.4a-e. Possible discontinuity configurations. (**a**) Interaction of two rarefaction fans, (**b**) two shock waves, (**c**) rarefaction fans and shock wave, (**d**) shock wave and rarefaction fans, and (**e**) two rarefaction fans with regions (*3*) of vacuum

scripts $n-3/2$, $n-1/2$, $n+1/2$. With the known values of U, V, P the system (3.56) is solved by means of iterations, while the terms with the superscript $n-1/2$ on the right side of (3.56) are taken from the previous iteration. In the course of the first iteration the superscript $n-1/2$ is changed for the subscript.

Let us consider the problem of the interaction of two flows in order to determine the "large" values at the upper and lower boundaries of an elementary cell. The interaction of the flows starts at the point $x=x_0$, $y=y_j$, $j=n-1, n$. Depending on the values of the parameters, there may develop one of the flow patterns shown in Fig.3.4. Here the solid lines correspond to shock waves, the dashed lines to tangential discontinuities and the dot-dash lines to the boundaries of rarefaction waves. The flow parameters are constant in the regions 1-4 in Fig.3.4a-d and in the region 1-3 in Fig.3.4e. A cell boundary may happen to lie in any of the regions.

Let us consider the situation shown in Fig.3.4a. In the rarefaction wave the flow parameters are interrelated by

$$d\zeta \mp \frac{1+\zeta^2}{\rho W^2}\sqrt{M^2-1}\,dp = 0 \quad . \tag{3.58}$$

We shall average the coefficient in front of dp and integrate (3.58). We get

$$\zeta - \zeta_j \mp \alpha(p - p_j) = 0 \quad ,$$

$$\alpha = \frac{1}{2}\left[\left(\frac{1+\zeta^2}{\rho W^2}\sqrt{M^2-1}\right)_{n-\frac{1}{2}} + \left(\frac{1+\zeta^2}{\rho W^2}\sqrt{M^2-1}\right)_{n+\frac{1}{2}}\right] \quad , \tag{3.59}$$

where the minus-sign and $j = n + 1/2$ correspond to the upper wave, and the plus-sign and $j = n - 1/2$ to the lower one. Since p and ξ are continuous at a tangential discontinuity, we get from (3.59) the following expression for p and ζ in the regions 3 and 4:

$$p = \frac{1}{2}(p_{n-\frac{1}{2}} + p_{n+\frac{1}{2}}) + \frac{1}{2\alpha}(\zeta_{n-\frac{1}{2}} - \zeta_{n+\frac{1}{2}}) \quad ,$$
$$\zeta = \zeta_{n-\frac{1}{2}} - \alpha(p - p_{n-\frac{1}{2}}) = \zeta_{n+\frac{1}{2}} + \alpha(p - p_{n+\frac{1}{2}}) \quad . \tag{3.60}$$

Now the remaining parameters can be calculated in the regions 3, 4:

$$\rho = \rho_j\left(\frac{p}{p_j}\right)^{1/\gamma} \quad , \quad W = \left[\frac{\gamma+1}{\gamma-1} - \frac{2\gamma p}{(\gamma-1)\rho}\right]^{\frac{1}{2}} \quad , \tag{3.61}$$

$$u = W(1+\zeta^2)^{-\frac{1}{2}} \quad , \quad v = \zeta u \quad , \quad M = W(\gamma p/\rho)^{-\frac{1}{2}} \quad .$$

Here $j = n - 1/2$ and $j = n + 1/2$ for the regions 3 and 4 respectively.

If the calculated values of p and ζ deviate considerably from the analogous values with indices $n - 1/2$, $n + 1/2$, they should be calculated anew:

$$p = \frac{1}{\alpha_{n-\frac{1}{2}} + \alpha_{n+\frac{1}{2}}}(\beta_{n-\frac{1}{2}}\zeta_{n-\frac{1}{2}} - \beta_{n+\frac{1}{2}}\zeta_{n+\frac{1}{2}} + \alpha_{n-\frac{1}{2}}p_{n-\frac{1}{2}} + \alpha_{n+\frac{1}{2}}p_{n+\frac{1}{2}}) \quad ,$$
$$\zeta = \beta_{n-\frac{1}{2}}\zeta_{n-\frac{1}{2}} - \alpha_{n-\frac{1}{2}}(p - p_{n-\frac{1}{2}}) = \beta_{n+\frac{1}{2}}\zeta_{n+\frac{1}{2}} + \alpha_{n+\frac{1}{2}}(p - p_{n+\frac{1}{2}}) \quad , \tag{3.62}$$

where

$$\alpha_j = \frac{1}{2}\left[\left(\frac{1+\zeta^2}{\rho W^2}\sqrt{M^2-1}\right)_j + \frac{1+\zeta^2}{\rho W^2}\sqrt{M^2-1}\right] \quad , \quad \beta_j = 1 \quad . \tag{3.63}$$

The second addend in α_j is calculated from the previous approximation; the β_j have been introduced for the convenience of further description.

If the shock waves are weak, then p and ζ are defined from (3.60) as before, while for the density we have

$$\rho = \frac{(\gamma-1)p_j + (\gamma+1)p}{(\gamma+1)p_j + (\gamma-1)p}\rho_j \quad , \quad p > p_j \quad . \tag{3.64}$$

Here $j = n - 1/2$ for the calculation of ρ in region 3 in Figs.3.4b,c and $j = n + 1/2$ when ρ is calculated in region 4 in Figs.3.4b,d. The remaining parameters are defined by (3.61). Equations (3.60,61,64) are used in the first approximation and in the case of shock waves of an arbitrary intensity. To calculate p and ζ in the subsequent approximations, one has to use relations for an oblique shock wave.

Let us consider the situation shown in Fig.3.4b. It follows from the conditions of coincidence of flow directions in the regions 3 and 4 and conditions at the shocks that

$$\frac{\zeta_{n-\frac{1}{2}} - \tan\delta_{n-\frac{1}{2}}}{1 + \zeta_{n-\frac{1}{2}}\tan\delta_{n-\frac{1}{2}}} = \frac{\zeta_{n+\frac{1}{2}} + \tan\delta_{n+\frac{1}{2}}}{1 - \zeta_{n+\frac{1}{2}}\tan\delta_{n+\frac{1}{2}}} \quad , \tag{3.65}$$

$$\tan\delta_j = \gamma_j(p - p_j); \quad \gamma_j = \left[\frac{2\gamma M_j^2 p_j}{(\gamma - 1)p_j + (\gamma + 1)p} - 1\right]^{\frac{1}{2}} \frac{1}{\rho_j W_j^2 + p_j - p} \quad ,$$

$$j = n \pm \frac{1}{2} \quad . \tag{3.66}$$

Substituting (3.66) into the numerators of (3.65), we once again obtain (3.62) for p with the coefficients

$$\alpha_j = \gamma_j\beta_j \quad , \quad \beta_j = (1 \pm \zeta_j \tan\delta_j)^{-1} \quad , \tag{3.67}$$

where the upper sign corresponds to $j = n - 1/2$ and the lower one to $j = n + 1/2$. To calculate α_j, β_j the values of p obtained in the previous approximation are used. For α_j, β_j found from (3.67), equation (3.62) for ζ is used again.

Equations (3.62) are valid also for the situations shown in Figs.3.4c,d. In the case of the first situation ($p_{n-\frac{1}{2}} < p < p_{n+\frac{1}{2}}$), (3.67) is used for $j = n - 1/2$ and (3.63) is used for $j = n + 1/2$, and vice versa for the second situation. To calculate ρ in the regions 3 and 4, (3.61) is used in the case of a rarefaction wave, and (3.64) in the case of a shock. The occurrence of the situation shown in Fig.3.4e corresponds to a large discrepancy in the parameters with subscripts $n - 1/2$, $n + 1/2$, i.e. a smaller grid spacing is necessary.

The "large" quantities in (3.56) are put equal to the values of the parameters in the region where the cell boundary happens to lie. If it lies in the region occupied by a rarefaction wave, the "large" quantities are obtained by means of linear interpolation between the values of parameters at the boundaries of the wave.

For the boundary cells (whose side facets are numbered $n = 1/2$ and $n = N - 1/2$) the "large" quantities are calculated in a different manner. If the boundaries are formed by nozzle walls, for which angles of inclination to the x axis, $\theta_\pm(x)$ (or $\zeta_\pm$) are known, then the "large" quantities can be obtained from the solution of the similar problem of uniform flow past a flat wall inclined to the x axis at an angle equal to $\theta_+(x_0+0.5\tau)$ or $\theta_-(x_0+0.5\tau)$. In this case (as in the case of the axis of symmetry) the following formula for determination of the pressure at the wall is valid:

$$p = p_j \pm (\beta_j\zeta_j - \zeta)/\alpha_j \quad , \quad \zeta \equiv \zeta(x_0 + 0.5\tau)$$

$$\approx [y_\pm(x_0 + h) - y_\pm(x_0)]\tau^{-1} \quad . \tag{3.68}$$

The coefficients α_j, β_j are calculated from (3.63) if $p < p_j$, and from (3.67) if $p > p_j$. The upper signs in (3.67,68) and the subscript $j = N - 1/2$ correspond to the upper boundary, while the lower signs and the subscript $j = 1/2$ correspond to the lower boundary. If the boundary is a free surface with a given p, then for calculation of ζ we have

$$\zeta = \beta_j\zeta_j \mp \alpha_j(p - p_j) \quad .$$

The choice and determination of the coefficients, signs and indices in the formula is analogous to (3.68). To ensure stability of calculation with the help of the described difference scheme, the size of the step τ must be chosen in such a way that all waves generated by interaction of flows would reach its right boundary at $x = x_0 + \tau$.

Concluding the section, we draw the reader's attention to the fact that the averaging of the coefficient of dp in (3.58) may sometimes result in poor accuracy. Then transformation to a new variable $\bar{p} = p^{(\gamma-1)/2\gamma}$ is advisable, when the approximate integration of (3.58) is better justified.

3.3 Methods of Calculating Three-Dimensional Supersonic Flows

A gas flow in a symmetric nozzle may be asymmetric, e.g. due to asymmetric disturbances of the flow. Sometimes it is desirable to have a nozzle with an asymmetric exit section or a nozzle with a curvilinear axis. In all these cases the gas flow is three-dimensional, i.e., the gasdynamic parameters depend on three space coordinates. Such flows may be successfully calculated by the method of characteristics and by grid methods.

3.3.1 Method of Characteristics

The schemes of the method of characteristics are subdivided into two classes – semicharacteristic and bicharacteristic. In the first case the initial system of differential equations is, first of all, by means of approximation in one of the variables, reduced to a system with two independent variables [3.32]. To solve the resulting two-dimensional system, the method of characteristics is used. In the case of bicharacteristic methods (among them the methods developed in [3.27,33]) the compatibility equations for a three-dimensional system of equations are approximated along the generatrices of the Mach conoid (i.e. along the bicharacteristics, see Sect.2.1.2) and the streamlines. The reader will find a survey of various characteristic schemes and a bibliography in [3.1,33].

Let us consider one of the bicharacteristic methods developed in [3.33] and based on the ideas of [3.34]. The characteristic equations for the system (2.23,25-27) describing a stationary equilibrium gas flow have the form of (2.34-40). The equations of the streamlines and the characteristic conoid [see (2.32)] may be written as

$$\frac{dx}{dt} = u \quad , \quad \frac{dy}{dt} = v \quad , \quad \frac{dz}{dt} = w \quad , \tag{3.69}$$

$$\begin{gathered}[u^2 - (W^2 - a^2)]\left(\frac{dx}{dt}\right)^2 + [v^2 - (W^2 - a^2)]\left(\frac{dy}{dt}\right)^2 \\ + [w^2 - (W^2 - a^2)]\left(\frac{dz}{dt}\right)^2 + 2uv\,\frac{dx}{dt}\,\frac{dy}{dt} + 2uw\,\frac{dx}{dt}\,\frac{dz}{dt} + 2uw\,\frac{dy}{dt}\,\frac{dz}{dt} = 0 \quad ,\end{gathered} \tag{3.70}$$

where t is the length of arc along a streamline and a bicharacteristic. Equation (2.40), upon writing the derivatives along bicharacteristics, acquires the following form:

$$\begin{aligned}&\rho a n_x \frac{du}{dt} + \rho a n_y \frac{dv}{dt} + \rho a n_z \frac{dw}{dt} - \frac{dp}{dt} \\ &\quad + \rho a^2\Big[(n_x^2 - 1)\,\frac{\partial u}{\partial x} + (n_y^2 - 1)\,\frac{\partial v}{\partial y} + (n_z^2 - 1)\,\frac{\partial w}{\partial z} \\ &\quad + \left(\frac{\partial u}{\partial y} + \frac{\partial v}{\partial x}\right)n_x n_y + \left(\frac{\partial u}{\partial z} + \frac{\partial w}{\partial x}\right)n_x n_z + \left(\frac{\partial v}{\partial z} + \frac{\partial w}{\partial y}\right)n_y n_z\Big] = 0 \quad .\end{aligned} \tag{3.71}$$

The term in square brackets contains derivatives along a shock wave surface with respect to the direction normal to the bicharacteristics (the lateral derivatives). The most important feature of the method is that there is no

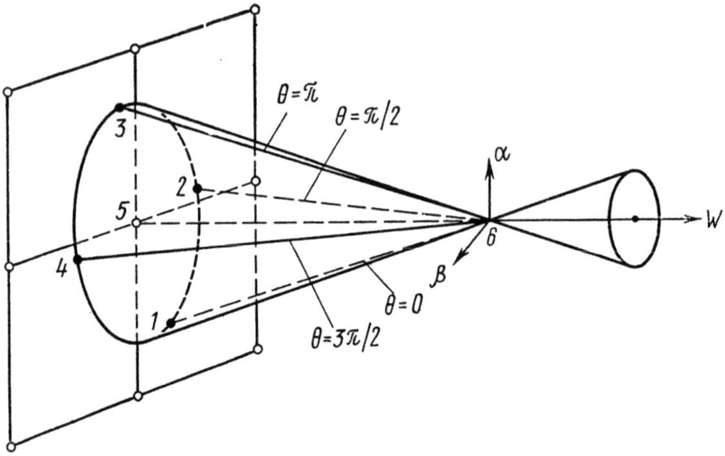

Fig.3.5. Arrangement of the characteristic grid nodes. Four bicharacteristics ($\theta=0$, $\pi/2$, π, $3\pi/2$) have been used to construct the difference scheme

need to calculate the lateral derivatives at the point where the solution is being sought. As a result the method ensures second order of accuracy[1].

Let the flow parameters be known at all the nodes of the layer $x=x_0$. We shall obtain relations for determination of the unknown values at the internal nodes of the grid at the layer $x=x_0+\Delta x$. In Fig.2.5 point 5 belongs to the layer $x=x_0$ in which all the values are known, point 6 is the point of intersection of the streamline passing through point 5 with the plane $x=x_0+\Delta x$, where the solution is sought and 1, 2, 3, 4 are the points of intersection of bicharacteristics passing through point 6 with the plane $x=x_0$. Below, for the sake of brevity, we use subscript notations: $x_1=x$, $x_2=y$, $x_3=z$, $u_1=u$, $u_2=v$, $u_3=w$, and repeated indices mean summation.

In the references above the parametric form of bicharacteristics is used:

$$dx_i = (u_i + c\alpha_i \cos\theta + c\beta_i \sin\theta)dt \quad , \quad (i = 1,2,3) \quad . \tag{3.72}$$

Here θ is an angle characterizing a given bicharacteristic, α_i, β_i are unit vectors such that u_i/W, α_i, β_i form an orthogonal system of vectors and $c^2=a^2W^2/(W^2-a^2)$ (Fig.3.5). The compatibility equation along the bicharacteristic (3.71) is written in the form

1 In this connection, [3.27] is to be mentioned, where a numerical scheme is proposed using derivatives with respect to bicharacteristic directions only

$$dp + \rho c(\alpha_i \cos\theta + \beta_i \sin\theta)du_i$$

$$= -\rho c^2(\alpha_i \sin\theta - \beta_i \cos\theta)(\alpha_j \sin\theta - \beta_j \cos\theta) \frac{\partial u_i}{\partial x_j} dt \tag{3.73}$$

[the components of the vector of the normal are $n_i = (a/c)\ (cu_i W^{-2} - \alpha_i \cos\theta - \beta_i \sin\theta)$]. The following noncharacteristic relationship which is valid along a streamline is also introduced and is obtained with the help of the continuity equation:

$$dp = -\rho c^2(\alpha_i\alpha_j + \beta_i\beta_j) \frac{\partial u_i}{\partial x_j} dt \quad . \tag{3.74}$$

It is seen that if, in (3.73), one puts $\theta = 0, \pi/2, \pi, 3\pi/2$, then these four equations in the manner of (3.74) will contain partial derivatives in the form of the two complexes $\alpha_i\alpha_j(\partial u_i/\partial x_j)$ and $\beta_i\beta_j(\partial u_i/\partial x_j)$ which may be considered as additional unknowns. They may be excluded from the five equations written in the form of finite differences. Then we obtain three independent relations.

Finally, we shall consider (3.73) along the four bicharacteristics corresponding to $\theta = 0, \pi/2, \pi, 3\pi/2$ together with (3.74) and (2.35-37), the latter three in the form

$$dp_0 = 0 \quad , \quad dH_0 = 0 \quad , \quad dp + \rho u_i du_i = 0 \quad . \tag{3.75}$$

We shall write (3.73) in the form of finite differences for $\theta = 0, \pi/2, \pi, 3\pi/2$ and designate the resulting equations by I-V respectively.

To exclude the derivatives $\partial u_i/\partial x_j$ at the point 6. III is subtracted from I, IV from II and V from the sum of I and II. This results in the following three independent relations:

$$\frac{p_6 - p_1}{t_6 - t_1} + (\rho c)_{16}\alpha_{i6} \frac{u_{i6} - u_{i1}}{t_6 - t_1} + \frac{p_6 - p_3}{t_6 - t_3} + (\rho c)_{36}\alpha_{i6} \frac{u_{i6} - u_{i3}}{t_6 - t_3}$$

$$= \left[- \left(\rho c^2 \frac{\partial u_i}{\partial x_j}\right)_1 + \left(\rho c^2 \frac{\partial u_i}{\partial x_j}\right)_3 \right] (\beta_i\beta_j)_6 \quad , \tag{3.76}$$

$$\frac{p_6 - p_2}{t_6 - t_2} + (\rho c)_{26}\beta_{i6} \frac{u_{i6} - u_{i2}}{t_6 - t_2} + \frac{p_6 - p_4}{t_6 - t_4} + (\rho c)_{46}\beta_{i6} \frac{u_{i6} - u_{i4}}{t_6 - t_4}$$

$$= \left[- \left(\rho c^2 \frac{\partial u_i}{\partial x_j}\right)_2 + \left(\rho c^2 \frac{\partial u_i}{\partial x_j}\right)_4 \right] (\alpha_i\alpha_j)_6 \quad , \tag{3.77}$$

$$\frac{p_6 - p_1}{t_6 - t_1} - (\rho c)_{16}\alpha_{i6}\frac{u_{i6} - u_{i1}}{t_6 - t_1} + \frac{p_6 - p_2}{t_6 - t_2} + (\rho c)_{26}\beta_{i6}\frac{u_{i6} - u_{i2}}{t_6 - t_2}$$

$$- \frac{p_6 - p_5}{t_6 - t_5} = \left[-\left(\rho c^2 \frac{\partial u_i}{\partial x_j}\right)_2 + \left(\rho c^2 \frac{\partial u_i}{\partial x_j}\right)_5\right](\alpha_i\alpha_j)_6$$

$$+ \left[-\left(\rho c^2 \frac{\partial u_i}{\partial x_j}\right)_1 + \left(\rho c^2 \frac{\partial u_i}{\partial x_j}\right)_5\right](\beta_i\beta_j)_6 \quad . \qquad (3.78)$$

Below, the general algorithm of solution is described. Coordinates of the point 5 (Fig.3.5) may be considered fixed. Approximating (3.69) gives three relations for the definition of x_{26}, x_{36}, t_6:

$$x_{i6} - x_{i5} = u_{i56}(t_t - t_5) \quad , \qquad (i = 1,2,3) \quad . \qquad (3.79)$$

To find the coordinates x_{2k}, x_{3k} of the points 1-4 at the plane $x = x_0$ and the values of parameter t_k, $(k = 1,2,3,4)$, the difference equations approximating (3.72) are used:

$$x_{i6} - x_{ik} = [u_{ik6} + (c\cos\theta)_{k6}\alpha_{i6} + (c\sin\theta)_{k6}\beta_{i6}]$$

$$\times (t_6 - t_k) \quad , \qquad (i = 1,2,3) \quad . \qquad (3.80)$$

Then, with the help of the second-order polynomials obtained using the values at the node 5 and eight adjacent nodes of the layer $x = x_0$ (Fig.3.5), the values of parameters at the points 1-4 are found. The vector α_i and β_i lie in the plane perpendicular to W. Numerical experiments have shown that they should be oriented in such a way that the angle between their projections onto the plane $x = x_0$ is divided in halves by the projection of the vector of pressure gradient. Now u_{i6}, p_6, p_{06}, H_{06} may be obtained from (3.75-78). In the first approximation for any coefficient $g_{\ell 6} = 0.5(g_\ell + g_6)$, $\ell = 1,2,\ldots,5$, g_6 is put equal to g_ℓ.

If the point 6 belongs to a solid wall, then (3.72) at $\theta = 3\pi/2$ is replaced by the condition $m_{i6}u_{i6} = 0$, where the m_i are components of an outer normal to the wall. The three remaining bicharacteristics are oriented in such a way that two of them $(\theta = 0,\pi)$ lie in the plane, tangential to the wall, while the bicharacteristic $\theta = \pi/2$ lies inside the region of flow. If point 6 belongs to a solid wall, then the parameters at symmetrical points on the plane $x = x_0$ are defined by the symmetry conditions and after that the values of parameters at an internal point or at a boundary are determined by means of the algorithms described above.

The step Δx is determined from the Courant-Friedrichs-Levy conditions and must not exceed

$$\frac{u^2}{cW}\left[1 - \frac{c}{W}\left(\frac{W^2}{u^2} - 1\right)^{\frac{1}{2}}\right]R_{min} \quad , \tag{3.81}$$

where R_{min} is the distance from point 5 to the nearest node used in the difference scheme. However, calculations have shown that this is not enough. To ensure stability one should use in the calculation formulas not the given values of parameters at the point 5, but the interpolated values (as at the points 1-4).

3.3.2 Difference Methods

The methods described in Sect.3.2.2 can be easily generalized for the case of three-dimensional flows of a perfect gas. Let us consider a three-dimensional modification of the difference method described in Sect.3.2.2. Let us once again write down the equations of gas dynamics in the form of integral laws of conservation, using cylindrical coordinates x, r, φ:

$$\frac{d}{dx}\iint_S \mathbf{a}\, dr\, d\varphi = \oint_\Gamma [(\mathbf{c} - \mathbf{a}\xi_\varphi)dr - (\mathbf{b} - \mathbf{a}\xi_r)d\varphi] + \iint_S \mathbf{f}\, dr\, d\varphi \quad . \tag{3.82}$$

Here ξ_r, ξ_φ are projections of the vector $\boldsymbol{\xi} = \mathbf{dn}/dx$ perpendicular to the axis x on the axes r, φ, where dn is the projection of the displacement of the contour $\Gamma(x)$, limiting the area S on its outer normal.

The column vectors **a**, **b**, **c**, **f** have the form

$$\mathbf{a} = \begin{pmatrix} \rho u \\ p + \rho u^2 \\ \rho uv \\ \rho uw \end{pmatrix}, \quad \mathbf{b} = \begin{pmatrix} \rho v \\ \rho uv \\ p + \rho v^2 \\ \rho vw \end{pmatrix}, \quad r\mathbf{c} = \begin{pmatrix} \rho w \\ \rho uw \\ \rho vw \\ p + \rho w^2 \end{pmatrix},$$

$$\mathbf{f} = -\frac{\rho}{r}\begin{pmatrix} v \\ uv \\ v^2 - w^2 \\ 2vw \end{pmatrix} \quad . \tag{3.83}$$

Let us consider a flow with two planes of symmetry and construct a grid in the region $x \geqslant x_0$, $0 \leqslant r \leqslant F(x,\varphi)$, $0 \leqslant \varphi \leqslant \pi/2$. The region of flow at x = const will be denoted by D. We shall divide it into K vertical strips along φ and number them $k = 1/2, \ldots, K - 1/2$. Boundaries of the strips are ascribed numbers,

$k=0,1,\dots,K$ ($k=0$ corresponds to $\varphi=0$). The segment $\varphi=\varphi_k$ is divided into N equal parts which are numbered from $n=1/2$ to $n=N-1/2$ while their end points are given numbers from $n=0$ to $n=N$. The points of two neighbouring segments $\varphi=\text{const}$ having the same numbers n are joined with straight-line segments. Thus obtained, elementary rectangles (cells) are ascribed two indices $n-1/2$, $k-1/2$, ($n=1,\dots,N$, $k=1,\dots,K$). The mean for the rectangle values of parameters in the plane $x=x_0$ are denoted by subscripts (e.g. $u_{n-\frac{1}{2},\,k-\frac{1}{2}}$); in the plane $x=x_0+\tau$ the same numbers are used as superscripts. The vertices of rectangles in the planes $x=x_0$ and $x=x_0+\tau$ having the same indices are joined with straight-line segments. Thus elementary volumes of the grid are obtained. Evidently, the side faces of elementary volumes will generally not be flat. Therefore, for calculation of "large" quantities — the mean values of parameters at each side face — a flat face is used passing through the edge of a cell at $x=x_0$ and through the middle of the edge at $x=x_0+\tau$.

Let us summarize the rules of numbering vertices, edges and faces of an elementary volume. The vertices are ascribed two integer numbers, for instance, $n-1$, $k-1$. The edges are given one integer and one semiinteger number: $n-1/2$, $k-1$ for the vertical edges and $n-1$, $k-1/2$ for the horizontal ones. Faces lying in the planes $x=\text{const}$ are numbered by two semiinteger numbers ($n-1/2$, $k-1/2$), and the side faces are numbered by one integer and one semiinteger number, e.g. $n-1/2$, k for the face separating volumes along φ, and n, $k-1/2$ for faces separating them along r.

Let us introduce the following notation:

$$(h_r)_{n-\frac{1}{2},j} = r_{nj} - r_{(n-1)j} \;, \qquad h_r^{n-\frac{1}{2},j} = r^{nj} - r^{n-1,j} \;,$$

$$(h_\varphi)_{k-\frac{1}{2}} = \varphi_k - \varphi_{k-1} \;, \qquad (h_r)_\varphi^{k-\frac{1}{2}} = \varphi^k - \varphi^{k-1} \;, \tag{3.84}$$

where $j=k-1,\ k$, and define the mean heights

$$(h_r)_{n-\frac{1}{2},k-\frac{1}{2}} = \frac{1}{2}\left[(h_r)_{n-\frac{1}{2},k-1} + (h_r)_{n-\frac{1}{2},k}\right] \;,$$

$$h_r^{n-\frac{1}{2},k-\frac{1}{2}} = \frac{1}{2}\left(h_r^{n-\frac{1}{2},k-1} + h_r^{n-\frac{1}{2},k}\right) \tag{3.85}$$

and the quantities

$$(\Delta_1)_{i,k-\frac{1}{2}} = \frac{1}{2}\,(r^{i,k-1} + r^{ik} - r_{i,k-1} - r_{ik}) \;, \tag{3.86}$$

$$(\Delta_2)_{i,k-\frac{1}{2}} = \frac{1}{2}\,(r^{ik} - r^{i,k-1} + r_{ik} - r_{i,k-1}) \quad , \qquad (i = n-1,n) \quad . \quad (3.86)$$

Let us integrate (3.83) with respect to x from x_0 to $x_0+\tau$ considering S an elementary cell. We have

$$\begin{aligned}(\mathbf{a}h_r)^{n-\frac{1}{2},k-\frac{1}{2}} &= (\mathbf{a}h_r)_{n-\frac{1}{2},k-\frac{1}{2}} + \left(\mathbf{A}\Delta_1 - \mathbf{B}\tau + \mathbf{C}\Delta_2\,\frac{\tau}{h_\varphi}\right)_{n,k-\frac{1}{2}}\\ &- \left(\mathbf{A}\Delta_1 - \mathbf{B}\tau + \mathbf{C}\Delta_2\,\frac{\tau}{h_\varphi}\right)_{n-1,k-\frac{1}{2}} - \left(\mathbf{C}\,\frac{\tau h_r}{h_\varphi}\right)_{n-\frac{1}{2},k} \qquad (3.87)\\ &+ \left(\mathbf{C}\,\frac{\tau h_r}{h_\varphi}\right)_{n-\frac{1}{2},k-1} + \frac{\tau}{2}\left[(\mathbf{f}h_r)_{n-\frac{1}{2},k-\frac{1}{2}} + (\mathbf{f}h_r)^{n-\frac{1}{2},k-\frac{1}{2}}\right] .\end{aligned}$$

The unknown values are found from (3.87) similarly to the two-dimensional case (Sect.3.2.2). To find "large" quantities participating in (3.87) we consider the problem of impact of two flows which meat along the line corresponding to the cell's side at $x=x_0$. The velocity vector of each of the flows may be represented as the vector sum of a component parallel to the line of impact of the flows and another component perpendicular to the latter. Then the problem reduces to the plane problem of interaction of flows, because the velocity component parallel to the line of interaction does not influence the interaction of the flows.

If the upper face coincides with the wall, and the "large" quantities are then defined there, the problem can be solved in the same manner as the two-dimensional case. If the boundary is a free surface, then in the process of transfer to the layer $x=x_0+\tau$ the "large" quantities are at first defined at the upper face and the ordinates $r^{N,k-\frac{1}{2}}$ of the middle points, then the ordinates of the vertices of the boundary cells are calculated:

$$r^{Nk} = 0.5(r^{N,k-\frac{1}{2}} + r^{N,k+\frac{1}{2}}) \quad , \qquad (k = 1,\dots,K-1) \quad ,$$

$$r^{N0} = r^{N,\frac{1}{2}} \quad , \qquad r^{NK} = r^{N,K-1} \quad .$$

If a side face lies in the plane of symmetry, then the "large" quantities on it are found from the solution of the problem of flow past a flat wall parallel to the axis x.

The scheme considered is, in the case of smooth solutions, of the first order of accuracy. To ensure stability, the step τ should be chosen in such a way that inside each elementary volume the waves generated at its edges in the plane $x=x_0$ can reach its right face.

3.4 Methods of Solving the Inverse Problem of the Theory of Nozzles

In Sect.2.1 we formulated the inverse problem of the theory of nozzles. For equations of gas dynamics (2.96-104) in the general case of a three-dimensional flow in the variables x, ψ, θ we proceed as follows. At the plane $\psi=\psi_0$ a function $r=r_0(s,\theta)$ and a component of velocity $u=u_0(s,\theta)$ (or pressure, or density) are specified. At the plane $s=s_0$ one specifies the components of velocity $w=w_0(\theta,\psi)$, a function $\varphi=\varphi_0(\theta,\psi)$, density $\rho=\rho_0(\theta,\psi)$ and $\alpha_i=\alpha_{i0}(\theta,\psi)$. One wants to find a family of stream surfaces and the flow parameters in the vicinity of the initial stream surface.

Below we describe a difference method which proved to be quite efficient for solving this problem for the system (2.96-104) [3.35]. Let us introduce a grid with nodes

$$s_i = s_0 + i\Delta s \quad , \quad \psi_j = \psi_0 + j\Delta\psi \quad , \quad \theta_k = k\Delta\theta \quad ,$$
$$i = 0,1,\dots,I \quad , \quad j = 0,1,\dots,J \ , \quad k = 0,1,\dots,K \quad .$$

and denote $f_{ijk}=f(s_i, \psi_j, \theta_k)$. The step sizes Δs and $\Delta\theta$ generally vary and depend on i and k respectively. Let all the flow parameters, namely u, v, w, p, ρ, T, α_i, r, φ and $\partial\varphi/\partial\psi$, be known at the grid nodes of the layer ψ_i = const. The parameters at the layer ψ_{j+1} = const are calculated iteratively in the following manner.

An iteration, numbered ℓ, begins with the calculation of $p^{(\ell)}_{i,j+1,k}$, $r^{(\ell)}_{i,j+1,k}$, $v^{(\ell)}_{i,j+1,k}$ from equations

$$p^{(\ell)}_{i,j+1,k} = p_{ijk} + \frac{\Delta\psi}{2}\left[(f_1)_{ijk} + (f_1)^{(\ell-1)}_{i,j+1,i}\right] ,$$

$$r^{(\ell)}_{i,j+1,k} = \left\{r^{1+\nu}_{ijk} + \frac{\Delta\psi}{2}\left[(f_2)_{ijk} + (f_2)^{(\ell-1)}_{i,j+1,k}\right]\right\}^{1/(1+\nu)} , \tag{3.88}$$

$$v^{(\ell)}_{i,j+1,k} = \left(\frac{\partial r}{\partial s}\right)^{(\ell)}_{i,j+1,k} \left(\frac{u}{H_1}\right)^{(\ell-1)}_{i,j+1,k} ,$$

which approximate (2.96,97,100). Here f_1, f_2 denote the right sides of (2.96, 97). In the first iteration ($\ell=1$) the values of variables with the superscript $\ell=0$ are taken from the lower layer j. The calculated values $p^{(\ell)}_{i,j+1,k}$, $r^{(\ell)}_{i,j+1,k}$, $v^{(\ell)}_{i,j+1,k}$ are used in the subsequent numerical integration of (2.98,99,101-104) in the direction s in order to obtain the remaining unknowns u, w, ρ, T, α_n, ($n=1,2,\dots,N$), at the layer $j+1$ in the ℓth iteration.

Let $f^{(\ell)}_{i,j+1,k}$, $(k=1,\dots,K)$, be known for some i, where f is any of the above-mentioned quantities. We shall try to find $f^{(\ell)}_{i+1,j+1,k}$. This process is also an iterative one. Let the iteration number of this internal iteration cycle be m. Approximating (2.98,101) similarly to (3.88), we have

$$(wr)^{(\ell,m)}_{i+1,j+1,k} = (wr)^{(\ell)}_{i,j+1,k} + \frac{\Delta s}{2}\left[(f_3)^{(\ell)}_{i,j+1,k} + (f_3)^{(\ell,m-1)}_{i+1,j+1,k}\right] \quad , \tag{3.89}$$

$$\varphi^{(\ell,m)}_{i+1,j+1,k} = \varphi^{(\ell)}_{i,j+1,k} + \frac{\Delta s}{2}\left[(f_4)^{(\ell)}_{i,j+1,k} + (f_4)^{(\ell,m-1)}_{i+1,j+1,k}\right] \quad ,$$

where f_3, f_4 are the right sides of (2.98,101) and $f^{(\ell,0)}_{i+1,j+1,k} = f^{(\ell)}_{i,j+1,k}$. To determine the velocity component u we shall use (2.99). Then we have

$$\begin{aligned}(u^2)^{(\ell,m)}_{i+1,j+1,k} = {} & (W^2)^{(\ell)}_{i,j+1,k} - (v^2)^{(\ell)}_{i+1,j+1,k} - (w^2)^{(\ell,m)}_{i+1,j+1,k} \\ & - \frac{\Delta s}{\gamma}\left[(\rho^{-1})^{(\ell)}_{i,j+1,k} + (\rho^{-1})^{(\ell,m-1)}_{i+1,j+1,k}\right] \\ & \times \left(p^{(\ell)}_{i+1,j+1,k} - p^{(\ell)}_{i,j+1,k}\right) \quad .\end{aligned} \tag{3.90}$$

Equations (2.103) form a system of relaxation equations and are solved simultaneously with (2.102) with the help of the implicit difference scheme described in Sect.3.1. Thus $T^{(\ell,m)}_{i+1,j+1,k}$ and $(\alpha_n)^{(\ell,m)}_{i+1,j+1,k}$, $(n=1,\dots,K)$ are calculated. Then $\rho^{(\ell,m)}_{i+1,j+1,k}$ is defined from (2.104). The iterative cycle in m is terminated after the necessary accuracy is obtained; then the values of u, v, w, p, ρ, T, α_n are found in the same way at the nodes $(i+2, j+1, k)$, $(k=0,1,\dots,K)$, and so on. The ℓth iteration terminates after these values are defined at all nodes of the $(j+1)^{th}$ layer. Then, if the required accuracy of the values of parameters at the $(j+1)^{th}$ layer is not achieved, the $(\ell+1)^{th}$ iteration is started.

An important feature of any difference scheme is the method of calculating the derivatives $\partial v/\partial s$, $\partial r/\partial s$, $\partial r/\partial\theta$ and $\partial p/\partial\theta$ on the right sides of (3.88,89). Let us consider the difference formulas of the derivatives $\partial v/\partial s$ and $\partial r/\partial s$. A typical dependence of v on nozzle length at the contour in the subsonic part of the nozzle is shown in Fig.3.6. In the region I corresponding to slow subsonic flow, the function v changes only slightly and its derivatives are small. Conversely, in the regions II and IV the gradients of the function v are large. Therefore, replacing the derivatives by divided finite differences we may vary the spacing of the difference grid: it may be coarse in the region I where the gradients are small in the axis direction and hence the errors of approximation will be small too. At the same

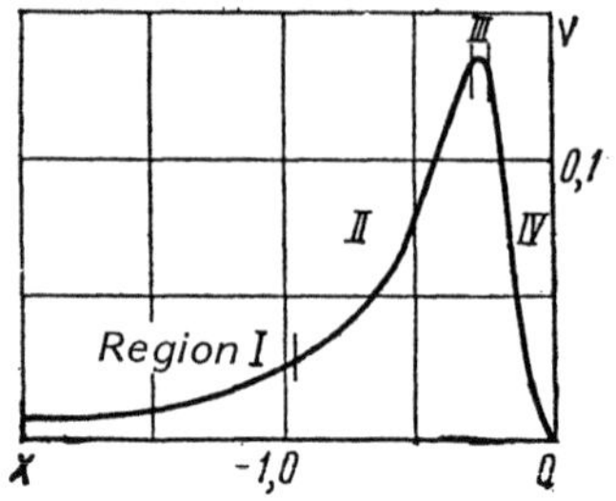

Fig.3.6. Dependence of the velocity component v on x

time the round-off error will also be small because the step size is large. On the other hand, the grid must be dense enough in the regions II and IV in order to make the approximation errors admissibly small. In this case round-off errors are small as well, because the gradients are large. In the neighbourhood of the maximum, just as in the region I, the step size of the difference grid must be large enough.

Thus, for calculating flow in an elliptic region, one should use a difference grid with variable spacing. The use of large step sizes in the regions with small gradients results in a negligible growth of the round-off errors in the process of numerical solution of an initial-value problem for elliptic equations. At the same time the round-off errors do not influence the stability of the numerical solution.

To calculate the derivatives $\partial v/\partial s$ and $\partial r/\partial s$, various difference schemes may be used. Special calculations have shown that the three-point scheme,

$$\left(\frac{\partial f}{\partial s}\right)_i = f_{i-1}\frac{2s-(s_{i+1}+s_i)}{(s_{i-1}-s_i)(s_{i-1}-s_{i+1})} + f_i\frac{2s-(s_{i+1}+s_{i-1})}{(s_i-s_{i-1})(s_i-s_{i+1})} + f_{i+1}\frac{2s-(s_i+s_{i-1})}{(s_{i+1}-s_{i-1})(s_{i+1}-s_i)}\ , \tag{3.91}$$

is the most stable one. While calculating derivatives at the extreme points of a layer, s is put equal to s_{i-1} or s_{i+1} for the right and the left end points respectivley. At the other points of a layer $s=s_i$. Derivatives $\partial p/\partial\theta$ and $\partial r/\partial\theta$ are calculated in a similar manner. However, when doing so one should use, evidently, a three-point scheme with a constant step $\Delta\theta$ at the plane s = const, which nevertheless, may be different for different planes. The three-point difference scheme together with a variable grid is a kind of regularizing operator for the process of calculating derivatives [3.36], allowing a noncorrect initial value problem to be solved successfully. Derivatives $\partial\varphi/\partial\theta$ on the right sides of the first two equations (2.88) are approximated by the expression $(\varphi_{i,j+1,k} - \varphi_{i,j,k})/\Delta\psi$. The finite difference scheme described above has second-order accuracy in $\Delta\psi$ and, in the case of a uniform spacing in the directions of θ and s, is also accurate to second order in $\Delta\theta$ and Δs. The second-order accuracy is attained due to iterations in ℓ and m.

In the cases of plane and axisymmetric flows at $w=0$ and $\varphi=0$ with a constant adiabatic exponent, the calculation is especially simple. In these cases the system of equations coincides with (2.107-110). Let us write down the difference equations replacing s by x:

$$r^{(\ell)}_{i,j+1} = \left\{r^{1+\nu}_{ij} + \frac{1+\nu}{2}\left[(1/\rho u)_{ij} + (1/\rho u)^{(\ell-1)}_{i,j+1}\right]\right\}^{1/(1+\nu)}\Delta\psi \quad , \tag{3.92}$$

$$p^{(\ell)}_{i,j+1} = p_{i,j} - \frac{\gamma}{2}\left[\left(\frac{1}{r^{\nu}}\frac{\partial v}{\partial x}\right)_{ij} + \left(\frac{1}{r^{\nu}}\frac{\partial v}{\partial x}\right)^{(\ell-1)}_{i,j+1}\right]\Delta\psi \quad , \tag{3.93}$$

$$v^{(\ell)}_{i,j+1} = \left\{\frac{\partial r/\partial x}{\sqrt{1+(\partial r/\partial x)^2}}\left[\frac{\gamma+1}{\gamma-1} - \frac{2}{\gamma-1}p^{(\gamma-1)/\gamma}\right]^{\frac{1}{2}}\right\}^{(\ell)}_{i,j+1} \quad , \tag{3.94}$$

$$\rho^{(\ell)}_{i,j+1} = (p^{1/\gamma})^{(\ell)}_{i,j+1} \quad , \quad u^{(\ell)}_{i,j+1} = \left\{\left[\frac{\gamma+1}{\gamma-1} - \frac{2}{\gamma-1}p^{(\gamma-1)/\gamma} - v^2\right]^{\frac{1}{2}}\right\}^{(\ell)}_{i,j+1} \quad . \tag{3.95}$$

In the first approximation (3.92,93) are used to define $r^{(\ell)}_{i,j+1}$, $p^{(\ell)}_{i,j+1}$, $(i=0,1,\dots,I)$, while it is assumed that $f^{(0)}_{i,j+1} = f_{i,j}$. Then, the values of r obtained are used to calculate $(\partial r/\partial x)^{(\ell)}_{i,j+1}$ from (3.91) and to find $v^{(\ell)}_{i,j+1}$, $\rho^{(\ell)}_{i,j+1} u^{(\ell)}_{i,j+1}$. In the subsequent iterations the parameters are calculated in a similar way.

An important feature of the proposed scheme is the almost complete absence of any stability limitations for the step $\Delta\psi$ whose size is defined by the approximation error only. The size of step Δs is limited only in the elliptic region. Also due to the evident simplicity of the numerical algorithm, especially in the plane and axisymmetric cases, the computer time needed is extremely small.

If the initial data surface $\psi=\psi_0$ coincides with the symmetry axis, the above method cannot be used to depart from the axis due to a singularity in (2.96-104). In order to determine the unknown values at a surface $\psi=\text{const}$ close to the axis of symmetry, expansion of the solution in the stream function series may be used. This question will be discussed in detail in Sect.4.2. From the stream surface thus obtained, the calculation can be carried on by a difference method.

For solving the inverse problem for gas flows in complicated channels, when the streamlines are multivalued functions of Cartesian coordinates, one can conveniently use (2.117-120) in ψ, σ, φ. The system (2.117-120) takes into account the swirl of flow. It can be solved by the method described above [3.37-39]. While doing so, the sequence of calculation of the flow parameters may be chosen in such a way that the calculation can be started from the axis of symmetry.

A method similar to the one discussed in this section, has been used to solve the inverse problem in [3.40]. The method of integral relations can also be used [3.41].

3.5 Methods of Solving the Direct Problem of the Theory of Nozzles

The majority of researchers solving the direct problem use the method of stabilization. The idea is to solve a stationary problem using nonstationary equations of gas dynamics. A boundary-value problem with boundary conditions corresponding to those of a stationary problem and independent of the additional time coordinate is solved for the nonstationary equations. The required stationary solution is obtained as a limit approached by the nonstationary process with growth of t. Despite an additional dimension, such a procedure is often justified. It facilitates calculation of gas flows in nozzles and jets and flows past bodies, when the motion of a gas is described by equations of a mixed elliptic-hyperbolic type, and in some other cases. Introduction of the time coordinate reduces the problem to the solution of hyperbolic equations.

In its application to the calculation of gas flows in nozzles with a fixed contour, the method of stabilization has been developed by a number of authors [3.42-45]. Here we briefly describe some of its modifications. We shall consider the flow of a perfect gas in a nozzle with a fixed contour $y = F(x)$, where x, y are cylindrical coordinates and the x axis coincides with the axis of symmetry. The contour of the central body is described by $y = G(x)$ [in the case of Laval nozzles $G(x) = 0$]. We shall assume that far enough from the nozzle's throat, in the subsonic part of the flow, the contour is cylindrical. When $x \to -\infty$, the flow becomes uniform, its parameters being defined in the process of calculation. The numerical solution will be sought in the region $x_0 \leqslant x \leqslant x_k$, where x_k is a cross-section downstream of the throat in the supersonic part of the flow. The usual impermeability conditions are put at the boundaries $y = F(x)$ and $y = G(x)$. At $x = x_0$ the flow is assumed uniform, or an equivalent (in the case of constant entropy and stagnation enthalpy) condition $\partial u/\partial x = 0$ is required [3.43]. At $x = x_k$ there is no need of a boundary condition (see Sect.2.1.3).

The method contained in [3.43] is based upon the known problem of disintegration of an arbitrary discontinuity [3.31]. It is similar to its stationary analogue described in Sect.3.3. The difference grid is constructed in just the same way. Along the x axis the region is divided into N layers numbered $n - 1/2$, $(n = 1,2,\dots,N)$; in the vicinity of the throat the spacing is made smaller. The boundaries of the layers are numbered n, $(n = 0,1,\dots,N)$. They are divided into K parts numbered $k - 1/2$, $(k = 1,2,\dots,K)$. The nodes of neighbouring vertical boundaries having equal numbers are joined with segments. The elementary cells thus obtained are ascribed two numbers each: $n - 1/2$ and $k - 1/2$. Parameters, averaged inside a cell, are given subscripts $(n - 1/2,\ k - 1/2)$ at the known layer $t = t_0$; at the next layer $t = t_0 + \tau$, where the solution is sought, they are supplied with equal superscripts (τ is the time step size). The averaged parameters at the side faces of the elementary volumes (denoted by capital letters) are denoted by an integer and a semiinteger index.

Let us write down the system of equations describing nonstationary flow of a gas in the divergent form:

$$
\begin{aligned}
&\frac{\partial}{\partial t}(\rho y^{\nu}) + \frac{\partial}{\partial x}(\rho u y^{\nu}) + \frac{\partial}{\partial y}(\rho v y^{\nu}) = 0 \quad ,\\
&\frac{\partial}{\partial t}(\rho u y^{\nu}) + \frac{\partial}{\partial x}[(p + \rho u^2)y^{\nu}] + \frac{\partial}{\partial y}(\rho u v y^{\nu}) = 0 \quad , \qquad (3.96)\\
&\frac{\partial}{\partial t}(\rho v y^{\nu}) + \frac{\partial}{\partial x}(\rho u v y^{\nu}) + \frac{\partial}{\partial y}[(p + \rho v^2)y^{\nu}] = \nu p y^{\nu-1} \quad ,\\
&\frac{\partial}{\partial t}\left[\rho\left(h + \frac{1}{2}W^2\right)\right] + \frac{\partial}{\partial x}\left[\rho u\left(h + \frac{1}{2}W^2\right) + \frac{\partial}{\partial y}\left[\rho v\left(h + \frac{1}{2}W^2\right)\right] = 0 \quad .
\end{aligned}
$$

Upon integrating over the elementary volumes we get relationships expressing the required superscripted values in terms of the values at the layer $t = t_0$ and the quantities in capital letter. Because the relationships are quite cumbersome, here we write down only one of them, the formula for calculating $\rho^{n-\frac{1}{2},k-\frac{1}{2}}$, obtained from the first of the equations (3.96):

$$
\begin{aligned}
\rho^{n-\frac{1}{2},k-\frac{1}{2}} = \rho_{n-\frac{1}{2},k-\frac{1}{2}} + \frac{\tau}{y^{\nu}_{n-\frac{1}{2},k-\frac{1}{2}}\Delta_{n-\frac{1}{2}}} \Bigg[& (RUy^{\nu})_{n-1,k-\frac{1}{2}} \frac{\Delta_{n-1}}{h_{n-\frac{1}{2}}} \\
& - (RUy^{\nu})_{n,k-\frac{1}{2}} \frac{\Delta_n}{h_{n-\frac{1}{2}}} + (RUy^{\nu})_{n-\frac{1}{2},k-1} \frac{y_{n-1,k-1} - y_{n,k-1}}{h_{n-\frac{1}{2}}}
\end{aligned}
$$

$$- (RUy^{\nu})_{n-\frac{1}{2},k} \frac{y_{n-1,k} - y_{n,k}}{h_{n-\frac{1}{2}}} + (RVy^{\nu})_{n-\frac{1}{2},k-1} - (RVy^{\nu})_{n-\frac{1}{2},k}\Bigg] \quad , \qquad (3.97)$$

where

$$\Delta = (F - G)/K \quad , \quad h_{n-\frac{1}{2}} = x_n - x_{n-1} \quad .$$

To determine the "capital letter" quantities we shall consider disintegration of an arbitrary discontinuity at every boundary of an elementary cell at the layer $t = t_0$. The initial values of the flow parameters on either side of a discontinuity are equal to the values of the parameters in the two cells bordering on the boundary under consideration. To calculate the quantities at the faces lying at the boundary $y = F(x)$, auxiliary cells, symmetrical relative to the boundary, are introduced. Here the flow parameters are chosen in such a way that the velocity at a side face, obtained as a result of a discontinuity disintegration, satisfied the impermeability condition.

The scheme is accurate to first order. The step size is chosen in such a way that for all elementary cells, the waves generated by disintegration of an arbitrary discontinuity at $t = t_0$ do not have time to reach the opposite faces of cells. The calculation is carried on from layer to layer until a stationary regime is established within the prescribed accuracy. The initial distributions of parameters may be obtained, for instance, from the one-dimensional theory.

In [3.44] implicit difference schemes are used to solve the direct problem for a Laval nozzle. The initial system of equations has the following form:

$$\frac{\partial \mathbf{X}}{\partial t} + A\frac{\partial \mathbf{X}}{\partial x} + B\frac{\partial \mathbf{X}}{\partial y} + \mathbf{F} = 0 \quad , \quad p = \rho^{\nu} \quad , \quad \text{where} \qquad (3.98)$$

$$\mathbf{X} = \begin{pmatrix} p \\ u \\ v \end{pmatrix}, \quad A = \begin{pmatrix} u & \gamma\rho a^2 & 0 \\ 1/\gamma\rho & u & 0 \\ 0 & 0 & u \end{pmatrix}, \quad B = \begin{pmatrix} v & 0 & \gamma\rho a^2 \\ 0 & v & 0 \\ 1/\gamma\rho & 0 & v \end{pmatrix} ,$$

$$\mathbf{F} = \begin{pmatrix} \nu\gamma v\rho a^2/y \\ 0 \\ 0 \end{pmatrix}$$

and the unknown quantities are referred to their critical values. We shall transform to the coordinates

$$t = t \quad , \quad \xi = y/F(x) \quad , \quad \eta = \eta(x,y) \quad ,$$

where $\eta = \eta(x,y)$ is chosen in such a way that the grid spacing becomes smaller in the vicinity of the throat. The resulting system may be reduced to the characteristic form by multiplying it on the left by the matrix

$$Q = \begin{pmatrix} 1 & \gamma\rho a \sin\varphi & \gamma\rho a \cos\varphi \\ 1 & -\gamma\rho a \sin\varphi & -\gamma\rho a \cos\varphi \\ 0 & \cos\varphi & -\sin\varphi \end{pmatrix} . \tag{3.99}$$

We have

$$Q \frac{\partial \mathbf{X}}{\partial t} + \Lambda Q \frac{\partial \mathbf{X}}{\partial \xi} = -Q(\eta_x A + \eta_y B)\frac{\partial \mathbf{X}}{\partial \eta} - QF \quad . \tag{3.100}$$

The equations of the system are at the same time the compatibility conditions for characteristic surfaces passing through the line $t = \text{const}$, $\xi = \text{const}$. In (3.99,100) the components of normals to characteristic surfaces are denoted by $\sin\varphi = \xi_x(\xi_x^2 + \xi_y^2)^{-1/2}$ and $\cos\varphi = \xi_y(\xi_x^2 + \xi_y^2)^{-1/2}$. The matrix Λ is diagonal. Its elements

$$\lambda_{11} = u\xi_x + v\xi_y + a(\xi_x^2 + \xi_y^2)^{1/2} \quad ,$$

$$\lambda_{22} = u\xi_x + v\xi_y - a(\xi_x^2 + \xi_y^2)^{1/2} \quad ,$$

$$\lambda_{33} = u\xi_x + v\xi_y$$

are time components of the characteristic normals.

To approximate the first two equations, the implicit difference scheme of [3.46] is used. The scheme cannot be used to approximate the third equation of the system (3.100) due to the fact that the scheme employing a four-point implicit approximation of the left side of (3.100) loses stability if any of the elements λ_{ii} changes sign, and has a low stability for small absolute values of λ_{ii}. Meanwhile, due to the boundary conditions having the form

$$u\xi_x + v\xi_y = 0 \quad \text{for} \quad \xi = 0 \quad \text{and} \quad \xi = 1 \quad , \tag{3.101}$$

λ_{33} is small near boundaries. Therefore, to approximate the left side of the third equation of the system (3.100) a six-point implicit scheme is used.

Let us introduce a grid with nodes t^n, ξ_m, η_ℓ in the region $t \geqslant 0$, $0 \leqslant \xi \leqslant 1$, $\eta_0 \leqslant \eta \leqslant \eta_k$. Denoting $t^n = t^{n-1} + \tau$, $\xi_m = \xi_{m-1} + h_1$, $\eta_\ell = \eta_{\ell-1} + h_2$, where $n = 1,2,\dots,N$, $m = 0,1,\dots,M$, $\ell = 1,2,\dots,L$, $f(t^n, \xi_m, \eta_\ell) = f^n_{m\ell}$, we get for the first two equations (3.100),

$$\left(\frac{\partial f}{\partial t}\right)^{n+1/2}_{m+1/2,\ell} = \frac{1}{2\tau}\left(f^{n+1}_{m+1,\ell} - f^n_{m+1,\ell} + f^{n+1}_{m,\ell} - f^n_{m,\ell}\right)$$

$$- \frac{\sigma_1\kappa_2}{2\tau}\left(f^n_{m+1/2,\ell+1} - 2f^n_{m+1/2,\ell} + f^n_{m+1/2,\ell-1}\right) \quad , \tag{3.102}$$

$$\left(\frac{\partial f}{\partial \xi}\right)^{n+\frac{1}{2}}_{m+\frac{1}{2},\ell} = \frac{\kappa_1}{\tau}\left[\alpha\left(f^{n+1}_{m+1,\ell} - f^{n+1}_{m,\ell}\right) + \beta\left(f^{n}_{m+1,\ell} - f^{n}_{m,\ell}\right)\right] \quad ,$$

$$\left(\frac{\partial f}{\partial \eta}\right)^{n+\frac{1}{2}}_{m+\frac{1}{2},\ell} = \frac{\kappa_2}{2\tau}\left[\alpha\left(f^{n+1}_{m+\frac{1}{2},\ell+1} - f^{n+1}_{m+\frac{1}{2},\ell-1}\right) + \beta\left(f^{n}_{m+\frac{1}{2},\ell+1} - f^{n}_{m+\frac{1}{2},\ell-1}\right)\right] \quad ,$$

where

$$f^{n}_{m+\frac{1}{2},\ell} = \frac{1}{2}\left(f^{n}_{m+1,\ell} + f^{n}_{m,\ell}\right) \quad ,$$

$$(m = 0,1,\dots,M-1 \quad , \quad \ell = 1,2,\dots,L) \quad .$$

For the third equation of the system (3.100) we have

$$\left(\frac{\partial f}{\partial t}\right)^{n+\frac{1}{2}}_{m\ell} = \frac{1}{\tau}\left(f^{n+1}_{m\ell} - f^{n}_{m\ell}\right) - \frac{\sigma_2\kappa_2}{\tau}\left(f^{n}_{m,\ell+1} - 2f^{n}_{m\ell} + f^{n}_{m,\ell-1}\right) \quad ,$$

$$\left(\frac{\partial f}{\partial \xi}\right)^{n+\frac{1}{2}}_{m\ell} = \frac{\kappa_1}{2\tau}\left[\alpha\left(f^{n+1}_{m+1,\ell} - f^{n+1}_{m-1,\ell}\right) + \beta\left(f^{n}_{m+1,\ell} - f^{n}_{m-1,\ell}\right)\right] \quad , \tag{3.103}$$

$$\left(\frac{\partial f}{\partial \eta}\right)^{n+\frac{1}{2}}_{m\ell} = \frac{\kappa_2}{2\tau}\left[\alpha\left(f^{n+1}_{m,\ell+1} - f^{n+1}_{m,\ell-1}\right) + \beta\left(f^{n}_{m,\ell+1} - f^{n}_{m,\ell-1}\right)\right] \quad ,$$

$$(m = 0,1,\dots,M-1. \quad \ell = 1,\dots,L) \quad .$$

In (3.102,103), $\sigma_i > 0$, $\alpha > \beta$, $\alpha + \beta = 1$, $\kappa_i = \tau/h_i$. Substituting (3.102,103) into (3.100), we get a system of difference equations, which can be written in the form (the obvious indices are omitted):

$$a_m\mathbf{X}_{m-1} + b_m\mathbf{X}_m + c_m\mathbf{X}_{m+1} = \mathbf{d}_m \quad , \quad (m = 1,\dots,M-1) \quad . \tag{3.104}$$

The coefficients of the system depend on the values of functions at the layers n and n + 1, and it is therefore solved by means of iteration. In the first iteration, while calculating coefficients, the values of all parameters are put equal to their values at the previous time interval. The vector x^{n+1} thus obtained is used in the next iteration.

Equations for each fixed ℓ which are absent in (3.104) form, together with (3.101), boundary conditions for (3.104):

$$\begin{aligned} g_0\mathbf{X}_0 + g_1\mathbf{X}_1 &= \mathbf{d}_0 \quad , \\ g_{M-1}\mathbf{X}_{M-1} + g_M\mathbf{X}_M &= \mathbf{d}_M \quad . \end{aligned} \tag{3.105}$$

The system (3.104) with the boundary conditions (3.105) is solved by means of a tridiagonal algorithm which is stable if $\lambda_{11} > 0$, $\lambda_{22} > 0$ for all values of ξ and η. The inequalities are satisfied if the flow along the coordinate η is subsonic, which can always be achieved by an appropriate choice of the transform $\xi(x,y)$.

Concluding the chapter, we note that in [3.41] the method of integral relations has been used to solve the direct problem. One can also use the algorithm of solving the inverse problem described in the previous section. This method of solving the direct problem is based on the assumption that the actual distributions of velocity at the axes of symmetry for two similar contours differ in proportion to the difference in the corresponding distributions of velocity in the one-dimensional approximation. Then the velocity distribution at the axis corresponding to the contour required is found from the velocity distribution for a known contour and known distributions in the one-dimensional approximation. The calculations show that a solution converges to the one sought after a number of approximations of the solution of the inverse problem [3.47].

4. Asymptotic Methods in the Theory of Nozzles

Before the advent of electronic computers, asymptotic methods were the main means of analytical investigation of nozzle flows. These methods have maintained their importance up to date and they allow, on the one hand, the accuracy of numerical solutions to be evaluated where convergence of the methods has been proved and, on the other hand, the study of solutions near singularities, which are often too complicated for numerical methods. Finally, asymptotic methods prove, sometimes, to be preferable to qualitative and even quantitative ones. The main asymptotic methods of the theory of nozzles are presented below: the source-sink method, series expansion in sub-, super- and transonic regions of flow, the method of small perturbations for flows close to that from a source and some others.

4.1 Source-Sink Method and Inverse Problem for Incompressible Fluid

The source-sink method is widely used in gas dynamics to solve various linear problems. Superposing flow fields corresponding to sources and sinks of different intensities, one can obtain flow patterns corresponding to flows past bodies and in channels of various configurations. This method can be used in the case of incompressible fluid only, when the potential and the stream function equations are linear, so that the principle of superposition may be applied. Choosing a system of sources and sinks and adjusting their intensities, one can construct a flow in a given channel. However, the problem is quite complicated. The inverse problem is much simpler. Starting from a fixed system of sources and sinks, surfaces can be found which may be looked upon as a nozzle's walls. Below we discuss application of the method to plane, axisymmetric and three-dimensional flows.

4.1.1 Plane Flow

Let us distribute 2N + 1 doublets along the r axis of a Cartesian coordinate system. We shall put a doublet characterized by the moment M_0 at the origin and place 2N doublets, having moments M_1, M_2, ..., M_N directed against the x axis, symmetrically about the origin at distances a_1, a_2, ..., a_N from it, Fig.4.1. If a uniform flow which has velocity u_∞ parallel to the x axis is superposed upon the flow produced by the symmetric system of doublets, then a new flow results which is symmetric about the x axis. Any streamlines of the resulting flow, e.g. AB and A'B', may be looked upon as the nozzle's walls. Evidently, at zero intensity M_0 of the central doublet, contours AB and A'B' form a Laval nozzle, while at $M_0 \neq 0$ they provide a nozzle with a central body. Curvilinear channels are obtained if the walls are formed by streamlines nonsymmetric with respect to the longitudinal axis.

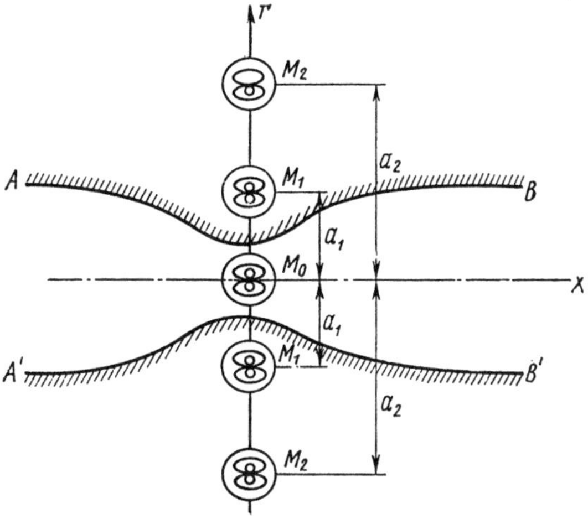

Fig.4.1. Construction of a nozzle contour by the source-sink method. At the points 0, $a_1, a_2, \ldots, a_N$ doublets with intensities $M_0, M_1, M_2, \ldots, M_N$ are placed

It can easily be shown that the potential and the stream function of the flow are

$$\varphi = x\left[u_\infty + \frac{M_0}{2\pi(x^2 + r^2)} + \frac{1}{\pi}\sum_{i=1}^{N} \frac{M_i(x^2 + r^2 + a_i^2)}{(x^2 + r^2)^2 + 2a_i^2(x^2 - r^2) + a_i^4}\right] , \qquad (4.1)$$

$$\psi = r\left[u_\infty - \frac{M_0}{2\pi(x^2 + r^2)} - \frac{1}{\pi}\sum_{i=1}^{N} \frac{M_i(x^2 + r^2 - a_i^2)}{(x^2 + r^2)^2 + 2a_i^2(x^2 - r^2) + a_i^4}\right] . \qquad (4.2)$$

This solution for an infinite number of equidistant doublets with the same moment is presented in [4.1].

4.1.2 Axisymmetric Flow

As in the case of a plane flow let us place a doublet with a moment M_0 at the centre of a cylindrical coordinate system x, r. Then we shall place uniformly distributed doublets with moments M_i at N concentric circles with the centre at the origin of the coordinate system and of radii a_i, $(i=1,2,\dots,N)$. Superposing a uniform flow upon the flow formed by this system of doublets, we get the following velocity components and stream function for the resulting flow:

$$u = u_\infty + \frac{M_0}{4\pi(x^2+r^2)^{3/2}}\left(1 - \frac{3x^2}{x^2+r^2}\right)$$

$$+ \frac{1}{\pi}\sum_{i=1}^{N}\frac{M_i a_i}{r_{1i}^2 r_{2i}}\left[1 - 2x^2\left(\frac{1}{r_{1i}^2} - \frac{1}{r_{2i}^2}\right)E_i + \frac{2x^2}{r_{2i}^2}K_i\right] , \tag{4.3}$$

$$v = -\frac{3M_0 xr}{4\pi(x^2+r^2)^{5/2}} + \frac{2}{\pi}\sum_{i=1}^{N}\frac{M_i a_i x}{r_{1i}^2 r_{2i}}\left\{\left(\frac{a_i - r}{r_{1i}^2} - \frac{a_i + r}{r_{2i}^2}\right)E_i\right.$$

$$\left. + \frac{1}{4r}\left[E_i - \left(1 - \frac{r+a_i}{2a_i}k_i^2\right)K_i\right]\right\} , \tag{4.4}$$

$$\psi = -\frac{r^2}{2}u_\infty + \frac{M_0 r^2}{4\pi(x^2+r^2)^{3/2}}$$

$$- \frac{1}{2\pi}\sum_{i=1}^{N}\frac{M_i a_i}{r_{2i}}\left[\frac{1}{1-k_i^2}\left(1 - \frac{r+a_i}{2a_i}k_i\right)E_i - K_i\right] , \tag{4.5}$$

where

$$r_{1i} = [x^2 + (r-a_i)^2]^{1/2} , \quad r_{2i} = [x^2 + (r+a_i)^2]^{1/2} , \quad k_i = 2\sqrt{ra_i}/r_{2i} ,$$

and

$$K_i = \int_0^{\pi/2}(1 - k_i^2\sin^2\varphi)^{-1/2}d\varphi , \quad E_i = \int_0^{\pi/2}(1 - k_i^2\sin^2\varphi)^{1/2}d\varphi$$

are the complete elliptic integrals of the first and second types. Equations (4.3-5) fully define the flow. Similarly to the plane case, one gets a Laval nozzle at $M_0=0$ and a nozzle with a central body at $M_0 \neq 0$.

4.1.3 Three-Dimensional Flow

Let us place uniformly distributed doublets with moments M at the plane $x = 0$ which has circular orifices of radii R_{ij} whose centres are placed at concentric circles of radii a_i. Superposing a uniform flow, perpendicular to the plane, upon the flow field produced by the system of doublets, we get a flow in a three-dimensional system similar to a multinozzle construction. By the principle of superposition the solution of the problem is reduced to summing the potential generated by doublets M, placed at a continuous plane, and the potentials generated by doublets with moments (-M) placed at circles of radii R_{ij} whose centres have cylindrical coordinates 0, α_i, θ_j. The potential generated by a circle in cylindrical coordinates x, r, θ is

$$\varphi = \frac{Mx}{2\pi(r_1^2 + R_{ij}^2)^{\frac{1}{2}}(1+b)^{\frac{1}{2}}}\left[\frac{\Pi(h_2,k)}{1+d}\left(1 - \frac{R_{ij}}{r_1}\right) + \frac{\Pi(h_1,k)}{1-d}\left(1 + \frac{R_{ij}}{r_1}\right)\right] ,$$

where

$$r_1 = [x^2 + r^2 + a_i^2 - 2ra_i \cos(\theta - \theta_j)]^{\frac{1}{2}} ,$$

$$d = (1 - x^2/r_1^2)^{\frac{1}{2}} , \quad b = 2(r_1^2 - x^2)^{\frac{1}{2}}(r_1^2 + R_{ij}^2)^{-1}R_{ij} ,$$

$$k = [2b/(1+b)]^{\frac{1}{2}} , \quad h_1 = 2d/(1-d) , \quad h_2 = -2d/(1+d)$$

and

$$\Pi(h,k) = \int_0^{\pi/2} (1 + h \sin^2\varphi)^{-1}(1 - k^2 \sin^2\varphi)^{-\frac{1}{2}} d\varphi$$

is a complete elliptic integral of the third type. The velocity components are defined by

$$u = \frac{\partial\varphi}{\partial x} , \quad v = \frac{\partial\varphi}{\partial r} , \quad w = \frac{\partial\varphi}{\partial z} . \tag{4.6}$$

4.1.4 Solution of the Inverse Problem for Incompressible Fluid

The irrotational stationary flow of an incompressible fluid is described by

$$\frac{\partial u}{\partial r} - \frac{\partial v}{\partial x} = 0 , \tag{4.7}$$

$$\frac{\partial^2 u}{\partial x^2} + \frac{\partial^2 u}{\partial r^2} + \frac{\nu}{r}\frac{\partial u}{\partial r} = 0 . \tag{4.8}$$

In order to solve the inverse problem boundary conditions

$$v = 0 \quad , \quad u = u_0(x) \tag{4.9}$$

are given at the axis of symmetry (at $r = 0$). In the case of plane flow ($\nu = 0$), (4.8) is Laplace's equation and its solution is an arbitrary function of a complex argument $F(x + ir)$. It follows from the boundary condition (4.9) that in the real plane the solution of the system (4.7,8) has the form

$$u = \mathrm{Re}\{u_0(x + ir)\} \quad , \quad v = \mathrm{Re}\{iu_0(x + ir)\} \quad . \tag{4.10}$$

In the axisymmetric case, substitution of $r_1 = ir$ reduces (4.8) to the Darboux equation and the system (4.7,8) has the following solution in the real plane [4.2]:

$$\begin{aligned} u &= \mathrm{Re}\left\{\frac{1}{\pi} \int_{-\pi/2}^{+\pi/2} u_0(x - ir \sin t)\, dt\right\} \quad , \\ v &= \mathrm{Re}\left\{-\frac{i}{\pi} \int_{-\pi/2}^{+\pi/2} u_0(x - ir \sin t) \sin t\, dt\right\} \quad . \end{aligned} \tag{4.11}$$

An axisymmetric flow of incompressible fluid with velocity distributed along the axis according to

$$u_0(x) = u_\infty + \frac{1 - u_\infty}{1 + Ax^2} \tag{4.12}$$

is discussed in [4.3]. It is evident that for this distribution the point with coordinates $x = 0$, $r = A^{-\frac{1}{2}}$ is a corner point because here the velocity tends to infinity. If the velocity is distributed along the axis arbitrarily, the corner point appears at the poles of the initial function extended into the complex plane. In the case of a compressible fluid the coordinates of the corner point cannot be found so easily, though there always exists such a point. To get the velocity distribution (4.12) by the source-sink method, it is necessary to place at the point $x = 0$, $r = A^{-\frac{1}{2}}$ the origin of the semiplane consisting of doublets and the normal to the axis.

4.2 Expansion in a Stream Function Series

It has been shown in Chap.2 that to solve the inverse problem of the theory of nozzles in the case of an isentropic three-dimensional flow of a nonreactive gas, the functions $r = r_0(s,\theta)$ and $u = u_0(s,\theta)$ must be specified at the initial surface $\psi=\psi_0$ and functions $w=w_0(\theta,\psi)$ and $\varphi=\varphi_0(\theta,\psi)$ at the initial plane $s=s_0$. The solution of the corresponding initial-value problem can be obtained in the form of a series. Various methods of series representation may be used: expansion in powers of Cartesian coordinates [4.4,5], expansion in negative powers of the throat's curvature radius [4.6,7], expansion in powers of the stream function [4.8]. It is characteristic that in all these references the series expansion is done in the transonic region of flow only. In [4.3,9] a solution is sought in the form of a series in powers of the stream function in the neighbourhood of the initial surface for sub-, trans- and supersonic regions of flow. The solution in [4.9] obtained for a three-dimensional flow is the most general one.

4.2.1 Three-Dimensional Flow

Let us construct a solution for the case of an ideal gas with a constant adiabatic exponent in the vicinity of the initial plane $\psi=0$ in the form of the stream function series [4.9]. We shall use the system (2.96-106) which for an ideal gas with a constant adiabatic exponent acquires the form

$$\frac{\partial p}{\partial \psi} = -\gamma \frac{G(s,\theta,\psi)}{r^{\nu} u (\partial\varphi/\partial\theta)^{\nu}} + \nu \frac{\partial p/\partial\theta}{\partial\varphi/\partial\theta} \frac{\partial \varphi}{\partial \psi} \quad , \tag{4.13}$$

$$\frac{\partial r^{1+\nu}}{\partial \psi} = \frac{1}{(\partial\varphi/\partial\theta)^{\nu}} \left(\frac{1+\nu}{\rho u} + \nu \frac{\partial\varphi}{\partial\psi} \frac{\partial r^{1+\nu}}{\partial\theta} \right) \quad , \tag{4.14}$$

$$\frac{\partial (wr)}{\partial s} = \pm \frac{ur}{R} \sin\varphi - \frac{1 \pm rR^{-1} \cos^{\nu}\varphi}{u(\partial\varphi/\partial\theta)} \left[\frac{1}{\gamma\rho} \frac{\partial p}{\partial\theta} + G(s,\theta,\psi) \frac{\partial r}{\partial\theta} \right] \quad , \tag{4.15}$$

$$\frac{\partial r}{\partial s} = \frac{v}{u} \left(1 \pm \frac{r}{R} \cos^{\nu}\varphi \right) \quad , \tag{4.16}$$

$$\frac{\partial \varphi}{\partial s} = \frac{w}{ru} \left(1 \pm \frac{r}{R} \cos^{\nu}\varphi \right) \quad , \tag{4.17}$$

$$p = \rho^{\nu} \quad , \tag{4.18}$$

$$u = \left[\frac{\gamma+1}{\gamma-1} - \frac{2}{\gamma-1} p^{(\gamma-1)/\gamma} - v^2 - w^2 \right]^{\frac{1}{2}} \quad , \tag{4.19}$$

where

$$G(s,\theta,\psi) = \left(u \frac{\partial v}{\partial s} \pm \frac{u^2}{R} \cos^{\nu}\varphi \right) \left(1 \pm \frac{r}{R} \cos^{\nu}\varphi \right)^{-1} - \nu \frac{w^2}{r} \quad .$$

Let us present the unknown functions u, v, w, p, ρ, φ, r in the form

$$f(s,\psi,\theta) = \sum_{n=0}^{N} f_n(s,\theta)\psi^n + \sqrt{\psi} \sum_{n=0}^{N} f_n^0(s,\psi)\psi^n \quad , \tag{4.20}$$

where $f(s,\psi,\theta)$ is any of the functions. Substituting series (4.20) into the system (4.13-19) and equating coefficients at equal powers of ψ, we obtain another system of partial differential equations in two independent variables s,θ, from which the functions $f_n(s,\theta)$ and $f_n^0(s,\theta)$ can be obtained.

If $r_0(s,\theta) \neq 0$, then to find the functions $f_0(s,\theta)$ we have the system (4.15-19) where now all the dependent variables are denoted by the subscript 0. We recall that the functions $r_0(s,\theta)$ and $u_0(s,\theta)$ are given at $\psi = 0$, while the functions $w_0(\psi,\theta)$ and $\varphi_0(\psi,\theta)$ are specified at $s = s_0$. Due to such a specification of the initial data all the functions sought are known at the surface $\psi = 0$ at $s = s_0$. It is evident that the right sides of (4.15,17) contain functions φ and w, and their derivatives with respect to θ, and the known functions $u_0(s,\theta)$ and $r_0(s,\theta)$ and their derivatives with respect to s and θ. Then, integrating the system (4.15,17) by some numerical method starting from $s = s_0$, we determine the functions $w_0(s,\theta)$ and $\varphi_0(s,\theta)$ at the surface $\psi = 0$, and later, from (4.16,18,19), the functions v_0, p_0, ρ_0.

In the case of axisymmetric and plane flows when $w = 0$ and $R = \infty$, the problem simplifies considerably, because there is no need to integrate the system (4.15,17) numerically, and the functions v_0, p_0, ρ_0 are obtained directly from (4.16,18,19). It is also important that at $u = u_0(s)$, $r = r_0(s)$ and $R(s) \neq \infty$, even if the functions w_0 and φ_0 are equal to zero at the initial section $s = s_0$, at other values of s they become nonzero, since for $R \neq \infty$ the system (4.15-17) is nonuniform. This means that if the axis is curvilinear, then the circumferential velocity component is nonzero.

It can be shown by means of simple, but cumbersome, calculations that the system of equations for the determination of the functions $f_0^0(s,\theta)$ is uniform in all these functions, and specification of $w_0^0(s_0,\theta)$ and $\varphi_0^0(s_0,\theta)$ at the initial plane defines all the remaining functions $f_0^0(s_0,\theta)$ at the plane. Evidently, at $w_0^0(s_0,\theta) = \varphi_0^0(s_0,\theta) = 0$, all the functions $f_0^0(s_0,\theta)$, due to uniformity of the corresponding system, are identically zero. Specification of the condition $w_0^0(s_0,\theta) = \varphi_0^0(s_0,\theta) = 0$ at $s = s_0$ is physically justified, because one can hardly imagine flows with infinite derivatives $\partial w/\partial\psi$, $\partial\varphi/\partial\psi$ at the initial plane in the regions lying far from the symmetry axis.

Let us consider a method of determining the function $f_1(s,\theta)$. Any of the required functions can be presented in the form [4.9]:

$$f_1(s,\theta) = \alpha_0(s,\theta) + \alpha_1(s,\theta)w_1 + \alpha_2(s,\theta)\varphi_1 \quad ,$$

where $\alpha_i(s,\theta)$ are functions of the known parameters $f_0(s,\theta)$ and their derivatives with respect to s and θ. Then the corresponding system of linear equations forms a Cauchy system, under the condition that $w_1(s_0,\theta)$ and $\varphi_1(s_0,\theta)$ are given at the initial plane $s = s_0$.

Now let us consider a practically important situation when $r_0(s,\theta) \equiv 0$ and $R = \infty$. We recall that $R = R(s)$ is the curvature radius of a curve in the plane x, y, where a curvilinear coordinate system s, r, φ is given. The case $R = \infty$ corresponds to the rectilinear axis of the coordinate system. To define the functions $f_0^0(s,\theta)$ and $f_1(s,\theta)$ we have the following system of equations:

$$v_0^0 = u_0 \frac{\partial r_0^0}{\partial s} \quad , \quad \frac{\partial \varphi_0}{\partial s} = \frac{w_0^0}{r_0^0 u_0} \quad , \quad r_0^0 = \left[\frac{2}{\rho_0 u_0 (\partial\varphi_0/\partial\theta)}\right]^{1/2} \quad , \tag{4.21}$$

$$\frac{\partial w_0^0}{\partial s} = -\frac{w_0^0 v_0^0}{u_0 r_0^0} - \frac{r_0^0}{2\gamma}\frac{\partial p_1}{\partial\theta} + \frac{p_1}{\gamma}\frac{\partial r_0^0}{\partial\theta} \quad , \tag{4.22}$$

$$p_1 = -\frac{\gamma}{r_0^0(\partial\varphi_0/\partial\theta)}\left[\frac{\partial v_0^0}{\partial s} - \frac{(w_0^0)^2}{u_0 r_0^0}\right] \quad . \tag{4.23}$$

It can be easily shown that only the functions f_n^0 are present in the expansions of the functions r, v, w, while $f_n = 0$. Conversely, only the functions f_n are present in the expansions of the functions p, ρ, u, and $f_n^0 = 0$, because the normal and circumferential velocity components are odd functions of r in the vicinity of the axis, while the pressure, the density and the longitudinal velocity component are even functions.

Construction of a solution for the system (4.21-23) for a three-dimensional flow proves to be quite a nontrivial task. It can be shown that (4.22) together with (4.21) and (4.23) can be reduced to

$$\frac{\partial^2}{\partial\theta^2}\left(\frac{\partial w_0^0}{\partial s}\right) + g_0(s,\theta)\frac{\partial}{\partial\theta}\left(\frac{\partial w_0^0}{\partial s}\right) + g_1(s,\theta)\frac{\partial w_0^0}{\partial s} + g_2(s,\theta) = 0 \quad , \tag{4.24}$$

where $g_i(s,\theta)$ are the known functions which can be calculated at the plane $s = \text{const}$ if w_0^0 is known and the function $u_0(s)$ is given. To find the derivative $\partial w_0^0/\partial s$ at a plane $s = \text{const}$, one has to solve a boundary-value problem for the differential equation (4.24), putting at the symmetry planes boundary conditions $\partial w_0^0/\partial s = 0$ (consistent with periodic behaviour of the function w_0^0).

In the general case of a three-dimensional flow a solution of the system of nonlinear partial differential equations (4.21-23) can be obtained only by numerical integration. However, sometimes it can be obtained in a closed form.

Let us construct a solution of the system using perturbation theory. We shall present the functions φ_0, r_0^0, v_0^0, w_0^0, p_1 in the form

$$\varphi_0 = \theta + \varepsilon\varphi_{01}(s,\theta) + \varepsilon^2\varphi_{02}(s,\theta) + \dots ,$$

$$r_0^0 = r_{00}^0(s) + \varepsilon r_{01}^0(s,\theta) + \varepsilon^2 r_{02}^0(s,\theta) + \dots ,$$

$$w_0^0 = \varepsilon w_{01}^0(s,\theta) + \varepsilon^2 w_{02}^0(s,\theta) + \dots , \tag{4.25}$$

$$v_0^0 = v_{00}^0(s) + \varepsilon v_{01}^0(s,\theta) + \varepsilon^2 v_{02}^0(s,\theta) + \dots ,$$

$$p_1 = p_{10}(s) + \varepsilon p_{11}(s,\theta) + \varepsilon^2 p_{12}(s,\theta) + \dots ,$$

where ε is a small parameter. This form of representation of the required functions is based upon the physically justified assumption that at the initial plane the values of flow parameters of a three-dimensional flow and of an axisymmetric flow with the same axial velocity distribution $u_0(s)$ differ only slightly. Substituting (4.25) into the system (4.21-23) and linearizing the latter, we arrive at the following equations:

$$r_{00}^0 = \left(\frac{2}{\rho_0 u_0}\right)^{\frac{1}{2}} , \quad v_{00}^0 = u_0 \frac{dr_{00}^0}{ds} , \quad p_{10} = -\frac{\gamma}{r_{00}^0}\frac{dv_{00}^0}{ds} , \tag{4.26}$$

$$r_{00}^0 \frac{\partial\varphi_{01}}{\partial\theta} = -2r_{01}^0 , \quad v_{01}^0 = u_0 \frac{\partial r_{01}^0}{\partial s} , \quad u_0 r_{00}^0 \frac{\partial\varphi_{01}}{\partial s} = w_{01}^0 , \tag{4.27}$$

$$\frac{\partial}{\partial s}(r_{00}^0 w_{01}^0) = \frac{1}{2}\left(\frac{\partial r_{01}^0}{\partial\theta}\frac{dv_{00}^0}{ds} - r_{00}^0\frac{\partial^2 v_{01}^0}{\partial s\,\partial\theta}\right) .$$

In a similar manner, the system of equations for coefficients of ε^2 can be obtained.

An important feature of the system of partial differential equations (4.27) is its linearity, which means that it can be solved by Fourier's method of separating variables. As a result we get the following approximate solution of the system (4.21-23) accurate to terms of order ε^2:

$$r_0^0(s,\theta) = r_{00}^0(s)\left\{1 - \varepsilon r_{00}^0(s_0)\sum_{k=0}^{K}\frac{k}{2}\left[\frac{w_{01k}^0(s_0)}{2}\int_{s_0}^{s}\rho_0\,ds + \frac{\varphi_{01k}(s_0)}{r_{00}^0(s_0)}\right]\cos k\theta\right\} ,$$

$$\varphi_0(s,\theta) = \theta + \varepsilon\sum_{k=0}^{K}\left[\frac{1}{2}r_{00}^0(s_0)w_{01k}^0(s_0)\int_{s_0}^{s}\rho_0\,ds + \varphi_{01k}(s_0)\right]\sin k\theta ,$$

$$p_1(s,\theta) = p_{10}(s)\left\{1 - \varepsilon r_{00}^0(s_0)\sum_{k=0}^{K}k\left[\frac{w_{01k}^0(s_0)}{2}\int_{s_0}^{s}\rho_0\,ds + \frac{\varphi_{01k}(s_0)}{r_{00}^0(s_0)}\right]\cos k\theta\right\} . \tag{4.28}$$

$$v_0^0(s,\theta) = v_{00}^0(s) - \frac{1}{2}\varepsilon r_{00}^0(s_0)\sum_{k=0}^{K}k\left\{v_{00}^0(s)\left[\frac{w_{01k}^0(s_0)}{2}\int_{s_0}^{s}\rho_0\,ds + \frac{\varphi_{01k}(s_0)}{r_{00}^0(s_0)}\right] + \frac{w_{01k}^0(s_0)}{r_{00}^0(s_0)}\right\}\cos k\theta ,$$

$$w_0^0(s,\theta) = \varepsilon\frac{r_{00}^0(s_0)}{r_{00}^0(s)}\sum_{k=0}^{K}w_{01k}^0(s_0)\sin k\theta ,$$

where $w_{01k}^0(s_0)$ and $\varphi_{01k}(s_0)$ are Fourier coefficients in the series expansions of the initial functions $w_0^0(s_0,\theta)$ and $\varphi_0(s_0,\theta)$ in terms of $\sin k\theta$. It follows from (4.28) that the configuration of a cross-section can be changed by variation of the initial data at $s = s_0$, i.e. by varying the values of $w_{01k}^0(s_0)$ and $\varphi_{01k}(s_0)$. If we put $\varphi_{01k}(s_0) = 0$ at $s = s_0$, then the nozzle's cross-section in the plane is circular; however, with the growth of s it acquires a form defined by $w_{01k}^0(s)$. At the same time $\partial r_0^0/\partial s|_{s=s_0}$ is, unlike the axisymmetric case, a function of θ. It is evident that if all $w_{01k}^0(s_0) = 0$, except $w_{012}^0(s_0)$, then the three-dimensional flow has two planes of symmetry, and at $w_{011}^0(s) \neq 0$, one plane of symmetry. For the case $r_0 \equiv 0$ at $R \neq \infty$ one can construct an analogous solution considering R^{-1} as ε. It can be shown that in this case, a nozzle which is circular at the initial section preserves its circular cross-section to an accuracy of ε^2; however, its axis becomes curvilinear in accordance with the relation $R = R(s)$.

A system of equations, similar to (4.21-23), can be written to determine other terms of the series (4.20).

4.2.2 Plane and Axisymmetric Flows

Let R tend to infinity and r_0 be identically equal to zero. Then the initial (Cauchy) data at the symmetry axis ($\psi = 0$) have the form $u = u_0(s)$, $r_0 = 0$ and hence $v_0 = 0$. For the case of an axisymmetric irrotational flow ($w = 0$, $\varphi = \theta$) the system (4.21-23) is much simplified and acquires the form:

$$r_0^0 = \left(\frac{2}{\rho_0 u_0}\right)^{1/2} , \quad v_0^0 = u_0 \frac{dr_0^0}{ds} , \tag{4.29}$$

$$p_1 = -\frac{\gamma}{r_0^0} \frac{\partial v_0^0}{\partial s} . \tag{4.30}$$

One can easily obtain expressions for the subsequent terms of the series. We have

$$r_1^0 = -\frac{1}{4} r_0^0\left(\frac{\rho_1}{\rho_0} + \frac{u_1}{u_0}\right) , \quad v_1^0 = u_1 \frac{dr_0^0}{ds} + u_0 \frac{dr_1^0}{ds} , \tag{4.31}$$

$$2r_0^0 p_2 = -\left(\gamma \frac{dv_1^0}{ds} + r_1^0 p_1\right) , \quad \frac{\rho_1}{\rho_0} = \frac{p_1}{\gamma p_0} ,$$

$$u_1 = -\frac{1}{2u_0}\left[\frac{2p_1}{\gamma\rho_0} + (v_0^0)^2\right] .$$

In a plane flow, all functions $f_n^0 = 0$ and by symmetry, functions f_n with odd n are equal to zero if $f = p, \rho, u$, and functions f_n with even n are equal to zero if $f = v, r$. We have

$$r_1 = \frac{1}{\rho_0 u_0} , \quad v_1 = u_0 \frac{dr_1}{ds} , \tag{4.32}$$

$$p_2 = -\frac{\gamma}{2} \frac{dv_1}{ds} , \quad u_2 = -\frac{1}{2u_0}\left(\frac{2p_2}{\gamma\rho_0} + v_1^2\right) , \quad \frac{\rho_2}{\rho_0} = \frac{p_2}{\gamma p_0} , \tag{4.33}$$

$$r_3 = -\frac{r_1}{3}\left(\frac{\rho_2}{\rho_0} + \frac{u_2}{u_0}\right) , \quad v_3 = u_2 \frac{dr_1}{ds} + u_0 \frac{dr_3}{ds} , \quad p_4 = -\frac{\gamma}{4} \frac{dv_3}{ds} . \tag{4.34}$$

Since the function $u = u_0(s)$ is given and hence p_0 and $\rho_0 = p_0^{1/\gamma}$ are known, (4.29-34) allow the functions sought to be defined from finite relations or by differentiation of the known functions with respect to s.

The formulas thus obtained can be used to calculate the entire flow field including the sub-, trans- and supersonic regions. Expansions constructed in [4.4,5,7,8] for the transonic region can be obtained from them as special cases

In fact, if, following [4.4], we represent the velocity at the symmetry axis (in the vicinity of the nozzle's centre) of the Meyer flow in the form

$$u_0(x) = 1 + Ax + Bx^2 + Cx^3 + Dx^4 + \dots \; , \tag{4.35}$$

(where $s = x$, since $R = \infty$), then it is evident that

$$r_0^0 = \sqrt{2}\left[1 + \frac{1}{2}(\gamma + 1)x^2(A + Bx)^2\right]^{\frac{1}{2}} \; . \tag{4.36}$$

Using (4.36) and subsequent differentiation, one can find all the required functions in (4.29-34) and, in accordance with the solution of [4.4], deduce that in a plane flow,

$$u = u_0 + u_2\psi^2 = u_0 + \frac{1}{2}(\gamma + 1)A^2r^2 + \dots \; , \tag{4.37}$$

$$v = v_1\psi + v_3\psi^3 = (\gamma + 1)A^2xr + \frac{1}{6}(\gamma + 1)^2A^3r^3 + \dots \; ,$$

while in an axisymmetric one,

$$u = u_0 + u_1\psi = u_0 + \frac{1}{4}(\gamma + 1)A^2r^2 + \dots \; ,$$

$$v = \sqrt{\psi}(v_0^0 + v_1\psi) = \frac{1}{2}(\gamma + 1)A^2xr + \frac{1}{16}(\gamma + 1)^2A^3r^3 + \dots \; . \tag{4.38}$$

From (4.37,38) we obtain the equations of the sonic line ($M = 1$), the line $\theta = 0$ (where θ is the angle between the velocity vector and the nozzle axis) and characteristics of the first (C_+^0) and the second (C_-^0) families issuing from the centre of the nozzle (the limiting characteristics). We have

$$\begin{array}{lll} & \nu = 0 & \nu = 1 \\ M = 1: & x = -\frac{1}{2}(\gamma + 1)Ar^2 \; , & x = -\frac{1}{4}(\gamma + 1)Ar^2 \; , \\ \theta = 0: & x = -\frac{1}{6}(\gamma + 1)Ar^2 \; , & x = -\frac{1}{8}(\gamma + 1)Ar^2 \; , \\ C_+^0: & x = \frac{1}{2}(\gamma + 1)Ar^2 \; , & x = \frac{1}{8}(\gamma + 1)(1 + \sqrt{5})Ar^2 \; , \\ C_-^0: & x = -\frac{1}{4}(\gamma + 1)Ar^2 \; , & x = \frac{1}{8}(\gamma + 1)(1 - \sqrt{5})Ar^2 \; . \end{array} \tag{4.39}$$

It follows that the sonic line lies at the symmetry axis upstream from the point $x = 0$; at the nozzle contour it is also displaced upstream from the throat. At the point where the symmetry axis lies in the plane of the throat, the velocity is subsonic, Fig.4.2.

Let us consider some extremum properties of the characteristic C_+^0 in the case of a plane flow [4.10]. It follows from the analysis of mapping the

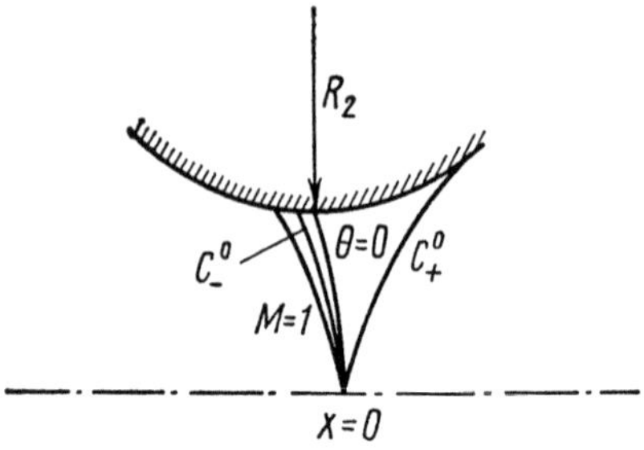

Fig.4.2. Transonic region of a Laval nozzle showing the sonic line, the line $\theta = 0$ and the limiting characterisitcs C_+^0, C_-^0

physical plane upon the plane of the hodograph that each characteristic of the second family has a point of inflection at the characteristic C_+^0. Also, the characteristic C_+^0 is the locus of the extreme values of u and v reached while moving along characteristics belonging to another family. The latter property is due to passing twice – in the direct and opposite directions – the corresponding characteristics of one of the families in the process of mapping the physical plane upon the plane of the hodograph.

Let us compare solutions given by (4.29-34) with those in [4.7,8] where functions are expanded in negative powers of the throat curvature radius R_2, Fig.4.2. We shall express R_2 in terms of velocity derivatives by means of relations from one-dimensional theory. Differentiating twice the continuity equation $r^{1+\nu} q(\lambda) = \text{const}$, where $q(\lambda)$ is defined by (2.126), we get, at the throat,

$$\frac{1+\nu}{r}\frac{d^2r}{dx^2} = -\left[\frac{d^2\lambda}{dx^2}\frac{d\ln q(\lambda)}{d\lambda} + \left(\frac{d\lambda}{dx}\right)^2 \frac{d^2 \ln q(\lambda)}{d\lambda^2}\right] \tag{4.40}$$

or

$$R_2 = -(1+\nu)\left[\frac{d^2\lambda}{dx^2}\frac{d\ln q(\lambda)}{d\lambda} + \left(\frac{d\lambda}{dx}\right)^2 \frac{d^2 \ln q(\lambda)}{d\lambda^2}\right]^{-1} \quad .$$

Let us consider the Meyer flow and the Taylor flow separately. For the former at $x = 0$ the velocity is equal to the speed of sound and

$$\frac{d\ln q(\lambda)}{d\lambda} = 0 \quad , \quad \frac{d^2 \ln q(\lambda)}{d\lambda^2} = -(\gamma+1) \quad , \quad R_2 = \frac{1+\nu}{\gamma+1}\left(\frac{d\lambda}{dx}\right)^{-2} \quad ,$$

while the velocity distribution $u_0(x) = \lambda(x)$ is given by (4.35). Then from (4.37,38) with the help of (4.40) we deduce that in a plane flow

$$\begin{aligned} u &= 1 + (\gamma+1) - \frac{1}{2}\, xR_2^{-\frac{1}{2}} + \frac{1}{2}\left(r^2 - \frac{1}{3}\right)R_2^{-1} + \ldots \;, \\ v &= xrR_2^{-1} + \frac{1}{6}(\gamma+1)^{\frac{1}{2}} r(r^2-1)R_2^{-3/2} + \ldots \;, \end{aligned} \tag{4.41}$$

while in an axisymmetric flow

$$u = 1 + \sqrt{2}(\gamma + 1)^{-\frac{1}{2}} x R_2^{-\frac{1}{2}} + \frac{1}{2}\left(r^2 - \frac{1}{2}\right) R_2^{-1} + \dots , \tag{4.42}$$

$$v = x r R_2^{-1} + \frac{1}{4}\left(\frac{\gamma + 1}{2}\right)^{\frac{1}{2}} r(r^2 - 1) R_2^{-3/2} + \dots .$$

These formulas have been obtained under the assumption that the plane $x = 0$ passes through the throat cross-section, whose ordinate $r_* = 1$. [Deriving (4.37,38) we assumed that the plane $x = 0$ passes through the point of the symmetry axis where the flow velocity is equal to the sound speed.] The equations (4.41,42) coincide exactly with the ones obtained in [4.7,8].

Let us now consider the Taylor flow. In the case of this flow the velocity at the symmetry axis is subsonic throughout; however, local supersonic zones may form in some parts of the flow field. To calculate the flow (4.29-34) can also be used. The velocity distribution at the axis in this case has the form

$$u_0 = \lambda_0 + a x^2 \quad , \quad \lambda_0 < 1 \quad , \quad a < 0 \quad , \tag{4.43}$$

because $du_0/dx = 0$ in the Taylor flow, if no supersonic zones contact the symmetry axis. Then, keeping in mind that

$$r_0^0 = \sqrt{2} q^{-\frac{1}{2}}(\lambda_0)\left[1 - \frac{1}{2} \left.\frac{d \ln q(\lambda)}{d\lambda}\right|_{\lambda=\lambda_0} a x^2\right] ,$$

$$r_1 = q^{-1}(\lambda_0)\left[1 - \left.\frac{d \ln q(\lambda)}{d\lambda}\right|_{\lambda=\lambda_0} a x^2\right] , \tag{4.44}$$

one can obtain formulas similar to (4.37,38). Here we shall write down to an accuracy of R_2^{-2} only the formulas analogous to (4.41) and (4.42). For plane and axisymmetric flows we have

$$u = \lambda_0 - \frac{(1 + \nu)\lambda_0}{2(1 - \lambda_0^2)}\left(1 - \frac{\gamma - 1}{\gamma + 1}\lambda_0^2\right) x^2 R_2^{-1} + \frac{1}{2}\lambda_0 r^2 R_2^{-1} + \dots ,$$

$$v = \lambda_0 x r R_2^{-1} + \dots , \tag{4.45}$$

where

$$R_2 = -(1 + \nu)\left[\frac{d^2\lambda}{dx^2}\frac{d \ln q(\lambda)}{d\lambda}\right]_{x=0}^{-1} .$$

While solving the inverse problem numerically, the expansions in the ψ series in the neighbourhood of the symmetry axis are used to calculate a streamline close to the axis and serving as an initial line for numerical integration, because equations (2.96) have a singularity at the symmetry axis. In the case of an axisymmetric flow this singularity can be excluded by transforming the equations; however, in the case of a three-dimensional flow the necessity of using an asymptotic expansion to obtain the solution at a streamline close to the axis cannot be circumvented.

The above relations were used to calculate nozzle gas flows before the advent of numerical methods. There is no proof that the series converge and the radii of convergence remain undetermined. Therefore, the possibility of their application is deduced by comparison with numerical solutions and experimental data which show that expansions in the series retaining 3 or 4 terms can be used to describe flow in the transonic region at $R_2 \geqslant 0.5r_*$ as well as in sub- and supersonic regions when the transverse gradients of parameters are not large. Comparison with a numerical solution of the inverse problem of the theory of nozzles is presented in the next section.

The equations (4.37,38,41,42) are applicable only to the calculation of flows in slender nozzles when $R_2 > 2r_*$. However, they allow a general impression of the structure of a transonic flow in the vicinity of the throat to be obtained. Also, in the case of plane and axisymmetric flows, (4.39) allow coordinates of all characteristic lines to be defined in the transonic region accurately enough (to 5-10%) even at small R_2, ($R_2 \geqslant 0.2r_*$), though in the latter case the accuracy is poor, Fig.4.3.

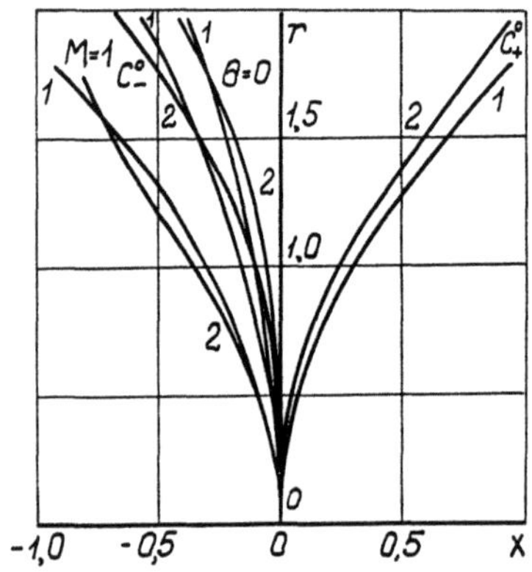

Fig.4.3. Configurations of the sonic line, the line $\theta = 0$ and limiting characteristics C_+^0, C_-^0 (*1*) calculated by (4.39) at $\gamma = 1.4$, $A = 0.24$, $\nu = 0$, (*2*) numerical solution [4.11]

4.2.3 Two-Phase Flows

Numerical solution of the inverse problem of the theory of nozzles for a two-phase flow with the initial data specified at the symmetry axis requires construction of an asymptotic expansion because the equations of the system (2.112) have a singularity at the axis. The velocity distribution is given at the symmetry axis, while u_s, v_s, T_s, ρ_s are specified at the plane $s = s_0$. Representing the functions u, u_s, ρ, ρ_s, T, T_s in the vicinity of the symmetry axis in the form

$$f = f_0 + f_{2-\nu}\psi^{2-\nu} \quad , \quad f_s = f_{0s} + f_{(2-\nu)s}\psi^{2-\nu} \quad ,$$

and the functions v, v_s, r in the form

$$f = f_0\psi^{1-\nu/2} \quad , \quad f_s = f_{0s}\psi^{1-\nu/2} \quad ,$$

we get, upon substituting these relations into the system (2.111,112),

$$r_0^{1+\nu} = \frac{1+\nu}{\rho_0 u_0} \quad , \quad v_0 = u_0 \frac{\partial r_0}{\partial s} \quad ,$$

$$p_{2-\nu} = - \frac{\gamma}{(2-\nu) r_0^{\nu}} \left(\frac{\partial v_0}{\partial s} + \frac{\rho_{0s}}{\rho_0 u_0} f_{0r} \right) \quad ,$$

$$\frac{\partial p_0}{\partial s} = - \left[\gamma \rho_0 \frac{\partial}{\partial s} \left(\frac{u_0^2}{2} \right) + \gamma \frac{\rho_{0s}}{u_0} (\mathbf{fW})_0 \right] \quad ,$$

$$\frac{\partial h_0}{\partial s} = - \left[\frac{\partial}{\partial s} \left(\frac{u_0^2}{2} \right) - \frac{\rho_{0s}}{\rho_0 u_0} (\mathbf{W}_s \mathbf{f} + q)_0 \right] \quad ,$$

$$u_{0s} \frac{\partial}{\partial s} (\ln \rho_{0s}) = -(1+\nu) \frac{v_{0s}}{r_0} - \frac{\partial u_{0s}}{\partial s} \quad ,$$

$$u_{0s} \frac{\partial u_{0s}}{\partial s} = f_{0x} \quad , \quad u_{0s} \frac{\partial v_{0s}}{\partial s} = f_{0r} \quad , \quad u_{0s} \frac{\partial e_{0s}}{\partial s} = q_0 \quad ,$$

$$h_0 = h(p_0, T_0) \quad , \quad p_0 = \rho_0 T_0 \quad , \quad e_{0s} = e_s(T_{0s}) \quad .$$

Integrating this system numerically from $s = s_0$, we find subsequently p_0, h_0, T_0, ρ_0, u_{0s}, e_{0s}, T_{0s}, v_0, v_{0s}, ρ_{0s}, r_0 and $p_{2-\nu}$ at all points.

4.3 Asymptotic Methods in the Transonic Region

Gas flow in the transonic part of a nozzle may be studied by means of the small perturbations method, allowing a solution to be obtained for the velocity potential, and by means of a method based upon series expansion in the vicinity of a rectilinear sonic line. The two methods are briefly discussed below.

4.3.1 Method of Small Perturbations

This method is dealt with in a number of references, particularly [4.10, 12-15]. The monograph [4.10] gives the solution of equations obtained by the method of small perturbations for an arbitrary three-dimensional nonstationary flow in the transonic region and formulates conditions under which shock waves form. We direct the interested reader to this monograph and here present only the most significant results obtained by the method of small perturbations for plane and axisymmetric flows.

Supposing that the gas velocity is close to the sound speed and that the angle between the velocity vector and the x axis is small we get the following equation for the potential φ of the perturbed flow:

$$-(\gamma + 1) \frac{\partial\varphi}{\partial x} \frac{\partial^2\varphi}{\partial x^2} + \frac{\partial^2\varphi}{\partial r^2} + \frac{\nu}{r} \frac{\partial\varphi}{\partial r} = 0 \quad . \tag{4.46}$$

This equation is one of the simplest equations of the mixed type: it is elliptic in the semiplane corresponding to the subsonic flow and hyperbolic in the semiplane where the flow is supersonic.

It is characteristic of the method of small perturbations in the transonic region that the resulting equations are nonlinear in the physical plane. In this connection, it would be interesting to study some partial solutions of these equations which are asymptotic in the vicinity of the nozzle centre. Existence of such solutions is explained by the invariance of transonic flow equations with respect to a two-parameter group of similarity transforms. Partial solutions of (4.46) have the form

$$\varphi = \frac{1}{2} Ax^2 + \frac{1}{2} (\gamma + 1)A^2xr^2 + \frac{1}{24} (\gamma + 1)^2A^3r^4 \quad , \quad (\nu = 0) \quad , \tag{4.47}$$

$$\varphi = \frac{1}{2} Ax^2 + \frac{1}{4} (\gamma + 1)A^2xr^2 + \frac{1}{64} (\gamma + 1)^2A^3r^4 \quad , \quad (\nu = 1) \quad . \tag{4.48}$$

where A is, as earlier, the velocity derivative at the nozzle centre. It is evident that the components of the perturbation velocity $u = \partial\varphi/\partial x$, $v = \partial\varphi/\partial y$

found from these relations exactly correspond to the first terms of the series constructed by Meyer [see (4.37,38)]. All conclusions made while analyzing these formulas in the previous section follow, evidently, from the equations of the method of small perturbations as well.

4.3.2 Series Expansion in the Vicinity of a Rectilinear Sonic Line

The most complete investigation of a plane flow in the vicinity of a rectilinear sonic line has been carried out in [4.16]. There it was shown that there may exist five different types of flow, including sub- and supersonic flows containing corner points. Axisymmetric flows are considered in [4.10]. In [4.3,10,14,16] a solution has been obtained for the vicinity of a rectilinear sonic line in the form of series expansions in terms of x, and in [4.17], in terms of a characteristic coordinate series.

Let us expand the solution into a series in x in the vicinity of a rectilinear sonic line for one of the possible types of flow of the greatest practical significance [4.3]. We shall take equations describing the irrotational flow of an ideal gas in the form

$$\frac{\partial u}{\partial r} = \frac{\partial v}{\partial x} \quad ,$$

$$\left[\frac{\gamma+1}{2}(1-u^2) - \frac{\gamma-1}{2}v^2\right]\frac{\partial u}{\partial x} + \left[\frac{\gamma+1}{2}(1-v^2) - \frac{\gamma-1}{2}u^2\right]\frac{\partial v}{\partial r}$$
$$+ 2uv\frac{\partial v}{\partial x} + \nu\left[\frac{\gamma+1}{2} - \frac{\gamma-1}{2}(u^2+v^2)\right]\frac{v}{r} = 0 \quad . \qquad (4.49)$$

In the vicinity of a rectilinear sonic line at which $u=1$, $v=0$, it is natural to seek a solution of the system (4.49) in the form

$$u = 1 + u_1(r)x + u_2(r)x^2 + u_3(r)x^3 + \dots \quad ,$$
$$v = v_1(r)x + v_2(r)x^2 + v_3(r)x^3 + \dots \quad . \qquad (4.50)$$

Substituting the series (4.50) into (4.49) and equating coefficients of the zeroth and first powers of x on the left and right sides of these equations, we get $u_1 = v_1 = v_2 = 0$. These are the Görtler conditions necessary and sufficient for a sonic line to be rectilinear [4.14]. Equating coefficients of other powers of x, we get

$$\frac{du_2}{dr} = 3v_3 \quad , \qquad \frac{d^2u_2}{dr^2} + \frac{\nu}{r}\frac{du_2}{dr} - 6(\gamma+1)u_2^2 = 0 \quad , \qquad (4.51)$$

$$\frac{du_3}{dr} = 4v_4 \quad , \qquad \frac{d^2u_3}{dr^2} + \frac{\nu}{r}\frac{du_3}{dr} - 20(\gamma + 1)u_2u_3 = 0 \quad , \tag{4.52}$$

$$\frac{du_4}{dr} = 5v_5 \quad , \qquad \frac{dv_5}{dr} = 6(\gamma + 1)u_2u_4 + 3(\gamma + 1)u_3^2$$
$$+ (\gamma + 1)(2\gamma - 1)u_2^3 + 6v_3^2 - \frac{\nu}{r}v_5 \quad . \tag{4.53}$$

Let the velocity distribution at the symmetry axis on the side of the subsonic part be given in the form (4.35), where $A = 0$. Then the boundary conditions for the system (4.51-53) at $r = 0$ will have the form

$$\begin{aligned} du_2/dr &= 0 \quad , & u_2 &= B < 0 \quad , \\ du_3/dr &= 0 \quad , & u_3 &= C \quad , \\ du_4/dr &= 0 \quad , & u_4 &= D \quad . \end{aligned} \tag{4.54}$$

The systems (4.51-53) in a general case cannot be integrated in a closed form. Their numerical integration has shown that at a certain $r = r^0$ the function $u_2(r)$ reaches zero, then changes its sign, becomes positive, and at a certain $r = r_0$ tends to infinity [4.3]. The change of sign of the function $u_2(r)$ means that another sonic line originates at the point $r = r^0$ directed upstream to the subsonic region, Fig.4.4. The flow between the rectilinear and the curvilinear sonic lines is supersonic. In the vicinity of the point $r = r_0$ the solution of (4.51) has the form

$$u_2 = d_0(r - r_0)^{-2} + d_1(r - r_0)^{-1} + d_2 + d_3(r - r_0) + d_4(r - r_0)^2$$
$$+ d_5(r - r_0)^3 + d_6(r - r_0)^4 \ln(r - r_0) + c_2(r - r_0)^4 \ldots \quad . \tag{4.55}$$

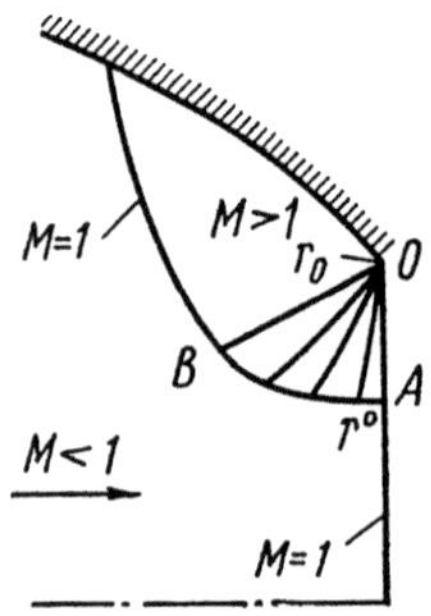

Fig.4.4. Diagram of flow in the vicinity of a rectilinear sonic line, with formation of a local supersonic line

Substituting (4.55) into (4.51) and equating powers of $r - r_0$, we define coefficients d_i. In particular,

$$d_0 = (\gamma + 1)^{-1} \quad , \quad d_1 = \frac{\nu}{5r_0} d_0 \quad . \tag{4.56}$$

The values of r_0 and c_2 are defined from the condition of matching the asymptotic solution (4.55) and the solution obtained by numerical integration of (4.51). A solution of the form

$$\begin{aligned} u &= 1 + \frac{x^2}{(r - r_0)^2} [d_0 + d_1(r - r_0) + \ldots] \quad , \\ v &= - \frac{x^3}{3(r - r_0)^3} [2d_0 + d_1(r - r_0) + \ldots] \end{aligned} \tag{4.57}$$

means that the point $x = 0$, $r = r_0$ is a corner point and hence the Prandtl-Meyer flow takes place here. The local supersonic zone contains a compression wave limited by a characteristic OB and the characteristic OA coinciding with the sonic line (Fig.4.4).

If the velocity distribution at the x axis is such that $A = B = 0$, then $u_2(r) = v(r) = 0$ and all the subsequent terms of the series can be obtained in an analytic form. We have

$$\begin{aligned} &u_3 = C = \text{const} \quad , \quad v_4 = 0 \quad , \\ &u_4 = \frac{15(\gamma + 1)}{2(1 + \nu)} C^2 r^2 + D \quad , \quad v_5 = \frac{3(\gamma + 1)}{1 + \nu} C^2 r \quad , \end{aligned} \tag{4.58}$$

hence

$$u = 1 + Cx^3 + \left[\frac{15(\gamma + 1)}{2(1 + \nu)} C^2 r^2 + D\right] x^4 + \ldots \quad . \tag{4.59}$$

Thus, another sonic line appears in the vicinity of the rectilinear one. At $D = 0$ the new sonic line and the rectilinear one have an infinite common point, unlike the case $A = 0$, $B \neq 0$, when the point lies at a finite distance.

Concluding the section, let us compare solutions obtained by asymptotic methods described above with the numerical results of [4.18]. Figures 4.5,6 show the numerical solution (curve 1) and analytic solutions obtained by expansion into a series in ψ in the vicinity of the symmetry axis (curve 2), into a series in x in the vicinity of a rectilinear sonic line (curve 3) and the solution of the inverse problem for incompressible fluid (curve 4). Comparing the solutions, one gets some impression of the regions of applicability of the approximate solutions. Series expansion of the solution in terms

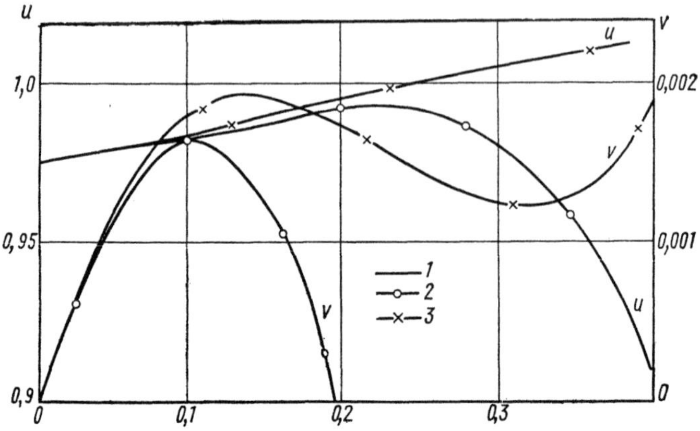

Fig.4.5

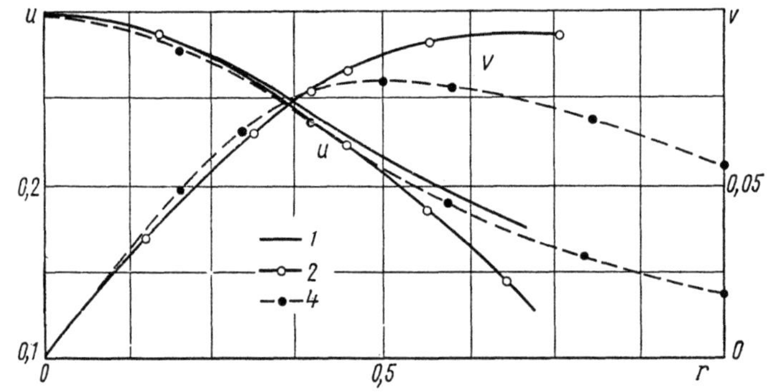

Fig.4.6

Fig.4.5. Comparison of numerical and asymptotic solutions ($x = -0.05$). Velocity components u, v in the transonic region of a nozzle. Expansion into ψ series (*2*) and x series (*3*) in the vicinity of the sonic line. Numerical solution of the inverse problem for velocity distribution (5.1)

Fig.4.6. Comparison of numerical and asymptotic solutions ($x = -0.6$). Velocity components u, v in the subsonic region of a nozzle. (*2*) expansion into ψ series; (*3*) solution for incompressible fluid. Numerical solution of the inverse problem for velocity distribution (5.1)

of ψ near the throat is justified for small values of r only, Fig.4.5. When $|x|$ grows, the results of the exact and the approximate solutions practically coincide. The results show that at small values of $|x|$ when compressibility is significant, the solution of the inverse problem for incompressible fluid cannot be used to describe flows of compressible fluid. At moderate gas flow velocities ($W < 0.3$), the flow parameters for compressible and incompressible fluids practically coincide. Comparison of the exact solution with the solution obtained by expanding the required functions into a series in x in the vicinity of a rectilinear sonic line shows that at $x = -0.05$ the exact and the approximate solutions give practically the same results. At $x < -0.2$ the series diverge.

4.4 Solution in the Neighbourhood of the Infinite Point in the Subsonic Regions

Let us construct an asymptotic solution for the system (2.107-110) in the vicinity of the infinite point for the direct and the inverse problems in the cases of plane and axisymmetric flows. We shall present the unknown functions in the form

$$g = \sum_{n=0}^{\infty} g_n(\psi)x^{-n} \quad . \tag{4.60}$$

The boundary conditions are: $r = v = 0$, $u = u_\infty + \Sigma_{n=1}^{N} u_n x^{-n}$ at $\psi = 0$ for the inverse problem and $r = v = 0$ at $\psi = 0$, $r = r_\infty + \Sigma_{n=1}^{N} f_n x^{-n}$, $v/u = -\Sigma_{n=1}^{N} n f_n x^{-(n+1)}$ at $\psi = \psi_k$ for the direct problem. Here ψ_k is the flow rate at the nozzle contour which depends on the flow regime and is always less than the critical value, at which the flow in the subscritical part does not depend on the external pressure. Substituting the series (4.60) into (2.107-110) and equating the powers of x, we get a system of equations for the determination of the functions g_n which can be solved in a closed form. In particular, we have

$$v_0(\psi) = v_1(\psi) = 0 \ , \quad p_0(\psi) = \text{const} \ , \quad p_1(\psi) = \text{const} \ , \quad p_2(\psi) = \text{const} \ ,$$

$$r_0 = \left[\frac{(1+\nu)\psi}{\rho_\infty u_\infty}\right]^{1/(1+\nu)} \ , \quad r_1 = -\frac{1}{\gamma}\left[\frac{\psi}{(1+\nu)\rho_\infty u_\infty}\right]^{1/(1+\nu)} \frac{p_1(M_\infty^2 - 1)}{p_\infty M_\infty} \ . \tag{4.61}$$

For the inverse problem the unknown constants p_0, p_1, p_2 are defined from the boundary conditions with the help of the known u_∞, u_1.

We have

$$p_0 = p_\infty = \frac{\pi(u_\infty)}{\pi(1)} \ , \quad \pi(u_\infty) = \left(1 - \frac{\gamma-1}{\gamma+1}u_\infty^2\right)^{\gamma/(\gamma-1)} \ ,$$

$$p_1 = -\gamma\rho_\infty u_\infty u_1 \ , \quad p_2 = \frac{p_1^2}{\gamma p_\infty} - \frac{\gamma\rho_\infty}{2}(2u_\infty u_2 + u_1^2) \ , \tag{4.62}$$

$$p = p_\infty + p_1 x^{-1} + p_2 x^{-2} + \dots \ .$$

It follows from (4.61,62) that the first three terms of the asymptotic expansion for the function p are independent of ψ (the dependence on ψ appears in the subsequent terms of the expansion which can be obtained in a similar way), and that for diminishing x, i.e. in the direction of gas flow in the vicinity of the infinite point, the pressure at all streamlines can only diminish ($p_1 < 0$; $p_2 < 0$ if $p_1 = 0$ and so on).

In the direct problem the constants p_0, p_1 are determined as follows. At a fixed ψ_k we get, from the boundary conditions and (4.61),

$$\rho_\infty u_\infty = q(u_\infty) = (1+\nu)\psi_k r_\infty^{-(1+\nu)} \quad ,$$

$$q(u_\infty) = u_\infty\left[\frac{1}{2}(\gamma+1) - \frac{1}{2}(\gamma-1)u_\infty^2\right]^{1/(\gamma-1)} \quad , \tag{4.63}$$

$$p_\infty = \frac{\pi(u_\infty)}{\pi(1)} \; , \quad p_1 = -\frac{\gamma f_1 p_\infty M_\infty^2}{M_\infty^2 - 1}\left[(1+\nu)\rho_\infty u_\infty \psi_k^{-1}\right]^{1/(1+\nu)} \quad .$$

It follows from (4.63) that in the direct problem, in the case of a given equation of the nozzle contour, the solution is uniquely defined by the parameter ψ_k only. This conclusion remains valid if all the subsequent terms of expansion are taken into consideration.

Asymptotic expansions in the vicinity of the infinite point in the subsonic part of a nozzle, for a viscous gas flow, have been constructed in [4.19].

4.5 Method of Small Perturbations for Flows Close to Radial

Radial flows or flows generated by a source/sink have been considered in Sect.2.2. The existence of a simple analytical dependence (2.149) between the flow velocity and the radial coordinate r substantially facilitates investigation of flows close to radial. Flows similar to those from a sink/source take place in the subsonic and supersonic parts of conical nozzles. The difference between flows in such nozzles and the strictly radial ones is due to the impossibility of realizing a spherical (circular) sonic surface, because in this case the sonic surface is a limiting line [4.20]. In the supersonic part of a nozzle a specially profiled contour may help to realize a strictly radial flow starting from a certain characteristic. However, the presence of nonisentropic and three-dimensional perturbations associated with the processes going on in the combustion chamber and the contour deformations may destroy a strictly radial flow.

Flows close to radial ones should be, naturally, studied by the method of small perturbations, which produces linear equations [4.21]. Let small perturbations be superposed upon a radial flow and the resultant flow be only slightly different from the initial one. Then any given flow parameter will be sought in the form

$$f = f_0 + \varepsilon f_1 + \varepsilon^2 f_2 + \dots \quad , \tag{4.64}$$

where f_0 is the value of the function in a radial flow and ε is a small parameter. The method of small perturbations is based on the assumption that the functions and their derivatives are infinitesimals of the same order and their squares may be neglected. It is convenient to solve the problem in a spherical (or, in the two-dimensional case, polar) system of coordinates r, θ, φ with the origin at the centre of a source, Fig.4.7.

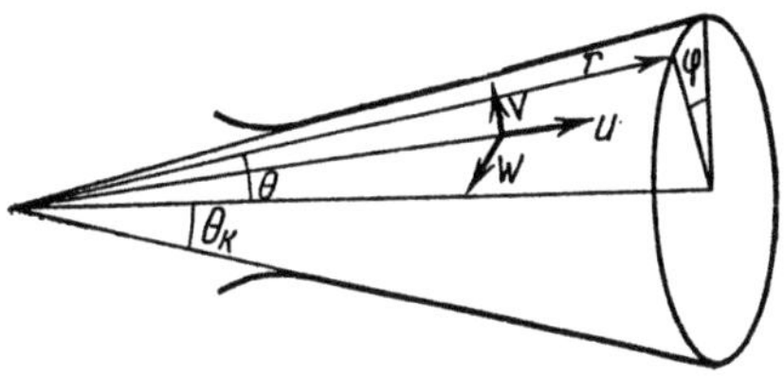

Fig.4.7. Spherical coordinate system. θ_k is semiangle at the vertex of the cone

Let us write down the equations of motion, continuity and entropy conservation in the form

$$v\left[\frac{\partial}{\partial\theta}(w\sin\theta) - \frac{\partial v}{\partial\varphi}\right] + u\left[\frac{\partial}{\partial r}(rw\sin\theta) - \frac{\partial u}{\partial\varphi}\right]$$

$$= \frac{a^2}{\gamma(\gamma-1)}\frac{\partial S}{\partial\varphi} - \frac{\partial H_0}{\partial\varphi} \quad , \tag{4.65}$$

$$\frac{w}{r\sin\theta}\left[\frac{\partial}{\partial\varphi}(rv) - \frac{\partial}{\partial\theta}(rw\sin\theta)\right] + u\left[\frac{\partial}{\partial r}(rv) - \frac{\partial u}{\partial\theta}\right]$$

$$= \frac{a^2}{\gamma(\gamma-1)}\frac{\partial S}{\partial\theta} - \frac{\partial H_0}{\partial\theta} \quad , \tag{4.66}$$

$$\frac{w}{r\sin\theta}\left[\frac{\partial u}{\partial\varphi} - \frac{\partial}{\partial r}(rw\sin\theta)\right] + \frac{v}{r}\left[\frac{\partial u}{\partial\theta} - \frac{\partial}{\partial r}(rv)\right]$$

$$= \frac{a^2}{\gamma(\gamma-1)}\frac{\partial S}{\partial r} - \frac{\partial H_0}{\partial r} \quad , \tag{4.67}$$

$$\frac{1}{r^2}\frac{\partial}{\partial r}(r^2\rho u)+\frac{1}{r\sin\theta}\frac{\partial}{\partial\theta}(\rho v\sin\theta)$$

$$+\frac{1}{r\sin\theta}\frac{\partial}{\partial\varphi}(\rho w)=0\quad, \tag{4.68}$$

$$u\frac{\partial S}{\partial r}+\frac{v}{r}\frac{\partial S}{\partial\theta}+\frac{w}{r\sin\theta}\frac{\partial S}{\partial\varphi}=0\quad, \tag{4.69}$$

where u, v, w are the velocity vector projections on the axes of the spherical coordinate system r, θ, φ, referred to the critical sound speed. The equations of motion (4.65-67) may be obtained from the Euler equations with the help of Bernoulli's integral and equations of entropy conservation and are often called equations in Crocco's form [4.22].

Let us present the required functions in the form

$$\begin{aligned} u &= u_0+\varepsilon u_1\ , & \rho &= \rho_0+\varepsilon\rho_1\ ,\\ v &= v_0+\varepsilon v_1\ , & S &= S_0+\varepsilon S_1\ ,\\ w &= w_0+\varepsilon w_1\ , & H_0 &= H_{00}+\varepsilon H_{01}\ ,\\ p &= p_0+\varepsilon p_1\ . \end{aligned} \tag{4.70}$$

Here values with the subscript 0 refer to parameters of a radial flow and the εf_1 are small additives whose squares may be neglected. In a radial flow $v_0=w_0=0$, while u_0, p_0, ρ_0 depend only on the coordinate r and s_0, H_{00} are constant throughout the flow.

Let us derive linearized equations for the additives. It follows from (4.69) that

$$\frac{\partial}{\partial r}(\varepsilon S_1)=0\qquad\text{or}\qquad \varepsilon S_1=\varepsilon S_1(\theta,\varphi)\quad. \tag{4.71}$$

These relations mean that entropy remains constant along any streamline of a radial flow, i.e. along rays issuing from the pole. Specification of entropy at the initial line $r=r_H=\text{const}$ fully defines entropy perturbations throughout the flow. It follows from (4.67) that entropy also stays constant along streamlines of a three-dimensional source $\varepsilon H_{01}=\varepsilon H_{01}(\theta,\varphi)$ and is fully defined by its value at the initial line $r=r_H=\text{const}$.

The linearized equations of motion (4.65,66) have the form

$$\sin\theta\frac{\partial}{\partial r}(rw)-\frac{\partial u}{\partial\varphi}=\frac{a_0^2}{u_0}\left[\frac{1}{\gamma(\gamma-1)}\frac{\partial S}{\partial\varphi}-\frac{1}{a_0^2}\frac{\partial H_0}{\partial\varphi}\right]\quad, \tag{4.72}$$

$$\frac{\partial}{\partial r}(rv)-\frac{\partial u}{\partial\theta}=\frac{a_0^2}{u_0}\left[\frac{1}{\gamma(\gamma-1)}\frac{\partial S}{\partial\theta}-\frac{1}{a_0^2}\frac{\partial H_0}{\partial\varphi}\right]\quad, \tag{4.73}$$

where ε and subscripts of the perturbation quantities are omitted.

Linearizing Bernoulli's equation and the entropy equation, we get

$$\frac{p}{p_0} = -\left(\gamma \frac{u_0 u}{a_0^2} + \frac{S}{\gamma - 1} - \frac{\gamma H_0}{a_0^2}\right) , \quad \frac{\rho}{\rho_0} = \frac{1}{\gamma}\left(\frac{p}{p_0} - S\right) . \tag{4.74}$$

Using (4.74) we reduce the continuity equation (4.68) to the form

$$\frac{\kappa}{\kappa - u_0^2}\left\{r \frac{\partial}{\partial r}\left[ru(u_0^2 - 1)\right] - ru + ruu_0^2\left(\frac{6}{\kappa} - 1\right)\right\}$$
$$= \frac{1}{\sin\theta}\left[\frac{\partial}{\partial\theta}(rv \sin\theta) + \frac{\partial}{\partial\varphi}(rw)\right] + \frac{4ru_0^3 H_0}{H_{00}(\gamma - 1)(u_0^2 - 1)(\kappa - u_0^2)} , \tag{4.75}$$

where $\kappa = (\gamma + 1)/(\gamma - 1)$ and u_0 is, as earlier, the velocity coefficient.

Let us write down the boundary conditions of the problem. If a nozzle contour is represented by $\theta = \theta_k + \varepsilon F(r,\varphi)$, then the boundary condition of impermeability at the contour can be written in the form

$$v = ru_0 \frac{\partial F}{\partial r} \quad \text{for} \quad \theta = \theta_k . \tag{4.76}$$

At the line $r = r_H$ or $u_0 = u_{0H}$ we specify

$$u = f_1(\theta,\varphi) , \quad v = f_2(\theta,\varphi) , \quad H_0 = f_3(\theta,\varphi) , \quad S = f_4(\theta,\varphi) . \tag{4.77}$$

In the case of axisymmetric perturbations the boundary condition $v = 0$ at $\theta = 0$ is given.

To solve the system (4.72,73,75) we shall use the Fourier method of separation of variables. Due to periodicity of the solution in φ, the unknown functions may be represented in the form

$$f = \sum_{k=0}^{K} [f_{1k}(r,\theta)\cos k\varphi + f_{2k}(r,\theta)\sin k\varphi] , \tag{4.78}$$

where f is any of the functions except w for which we have

$$w = \sum_{k=0}^{K} [w_{1k}(r,\theta)\sin k\varphi - w_{2k}(r,\theta)\cos k\varphi] .$$

Substitution of the series (4.78) into (4.72,73,75) gives equations for the determination of $f_{1k}(r,\theta)$, $f_{2k}(r,\theta)$. For instance, we have the following system of equations for the determination of the functions u_{1k}, v_{1k}, w_{1k}:

$$\sin\theta \frac{\partial}{\partial r}(rw_{1k}) + ku_{1k} = -\frac{ka_0^2}{u_0}\left[\frac{S_{1k}(\theta)}{\gamma(\gamma - 1)} - \frac{H_{01k}(\theta)}{a_0^2}\right] , \tag{4.79}$$

$$\frac{\partial}{\partial r}(rv_{1k}) - \frac{\partial u_{1k}}{\partial \theta} = \frac{a_0^2}{u_0}\left[\frac{1}{\gamma(\gamma - 1)}\frac{\partial S_{1k}(\theta)}{\partial \theta} - \frac{1}{a_0^2}\frac{\partial H_{01k}(\theta)}{\partial \theta}\right] , \tag{4.80}$$

$$\frac{\kappa}{\kappa - u_0^2}\left\{r\frac{\partial}{\partial r}\left[ru_{1k}(u_0^2 - 1)\right] - ru_{1k} + ru_0^2u_{1k}\left(\frac{6}{\kappa} - 1\right)\right\}$$

$$- \frac{1}{\sin\theta}\left[\frac{\partial}{\partial \theta}(rv_{1k}\sin\theta) + krw_{1k}\right] = \frac{4ru_0^3H_{01k}(\theta)}{(\gamma - 1)(\kappa - u_0^2)(u_0^2 - 1)H_{00}} . \tag{4.81}$$

A similar system may be obtained for u_{2k}, v_{2k}, w_{2k}. Differentiating (4.81) with respect to r and using (4.79,80) we get

$$r\frac{\partial}{\partial r}\left\{\frac{\kappa}{\kappa - u_0^2}\left[r\frac{\partial}{\partial r}[(u_0^2 - 1)Z_1] - Z_1 + u_0^2Z_1\left(\frac{6}{\kappa} - 1\right)\right]\right\}$$

$$- \frac{1}{\sin\theta}\frac{\partial}{\partial \theta}\left(\sin\theta\frac{\partial Z_1}{\partial \theta}\right) - \frac{k^2Z_1}{\sin^2\theta}$$

$$= \frac{ra_0^2}{u_0\sin\theta}\left\{\frac{\partial}{\partial \theta}\left[\frac{\sin\theta}{\gamma(\gamma - 1)}\frac{\partial S_{1k}}{\partial \theta}\right] - \frac{\kappa}{2a_0^2}\frac{\partial}{\partial \theta}\left[\frac{\sin\theta}{H_{00}}\frac{\partial H_{01k}}{\partial \theta}\right]\right\}$$

$$- \frac{rk^2a_0^2}{u_0\sin^2\theta}\left[\frac{S_{1k}}{\gamma(\gamma - 1)} - \frac{\kappa}{2a_0^2}\frac{H_{01k}}{H_{00}}\right] + r\frac{\partial}{\partial r}\left[\frac{4ru_0^3H_{01k}}{(\gamma - 1)(u_0^2 - 1)(\kappa - u_0^2)H_{00}}\right] , \tag{4.82}$$

where $Z_1 = ru_{1k}$.

The right side of (4.82) is known, because $S_{1k}(\theta)$, $H_{01k}(\theta)$ are known throughout the flow. Therefore, we shall consider the solution of a uniform system corresponding to potential perturbations ($S_{1k} = H_{01k} = 0$). After finding the general solution of the uniform equation, one can easily find the solution of the nonuniform equation. The solution of the uniform equation will be sought in the form

$$Z_1 = \Theta(\theta)R(r) . \tag{4.83}$$

Substituting (4.83) into (4.82) and separating the variables, we obtain the Legendre equation for $\Theta(\theta)$,

$$\frac{1}{\sin\theta}\frac{d}{d\theta}\left(\sin\theta\frac{d\Theta}{d\theta}\right) + \left[n(n + 1) - \frac{k^2}{\sin^2\theta}\right]\Theta = 0 , \tag{4.84}$$

and the Hein equation for $R(r)$,

$$r \frac{d}{dr}\left(\frac{\kappa}{\kappa - u_0^2}\left\{r \frac{d}{dr}\left[(u_0^2 - 1)R\right] - R + u_0^2 R\left(\frac{6}{\kappa} - 1\right)\right\}\right) + n(n+1)R = 0 \quad . \quad (4.85)$$

The associated Legendre functions are one of the partial solutions of (4.84):

$$P_n^{(k)}(\cos\theta) = (-1)^k \sin^k\theta \frac{d^k P_n(\cos\theta)}{d(\cos\theta)^k} \quad ,$$

where $P_n(\cos\theta)$ are Legendre polynomials of order n (integer).

The functions $Z_{1n}(r)$ and $Z_{2n}(r)$ which form a complete system, just as the associated Legendre functions [4.23], are solutions of (4.85) and they can be obtained by numerical integration of (4.85). Having partial solutions of (4.82), one can find its general solution. Finally, we have for the determination of u,

$$ru = \sum_{n=1}^{\infty} \sum_{k=1}^{\infty} \{[a_{1kn} Z_{1n}(r) + a_{2kn} Z_{2n}(r)]\cos k\varphi$$

$$+ [b_{1kn} Z_{1n}(r) + b_{2kn} Z_{2n}(r)]\sin k\varphi\} P_n^{(k)}(\cos\theta) \quad . \quad (4.86)$$

Using (4.79,80), we get

$$rv = \sum_{n=1}^{\infty} \sum_{k=1}^{\infty} \{[a_{1kn} F_{1n}(r) + a_{2kn} F_{2n}(r)]\cos k\varphi$$

$$+ [b_{1kn} F_{1n}(r) + b_{2kn} F_{2n}(r)]\sin k\varphi\} \frac{dP_n^{(k)}(\cos\theta)}{d\theta} \quad , \quad (4.87)$$

$$rw = - \sum_{n=1}^{\infty} \sum_{k=1}^{\infty} \{[a_{1kn} F_{1n}(r) + a_{2kn} F_{2n}(r)]k \sin k\varphi$$

$$- [b_{1kn} F_{1n}(r) + b_{2kn} F_{2n}(r)]k \cos k\varphi\} \frac{P_n^{(k)}(\cos\theta)}{\sin\theta} \quad . \quad (4.88)$$

Here a_{1kn}, b_{1kn} are constants found, together with the numbers k and n, from the initial and boundary conditions. The functions Z_{1n}, Z_{2n}, F_{1n}, F_{2n} are interrelated by means of

$$Z_{1n} = r \frac{dF_{1n}}{dr} \quad , \quad Z_{2n} = r \frac{dF_{2n}}{dr} \quad .$$

Thus, we have constructed a solution for the case of potential three-dimensional flows close to radial.

Let us consider in detail the case of axisymmetric perturbations. Equations (4.86,87) become

$$u = \frac{1}{r} \sum_{n=1}^{\infty} [a_{1n} Z_{1n}(r) + a_{2n} Z_{2n}(r)] P_n(\theta) \quad ,$$
$$v = \frac{1}{r} \sum_{n=1}^{\infty} [a_{1n} F_{1n}(r) + a_{2n} F_{2n}(r)] \frac{dP_n(\theta)}{d\theta} \quad , \tag{4.89}$$

where $P_n(\theta)$, $dP_n(\theta)/d\theta$ are the Legendre functions and their derivatives satisfying the equation

$$\frac{d^2 P_n}{d\theta^2} + \mathrm{cotan}\theta \frac{dP_n}{d\theta} + n(n+1)P_n = 0 \quad . \tag{4.90}$$

We shall introduce a new variable $x = u_0^2$ and rewrite the Hein equation in the form of the following system of two equations:

$$4x \frac{dZ_n}{dx} + \left[\frac{4x}{x-1} + \frac{x(6\kappa^{-1} - 1) - 1}{1 - x\kappa^{-1}}\right] Z_n + n(n+1)F_n = 0 \quad , \tag{4.91}$$

$$\frac{dF_n}{dx} - \frac{x - 1}{4x(1 - x\kappa^{-1})} Z_n = 0 \quad . \tag{4.92}$$

To find solutions Z_{1n}, Z_{2n}, F_{1n}, F_{2n}, the system (4.91,92) is numerically integrated at the initial conditions,

$$Z_{1n}(x_H) = r_H \quad , \qquad F_{1n}(x_H) = 0 \quad ,$$
$$Z_{2n}(x_H) = 0 \quad , \qquad F_{2n}(x_H) = r_H \quad , \tag{4.93}$$

i.e. the fundamental solutions are found and with their help any partial solution can be constructed.

The constants a_{1n}, a_{2n}, b_{1n}, b_{2n} are found with the help of boundary and initial solutions for u and v. For the case of axisymmetric potential perturbations given at the initial line, and in the absence of perturbations of the contour [$F = 0$ in (4.76)] we have

$$v = 0 \quad \text{at} \quad \theta = 0 \quad \text{and} \quad \theta = \theta_k \quad , \tag{4.94}$$

$$u = f_1(\theta) \quad , \quad v = f_2(\theta) \quad \text{at} \quad x = x_H \quad . \tag{4.95}$$

To satisfy the conditions (4.94), the numbers n are chosen in such a way that

$$dP_n/d\theta = 0 \quad \text{at} \quad \theta = 0 \quad , \quad \theta = \theta_k \quad . \tag{4.96}$$

Approximate values of n satisfying (4.96) can be found from the relation

$$n = \pi\left(m - \frac{3}{4}\right)\theta_k^{-1} - \frac{1}{2} \quad , \quad (m = 2,3,\dots) \quad . \tag{4.97}$$

This can be done because the following asymptotic expansions of the associated Legendre functions are valid at large n:

$$\begin{aligned} P_n(\cos\theta) &= \sqrt{\frac{2}{\pi n \sin\theta}}\left[\left(1 - \frac{1}{4n}\right)\sin\varphi - \frac{1}{8n}\cot\text{an}\theta \cot\text{an}\varphi\right] \quad , \\ P_n^{(k)}(\cos\theta) &= (-n)^k \sqrt{\frac{2}{\pi n \sin\theta}} \sin\left(\varphi + \frac{k\pi}{2}\right) \quad , \end{aligned} \tag{4.98}$$

if $u \geqslant k$, where $\varphi = (n + 1/2)\theta + \pi/4$.

It follows from (4.97) that while θ_k diminishes the order of the first Legendre function satisfying (4.96) grows. For example, at $\theta_k = \pi/12$, $n = 14$, while at $\theta_k = \pi/18$, $n = 22$.

Using the completeness of the system of the associated Legendre functions satisfying (4.96), we shall represent $f_1(\theta)$, $f_2(\theta)$ in the form

$$f_1(\theta) = \sum_{n=1}^{\infty} f_{1n}P_{1n}(\theta) \quad , \quad f_2(\theta) = \sum_{n=1}^{\infty} f_{2n}\frac{dP_n}{d\theta} \quad , \tag{4.99}$$

where

$$f_{1n} = \frac{\int_0^{\theta_k} f_1(\theta)P_n(\theta)\sin\theta\, d\theta}{\int_0^{\theta_k} P_n^2(\theta)\sin\theta\, d\theta} \quad , \tag{4.100}$$

$$f_{2n} = \frac{\int_0^{\theta_k} f_2(\theta)(dP_n/d\theta)\sin\theta\, d\theta}{\int_0^{\theta_k} (dP_n/d\theta)^2 \sin\theta\, d\theta} \quad .$$

Now we can define constants in (4.89) and write down the final formulas for the calculation of $u(r,\theta)$, $v(r,\theta)$:

$$\begin{aligned} u(r,\theta) &= \frac{1}{r}\sum_{n=1}^{\infty} [f_{1n}Z_{1n}(r) + f_{2n}Z_{2n}(r)]P_n(\theta) \quad , \\ v(r,\theta) &= \frac{1}{r}\sum_{n=1}^{\infty} [f_{1n}F_{1n}(r) + f_{2n}F_{2n}(r)]\frac{dP_n(\theta)}{d\theta} \quad . \end{aligned} \tag{4.101}$$

To solve the problem in the presence of perturbations of the nozzle contour and in the absence of perturbations at the initial line, we must satisfy the nonuniform boundary condition at the nozzle wall and the uniform one at the initial line. This is associated with the necessity of solving the complicated problem of finding eigenvalues of the Hein function. This may be avoided if one assumes that at the initial line there are perturbations produced by those of the contour. Then the boundary condition at the nozzle contour may be used to define its shape.

Let us analyze the properties of the solution obtained. In [4.21] an approximate solution of the system (4.91,92) has been constructed, based on substitution of the original system by a reference system with known partial solutions, and asymptotic expansions in terms of negative power of n have been obtained with the help of these partial solutions. It has been found that such partial solutions for (4.91,92) were Bessel functions having in the supersonic part an oscillatory character and an amplitude diminishing with the growth of u_0. Figures 4.8,9 show the dependence of the perturbed velocity components on u_0 in the supersonic part of conical nozzles with different angles θ_k expressed in terms of the functions Z_{1n}, Z_{2n}, F_{1n}, F_{2n}. The first of the conditions (4.93) is evidently equivalent to specifying at the initial line ($r = r_H$) perturbations of the radial velocity component u only, while the second one does the same for the normal velocity component v. It follows from Figs.4.8,9 that the perturbations propagate vibrationally and the amplitude of the vibrations diminishes along the nozzle's length. The number of zeros of functions grows as the angle θ_k becomes smaller. The oscillatory character of the propagation of the perturbations is due to reflections of compression/rarefaction waves from the walls and the axis of a nozzle. Evidently, the number of such reflections grows when the angle θ_k becomes smaller, thus leading to the growth of the frequency of vibrations.

The solution presented is valid for the subsonic part of conic nozzles. However, propagation of perturbations from subsonic to supersonic values of u_0 is not oscillatory but monotonically attenuating.

The method of small perturbations allows a solution to be found for the case $\theta_k = 0$, i.e. for flows in a cylindrical tube. In this case perturbations are superposed upon a translational flow with the Mach number M_0. In cylindrical coordinates z, r, φ the solution for a supersonic flow has the form

$$u = - \sum_{n=1}^{N} \sum_{k=1}^{N} \frac{n}{\beta_0} \left[\left(a_{1n} \cos \frac{nz}{\beta_0} - a_{2n} \sin \frac{nz}{\beta_0} \right) \cos k\varphi \right.$$

$$\left. - \left(b_{1n} \cos \frac{nz}{\beta_0} - b_{2n} \sin \frac{nz}{\beta_0} \right) \sin k\varphi \right] J_k(nr) \quad ,$$

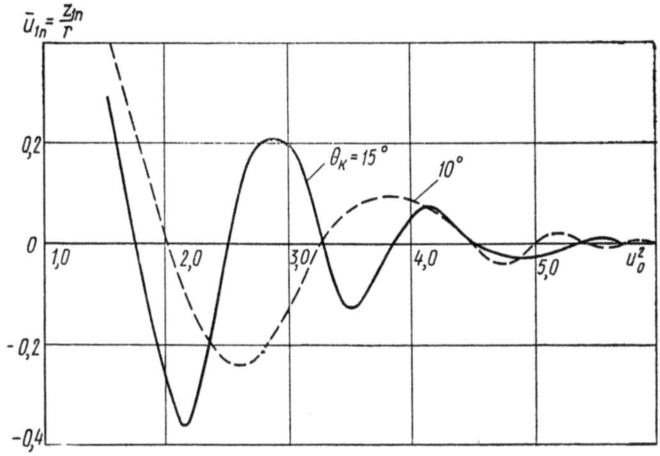

Fig.4.8

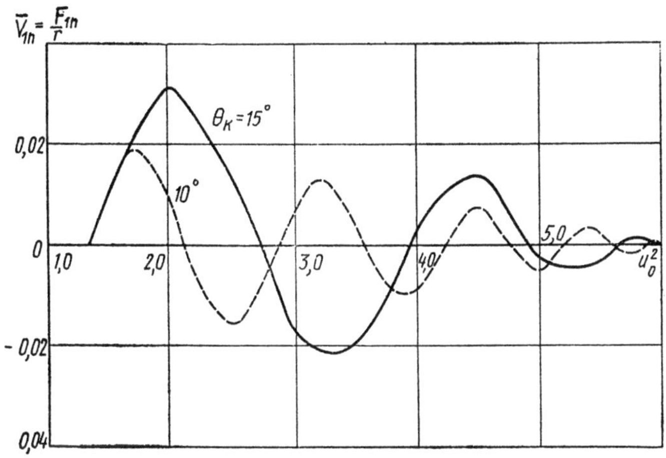

Fig.4.9

Fig.4.8. Variation of the perturbed velocity component $\bar{u}_{1n} = Z_{1n}/r$ in a conic nozzle for different θ_k. u_0 is the velocity coefficient

Fig.4.9. Variation of the perturbed velocity component $\bar{v}_{1n} = F_{1n}/r$ in a conic nozzle for different θ_k. u_0 is the velocity coefficient

$$v = -\sum_{n=1}^{N}\sum_{k=1}^{N}\left[\left(a_{1n}\sin\frac{nz}{\beta_0} + a_{2n}\cos\frac{nz}{\beta_0}\right)\cos k\varphi - \left(b_{1n}\sin\frac{nz}{\beta_0} + b_{2n}\cos\frac{nz}{\beta_0}\right)\sin k\varphi\right]\frac{dJ_k(nr)}{dr}\ , \qquad (4.102)$$

$$w = \frac{1}{r} \sum_{n=1}^{N} \sum_{k=1}^{N} k \left[\left(a_{1n} \sin \frac{nz}{\beta_0} + a_{2n} \cos \frac{nz}{\beta_0}\right) \sin k\varphi \right.$$

$$\left. + \left(b_{1n} \sin \frac{nz}{\beta_0} + b_{2n} \cos \frac{nz}{\beta_0}\right) \cos k\varphi \right] J_k(nr) \quad .$$

Here $\beta_0 = \sqrt{M_0^2 - 1}$ and $J_k(nr)$ is the kth-order Bessel function of the first type. The formulas also demonstrate the oscillatory character of the perturbations. The frequency of oscillations grows as M_0 tends to unity. However, there is no amplitude attenuation in this case.

The numbers n are defined from the boundary condition of impermeability of the wall, according to which $v = 0$ at $r = 1$. This means that the required values of n are roots of the equation

$$\left. \frac{dJ_k(nr)}{dr} \right|_{r=1} = 0 \quad , \tag{4.103}$$

which can be found approximately from the relationship

$$n = (k + 0.5 + 2i) \frac{\pi}{2} - \frac{7}{4\pi} (k + 0.5 + 2i)^{-1} \quad , \quad (i = 0,1,\ldots) \quad . \tag{4.104}$$

5. Nozzles of Jet Engines

A nozzle is a necessary part of any jet engine and its degree of perfection influences very much the performance of a jet vehicle as a whole. In fact, even in dense atmospheric layers, the jet thrust makes 90% of the total force acting upon a jet vehicle, while aerodynamic forces constitute only 10%. With the progress in aerospace technolgoy the cost of a unit of a nozzle's specific impulse increases: for a long-range intercontinental ballistic missile covering 10.000 km at the specific impulse of 310 s, the latter's increase by 1% results in an additional 500 km or so. For satellite lauchers an increase in the specific impulse means either a greater orbital altitude or an enlarged payload transferred to a fixed orbit. An increase of the specific impulse by 1 s (i.e., an 0.3% increment) results either in an increase of the orbital altitude from 550 to 600 km (i.e. almost a 10% increase) or in an enlargement of the payload by ~1.3%.

Therefore, the accuracy with which a nozzle's characteristics are determined must be high. One-dimensional approximations allows a nozzle's impulse and its mass flow to be calculated to an accuracy of about 5% (Sect.2.2). Thus, experimental and theoretical investigations of flows in nozzles of jet engines must ensure an accuracy an order of magnitude higher, i.e. the flow integral characteristics must be defined to an accuracy of 0.1% and the local characteristics to an accuracy of no less than 0.5%. Methods incapable of ensuring such an accuracy may be used only for a qualitative description of flow. Design and energy constraints make the task of choosing and calculating an appropriate nozzle quite a complicated search problem for the thrust extremum under a large number of constraints. When choosing an optimum nozzle, a specific trajectory of the missile and the nozzle's size and weight limitations must be taken into consideration. *Glushko* in 1930 was the first to proposed profiled nozzles for jet engines [5.1].

A typical nozzle of a modern jet engine is presented in Fig.5.1. For such nozzles considerable longitudinal and lateral gradients of the gasdy-

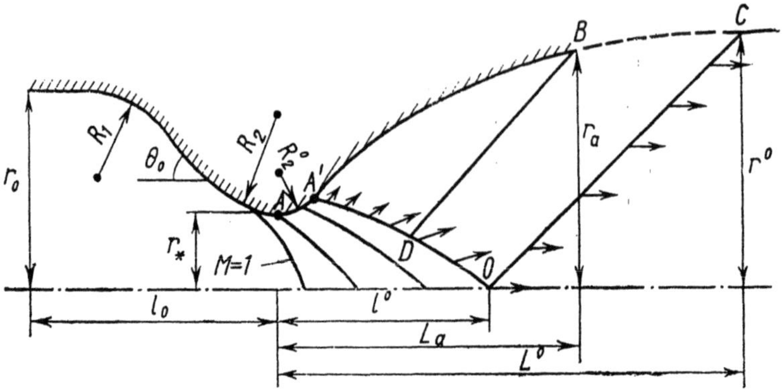

Fig.5.1. Diagram of a jet nozzle. AA'0 and A'0C are acceleration and straightening parts, respectively

namic parameters in the transonic and supersonic parts of a nozzle, necessary to accelerate the flow to high supersonic velocities in the shortest distance possible, are characteristic. A nozzle may be divided into three parts: subsonic, accelerating and straightening.

5.1 Flow Peculiarities in the Subsonic and Transonic Parts of a Nozzle

The geometry of sub- and transonic parts of modern nozzles may be characterized by the following parameters, Fig.5.1: the length of the subsonic part ℓ_0, the curvature radii R_2 and R_2^0 of the contour in the vicinity of the throat on its sub- and supersonic sides respectively, the angle of inclination of the conic part θ_0, the radius of matching of the conic and cylindrical parts R_1, the radius of the throat r_* and the radius of the initial cross-section r_0 (or the degree of compression $n = r_0/r_*$). If the nozzle generatrix has an inflection in the throat, then it is called a nozzle with a corner point. The character of the flow in the sub- and transonic parts of a nozzle is defined to a great extent by the values of these parameters. Evidently, the adiabatic exponent γ also belongs to the number of substantial parameters which characterize the gas properties. Its change in the sub- and transonic regions of flow due to physico-chemical processes is insignificant because the temperature of the combustion products changes only slightly between the initial cross-section of a nozzle and its throat. The results presented in this section have been mainly obtained by numerical solution of the inverse problem of the theory of nozzles [5.2].

5.1.1 Nozzle with a Rectilinear Surface of Transition

The practical interest in nozzles with a rectilinear sonic line is associated with profiling of nozzles for wind tunnels and jet engines. In this case profiling of the supersonic part does not depend on that of the subsonic part, because a rectilinear sonic line is simultaneously a characteristic of the first and the second families. It is practically impossible to specify a priori the contour of a nozzle which ensures straightness of the sonic line, because for that purpose the nozzle's contour and all the streamlines at the throat must have first, second and third derivatives all zero. On the other hand, calculation of a Laval nozzle with a rectilinear transition line in the framework of the inverse problem is rather a simple task. To calculate it in the case of plane and axisymmetric flows, it is necessary and sufficient to specify at the symmetry axis a velocity distribution with a zero first derivative at the sonic point.

Let us consider the results of solving the inverse problem with the velocity specified at the nozzle axis in the form

$$W = W_\infty + (1 - W_\infty)(1 + Ax^2)^{-1} \quad , \tag{5.1}$$

$$W = W_\infty + (1 - W_\infty)(1 + A|x|^3)^{-1} \quad , \tag{5.2}$$

where W is the absolute value of velocity referred to the critical sound speed a_* which is often denoted by λ and called the velocity coefficient. Some results of calculation are presented in Fig.5.2 (here and below $\gamma = 1.4$, if not specified otherwise). The figure shows streamlines and lines of $W = \text{const}$ and the change of W along the streamline $\psi = 0.06$ at the initial conditions (5.1) on the axis. It is seen that in the region where $W > 0.2$ and $\psi > 0.02$, the flow departs considerably from the one-dimensional calculation. In the vicinity of the rectilinear sonic line lying in the plane $x = 0$, at $r > 0.25$ a local supersonic zone and the second sonic line form in accordance with the solution of Sect.4.3. The flow in the zone is isentropic. As follows from Fig.5.2, the streamlines in the vicinity of the sonic line are almost parallel to the axis. If the velocity at the symmetry axis is specified by (5.2), then a local supersonic zone forms in the flow field; however, the second sonic line has a common point with the rectilinear sonic line at infinity.

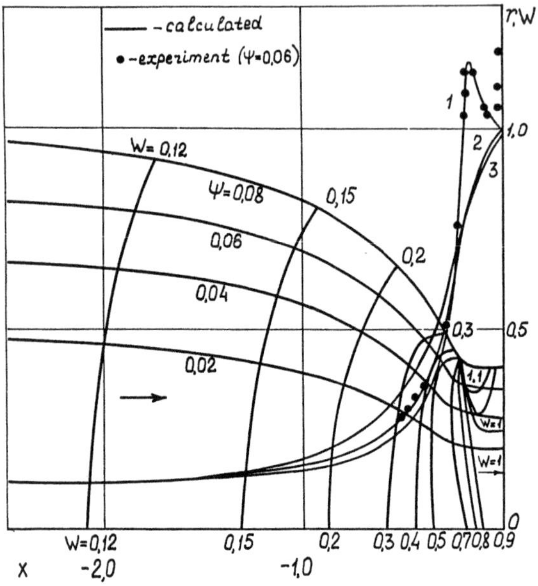

Fig.5.2. Flow in a nozzle with a rectilinear sonic line. Along the axis, velocity is distributed according to (5.1) at $A=10$, $W_{\infty}=0.1$. Velocity distribution: (*1*) streamline $\psi=0.06$, (*2*) one-dimensional theory, (*3*) symmetry axis

The existence of local supersonic zones has been confirmed for nozzles, corresponding to the streamlines $\psi=0.02$ and 0.06, Fig.5.2. The values of velocity presented in the figure have been obtained from calculations based on experimentally measured pressure. The calculated and experimental data agree well where the viscosity influence is negligible. Some discrepancy can be seen in almost cylindrical transonic regions where the influence of viscosity is strong. It should also be noted that despite the presence of a large positive pressure gradient in the transonic region, the boundary layer does not separate.

Numerical and experimental studies of flows in nozzles with a rectilinear sonic line show that the subsonic part of such nozzles computed in a framework of ideal fluid is rather long and the effects of viscosity prevent practical realization of a rectilinear sonic line. Therefore, neither these nozzles, nor the nozzles recommended in [5.3] are used in jet engines.

5.1.2 Nozzle with a Curvilinear Surface of Transition

If the derivative dW/dx is nonzero at the centre of a nozzle, then the transition surface is curvilinear and the subsonic and supersonic parts of a nozzle must be calculated simultaneously. Let us consider flows in a nozzle for the following velocity distribution at the symmetry axis:

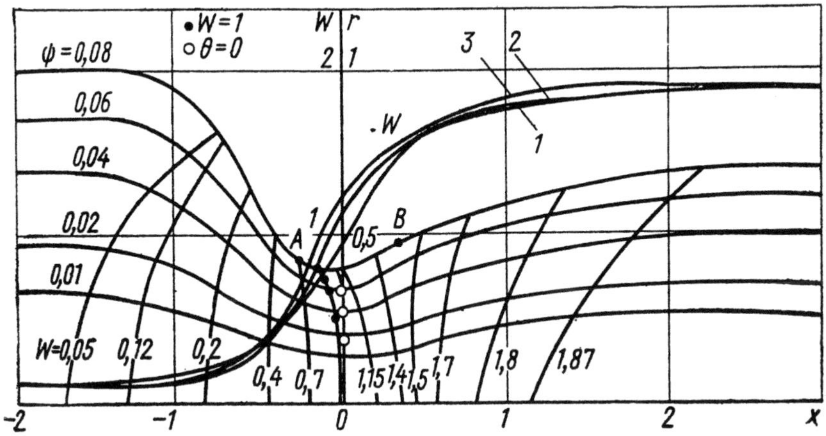

Fig.5.3. Flow field in a nozzle with a curvilinear sonic line. Along the axis the velocity is distributed according to (5.3) at $W_\infty = 0.1$, $\bar{W}_\infty = 1.9$, $b = 3.5$. Velocity distribution: (1) streamline $\psi = 0.06$, (2) one-dimensional theory, (3) symmetry axis

$$W = 1 + \frac{(1 - W_\infty)(\bar{W}_\infty - 1)(e^{bx} - 1)}{(1 - W_\infty)e^{bx} + (\bar{W}_\infty - 1)} , \tag{5.3}$$

where W_∞, $\bar{W}_\infty$ are asymptotic values of velocity at infinity in the subsonic and supersonic regions and the parameter b is proportional to the velocity derivative at the nozzle's centre.

Figure 5.3 shows a typical Laval nozzle flow with a curvilinear surface of transition. It is seen that at $\psi > 0.01$ the flow differs substantially from one-dimensional both in the subsonic and the supersonic regions, and the sonic line and the line $\theta = 0$ are displaced upstream from the nozzle centre in such a way that the sonic point at the contour moves upstream from the throat. The configuration of the lines $W = \text{const}$ is defined by the curvature of the streamlines and, in particular, of the nozzle contour. The inclination of these lines changes in the vicinity of the points where the streamline curvature changes (the points A and B in Fig.5.3). The segments with a negative curvature correspond to the lines $W = \text{const}$ inclined downstream from the axis, and here, as follows from (2.146), the pressure increases in the direction from the axis to the nozzle contour. In contrast, the segments with a positive contour curvature correspond to the lines $W = \text{const}$ inclined upstream from the axis and here the pressure decreases in the direction from the axis to the nozzle contour. The flow acceleration at

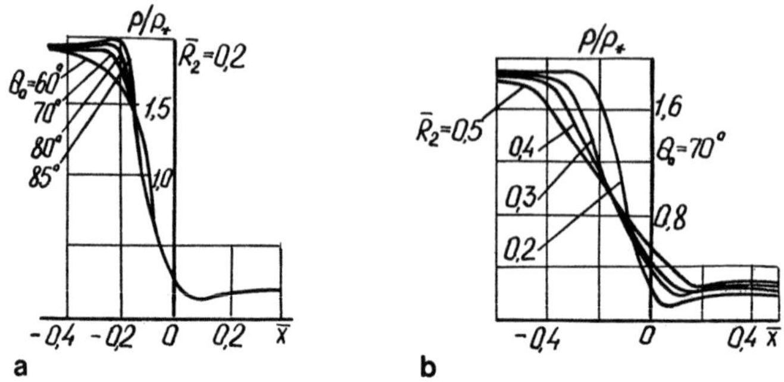

Fig.5.4a,b. Dependence of p/p_* on $\bar{x} = x/r_*$ at a nozzle contour at various $\bar{R}_2$ and θ_0

the streamlines increases with decrease of the curvature radius. At the axis the flow accelerates much more slowly.

Let us describe some results of the parametric study of flow in the sub- and transonic parts of axisymmetric nozzles [5.2,4-6].

It has been found that the value of the radius R_2 (Fig.5.1) has the greatest influence upon the flow parameters in the transonic region. For instance, at $R_2 > 0.2r_*$ and $\theta_0 < 85^\circ$ the flow parameters depend mainly on R_2 and only slightly on θ_0. Typical pressure distributions at the contour for different $\bar{R}_2 = R_2/r_*$ and θ_0 are presented in Fig.5.4.

Figure 5.5 demonstrates the influence of an adiabatic exponent upon the flow in the sub- and transonic regions. We see that for a fixed nozzle the

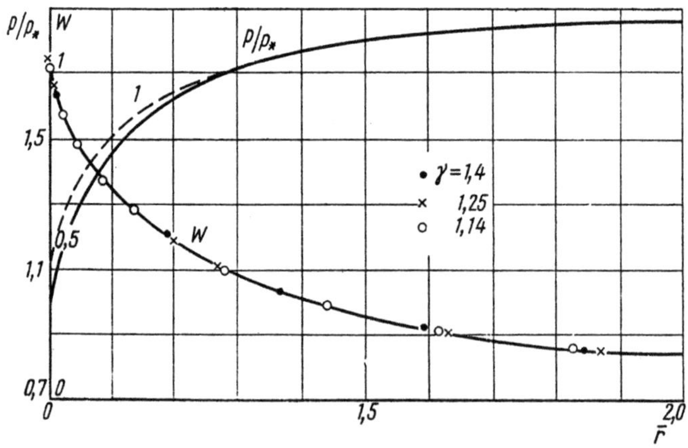

Fig.5.5. Dependence of $p/p_*(\gamma = 1.4)$ and W on $\bar{r} = r/r_*$ at a nozzle contour at $\bar{R}_2 = 2$:(1) one-dimensional theory

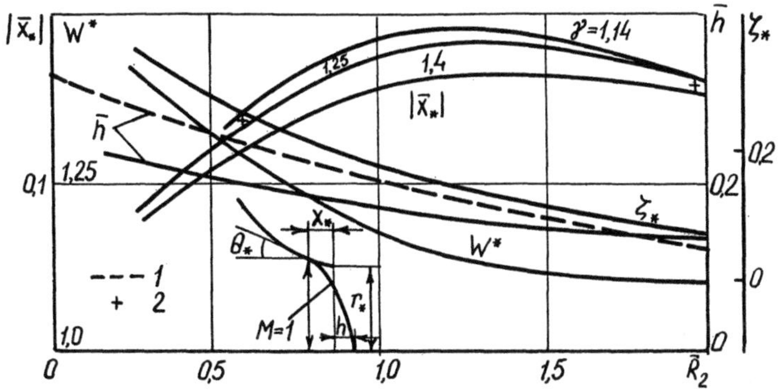

▲ Fig.5.6. Effects of $\bar{R}_2$ and γ upon the sonic line abscissas at the axis and the contour, the value of the velocity W^* at the contour at the throat and the wall angle at the sonic point ($\xi_* = \tan\theta_*$: (*1*) experiment [5.6], (*2*) experiment [5.7]

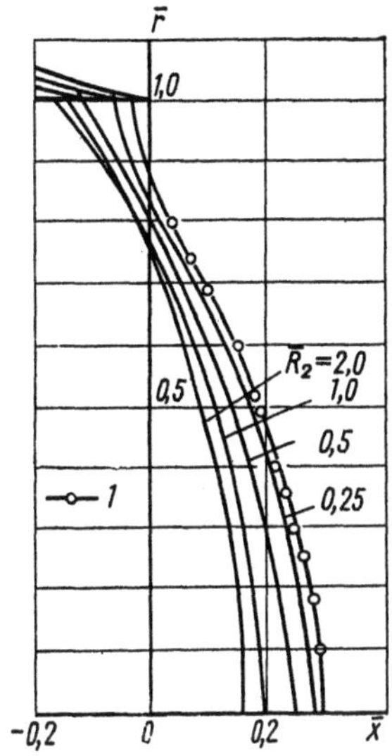

◄ Fig.5.7. Sonic lines in a nozzle with different $\bar{R}_2$ [5.2]: (*1*) calculated and experimental from [5.4] at $\bar{R}_2 = 0.1$

values of the velocity W in flows with different γ practically coincide, though the values of p/p_* may be different. This agrees with the one-dimensional theory. Thus, if a flow field has been calculated for a fixed nozzle geometry and γ, the results can also be used for other values of γ.

The dependence of some gasdynamic parameters on $\bar{R}_2$ and γ are presented in Fig.5.6. Figure 5.7 shows sonic lines for nozzles with different values of $\bar{R}_2$. It is seen that at small $\bar{R}_2$ the sonic lines have a point of inflection.

If a flow field has been obtained as a result of a numerical or theoretical solution, then the positions of characteristics can be obtained. Figure 5.8 shows the form of the second family characteristics issuing from a point at the throat contour and distributions of $\xi = \tan\theta$ and M along

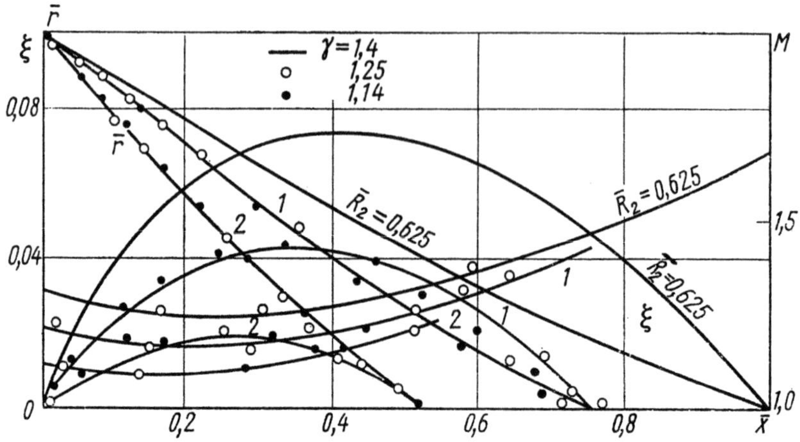

Fig.5.8. The second family characteristics issuing from the throat contour point and the distribution of ξ and M along them

them for different $\bar{R}_2$ and γ. Nonmonotonic change of ξ along characteristics ($\xi = 0$ at the axis and the contour) results in nonmonotonic change of the number M both in plane and in axisymmetric flows. For plane flows the fact that M changes nonmonotonically is evident and follows directly from the compatibility equation (2.150).

The velocity derivative $dW/d\bar{x}$ at the centre of a nozzle cannot exceed a certain limiting value $(dW/d\bar{x})_*$, corresponding to a flow with a corner point. Calculations show that this value is 0.6-0.7 for axisymmetric flows [5.2] and about 0.55 for plane flows [5.8]. The calculated value of $(dW/d\bar{x})_*$ in an axisymmetric case agrees well with experimental data obtained by measuring the velocity at the axis in a flow past a corner point [5.6].

The above results have been obtained for Meyer flows characterized by a transition surface crossing the flow from the nozzle wall to its axis. Let us follow transformation of flow in the convergent part of a nozzle while the pressure drop increases, i.e. transition from the Taylor flow to Meyer flow. Figure 5.9 shows sonic lines in a nozzle with $\bar{R}_2 = 2$ in the case of Taylor flow for velocity specified at the nozzle axis in the form

$$W = W_\infty + (W_0 - W_\infty)(1 + Ax^2)^{-1} \quad , \tag{5.4}$$

where W_0 is the velocity at the nozzle centre [5.9]. Such a specification of the velocity at the nozzle axis means that the contour will be symmetric about the plane $x = 0$. At $W_0 \approx 0.75$, a supersonic zone appears near the contour whose size increases with increasing pressure drop, i.e. with increase

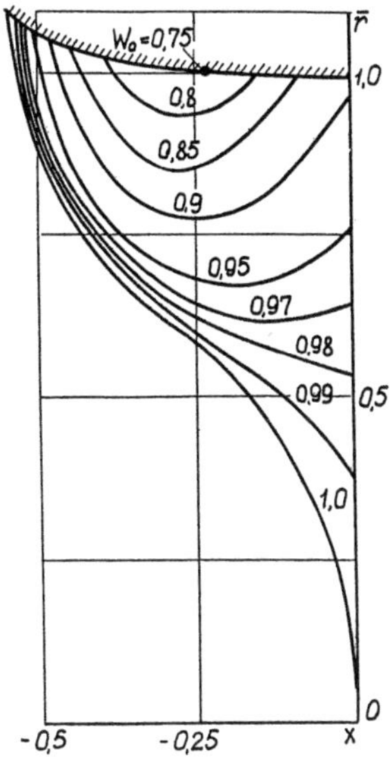

Fig.5.9. Sonic lines in an axisymmetric Taylor flow

of W_0 and hence W_∞. At $W_0 = 1$ the supersonic zone stretches from the nozzle contour to its axis.

Unlike the Taylor solution [see (4.45)], which is valid for large $\bar{R}_2$ ($\bar{R}_2 > 5$) only and according to which the supersonic zone always includes the nozzle throat, the local supersonic zones are at first isolated and symmetric about the throat and capture it at $W_0 > 0.88$ only. The value $W_0 = 0.75$ corresponds to $W_\infty = 0.094$, while $W_0 = 1$ corresponds to $W_\infty = 0.1014$. In the Meyer flow for the nozzle under consideration $W_\infty = 0.1015$, i.e. transition from the Taylor flow with a closed sonic line at $W_0 = 1$ to the Meyer flow occurs while the velocity changes by 0.1% only.

5.1.3 Local Deceleration Zones

Sometimes in a generally accelerating flow there appear local deceleration zones which may result in unwanted boundary layer separation and weakening of the nozzle's heat protection. A deceleration zone appears at small R_1 in the subsonic part of a nozzle in the neighbourhood of the matching point of the "radius" and cylindrical parts. Dependence of the pressure coefficient C_p on the nozzle length in this region is shown in Fig.5.10 comparing experimental [5.6] and numerical [5.2] data. At small R_1 the flow decelerates in the vicinity of matching of the cylindrical and convergent parts, while for $R_1 > 2r_0$ it accelerates monotonically. The region of constant C_p at

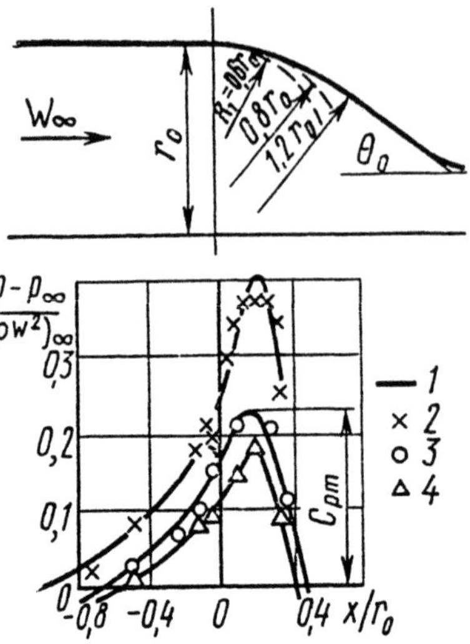

Fig.5.10. Pressure coefficient in the vicinity of the point of conjugation of cylindrical and conical segments of a nozzle in its subsonic part, obtained (*1*) numerically and (*2*) experimentally at $R_1=0.6\ r_0$, (*3*) $R_1=0.8\ r_0$ and (*4*) $R_1=1.2\ r_0$

$R_1=0.6r_0$ obtained experimentally corresponds to boundary layer separation. It has been found that the maximum positive value of C_{pm} depends mainly on R_1 and only slightly on θ_0, n and γ. At the same time, for any n, a short enough nozzle contour can be chosen containing no deceleration zones in the region.

Above, we have already given examples of the formation of local deceleration zones in the transonic region in the neighbourhood of a rectilinear sonic line. It is characteristic that they are local supersonic zones situated upstream from the throat. The study of flow in the zones carried out in the framework of the ideal fluid model has, in solving the direct problem for nozzles whose contours were obtained from solution of the inverse problem, demonstrated the stability of such flows with respect to small perturbations under the condition that the geometry of the nozzle contour be ensured with high precision. Experimental studies have also shown the existence of deceleration zones, though the presence of a boundary layer distorts the calculated flow pattern.

Upstream from the sonic point at the contour, local deceleration zones may also form if the angle of the contour inclination changes sharply over a short distance (Fig.5.4a). A local deceleration zone at large values of θ_0 appears upstream from the throat in the neighbourhood of the matching point of the contour's conic part with a circular arc [5.2,5]. Similar deceleration zones appear in the supersonic region if the gas has been accelerated at slightly curved parts of a nozzle contour in the neighbourhood

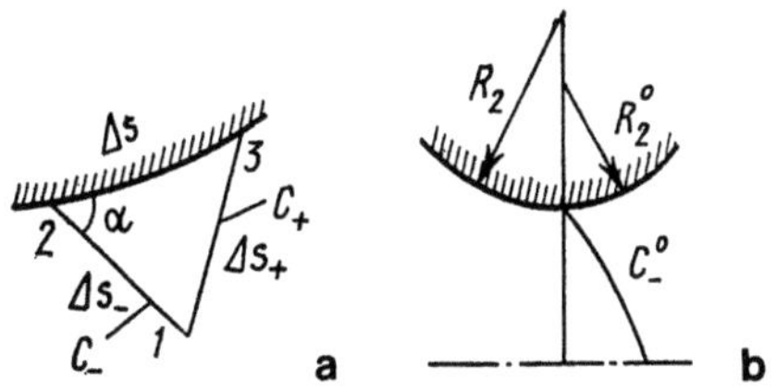

Fig.5.11a,b. Geometry of flow deceleration zones

of the point where a circular arc matches with a conic part of the contours (Fig.5.4b) [5.2,5,7].

The possibility of deceleration zones appearing in the supersonic flow near the point of curvature discontinuity and in the vicinity of a corner point can be predicted analytically. Let us consider the triangle 123 (Fig. 5.11a). Along the characteristics of the first family 1-3 and of the second family 1-2 compatibility equations (2.150) are satisfied, and can be written in the form

$$d\zeta \mp L(M,\zeta)dM \pm N(\zeta,M,r)ds = 0 \quad , \tag{5.5}$$

where

$$L(M,\zeta) = \frac{2\sqrt{M^2-1}(1+\zeta^2)}{M[2+(\gamma-1)M^2]} \quad , \tag{5.6}$$

$$N(\zeta,M,r) = \frac{\nu\zeta(1+\zeta^2)}{rM} \quad , \quad \zeta = \tan\theta \quad .$$

Parameters at the points 1, 2, 3 are related by

$$M_3 = M_2 + \left(\frac{dM}{ds}\right)_w \Delta s \quad , \quad \zeta_3 = \zeta_2 + \left(\frac{d\zeta}{ds}\right)_w \Delta s \quad , \tag{5.7}$$

$$\zeta_1 = \zeta_2 + \left(\frac{d\zeta}{ds}\right)_- \Delta s_- = \zeta_2 + \left[N - L\left(\frac{dM}{ds}\right)_-\right]\Delta s_- \quad , \tag{5.8}$$

$$M_1 = M_2 + \left(\frac{dM}{ds}\right)_- \Delta s_- \quad . \tag{5.9}$$

Here $(dM/ds)_w$ and $(dM/ds)_-$ are derivatives of M along the wall and the second family characteristics respectively. Using the relationships along the first family characteristics,

$$\zeta_3 - \zeta_1 - L(M_3 - M_1) + N\Delta s_+ = 0 \quad , \tag{5.10}$$

we get

$$\left(\frac{dM}{ds}\right)_w = \frac{M}{\sqrt{M^2-1}}\left(\frac{dM}{ds}\right)_- + \frac{1}{L}\left(\frac{d\zeta}{ds}\right)_w \quad . \tag{5.11}$$

Equation (5.11) allows $(dM/ds)_w$ to be defined through the known quantities $(dM/ds)_-$ and $(d\zeta/ds)_w$. In particular, if the second derivative is discontinous at a certain point at the contour, the discontinuity of the derivative dM/ds is defined by

$$\left[\frac{dM}{ds}\right]_w = \frac{1}{L}\left[\frac{d\zeta}{ds}\right]_w \quad . \tag{5.12}$$

Calculation of derivatives at the points of curvature discontinuities with the help of (5.11) shows that there can appear deceleration zones at small $\bar{R}_2^0$ in the supersonic region near the points of matching with the conic part of a nozzle (Sect.7.2).

It follows, in particular, from (5.12) that at the contour of a cylindrical part adjacent to the contour in the form of a circular arc the flow must decelerate, because $(dM/ds)_- < 0$ at the second family characteristics, issuing from the point of matching (Fig.5.8). Figure 5.12a presents the change of M along the cylindrical part. At the initial portion of the cylindrical part M decreases (compression flow), then it increases and further diminishes again. Such behaviour of M is in accordance with the oscillatory character of propagation of small perturbations in nozzles and the solution presented in Chap.4 [see (4.102)]. The presence of a compression wave results in the intersection of the second family characteristics issuing from the contour's points and the formation of shock waves. Considerable slow-down of the flow at the initial portion of the cylindrical part may result in boundary layer separation. The flow at the cylindrical part can also be calculated by the method of small perturbations with the help of (4.102). Then the number M_0 is calculated from the continuity equation $q(M_0) = \mu$, where μ is the flow rate coefficient. The Mach number M_0 increases with decreasing $\bar{R}_2$ and equals 1.05, 1.1, 1.2 at $\bar{R}_2 = 2$, 1, 0.5, respectively. If the nozzle contour in the vicinity of the throat is formed by matched circular arcs of radii R_2 and R_2^0, Fig.5.11b, then according to (5.12) the pressure gradient remains negative despite the discontinuity of the derivative dM/ds.

Let us now study the flow beyond the corner point in the nozzle throat. Consideration of the characteristic triangle in the vicinity of the corner point allows a differential equation to be obtained for derivatives along characteristics of the rarefaction wave issuing from the corner point, which makes it possible to define the values of these derivatives at any characteristic by means of the known values at a certain initial characteristic. Let us write this equation for plane and axisymmetric flows without derivation:

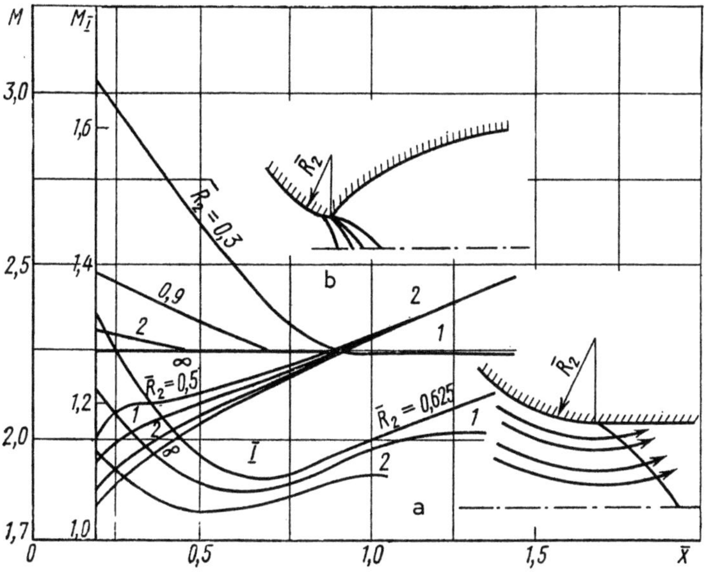

Fig.5.12. (**a**) Number M distribution in a nozzle with a cylindrical head and (**b**) at contours of plane and axisymmetric nozzles with a corner point at different $\bar{R}_2$: (*1*) $M_0 = 4$, $\gamma = 1.4$, $\nu = 0$, (*2*) $M_0 = 5.5$, $\gamma = 1.4$, $\nu = 1$

$$\frac{\partial}{\partial\alpha}\left(\frac{d\alpha}{dx}\right) = A\,\frac{d\alpha}{dx} + B \quad , \qquad \text{where} \tag{5.13}$$

$$A = \frac{\gamma + 1}{\gamma - \cos 2\alpha}\left[\frac{\cos(\theta + \alpha)}{\sin 2\alpha \cos(\theta - \alpha)} + \frac{2 \sin 2\alpha}{1 + \cos 2\alpha}\right] \quad ,$$

$$B = \frac{\nu \sin(\theta + \alpha)}{2 \cos(\theta - \alpha)} \quad .$$

This equation has the solution

$$\frac{d\alpha}{dx} = \exp\left(\int_{\alpha_0}^{\alpha} A\,d\alpha\right)\left[\left(\frac{d\alpha}{dx}\right)_0 + \int_{\alpha_0}^{\alpha} B \exp\left(-\int_{\alpha_0}^{\alpha} A\,d\alpha\right)d\alpha\right] \quad , \tag{5.14}$$

where $(d\alpha/dx)_0$ are the values of $d\alpha/dx$ at a certain initial characteristic of the second family.

As an initial characteristic, a characteristic of the second family is taken, issuing from a point of the contour in the throat of a nozzle without a corner point, whose contour up to the throat coincides with the contour of a nozzle with a corner point. Now we can calculate $(d\alpha/dx)_0$ at the characteristic [where $(d\alpha/dx)_0 > 0$ according to Fig.5.8]. Then, from (5.14) one can define $d\alpha/dx$ at various characteristics of the second family issuing from the corner point. It can be shown that the derivative $d\alpha/dx$ decreases sharply with increase of θ at the corner point. It reverses its sign at $\bar{R}_2 = 2$ and becomes negative at $\theta \approx 3^{\circ}$. For nozzles with a corner point and a uniform flow at the exit cross-section this corresponds to $M_0 \approx 1.5$, where M_0 is the Mach number at the exit. For such nozzles the function $(d\zeta/ds)_w$ is positive at the contour in the vicinity of the corner point; therefore, it follows from (5.11) that starting from certain values of $\theta = \theta_*$ the pressure gradient beyond the corner point must be negative. The value of θ_* depends on $\bar{R}_2$ and increases with decreasing $\bar{R}_2$. Derivatives $d\alpha/dx$ along characteristics of the second family calculated from (5.14) and along the contour with a corner point, from (5.11), show that at $\bar{R}_2 > 0.5$ and $M_0 > 1.5$ the derivative $d\alpha/dx$ (and hence dp/dx) is negative at the contours of axisymmetric nozzles with a corner point, while dM/dx is positive.

It follows from (5.14) that in the axisymmetric case the derivative $d\alpha/dx$ decreases quicker than in the plane case (due to the second term in the square brackets) and may even become negative at a positive value of $(d\alpha/dx)_0$. In the plane case, $(d\zeta/ds)_w \approx 0$ in the neighbourhood of the corner point, while in the axisymmetric case $(d\zeta/dx)_w > 0$. Thus, flow in a plane nozzle may decelerate in the neighbourhood of the corner point [see (5.12)] and this has been confirmed by straightforward calculations of flows in plane and axisymmetric nozzles. This was the purpose of the determination of gasdynamic parameters at characteristics of the rarefaction wave generated in the flow past a corner point, done with the help of the data obtained while calculating flow in a nozzle without a corner point. The data thus obtained at the characteristics of the rarefaction wave were later used to calculated flow in a given nozzle chosen from the family of nozzles with a corner point and a uniform characteristic at the exit, calculated under the condition of the straightness of the sonic line. Figure 5.12b shows some typical results. It is seen that the Mach number distributions for nozzles with curvilinear and rectilinear sonic lines differ significantly only in the nearest vicinity of the corner point ($\bar{x} < 1$), i.e. in accordance with the known fact of rapid attenuation of initial perturbations in supersonic nozzles. In an axisymmetric case (unlike the plane one) the presence of a curvilinear sonic line does not result in the appearance of a deceleration zone in the neighbourhood of the corner point.

The existence of a deceleration zone in a plane nozzle with a corner point and a uniform flow in the exit cross-section in the presence of a curvilinear sonic line can also be shown in the following way. Let us consider Fig.5.13

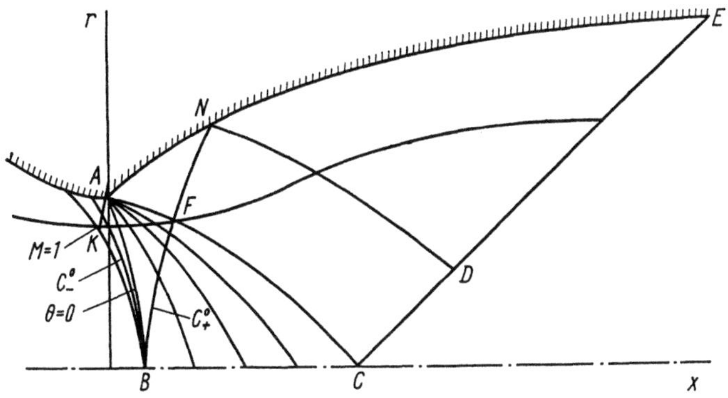

Fig.5.13. Flow in a plane nozzle with a corner point

which shows schematically the flow field of the nozzle. At the boundary characteristic of the rarefaction wave AC we have, from (2.150),

$$f(M) + \theta = f(M_0) \quad , \tag{5.15}$$

where M_0 is the Mach number at the rectilinear characteristic CE and f(M) is the Prandtl-Meyer angle. On the other hand, since the point A is a corner point, the flow parameters here satisfy the relationship for characteristics of the first family and

$$f(M_A) - \theta_A = - \theta_K > 0 \quad , \tag{5.16}$$

since the point K lies at the sonic line where the angle between the velocity vector and the symmetry axis is negative. From (5.15) and (5.16) we have

$$f(M_A) = \frac{1}{2} [f(M_0) - \theta_K] \quad . \tag{5.17}$$

At the first family characteristic BN issuing from the nozzle centre, $f(M) - \theta = 0$, while at the characteristic ND, $f(M) + \theta = f(M_0)$. Hence

$$f(M_N) = \frac{1}{2} f(M_0) \tag{5.18}$$

at the point N. It follows from (5.17,18) that $M_A > M_N$, i.e. the Mach number M decrease in the course of transition from the point A to the point N. If $\theta_k = 0$, as in the case of a nozzle with a rectilinear sonic line, then the point A coincides with the point N and the gasdynamic parameters at the nozzle contour beyond the corner point change monotonically.

The characteristic CE is rectilinear, therefore the flow in the region ACE is a simple wave and $M_N = M_F$. Thus, $M_A > M_F$, and since $M_F < M_0$, the Mach number changes nonmonotonically along the characteristic change of M at the nozzle contour. The minimum value of the Mach number at each characteristic of the fan is attained at the points of intersection with the characteristic BN. Note that the distance from the point A to the point N does not exceed $0.7r_*$. If the streamlines lying below the point F are used to form the nozzle contour, then there will be no deceleration zones at the lines.

5.1.4 Optimum Shape of the Subsonic Part of a Nozzle

The subsonic and transonic parts of a jet engine nozzle are profiled in such a manner as to ensure absence of flow separation throughout the flow field and to have minimum overall impulse losses for a minimum nozzle length. It follows from the analysis done in the previous subsection that at $\bar{R}_2 > 0.5$ no deceleration zones, and hence no separation zones, develop in the transonic part of an axisymmetric nozzle. On the other hand, the absence of separation near the matching point of the cylindrical and the convergent parts of the subsonic part of a nozzle is ensured if $R_1 > r_0$ and $\theta_0 \leqslant 40^\circ$, despite the presence of local deceleration zones. (It should be pointed out that in the case of a two-layer flow with a lower total pressure in the pre-wall layer compared to the total pressure in the flow core, the conditions for flow separation are more advantageous than in the case of a single-layer flow, and to ensure the absence of separation, R_1 must be enlarged by a factor of 1.5-2, while θ_0 must be somewhat smaller.)

The optimum values of R_2 in nozzles of jet engines are selected in the following way. The highest speed flows take place in the transonic region; therefore, increase of the length of the zone, say, by increasing R_2, results in an increase of friction losses. On the other hand, decrease of R_2 results (Fig.5.7) in a considerable curving of the sonic line and increases nonuniformity of the velocity distribution in the throat, which intensifies the flow nonuniformity at the nozzle exit cross-section and, as a consequence, increases the impulse losses. The losses evidently increase as the radius R_2 decreases.

Thus, the optimum value of R_2 is chosen under the condition that the overall losses due to friction in the subsonic part and the additional flow nonuniformity at the nozzle exit are a minimum. Since the propagation of small perturbations in the supersonic part of a nozzle has an attenuating character (Sect.4.5), increase of the degree of expansion results in a decrease of additional losses due to flow nonuniformity, and the optimum value of R_2

shifts to smaller values. The optimum value of $\bar{R}_2 = R_2/r_*$ for a modern nozzle is approximately equal to 0.8-1.2.

5.1.5 Flow Rate Coefficient and Critical Pressure Drop in Nozzles with Curvilinear Transition Surfaces

The flow rate coefficient μ is an important characteristic of a nozzle. It is defined as the ratio of the true gas flow rate through a throat of given cross-sectional area to the gas flow rate through the same section in the case of a one-dimensional flow with the same p_0 and T_0. The flow rate coefficient differs from unity due to the nonuniformity of flow parameters at the throat, the presence of a boundary layer and nonequilibrium physico-chemical processes.

If the nozzle contour in the vicinity of the throat is a circular arc, then the sonic line is curvilinear and the flow rate coefficient is less than 1 due to the flow field nonuniformity increasing as $\bar{R}_2$ decreases. If the transonic part has such a shape, then the flow rate coefficient depends mainly on $\bar{R}_2$ and only slightly on θ_0 and the adiabatic exponent γ. The dependence of μ on θ_0 is evident at small $\bar{R}_2$ ($\bar{R}_2 < 0.5$) only. The difference in the values of the flow rate coefficient for plane and axisymmetric flows at the same $\bar{R}_2$ is not large, Fig.5.14.

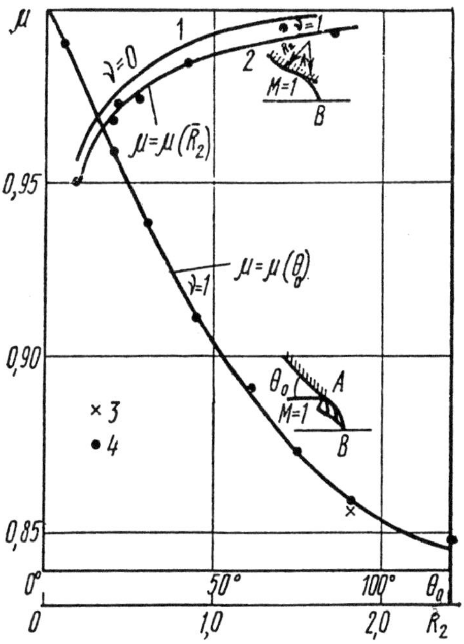

Fig.5.14. Dependence of the flow rate coefficient on θ_0 and $\bar{R}_2$: (*1*) calculation [5.8], (*2*) calculation [5.2], (*3*) calculation at $\nu = 0$ [5.10], (*4*) experiment at $\nu = 1$ [5.6]

In a convergent nozzle with a corner point at the throat, the sonic line has a point of inflection and flow nonuniformity may be substantial and grow with increase of the angle θ_0. The flow rate coefficient in an axisymmetric case may be found from experimental data presented in Fig.5.14. In a plane case at $\theta_0 = 90^\circ$ the calculated value of μ is equal to 0.856 and it remains almost the same in an axisymmetric case.

The data presented for the flow rate coefficient refer to the case of choked flow of a compressible gas when the gas flow rate does not depend on the external pressure. The quantity of numerical and experimental data referring to unchoked flow which depends on the external pressure p_H is much less. Note, in this connection, the numerical data of [5.11], according to which $\mu = 0.611$ in the case of efflux of an incompressible fluid from an orifice with plane walls at $\theta_0 = 90^\circ$ and $\gamma = 1.4$, and $\mu = 0.745$ for a compressible gas at the critical pressure ratio $p_H/p_0 = 0.528$.

The change of the flow rate coefficient due to the boundary layer can be computed from

$$\Delta\mu = 2\delta^{**} r_*^{-1} H \quad , \quad H = \delta^*/\delta^{**} \quad , \tag{5.19}$$

where δ^*, δ^{**} are the displacement and the momentum thickness respectively. The expression for δ^{**} will be presented below [see (5.45,47)]. For a turbulent boundary layer at a smooth surface, H can be defined from

$$H = \frac{9}{7}\left\{\left[\bar{T}_w\left(1 + \frac{\gamma - 1}{2} M^2\right) + \frac{7}{9}\frac{\gamma - 1}{2} M^2\right] - (1 - \bar{T}_w)\left(1 + \frac{\gamma - 1}{2} M^2\right)\left[\frac{1}{3} + \frac{5}{9}\bar{T}_w(1 + \bar{T}_w)\right]\right\} \quad , \tag{5.20}$$

where $\bar{T}_w$ is the wall temperature referred to the stagnation temperature. Usually the change of the flow rate coefficient due to the boundary layer is 0.1-0.2%. For a laminar boundary layer (at $\bar{T}_w = 1$), $H = 4.4$, 10, 20 at Mach number $M = 1,3$ and 5 respectively. An approximately twofold decrease of H corresponds to a decrease of $\bar{T}_w$ from 1 to 0.2. If, however, the characteristic Reynolds number Re_0 is small and the viscous layer occupies a considerable part of a nozzle cross-section, then the change of the flow rate coefficient due to the change of the Reynolds number may be significant. While $Re_0 = \rho_0 W_{max} r_* \eta_0^{-1}$ (ρ_0, η_0 are the density and the viscosity coefficient at the stagnation point, and W_{max} is the maximum efflux velocity) increases from 10^2 to 10^3, the flow rate coefficient changes from 0.6 to 0.9 [5.12]. The Reynolds number for nozzles of jet engines is usually 3 to 4 orders of magnitude higher.

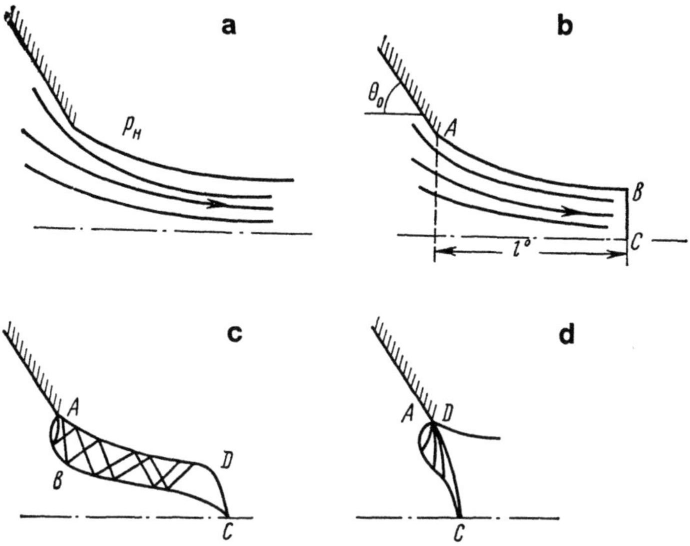

Fig.5.15a-d. Efflux from a plane orifice at various pressure drops

The presence of a curvilinear solic line results in a dependence of the critical pressure drop on the shape of the transonic part, i.e. on the value of R_2 (or θ_0 in the case of a conic convergent nozzle). To make clearer the physical essence of the phenomenon, let us consider efflux of a gas from a plane orifice with rectilinear walls, Fig.5.15. If the jet velocity is subsonic, then the cross-section where the streamlines become parallel and the pressure across the jet is constant lies at infinity, Fig.5.15a. If the velocity at the jet boundary is sonic, i.e. $p_H/p_0 = \pi(1)$, then the section lies at a finite distance ℓ^0 (at $\theta_0 = \pi/2$, $\ell^0 \approx 0.6 r_*$), and the sonic line coincides with the line ABC in Fig.5.15b; in this case the distance ℓ^0 increases with decreasing θ_0 [5.13]. If we now decrease the external pressure in such a way that p_H/p_0 becomes smaller than $\pi(1)$, then the boundary of the jet and sonic line ABC acquire the form presented in Fig.5.15c. The flow at the corner point A expands to the external pressure. Naturally, the waves issuing from the corner point are the rarefaction waves which reflect from the sonic line in the form of compression waves. If the external pressure is close to the critical one, i.e. $p_H/p_0 \approx \pi(1)$, then the Mach waves cross the space between the sonic line and the jet surface many times. At the jet surface, the compression waves issuing from the sonic line convert into rarefaction waves; hence, only rarefaction waves approach the sonic line. The action of the jet upon the sonic line ceases downstream from the characteristic DC issuing from the sonic point at the axis. In the course of a further drop in external pressure the qualitative picture does not change until the characteristic DC hits the jet boundary downstream from the corner point A. With decrease of the external pressure the sonic line AC becomes shorter and the sonic point at the symmetry axis moves upstream.

As long as the Mach waves issuing from the free surface of the jet reach the sonic line, the external pressure influences the subsonic part of flow. For the same reason the jet flow rate depends on the external pressure even for a supersonic efflux velocity. The dependence of the flow rate on the external pressure ceases only after the pressure drops to the level when the characteristic DC starts directly from the corner point (Fig.5.15d) and no Mach wave issuing from the free surface reaches the sonic line.

An analogous flow pattern can be observed at any nonzero θ_0. At $\theta_0 = 0$, i.e., in the case of efflux through an orifice whose walls are parallel to the axis, the sonic line is rectilinear and normal to the axis. It coincides with a characteristic and after the critical pressure $p_H = p_*$ is attained in the environment, the flow rate does not depend on any further decrease in p_H. Thus, the relative external pressure p_*^0/p_0, at which the gas flow rate ceases to depend on the pressure, is a function of the angle θ_0. For a plane case this dependence can be easily obtained from the compatibility equation. Actually, at the corner point A, according to the compatibility equation (2.150) (Fig.5.15),

$$f(M_A) - \theta_A = \theta_0 \quad . \tag{5.21}$$

At a second family characteristic DC issuing from a sonic point at the axis,

$$f(M) + \theta = 0 \quad . \tag{5.22}$$

If the characteristic DC enters the point A, we have

$$f(M_A^*) = \frac{\theta_0}{2} \quad . \tag{5.23}$$

With M_A^* known from (5.23), we can define $p_*^0/p_0 = \pi(M_A^*)$. Figure 5.16 shows the dependence of the critical pressure drop p_*^0/p_0 on θ_0 and γ. Note that in the case of a plane flow at $\theta_0 = \pi/2$ the flow rate increases from $0.725Q_m$ to $0.856Q_m$ while the external pressure drops from p_* to p_*^0 (here $Q_m = \rho_* a_* r_*$) [5.11].

If the external pressure is below p_*^0, the jet efflux becomes purely supersonic and can be calculated by the method of characteristics.

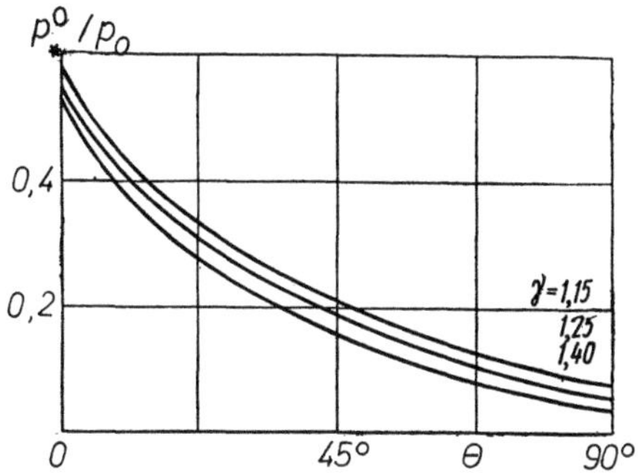

Fig.5.16. Dependence of the critical pressure drop on θ_0 and γ

5.1.6 Profiling of the Acceleration Part of a Jet Engine Nozzle

The shape of the slightly curved part of the nozzle contour AA' (Fig.5.1) defines the length of the acceleration part ℓ^0 and, in the case of a nozzle with a uniform flow field at the exit, the full length of the supersonic part of a nozzle $L^0 = \ell^0 + r_*[q(M_0)]^{-1+\nu/2}\sqrt{M_0^2 - 1}$, where M_0 is the value of the Mach number at the point 0. If the contour of a nozzle throat contains a corner point, the gas velocity increases suddenly. Flowing past a corner point a gas will accelerate to a given Mach number M_0 over a shorter length ℓ^0 compared to any segment AA' of finite curvature [5.14]. However, the use of a nozzle with a corner point is not always advantaneous and, therefore, the curvature radius R_2^0 is usually taken to be equal to $0.2r_*$-$0.5r_*$.

A flow past a corner point contains a rarefaction wave and is calculated by the classic method of characteristics. The initial characteristic of a rarefaction wave in the case of a curvilinear sonic line is defined by calculating the transonic flow region (Fig.5.8 presents typical characteristics and the change of parameters along them). In the case of a rectilinear transition surface the boundary characteristic coincides with the sonic line.

The flow past a corner point in an axisymmetric case in the presence of a plane transition surface has been calculated in [5.15] at various values of the adiabatic exponent γ. If the flow deflects through a large angle the second family characteristics possess maxima, where $\theta = \alpha$. The values of α and θ change along characteristics of a rarefaction wave monotonically: α increases in the direction from the corner point to the axis, while θ decreases. In an axisymmetric flow α and β change in a similar manner, starting from a certain characteristic in the case of a curvilinear transition surface, if $\bar{R}_2 > 0.5$ and $M_0 > 1.5$ (Sect.5.1). In a plane flow the parameters

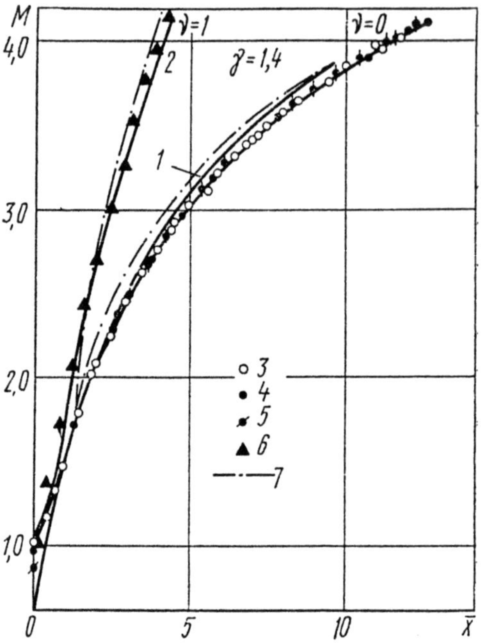

Fig.5.17. Variation of Mach number M at the acceleration part of a nozzle. Efflux from an orifice with rectilinear walls at $\theta_0=\pi/2$: (*1*) calculated, (*7*) calculated from (5.24). Flow past a corner point at various R_2: (*2*), (*3*) calculation at $\bar{R}_2=\infty$, (*4*) calculation at $\bar{R}_2=2$, (*5*) calculation at $\bar{R}_2=0.3$, (*6*) experiment at $\bar{R}_2=2$ [5.6]

at the characteristics of a rarefaction wave change monotonically in the case of a rectilinear sonic line, and nonmonotonically in the case of a curvilinear sonic line.

The intensity of flow acceleration may, generally, depend on the form of the transonic region. However, numerical solution of the plane problem of jet efflux from an orifice with rectilinear walls shows that this dependence is very weak [5.8]. An analogous result has been obtained for the axisymmetric case by means of experimental study of the velocity distribution developed at the axis in a flow past a corner point [5.6]. Figure 5.17 shows the dependence of the Mach number M on $\bar{x}$ for plane and axisymmetric cases. In the axisymmetric case flow is accelerated to a given number M in a shorter distance than in the plane case. In both cases, replacement of a corner point by an arc with a small radius of curvature does not result in a substantial change of length of the acceleration part. Thus, at $\bar{R}_2^0 \leqslant 0.5$, this length increases by only 2 or 3% compared to the case of flow past a corner point. The length of the acceleration part of a nozzle depends not only on the number M, but on the adiabatic exponent too—decrease of γ results in a length increase.

One can obtain an approximate analytical relationship for the dependence of the Mach number M on $\bar{x}$ along the axis of a plane or an axisymmetric jet. In the case of efflux into vaccum, the streamlines and characteristics merge at infinity and each characteristic and the corresponding streamline approaches a straight line inclined to the jet axis at a certain angle. It can be shown that the asymptotic behaviour of the flow parameters at a sufficiently large distance from the exit cross-section corresponds to the flow from a certain equivalent source whose intensity is different for different streamlines. In particular, for a streamline coinciding with the nozzle axis the following approximate dependence of M on the distance from the source is valid:

$$\bar{x} = C[q(M)]^{-1+\nu/2} \quad , \tag{5.24}$$

where C is a constant depending on γ and chosen by trial and error. For instance, for a plane flow $C = 1$ at $\gamma = 1.4$, $C = 1.1$ at $\gamma = 1.25$ and $C = 1.15$ at $\gamma = 1.14$. In the case of an axisymmetric flow for the same values of γ, C is equal to 1.17, 1.2, 1.25 respectively. The results of calculations with the help of (5.24) and by the method of characteristics are compared in Fig.5.17.

5.2 Profiling of Supersonic Parts of Jet Engine Nozzles and Impulse Losses

Profiling of the supersonic part of a nozzle is most important for solving the problem of an optimum nozzle. There are a number of approaches to its solution. Here, attention is concentrated on analyzing two families of nozzles from the point of view of obtaining the maximum thrust under given size restriction and studying various kinds of impulse losses. The question of calculating a nozzle at off-design regimes is considered and methods of numerical and experimental simulation of nozzle flows are described.

5.2.1 Profiling of Supersonic Parts of Jet Engine Nozzles

To profile the supersonic part of a nozzle one has to know the flow parameters in the region of the throat. Above, we described flow peculiarities in the acclerating part of a nozzle and pointed out that acceleration of a flow up to a given velocity over the shortest distance is achieved in the case of flow past a corner point. In practice, to decrease convective heat fluxes or to avoid separation in the case of a two-layer flow, the contour in the neighbourhood of the throat on its supersonic side is formed as a circular arc of small radius $\bar{R}_0^2 = (0.2\text{-}0.5)r_*$. Since the intensity of acceleration is practically independent of the shape of the sonic line, it is usually considered straight while calculating the supersonic contour.

Let the accelerating part of a nozzle be terminated by a characteristic of the rarefaction wave A'O, such that at the axis the given velocity characterized by the number M_0 is achieved, Fig.5.1. From any point of the characteristic A'O a characteristic DB may be drawn in such a way that the gas flow rate across A'D is equal to the gas flow rate across DB. If we solve now the Goursat problem by the method of characteristics and define the flow in the characteristic triangle A'DB, then the streamline A'B passing through the points A' and B may be looked upon as a nozzle contour.

There are a number of methods of specifying the closing characteristic DB [5.6]. Usually, a rectilinear characteristic OC, at which the velocity vector is constant and parallel to the axis, is chosen as the closing characteristic. Sometimes, this characteristic is called uniform. Having calculated a series of nozzle contours at a fixed adiabatic exponent with different values of M_0, one obtains a one-parametric family of nozzles.

As a closing characteristic starting at the symmetry axis, the characteristic OC may be chosen, symmetric with A'C and such that the values of velocity at symmetric points are equal, while the angles of inclination of the velocity vector have opposite signs, Fig.5.18a. The flow will be symmetric about the plane OO', with the gas flow accelerating in the region I and decelerating in the region II. In the plane of symmetry the angle of velocity inclination is zero, and the pressure increases in the direction from the axis. For jet engine nozzles only the segment AO' of the contour should be used. For the role of closing characteristic, a sink characteristic starting from the symmetry axis can be used, Fig.5.18b. In this case one will have a sink flow in the region II with the streamlines converging

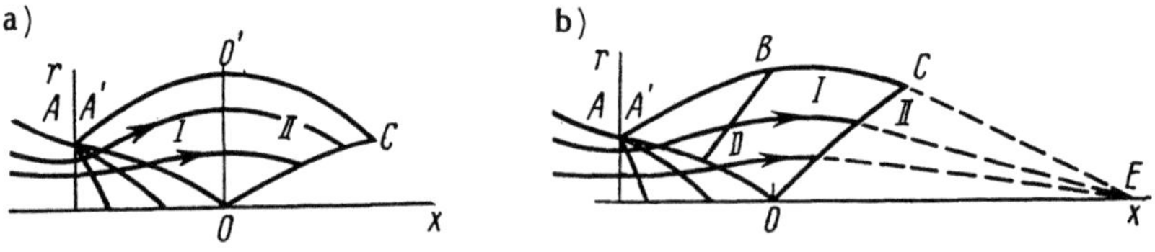

Fig.5.18a,b. Specification of a closing characteristic: (**a**) a symmetric characteristic, (**b**) sink flow characteristic

at the point E. Only a part of the contour, at the end point of which the angle of inclination of the velocity vector is zero, is used for supersonic nozzles of jet engines. The nozzle contours thus designed and passing through fixed points may prove to be the ones which do not develop the maximum thrust or impulse. It should be noted, however, that the nozzles whose contours have been designed on the basis of symmetric, sink or uniform closing characteristics and pass through the same points have practivally equal impulse values [5.6].

To design a nozzle contour, realizing the maximum thrust under given size constraints, one has to solve a variational problem. The problem of designing a contour providing the maximum thrust at a given external pressure has been solved in [5.16,17] in the framework of the ideal gas model with the use of the method of a control contour. In Sect.2.3 we presented the finite relationships which allow the variational characteristic and the corresponding values of the flow parameters to be defined [see (2.171-173)]. These relationships may be used for other problems too, when transfer to a control contour is possible (e.g. for the problem of finding the nozzle contour which realizes the maximum thrust at a given length or a fixed area of the exit cross-section).

The general method of Lagrange multipliers in the framework of the ideal gas model allows, in principle, an optimum nozzle to be constructed without transition to a control contour and, hence, without limitations superposed upon isoperimetric conditions and flow properties by such a transition [5.18, 19]. Using this method it proved possible to account for nonequilibrium physico-chemical processes [5.20,21]. However, its concrete realization is connected with considerable difficulties. At the same time, even in the case of the most complete consideration of all kinds of factors allowable in the framework of the general method of Lagrange multipliers, many important flow qualities remain beyond the scope of the method while solving concrete problems of optimum contour construction; therefore, an optimum contour so obtained proves to be optimum only for a hypothetical flow approximating the real one.

Up to now, it has been impossible to strictly optimize a nozzle as a whole, including sub-, trans- and supersonic flow regions, for a real flow of a viscous heat-conductive gas with diffusion of components and nonequilibrium physico-chemical processes. Therefore, none of the variational problems so far solved have the necessary generality. The natural way in which studies of the optimum profiling of jet engine nozzles develop nowadays is an a priori selection of a family of nozzles close to optimum and based on some phenomenological considerations. For this purpose either a one-parametric family of truncated nozzles with a uniform characteristic [5.6,22] or a two-parametric family of nozzles with a so-called variational (extremal) characteristic, whose coordinates and the corresponding values sought are defined according to [5.16,17,23], are usually chosen.

In practice, the contour of a nozzle with a uniform characteristic is truncated. In fact, shear friction stresses acting upon the walls of a nozzle result in a force opposite to the jet thrust and, since the end part of the contour is practically parallel to the axis, the increment of friction forces here exceeds the increment of the nozzle impulse due to the pressure forces. The length of a truncated nozzle is defined under the condition that the sum of increments of all the forces acting upon a nozzle is zero. Usually a nozzle is shortened even more in order to satisfy the requirements of the engine design as a whole and decrease its weight.

The family of nozzles with a variational characteristic is a two-parametric one because every nozzle belonging to it is characterized by two parameters — the coordinates of the point on a characteristic of the rarefaction wave from which the variational characteristic issues. A uniform characteristic always emanates from the symmetry axis and, therefore, the corresponding family of nozzles is one-parametric. In the case of a plane irrotational flow the two families are equivalent, because in accordance with (2.171-173) the variational characteristic is rectilinear, i.e. it is realized in a simple wave which represents a flow in a plane nozzle with a uniform characteristic. Therefore, every truncated nozzle with a uniform characteristic in the plane case coincides with a certain nozzle designed on the basis of a variational characteristic. In the axisymmetric case the two families of nozzles are identical only if the variational characteristic issues from a point at the symmetry axis and, thus, is rectilinear. Two nozzle contours, one which has been constructed on the basis of a variational characteristic and one which belongs to the family of truncated nozzles with a uniform characteristic, both coming into the same point, do not coincide in the axisymmetric case. To draw the variational characteristic DB passing through the fixed point B one has to vary the coordinates of the point D in the rarefaction wave, Fig.5.18. Therefore, construction of a nozzle contour with a variational characteristic is a more labour-consuming problem than the construction of a nozzle contour passing through the same points but belonging to the family of nozzles with a uniform characteristic.

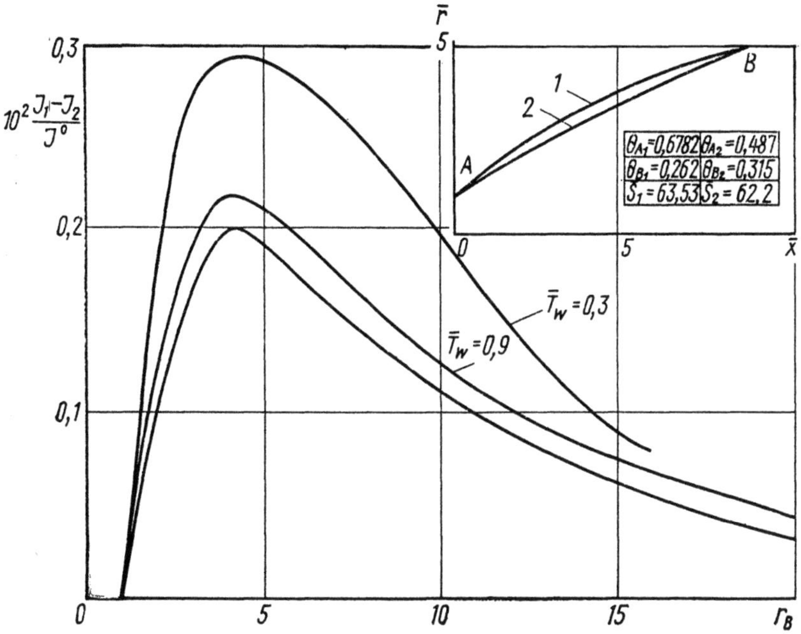

Fig.5.19. Comparison of the impulse J_1 with the impulse J_2 for a nozzle contour (*1*) with a variational characteristic, and for the contour (*2*) of a truncated nozzle with a uniform characteristic. J^0 is the impulse of an ideal nozzle with the same area ratio. The lower curve has been obtained ignoring friction losses

Below we compare nozzles with a variational and a uniform characteristic whose contours pass through the same points. The variational characteristics have been calculated from (2.171-173) and issue from the points of the fan lying to the right of the line defined by (2.176); thus, the flow in the nozzles is isentropic and shock-free throughout. Impulses of a nozzle with a variational characteristic and of a nozzle obtained by truncating one with a uniform characteristic, whose contours pass through the same end points A and B, are compared in Fig.5.19 which also shows the geometry of such nozzles. We see that the contour of the nozzle with a variational characteristic has a bigger angle of inclination at the point A and hence a smaller gas density at the wall. In the absence of friction, impulses of nozzles with a variational characteristic exceed those of truncated nozzles with a uniform characteristic by 0.1-0.2% only.

If friction of the gas on the walls is taken into account, then the advantages of nozzles with a variational characteristic are more pronounced and grow with increasing shear stresses at the walls (which increases as the dimensionless wall temperature $\bar{T}_w = T_w/T_0$ decreases). An additional gain is due to the lower density (and, hence, lower friction) of a gas in the boundary layer of a nozzle with a variational characteristic. Note that the gain does not follow from the variational problem formulation given in [5.16, 17] where optimization has been performed for an ideal gas. It will be smaller in the case of nonequilibrium physico-chemical processes, since the larger the angles the flow turns through at a corner point, the bigger the impulse losses due to the freezing of reactions.

It should be noted that the solutions known for the problem of designing an optimum nozzle contour by the general method of Lagrange multipliers have only an insignificant advantage in comparison with truncated nozzles belonging to the family of nozzles with a uniform characteristic or with the nozzles designed on the basis of variational characteristics, as in [5.16,17]. If the problem under consideration does not allow transition to a control contour, direct methods of the calculus of variations may sometimes prove to be simpler than the general method of Lagrange multipliers.

An example of calculating an optimum contour by direct methods taking into consideration flow nonuniformities and friction in a turbulent boundary layer is presented in [5.6]. Contours of such nozzles are more convex than those of nozzles with a variational characteristic. The impulse of optimum nozzles designed by direct methods is 0.1-0.2% higher than that of nozzles with a variational characteristic. This conclusion has been confirmed by straightforward calculations and special experiments. In every particular case the endpoints of an optimum nozzle should be chosen (a labour-consuming problem in itself) on the basis of a one-parametric family of nozzles with a uniform characteristic. Through the endpoints thus obtained one can draw contours calculated by one of the above methods and compare them by means of direct calculation of the flow in a nozzle with a fixed contour, taking into consideration friction, nonequilibrium physico-chemical processes and so on. Such a comparison allows the optimum nozzle contour to be defined. This is an approximate method of designing an optimum contour because it does not involve solution of the variational problem taking into consideration all isoparametric constraints and physical phenomena.

At the dawn of jet propulsion, conic and "radius" nozzles with a constant pressure at the contour, were used. The contour of the supersonic part of a "radius" nozzle is designed in the form of two matched circular arcs with radii R_2^0 and R_3, the latter being chosen under the condition that the inclination of the velocity vector at the nozzle exit cross-section should be a minimum in order to decrease flow nonuniformities. A nozzle with a constant pressure at the contour is profiled along the boundary of a jet outflowing from a sonic nozzle into a lower pressure environment. The structure of such jets includes a suspended shock wave formed in the vicinity of the corner point.

For very short nozzles whose variational characteristics issue from the points of the fan lying to the left of the line defined by (2.176) there are no shock-free solutions of variational problems. There are no published works containing results of calculations of flow in such nozzles and likewise for their shape. Evidently, this was the reason why the so-called affinely truncated nozzles were proposed to be used as optimum truncated nozzles. The contour of such a nozzle is obtained by dividing the x coordinate of the contour of a nozzle with a uniform characteristic by the same number while leaving the r coordinate unchanged [5.24].

Figure 5.20 shows results of experiments made to compare impulses of the above-mentioned types of nozzles with the same contour end points [5.6]. Here J^0 is the impulse of a truncated nozzle with length $\bar{L}_a = 4.25$ and with the same exit cross-section area and flow rate as those of the considered nozzles. Figure 5.20 also presents the results of comparative testing of two series of nozzles. The first series (I) consisted of truncated nozzles with a uniform characteristic. The second series (II) was composed of affinely truncated nozzles obtained by dividing the x coordinate of the contour of the nozzles of the first series with $\bar{L}_a = 10.5$ by a constant number, while r was left unchanged. We see that, independent of the supersonic part lengths, impulses of nozzles belonging to the first series exceed substantially those of the nozzles of the second series though the angle at the exit cross-section of the latter is smaller. The decrease of the impulse of the affinely

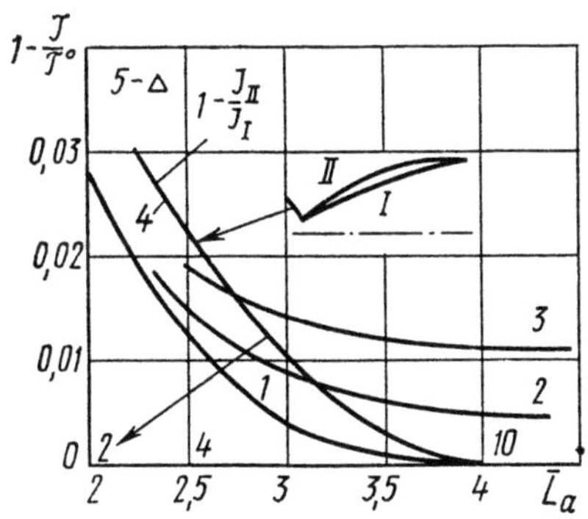

Fig.5.20. Impulse variation for nozzles with equal expansion ratios ($M_0 = 2.6$) as a function of the nozzle length $\bar{L}_a = L_a/r_*$: (*1*) truncated nozzle with a rectilinear characteristic, (*2*) conical nozzle, (*3*) "radius" nozzle, (*4*) divided difference of impulses of truncated nozzles with a uniform characteristic J_I, and of affinely shortened nozzles J_{II}, with equal lengths and exist cross-section areas at $M_0 = 3.25$ (the upper scale refers to $\bar{L}_a$), (*5*) nozzle with constant pressure at the contour

truncated nozzles is due to formation of shock waves observed experimentally. It is seen from Fig.5.20 that the nozzle with a constant pressure at the contour has the worst characteristics, while the truncated nozzle with a uniform characteristic is the best one. The impulse of the conic nozzle is higher than that of the "radius" nozzle.

5.2.2 Impulse Losses

The problem of choosing an optimum nozzle is divided into two stages. In the first stage, having found the initial one-parametric family of contours with known impulses and thrust characteristics, one determines the overall dimensions of an optimum nozzle. This stage is described in detail in Sect.5.3. Simultaneously one can define the optimum contour of a nozzle belonging to the family. In the second stage, having fixed the end points, one can correct the nozzle contour by means of rigorous variational principles. The basic ideas of constructing a nozzle contour passing through fixed points have been presented above.

In the first stage a family of truncated nozzles with a uniform characteristic should be chosen as the basic one. Below we describe the methods of determining impulse losses which are necessary to define the thrust and impulse characteristics. To construct a one-parametric family of nozzles one has to fix a priori either a constant adiabatic exponent γ or the properties of the working medium (if physico-chemical transformations are to be considered). In the latter case calculations of the corresponding family of nozzle contours become quite cumbersome and lose generality, because they have to be done anew for every different fuel.

In this connection, it would be advantageous to have an estimate of the influence of an equilibrium change of a gas property upon the nozzle's shape and the impulse losses. In this case the entropy of a gas remains constant throughout the flow field and, therefore, it becomes possible, with the help of the results of a thermodynamic calculation of the equilibrium expansion in the one-dimensional approximation, to obtain the dependence of the pressure on density $p = p(\rho)$ at $S = \text{const}$, presenting the latter in the form

$$\bar{p} = \bar{\rho}^{n} + a_0 + a_1\bar{\rho} + a_2\bar{\rho}^{-2} \ , \tag{5.25}$$

where $\bar{p} = p/p_0$, $\bar{\rho} = \rho/\rho_0$, $n = \ln \bar{p}_a/\ln \bar{\rho}_a$, and $\bar{p}_a$, $\bar{\rho}_a$ are the dimensionless pressure and density at the exit cross-section of the nozzle.

The dependence (5.25) was used to calculate a supersonic flow in axisymmetric nozzles with a corner point using the method of characteristics. Comparison of these results with the results obtained for a mixture with a constant adiabatic exponent $\gamma = n$ shows that it is necessary to take into account the change of properties of combustion products while calculating the gasdynamic parameters and the nozzle impulse. At the same time, calculation of the impulse loss coefficients and selection of the optimum nozzle sizes may be done with an accuracy sufficient for practical needs, if one uses an average value of the expansion isentrope exponent $n = \text{const}$. Thus, in the case of an equilibrium flow the influence of a change in the combustion products' properties upon the contour geometry and the impulse loss coefficients may be accounted for by an appropriate choice of a particular fuel composition.

Note that the average value of the isentrope exponent n, as defined by (5.25), changes together with $\bar{p}_a$. Evidently, n is not generally equal to the ratio of specific heats $\gamma = c_p/c_v$ though for a perfect gas without phyisco-chemical transformations $\gamma = n$. However, below we shall use γ everywhere, keeping in mind that for complex real compositions it is equal to n. At the same time it should be noted that the value of n cannot be used in the case of complicated reactive mixtures to calculate the impulse, the flow rate and the flow parameters at a nozzle's exit cross-section by means of the gasdynamic functions (2.137-143). In the case of complicated gas compositions the flow parameters at a nozzle's exit cross-section should be determined considering all physico-chemical processes (Chap.6).

Usually, gasdynamic and geometric characteristics of nozzle flows are presented in the form of tables calculated for practically important intervals of γ and M_0. There are some useful analytic formulas which approximate the tables. The work [5.25] contains some approximate formulas allowing the length and side surface to be calculated for nozzles with a corner point and uniform flow, parallel to the axis at the exit cross-section, if γ, the dimensionless exit cross-section radius r_a and the degree of truncation $m = (\bar{r}_a - 1)(\bar{r}_a^0 - 1)^{-1}$ are known. The formulas are

$$\bar{L}_a = \bar{L}^0\{0.32 + 0.68 \exp[-a_L(1 - m)^{\sigma}]\} \quad , \tag{5.26}$$

$$\bar{S}_a = \bar{S}^0 - \bar{r}_a^{2.6}(\bar{r}_a - 1)^{-0.5}[(1 - a_s)^{-0.5} - (m^{-1} - a_s)^{-0.5}] \quad ,$$

where

$$\bar{L}^0 = 3.16\bar{r}_a^{\,1.055+0.1\gamma^4} \quad , \quad a_L = 2.73 + 0.9(\gamma - 1)^{0.125} \quad ,$$

$$\sigma = 0.7 - 0.03\bar{r}_a^{0.5} \quad , \quad \bar{S}^0 = 5.62\bar{r}_a^{2.14+10(\gamma-1)^3} \quad ,$$

$$a_s = 0.984 + 5.5(\gamma - 1)^4 - [0.045 + 2.25(\gamma - 1)^6](\bar{r}_a - 1)^{-0.5} \quad . \qquad (5.27)$$

Here $\bar{L}_a$ is the length of the supersonic part of a nozzle referred to the radius of the critical cross-section and $\bar{S}_a$, $\bar{S}^0$ are the side surfaces of the supersonic parts of the truncated and nontruncated nozzles with a uniform flow at the exit cross-section at $\bar{r}_a = \bar{r}^0$, Fig.5.1. Equations (5.26,27) are valid in the intervals of $\gamma = 1.20\text{-}1.25$, $m = 0.4\text{-}1$ and $p_0/p_a = 20\text{-}10^4$ and values of $\bar{L}_a$, $\bar{S}_a$ can be calculated from them to an accuracy of 5-10%.

Let us proceed with the determination of the major parameters characterizing the integral gasdynamic characteristics of a nozzle. First of all these are the actual gas flow rate and the actual impulse in the direction of the nozzle axis defined by

$$Q = \int_S \rho(\mathbf{Wn})\, dS \quad , \qquad (5.28)$$

$$J = \int_S [\rho\mathbf{W}(\mathbf{Wn}) + p\mathbf{n}]\mathbf{e}_x\, dS \quad , \qquad (5.29)$$

where $\mathbf{n}$ is the unit vector normal to the surface and $\mathbf{e}_x$ is the unit vector parallel to the nozzle axis.

Generally speaking, the last formula defines the momentum flux through the exit cross-section directed along the axis; however, J is usually called the nozzle impulse. For axisymmetric and plane nozzles these formulas at $x = \text{const}$ reduce to

$$Q = 2\pi^\nu \int_0^r \rho W \cos\theta\, r^\nu\, dr \quad , \qquad (5.30)$$

$$J = 2\pi^\nu \int_0^r (\rho W^2 \cos^2\theta + p) r^\nu\, dr \quad . \qquad (5.31)$$

Equation (5.29) allows the impulse to be defined not only in the direction of the nozzle axis, but in any other direction too, if, instead of the unit vector $\mathbf{e}_x$, another unit vector corresponding to the particular direction is used. The actual thrust of the nozzle,

$$R = J - p_H F_a \quad , \qquad (5.32)$$

is defined as the resultant of all the forces acting upon a rocket in the direction of its axis if its outer surfaces are loaded with the external pressure p_H only. The specific impulse and the specific thrust are defined as $J_s = J/Q$ and $R_s = R/Q$ respectively. If $p_H = 0$, then $R_s = J_s$, and the latter is called the vacuum specific impulse. If $p_H = p_a$, the R_s is called the specific thrust at sea-level.

Usually dimensionless coefficients characterizing the flow rate, the impulse or the nozzle thrust are used, defined as the ratios of actual nozzle parameters to those of an ideal nozzle, the latter being a nozzle parameters to those of an ideal nozzle, the latter being a nozzle in which an ideal expansion takes place with entropy remaining constant along any streamline and in the absence of energy exchange with the environment through the nozzle walls. At any cross-section of an ideal nozzle the static pressure is constant and the velocity vector is parallel to the axis; however, the entropy, the total pressure and the total temperature may, generally speaking, vary over the section.

An actual nozzle can be compared with an ideal one in a variety of ways. For instance, initial nonuniform profiles of the total temperature, the pressure and the composition can be substituted by some equivalent averaged profiles conserving, say, constant energy and mass fluxes [5.26]. If one tries to conserve the entropy flux at the same time, then an averaged flow with uniform profiles may, generally, prove to be nonexistent. Also, in the transonic part of flow, the averaging may be ambiguous. Finally, the impulse of an averaged uniform flow may either exceed or be less than that of an nonuniform flow.

Therefore, comparing impulses of ideal and real nozzles, one should take for an ideal one the same nonuniform distribution of the flow parameters at the initial cross-section as in the case of a real nozzle, and use the model of multilayer one-dimensional equilibrium flow to define the ideal nozzle impulse. In this case the actual and the ideal flows will have equal mass, energy and entropy fluxes. Since mixing may sometimes result in an increase of the specific impulse, the impulse of an ideal nozzle may prove to be less than that of a real one. There are some other methods of comparison; however, it is most important that they be fully defined and allow the correct value of the actual impulse of a nozzle to be found by means of dimensionless coefficients.

Comparison should also be made under the condition of equality of the absolute areas of the exit cross-sections of an ideal and a real nozzle. Due to the requirement that the flow rates of the nozzles being compared must be equal, their throat areas may differ; therefore, the impulses must be compared at equal absolute exit cross-sections, not at equal dimensionless exit cross-sections. In this connection the Mach number of an ideal nozzle with a flow rate equal to that of a real nozzle is defined from the relation $q(M) = \mu F_*/F_a$, where μ is the flow rate coefficient of the real nozzle. If the flow rate coefficient is taken into consideration, then the Mach numbers of an ideal nozzle (and, hence, its impulse) prove to be higher compared to the case $\mu = 1$. In practice, another method of comparison is often used, when stagnation parameters, areas of exit cross-sections and those of the throats are assumed equal for the compared nozzles; in this case the flow rates are different.

Let us consider the two methods of comparison for the example of two nozzles with a flow at the exit cross-section which is uniform and parallel to the axis. First we shall require that the following equalities hold: $\gamma_1 = \gamma_2$, $p_{01} = p_{02}$, $T_{01} = T_{02}$, $F_{a1} = F_{a2}$, $Q_1 = Q_2$ ($F_{*1} \neq F_{*2}$ due to the difference in the flow rate coefficients). Evidently, in this case the impulse of the first nozzle J_1 is equal to that of the second one J_2; the same is true for the specific impulse $J_{1s} = J_1/Q_1 = J_{2s} = J_2/Q_2$. In the second case the condition $F_{*1} = F_{*2}$ ($\mu_1 \neq \mu_2$, $Q_1 \neq Q_2$) is used instead of $Q_1 = Q_2$. For the sake of simplicity we shall assume that $\mu_1 < 1$ and $\mu_2 = 1$, i.e. $Q_1 < Q_2$. Then we have $q(\lambda_1) = \mu_1 F_*/F_a$, $q(\lambda_2) = F_*/F_a$, i.e. $q(\lambda_1) < q(\lambda_2)$, $\lambda_1 > \lambda_2$. It can be easily shown that to an accuracy of $(1 - \mu_1)^2$

$$\frac{J_1}{J_2} = \mu_1 \frac{z(\lambda_1)}{z(\lambda_2)} = \mu_1 \left[1 + \frac{(\lambda_1^2 - 1)(\lambda_1 - \lambda_2)}{\lambda_1^2 z(\lambda_2)}\right]$$
$$= \mu_1 \left[1 + \frac{1 - \mu_1}{\lambda_1 z(\lambda_2)} \left(1 - \frac{\gamma - 1}{\gamma + 1}\right)\lambda_1^2\right] \tag{5.33}$$

where $z(\lambda) = \lambda \pm \lambda^{-1}$. Since, at $\lambda_1 \geqslant 1$, the latter term is less than $1 - \mu_1$, it follows that $J_1 < J_2$. It is also evident, that $J_{1s}/J_{2s} = z(\lambda_1)/z(\lambda_2) > 1$, i.e. the impulse of the first nozzle is less than that of the second one while its specific impulse exceeds the specific impulse of the latter.

The results obtained are obvious enough. The first nozzle has a larger value of λ at the exit cross-section and a lower flow rate, therefore, it has a smaller impulse (due to the lower flow rate), but a larger specific impulse (due to the larger λ) . Thus, in the case of the second method of comparison, the nozzles are put into unequal conditions. The first method is preferable, especially for comparing real nozzles, when the flow at the exit cross-section is nonuniform.

The differences between the actual values of parameters and their ideal values are characterized by various coefficients. Thus, the difference in the flow rates for the same throat areas is characterized by the flow rate coefficient μ. The nozzle impulse coefficient $\varphi = J/J^0$ is defined as the ratio of the actual nozzle impulse J to the impulse of the ideal nozzle J^0 with the same Q and F_a and the same distribution of the flow parameters at the initial cross-section as in the real nozzle.

The impulse of an ideal nozzle may be presented in the form

$$J^0 = 2\pi^\nu \int_0^{r_a} (\rho W^2 + p) r^\nu \, dr \quad . \tag{5.34}$$

For a perfect gas and uniform distribution of parameters over a cross-section we have

$$J^0 = a_* Q \frac{\gamma + 1}{2\gamma} z(\lambda^0) = \left(\frac{2}{\gamma + 1}\right)^{1/(\gamma-1)} \mu p_0 F_* z(\lambda^0) \quad , \tag{5.35}$$

where p_0 is the total pressure and λ^0 is defined by means of the gasdynamic function $q(\lambda^0) = \mu F_*/F_a$. The coefficients of loss of impulse, specific impulse and thrust are defined by

$$\zeta = 1 - \frac{J}{J^0} = 1 - \varphi \quad , \quad K_n = \frac{J}{p_0 F_*} \quad , \quad K_H = \frac{R}{p_0 F_*} \quad . \tag{5.36}$$

The values of K_n, φ for the case of an ideal gas with constant γ are related by

$$K_n = \left(\frac{2}{\gamma + 1}\right)^{1/(\gamma-1)} \mu\varphi z(\lambda^0) \quad . \tag{5.37}$$

For similar regimes of work for a supersonic nozzle with pressure, which any point of the flow field is independent of the external pressure p_H, the thrust coefficient K_H for a gas with $\gamma = \text{const}$ is related to K_n, φ by

$$K_H = K_n - r_a^{-2} p_H/p_0 \quad ,$$

$$K_H = \left(\frac{2}{\gamma + 1}\right)^{1/(\gamma-1)} \mu\varphi z(\lambda^0) - \frac{\mu p_H}{p_0 q(\lambda^0)} \quad . \tag{5.38}$$

In the course of a real gas flow through a supersonic nozzle, the process of expansion deviates considerably from the scheme adopted for an ideal process of expansion and, thus, results in losses. The difference between the actual thrust and the ideal one is produced by various gasdynamic and physico-chemical phenomena, the most important of which are listed below.

In the process of expansion, gasdynamic and thermodynamic parameters do not stay constant over a nozzle's cross section. In particular, the velocity vector is not parallel to the nozzle axis and the pressure and the Mach number may change substantially in the direction from the nozzle axis to its walls. Nonuniformity of the flow parameters at the exit cross-section results in divergence losses because at a fixed area of the exit cross-section the maximum thrust is developed by an ideal nozzle with a uniform flow parallel to the axis.

If the gas is viscous and heat conductive, then a boundary layer develops near the walls. The force of friction acting in the direction opposite to that of the thrust and the redistribution of pressure at the nozzle walls associated with the presence of the boundary layer result in friction losses. The heat exchange between the gas and the nozzle walls also influences the value of the thrust.

The process of expansion of a gas in a nozzle is accompanied by physico-chemical transformation (excitation of the vibrational degrees of freedom, reactions of recombination and dissociation and so on), whose characteristic times may be comparable to the time a fixed gas volume resides in a nozzle. If such is the case, then the physico-chemical processes are nonequilibrium and result in losses as compared to an ideal equilibrium efflux. The presence of condensate particles results in various nonequilibrium processes which are also sources of thrust loss. The losses of the specific impulse associated with the nonequilibrium character of physico-chemical processes will be considered in Chap.6.

Due to special features of the technological process of construction, the character of heat exchange and the erosion of a nozzle, the contour of a real nozzle differs from that of a calculated, specially profiled, optimum contour and this, as a rule, results in an increase of the thrust losses.

If the external pressure exceeds that at the nozzle contour at the exit cross-section, then the pressure distribution along the nozzle wall may change. This is possible both due to pressure transfer through the subsonic part of the boundary layer (at a small difference between the external pressure and the wall pressure) and due to flow separation at a large pressure drop. These phenomena also result in a change of thrust.

In accordance with the above reasons leading to impulse losses in an actual nozzle, the corresponding coefficient may be presented in the form of a sum of various losses, the most important of which are the coefficient, ζ_p, of impulse losses due to divergence associated with the nonuniformity of the flow parameters at the nozzle's exit cross-section, and the coefficient, ζ_T, of the impulse losses due to friction.

Let us consider the divergence impulse losses due to truncation of the supersonic part of a nozzle contour profiled to produce a uniform parallel flow at the exit. For such a nozzle the divergence impulse losses coefficient depends on the length of the supersonic part, the radius of the nozzle's exit cross-section and the adiabatic exponent γ. A considerable nonuniformity of the velocity vector values and directions may take place in optimum profiled nozzles. According to the law of momentum conservation, the divergence impulse losses of a nozzle are defined by the formula

$$\zeta_p = 1 - (J_* + \pi^\nu P)/J^0 \quad , \tag{5.39}$$

where J_* is the impulse of the convergent part of a nozzle and

$$P = 2 \int_{r_*}^{r_a} pr^\nu \, dr$$

is an integral of the pressure forces acting upon the supersonic part of a nozzle and projected on the x axis, defined up to the factor π^ν. In the special case of a nozzle with a rectilinear sonic line (5.39) becomes

$$\zeta_p = \left\{[z(\lambda^0) - 2] - \left(\frac{2}{\gamma + 1}\right)^{-1/(\gamma-1)} \bar{P}\right\} z^{-1}(\lambda^0) \quad , \quad \bar{P} = 2 \int_1^{\bar{r}} \frac{p}{p_0} \bar{r}^\nu \, d\bar{r} \quad .$$

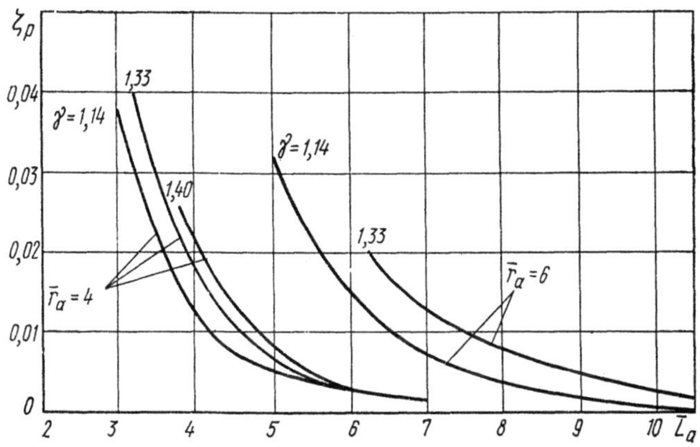

Fig.5.21. Dependence of the divergence losses coefficient ζ_p on the nozzle length at different values of $\bar{r}_a$ and γ

Figure 5.21 shows the dependence of ζ_p on the determining parameters. In the case of a fixed radius of the nozzle's exit cross-section, ζ_p decreases as the length increases, since the nonuniformity of the flow parameters in a cross-section decreases with the increase of a nozzle's length. If the nozzle's length is fixed, then ζ_p increases as the radius of the exit cross-section increases because this results in an increase in nonuniformity of the flow parameters. If the end points of a nozzle are fixed and the value of γ is changed, then different nozzle contours corresponding to different exponents γ enter the same fixed point. For these contours ζ_p increases with increase of γ. However, ζ_p changes only slightly: while γ grows from 1.14 to 1.40, ζ_p changes only by 0.5-0.7%. Nozzle contours corresponding to different γ and entering the same point also differ only slightly.

Various approximate formulas for defining the values of ζ_p are quite suitable for practical calculations. The following approximate formula allows the divergence losses to be determined for a family of truncated nozzles with a corner point and a uniform parallel flow over the exit cross-section at fixed γ, $\bar{r}_a$ and extent of truncation m:

$$\zeta_p = A_p(e^{n_1} - 1)^{-1} \exp\left[n_1\left(1 - \frac{\bar{r}_a}{\bar{r}^0}\right)\right]^{-1} \quad , \qquad (5.40)$$

where

$$A_p = 1.52\{\exp[-30(\gamma - 1)] + 0.1\} \quad ,$$

$$n_1 = 1.45(\bar{r}^0)^{0.25} - 0.005\bar{r}^0 \quad , \qquad \bar{r}^0 = 1 + (\bar{r}_a - 1)m^{-1} \quad ,$$

This formula is valid for $\gamma = 1.10$-1.25, $m = 0.4$-1, $p_0/p_a = 20$-10^4 and allows ζ_p to be defined to an accuracy of 5-10%.

Above, we have considered the divergence losses due to truncation of the supersonic part of a nozzle contour designed to produce a uniform and parallel flow. Besides the nonuniformities of the pressure and velocity fields caused by truncation of a nozzle, an additional nonuniformity may appear, generated by deviation of the shape of the sonic line at the throat of a real nozzle from an exactly straight one. However, the corresponding additional divergence losses are comparatively small and do not exceed 0.2% for $\bar{R}_2 > 0.5$ and $M > 1.5$. For a conic convergent nozzle, the flow nonuniformity at the throat associated with curvilinearity of the sonic surface results in a decrease of its impulse compared to that of an ideal nozzle with a uniform flow and the same flow rate and exit cross-section area. For example, for a plane nozzle with rectilinear walls, $\zeta_p = 0.0275$ at $\theta_0 = \pi/2$; for an axisymmetric nozzle ζ_p has approximately the same value. The ratio of the impulse J^0 of a corresponding ideal nozzle to the impulse J_* of a nozzle with a rectilinear sonic line and the same flow rate is equal to $0.5z(\lambda^0)$, where λ^0 is defined by the relationship $q(\lambda^0) = \mu$. For the plane nozzle under consideration $\mu = 0.856$ (Fig. 5.14), giving $0.5z(\lambda^0) = 1.048$.

Thus, comparison of the nozzles by the first method shows that the actual impulse J and the specific impulse J_s of a plane nozzle with $\theta_0 = \pi/2$ exceed J_* by 1.5% approximately, because $J/J_* = 0.5z(\lambda^0)(1 - \zeta_p)$. If the nozzles are compared by the second method, i.e. if, instead of equality of flow rates, it is required that the throat area of the considered nozzle be equal to that of a sonic one, then $J/J_* = 0.5\mu z(\lambda^0)(1-\zeta_p)$. By (5.33) $J < J_*$, i.e. using such a method of comparison the actual impulse is less than that of a sonic nozzle by approximately 12.5%. To obtain an impulse equal to that of a sonic nozzle, the throat area of the plane nozzle under consideration has to be increased μ^{-1} times approximately. The ratio of the specific impulses J_s/J_{*s} equals $0.5z(\lambda^0)(1-\zeta_p)$, the actual specific impulse exceeding that of a sonic nozzle by 1.5% approximately.

Relations between impulses and specific impulses, corresponding to different methods of comparison, are general enough and valid, at the critical flow regime, for all convergent nozzles with a nonuniform velocity field at the throat. Actually, according to (5.28) and (5.29),

$$dJ = \left[\mathbf{W} + \frac{p\mathbf{n}}{\rho(\mathbf{Wn})}\right]\mathbf{e}_x \, dQ$$

and at the line along which $\theta = 0$ (every nozzle gas flow contains such a line),

$$dJ = \frac{\gamma+1}{2\gamma} a_* z(\lambda) dQ \quad .$$

Integrating along the line and using the mean-value theorem, one gets

$$J = \frac{\gamma+1}{2\gamma} a_* \bar{z}(\lambda) Q \quad , \quad J_s = \frac{\gamma+1}{2\gamma} a_* \bar{z}(\lambda) \quad ,$$

where $\bar{z}(\lambda)$ is the flow rate averaged value of $z(\lambda)$ along the line $\theta = 0$. Since the line always lies in the supersonic part of a nozzle (Fig.5.3), $\lambda > 1$ and $\bar{z}(\lambda) > z(\lambda)$ along it. Hence, a convergent nozzle has a larger impulse and specific impulse than a sonic nozzle with the same flow rate. However, if a sonic nozzle has the same throat area as a convergent nozzle, and hence its flow rate is μ times more, then the impulse and specific impulse of the sonic nozzle are equal to

$$J_* = \frac{\gamma+1}{\gamma} a_* Q \mu^{-1} \quad , \quad J_{*s} = \frac{\gamma+1}{\gamma} a_* \quad ,$$

and thus, by (5.33), $J < J_*$ and $J_s > J_{*s}$.

Calculations show that for a nozzle whose transonic contour is formed by a circular arc of radius $\bar{R}_2 = 0.5$, J_* exceeds J by approximately 1.2%, while J_{*s} is about 1% less than J_s.

Usually, nozzles used in practice have deviations from the design, explained by the requirements of construction, peculiarities of technological processes and erosion. These deviations occur both in supersonic and subsonic parts of nozzles. For the supersonic part of a nozzle the additional impulse divergence losses due to geometric deviations are mainly defined by alteration of local angles of the contour inclination. The additional losses may be approximately calculated from

$$\Delta\zeta_p = 2\gamma\left(\frac{2}{\gamma+1}\right)^{-1/(\gamma-1)}(0.455\bar{r}_a + 1.7)^{-1}z^{-1}(\lambda^0)$$
$$\times \int_A^B \pi(M)M^2\bar{r}(M^2-1)^{-0.5}\left\{1 + \frac{[\gamma M^4 + (M^2-2)^2]\tan\theta}{4(M^2-1)^{3/2}}(\delta\theta)^2\right\}d\bar{x} \quad , \tag{5.41}$$

where $\delta\theta$ is the difference between a deformed and a designed contours at the same $\bar{x}$, and M is the Mach number at the designed contour. By (2.150) the Mach number at a deformed contour $\tilde{M}$ is determined by

$$\tilde{M} = M + \frac{dM}{d\theta}\delta\theta = M\left[1 + \left(1 + \frac{\gamma-1}{2}M^2\right)(M^2-1)^{-0.5}\delta\theta\right] \quad .$$

At $\delta\theta \approx 0.05$-$0.15$ additional losses due to local contour deformations do not exceed 0.2%, which is a consequence of the general property of the family of contours chosen close to the ones designed on the basis of a variational characteristic. For the latter, deviations from the optimum thrust are proportional to the squared variations of a nozzle contour; therefore, even relatively large variations of a contour do not result in considerable impulse losses.

If the contour near a corner point is rounded off by a circular arc of radius $\bar{R}_2^0$, then the associated additional impulse losses at $\bar{R}_2^0 < 1$ may be approximately defined from the formula

$$\Delta\zeta_p = 0.002\bar{R}_2^0 \quad . \tag{5.42}$$

To calculate friction impulse losses in the case of a viscous, heat-conductive gas, one has to introduce, besides the numbers M and γ, two dimensionless similarity parameters: the Reynolds number Re_{w0} and the temperature factor $\bar{T}_w$, which we define as follows:

$$Re_{w0} = W_m\rho_0 L_c\eta_w^{-1} \quad , \quad \bar{T}_w = T_w/T_0 \quad . \tag{5.43}$$

Here W_m is the velocity of efflux into vacuum, $L_c = \ell_0 + L_a$ is the overall length of a nozzle (Fig.5.1), ρ_0, T_0 are the stagnation density and temperature at the nozzle's initial cross-section and η_w is the dynamic viscosity of a gas at the wall temperature T_w. If there is no heat exchange in a nozzle, then $\bar{T}_w$ is put equal to 0.9, but otherwise $\bar{T}_w < 0.9$.

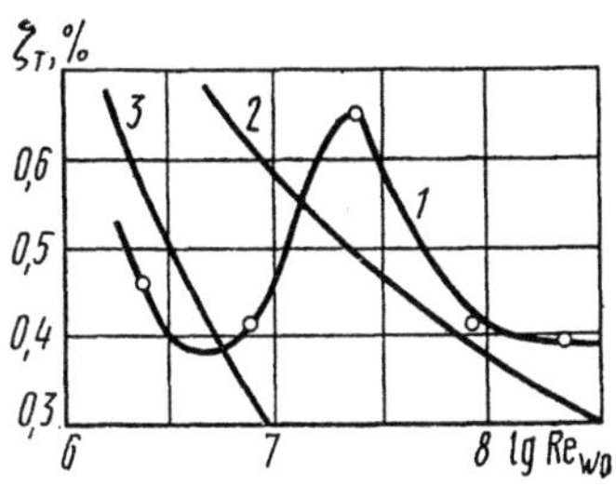

Fig.5.22. Dependence of the friction loss coefficient on the Reynolds number for a truncated nozzle belonging to the uniform flow family of nozzles at $M_a = 2.6$, $\gamma = 1.4$, $\bar{T}_w = 0.9$: (*1*) experiment, (*2*) calculation of turbulent boundary layer, (*3*) calculation of laminar boundary layer

It is common knowledge that the value of Re_{w0} defines one of the three possible regimes of flow in a boundary layer: laminar, turbulent and transitional. Experimental studies of boundary layers in optimum nozzles for air at $M_a = 2.5$-3.0, $\bar{T}_w = 0.9$ have shown that for $Re_{w0} < 10^7$ a boundary layer is laminar, while at $Re_{w0} > 3 \cdot 10^7$ it is turbulent [5.6]. In the interval $Re_{w0} = 10^7$-$3 \cdot 10^7$ a boundary layer is transitional. Results of experimental studies are presented in Fig.5.22 as impulse losses due to friction versus the Reynolds number. We see that for $Re_{w0} > 10^8$ the dependence of the losses on the number Re_{w0} is slight. At $\bar{T}_w < 0.9$ the dependence of the boundary layer flow regimes on the Reynolds number may be somewhat different due to the influence of heat exchange.

Appearance of the region where friction losses do not practically depend on the number Re_{w0} (the selfsimilarity region) is associated with the wall roughness. The value of Re^0 from which the friction losses cease to depend on the Reynolds number, is a function of the temperature factor and the degree of roughness.

All the three flow regimes may take place in boundary layers of modern jet engine nozzles. A laminar flow is possible in boundary layers of low thrust nozzles, while a turbulent flow is characteristic for nozzles developing large thrust. Estimates show that if the pressure at the nozzle's initial cross-section exceeds 1MPa ant the thrust is in excess of 5 kN, the boundary layer is turbulent.

The boundary layer in supersonic nozzles develops in a flow accelerated by large negative pressure gradients. Acceleration produces a stabilizing effect and may sometimes result in a reverse transition from a developed turbulent boundary layer to a laminar one. The phenomenon of the reverse transition has been discovered experimentally in nozzles with large negative pressure gradients [5.27]. In this experiment a fully developed turbulent boundary layer has been created at the initial cross-section of a nozzle; nevertheless, at the exit cross-section the boundary layer proved to be laminar. Thus, a reverse transition may be expected near throats of intensely accelerating nozzles. At large Reynolds numbers ($Re_{w0} > 3 \cdot 10^7$) a laminar boundary layer may become turbulent again.

The friction impulse losses in a turbulent or laminar boundary layer may be calculated with the help of the results obtained in [5.28,29], where formulas for a laminar boundary layer have been derived on the basis of exact solutions obtainable for certain forms of velocity distribution outside the boundary layer. Formulas for the calculation of turbulent boundary layers have been obtained by solving integral momentum and energy relations for such layers taking into consideration pressure gradient in the flow core. To derive these relationships, the Reynolds hydrodynamic analogy and numerous appropriately reduced experimental data on heat exchange and friction at a smooth flat plate were used. The formulas for the calculation of turbulent boundary layers have been tested experimentally only at $\gamma = 1.4$ and $M_0 < 10$, $\bar{T}_w = 0.5$-1, and they do not take into account the influence of the wall roughness.

Below we present formulas which allow friction impulse losses to be calculated in laminar and turbulent boundary layers of nozzle flows. If a transitional flow regime takes place in the boundary layer, then the friction losses should be calculated from the formulas for a turbulent boundary layer.

The coefficient of friction impulse losses is calculated from

$$\zeta_T = 2\bar{\delta}_a^{**}\left(1 + \frac{1}{\gamma M^2}\right)^{-1} , \tag{5.44}$$

where $\bar{\delta}_a^{**}$ is the impulse thickness referred to a current radius. This formula takes into account not only decrease of a nozzle impulse due to wall shear stresses, opposite to the thrust, but also its dependence on the wall pressure, which changes due to the displacement thickness (i.e. due to the change of the effective cross-section of a nozzle) and centrifugal forces acting upon gas particles moving along a curvilinear surface. The centrifugal forces change because the curvature of a contour corrected by the displacement thickness differs from that of the true contour. If the pressure changes are not taken into consideration, then the predicted friction impulse losses may contain substantial errors.

For a laminar boundary layer $\bar{\delta}_a^{**}$ is defined by

$$\bar{\delta}_a^{**} = \left[\pi(M_w)\right]^{-(\gamma+1)/2\gamma} \frac{\bar{\ell}^{-\frac{1}{2}}\left(\frac{2}{\gamma-1}\right)^{\frac{1}{4}}\bar{\nu}_T z(f,\bar{T}_w)}{\bar{r}_a Re_{w0}^{\frac{1}{2}}\bar{T}_w} , \tag{5.45}$$

where M_w is the wall Mach number at a current cross-section, $\bar{\ell}$ is the length of a nozzle contour generatrix referred to the throat radius, $\bar{\nu}_T$ is the energy thickness referred to a current cross-section radius and $z(f,\bar{T}_w)$ is a function characterizing velocity and temperature profiles in the boundary layer.

The values of $\bar{\nu}_T$ and f are defined by

$$\bar{\nu}_T^2 = \frac{A}{\bar{r}_a^2 M_w^b} \int_0^{\bar{\ell}} M^{b-1}\left[\pi(M)\right]^{(3\gamma-1)/2\gamma}\bar{r}^2 d\bar{\ell} \quad , \tag{5.46}$$

$$f = -\frac{A}{\gamma\bar{r}_a^2 M_w^{b+1}}\left[\pi(M_w)\right]^{-(7\gamma-3)/2\gamma}$$

$$\times \frac{d\pi(M)}{d\bar{\ell}} \int_0^{\bar{\ell}} M^{b-1}\left[\pi(M)\right]^{(3\gamma-1)/2\gamma}\bar{r}^2 d\bar{\ell} \quad ,$$

where $d/d\bar{\ell}$ is a derivative with respect to the direction of the contour generatrix at a fixed point. Table 5.1 presents the dependence of the coefficients A and b on the temperature factor, while Fig.5.23 shows the function $z(f,\bar{T}_w)$.

Table 5.1. Dependence of coefficients A and b on T_w

T_w	A	b
0.0	0.465	1.89
0.2	0.500	1.85
0.6	0.508	1.70
1.0	0.532	1.64

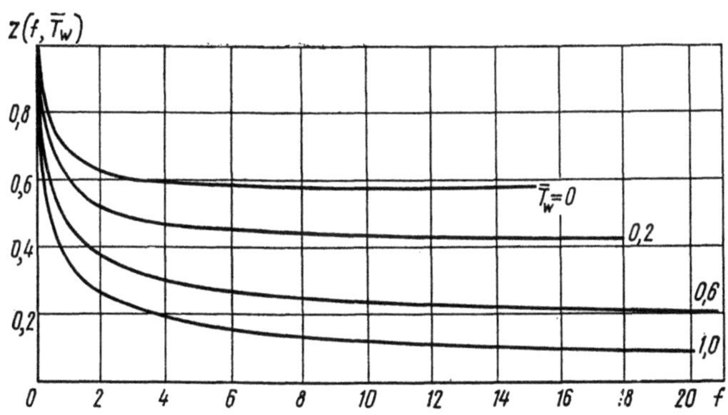

Fig.5.23. The function $z(f,\bar{T}_w)$ in (5.45)

It follows from the above formulas that the coefficient of laminar friction impulse losses grows as the nozzle's length increases and as $\bar{T}_w$, $\bar{r}_a$ and Re_{w0} decrease.

For a turbulent boundary layer the coefficient of friction impulse losses is calculated by means of (5.44), while the momentum thickness $\bar{\delta}_a^{**}$ referred to the exit cross-section radius may be calculated from the expansion

$$\bar{\delta}_a^{**} = \left(\frac{2}{\gamma - 1}\right)^{0.1} Re_{w0}^{-1/5}\left(\frac{0.015}{\bar{T}_w^{0.5}}\right)^{4/5}$$
$$\times\left(1 + \frac{\gamma - 1}{2} M_w^2\right)^{(\gamma+1)/2(\gamma-1)} \frac{\bar{\ell}^{0.2}}{\bar{r}_a^2 M_w^{q+1}} \tag{5.47}$$
$$\times \left[\int_0^{\bar{\ell}} \bar{r}^{5/4} M^{1+5/4q} \left(1 + \frac{\gamma - 1}{2} M^2\right)^{-(1.36\gamma-0.36)/(\gamma-1)} d\bar{\ell}\right]^{4/5} ,$$

where $q = 18/7\bar{T}_w - 2.7$. The formula allows friction losses to be calculated both in subsonic and supersonic parts of a nozzle. Note that if the momentum thikcness has been determined, then the displacement thickness δ^* may be calculated from (5.19,20). The formula and experimental results show that at small $\bar{T}_w$, δ^* may become negative because the density of a gas near a cooled wall may considerably exceed that at a flow core.

Usually the displacement thickness in nozzles of jet engines is small and equal to 5 to 10% of the nozzle radius. Having calculated the displacement thickness, one can find the wall pressure change due to the change both of the effective area of a nozzle cross-section and of the wall curvature, which differs from that of the true contour. Evidently, the two factors result in corrections of opposite signs.

At a fixed value of Re_{w0} the friction impulse losses depend on $\bar{L}_a$, $\bar{r}_a$, γ, $\bar{T}_w$ and the distribution of the Mach number along the nozzle axis. Figure 5.24 shows the characteristic dependence of turbulent friction impulse losses in a supersonic part of a nozzle for a family of truncated nozzles with a uniform flow at the exit cross-section. At a fixed exit cross-section radius the value of ζ_T grows as the nozzle length increases because the surface upon which the friction forces act increases. At a fixed nozzle length, an increase of the radius of the exit cross-section results in a decrease of ζ_T, because the density of the gas near the walls decreases. Also, at a fixed nozzle radius r_a and a fixed length the friction losses increase as γ, $\bar{T}_w$ decrease, because the density of the gas near the walls increases. Usually, in calculating $\bar{\delta}_a^{**}$ from (5.47), it is assumed that $Re_{w0} = Re^0 = 10^8$.

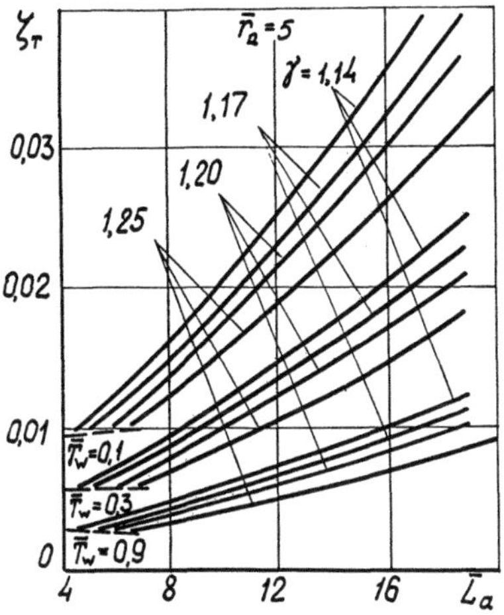

Fig.5.24. Dependence of the firction loss coefficient on the length of the supersonic part of a nozzle

If $Re_{w0} \neq 10^8$, then for a turbulent boundary layer the friction losses must be corrected in the following way:

$$\frac{\zeta_T \big| Re_{w0} = 10^8}{\zeta_T} = \left(\frac{Re_{w0}}{10^8}\right)^{0.2} \quad . \tag{5.48}$$

Figure 5.22 presents a comparison of the friction impulse losses obtained experimentally at $\bar{T}_w = 0.9$, $\gamma = 1.4$ and the results of calculations done with the helpf of the above formulas for laminar and turbulent boundary layers. The comparison shows that the results coincide to an accuracy of 20% for both turbulent and laminar boundary layers.

The following approximate formula allows turbulent friction losses to be defined for a family of truncated nozzles with a corner point and a uniform flow at the exit cross-section, at $Re_{w0} = 10^8$ and known values of γ, $\bar{r}_a$, m and $\bar{T}_w$ [5.25]:

$$\zeta_T = \zeta_T^0 m^{0.1}(0.3 + 0.035e^{3m^3}) \quad , \quad \text{where} \tag{5.49}$$

$$\zeta_T^0 = 8 \cdot 10^{-3}(2.62\gamma^2 \bar{T}_w^{-1/3} - 1)(\bar{r}_a - 1)^{½} \quad .$$

This formula is valid in the intervals $\gamma = 1.10\text{-}1.25$, $m = 0.4\text{-}1$, $p_0/p_a = 20\text{-}10^4$ and $\bar{T}_w = 0.1\text{-}0.9$ and allows ζ_T to be defined to an accuracy of 5-10%.

If the inner surface of the nozzle is rough, the turbulent friction impulse losses for a constant shape of the roughness depend mainly on the degree of roughness or on the dimensionless height of the roughness $\bar{K}_s = K_s/2r_*$, where

K_s is the absolute height of the roughness. The change of $\bar{K}_s$ in the interval from $4 \cdot 10^{-4}$ to $3.5 \cdot 10^{-2}$ results, for an optimum nozzle with $M_a = 3$, in the change of ζ_T by 1.5% approximately. In the case of gas flows through jet engine nozzles, the Reynolds numbers are rather large; therefore, the boundary layer displacement thickness is small and approximately equal to 5-10% of the nozzle radius. Hence, flows in the boundary layer and in the flow core can be calculated separately. However, their mutual influence is felt both in the case of solving the boundary layer equations—through the boundary conditions at the outer boundary—and while calculating the flow core—due to the necessity of correcting the nozzle contour by the displacement thickness.

At small Reynolds numbers ($Re_{w0} < 10^4$) such an approach is invalid due to the large thickness of the viscous layer. In this case, to calculate a flow, one has to use the full Navier-Stokes equations. Results of experimental and numerical studies of nozzle flows at small Reynolds numbers are presented in [5.12,30] and show that division of a nozzle gas flow, in the above sense, into an inviscid core flow and a boundary flayer is valid for $Re_{w0} > 5 \cdot 10^4$.

5.2.3 Thrust Changes Caused by Deviations from the Design Flow Regime

If the external pressure p_H is nonzero, then two characteristic flow regimes can be pointed out: a selfsimilar flow, when the ratio of the static wall pressure to the total pressure at any point of the supersonic part of a nozzle does not depend on the external pressure, and a nonselfsimilar flow, when the pressure distribution over a certain part of a nozzle does depend on the external pressure. In the case of the selfsimilar flow the nozzle thrust is defined by (5.32). To determine the nozzle thrust in the case of a nonselfsimilar flow regime, one has to know the force acting upon that part of a nozzle where distributions of the flow parameters depend on the external pressure. The thrust developed by the other part of a nozzle may be calculated from (5.32) as earlier. Generally, the thrust in the case of the nonselfsimilar flow regime, particularly in the case of flow separation, exceeds the one obtained from (5.32) for attached flows.

If the external pressure p_H exceeds the static wall pressure p_w near a nozzle's exit cross-section, the problem of predicting the force acting upon that part of a nozzle with a nonselfsimilar distribution of flow parameters becomes quite complicated because of the complex nature of interaction of the shock wave with the boundary layer. An extensive series of experiments at different values of γ allowed the determination of the dependence of the number M on the critical pressure ratio at a shock $(p_2/p_1)_*$ (p_1 and p_2 are the pressure values in front of and behind the shock wave respectively), which cannot be exceeded without causing separation of a turbulent boundary layer from a nozzle wall [5.31,32]. For $1 \leqslant M \leqslant 3$ the dependence can be approximated by

$$\left(\frac{p_2}{p_1}\right)_* = \frac{2\gamma}{\gamma + 1} [1.21 + 0.39(M - 1) + 0.055(M - 1)^2] \quad .$$

For $p_H > p_w$ the external pressure propagates upstream in the boundary layer even at $p_H/p_w < (p_2/p_1)_*$ at a fixed number M. As a result, a zone of elevated pressure forms near the nozzle's exit cross-section, some boundary layer thicknesses long, but the flow remains attached [5.33]. At $p_H/p_w \approx 0.5(p_2/p_1)_*$ and $M \approx 3.5$ the additional nozzle thrust augmentation due to the pressure propagation may reach 0.1-0.3%, while at $p_H/p_w \approx (p_2/p_1)_*$ and in the presence of a fully developed boundary layer ($\bar{\delta}^* \approx 10\%$), this augmentation may sometimes reach 0.4-0.8%. At $p_H/p_w > (p_2/p_1)_*$ the shock moves into the nozzle, fully developed separation zones form and the pressure at the nozzle wall downstream of the shock grows, especially in profiled nozzles. Figure 5.25 shows wall pressure distributions at various $\bar{p}_H = p_H/p_0$ obtained experimentally.

In the general case for all regimes when $p_H > p_w$, the nozzle thrust should be determined by means of empirical relations obtained either by direct thrust measurements or by means of drainage tests. As an example, Fig.5.26 presents an experimental dependence of the impulse coefficient K_n on p_a/p_H for optimum nozzles at $F_a = \text{const}$ (p_a, M_a are the pressure and the Mach number at the exit cross-section obtained on the basis of the one-dimensional theory). This figure shows a considerable impulse increase (up to 5-10%) in comparison with a fully attached flow. Figure 5.26 also presents a theoretical curve, to the right of which K_n does not depend on p_a/p_H.

In practice, the thrust coefficient $\bar{R}$ is often used, defined as the ratio of thrusts produced by an actual nozzle and an ideal one with $p_H = p_a$ at the exit cross-section. Evidently, such an ideal nozzle has a variable exit cross-sectional area which is a function of p_H, and produces a maximum thrust, because at a fixed external pressure the maximum thrust is developed by a nozzle with $p_H = p_a$. The last result can be easily obtained by differentiating (5.38) with respect to λ^0 (at $\varphi = \mu = 1$) and equating the first derivative to zero, which immediately results in $p_H = p_a$ at the point of the maximum. Note

Fig.5.25. Pressure distribution in a conical nozzle at $\theta_k = 15^\circ$, $M_a = 4$, $\gamma = 1.4$ at off-design regimes

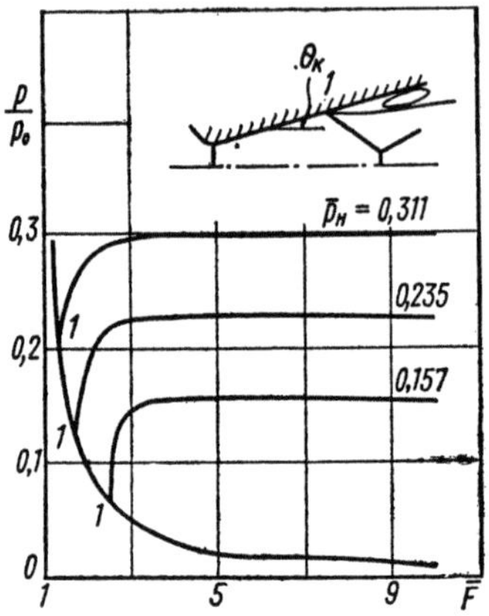

▲ Fig.5.25

Fig.5.26 ►

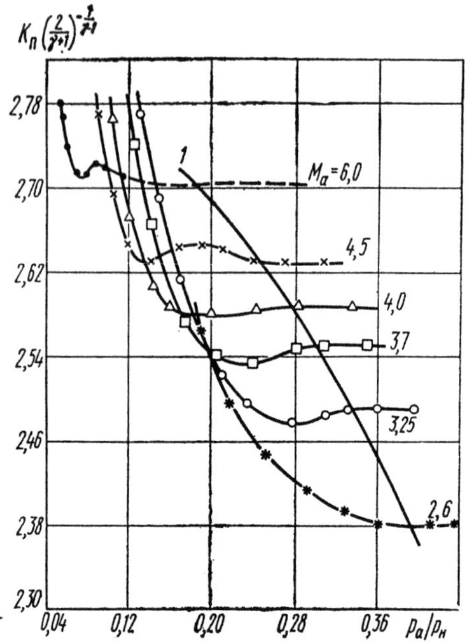

Fig.5.26. Experimental dependence of the specific impulse coefficient: (1) calculated curve to the right of which K_n does not depend on p_a/p_H and the outflow is selfsimilar

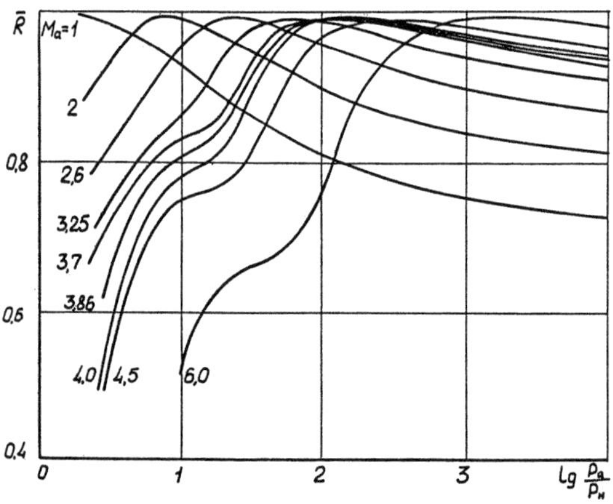

Fig.5.27. Dependence of the thrust coefficient $\bar{R}$ on p_a/p_H

that the equality is exact in the case of zero impulse losses only. Figure 5.27 shows the dependence of the thrust $\bar{R}$ on p_a/p_H for optimum nozzles at $F_a = \text{const}$.

5.2.4 Numerical and Experimental Simulation of Flows in Jet Engine Nozzles

The possibilities for testing life-size, large jet engines are, as a rule, limited. At the same time there are situation when is is necessary to simulate the flow at a nozzle's exit cross-section in order to study the jet characteristics, its dynamic and heat impact upon some parts of the structure and so on. Below we describe an approximate method of simulation, which allows some parameters of the flow to be reproduced at a nozzle's exit cross-section. In Chap.6 it will be shown that both in equilibrium and nonequilibrium flows, concentrations of the main components of the combustion products practically do not change at the low temperatures and pressures characteristic of jets in reactive motors. The effective isentropic exponent $n_1 = \ln(p/p_a)/\ln(\rho/\rho_a)$ (where p_a, ρ_a are respectively the pressure and density at a nozzle's exit cross-section, and p, ρ are pressure and density in the jet) equals 1.3-1.4 for a majority of combustion products, while their molecular weight changes from 27 to 28 only. Thus, physical properties of combustion products can be simulated by means of cold or slightly heated air. The adiabatic exponent n_1 differs from n used in (5.25) because the latter characterizes the process of expansion from the conditions of a combustion chamber to those at a nozzle's exit cross-section, while the former describes the process of expansion in a jet beginning from a nozzle's exit. Also, to simulate a flow, distribution of the Mach number and the inclination of the velocity vector at the exit cross-section of a life-size nozzle and of a model one must be exactly the same. If these requirements are satisfied, then simulation of the flow at the exit cross-section will be full enough.

The contour of a model nozzle is designed in the following way. Let the distributions of M and θ at the exit cross-section BC (Fig.5.28) be known as a result of calculation of the life-size nozzle flow. Using the same distributions of the flow parameters at the exit cross-section and an effective value of the isentropic exponent n_1, one solves the initial-value problem in the triangle BCO by the method of characteristics. Having, in addition, results of an independent calculation of the accelerating part of a nozzle,

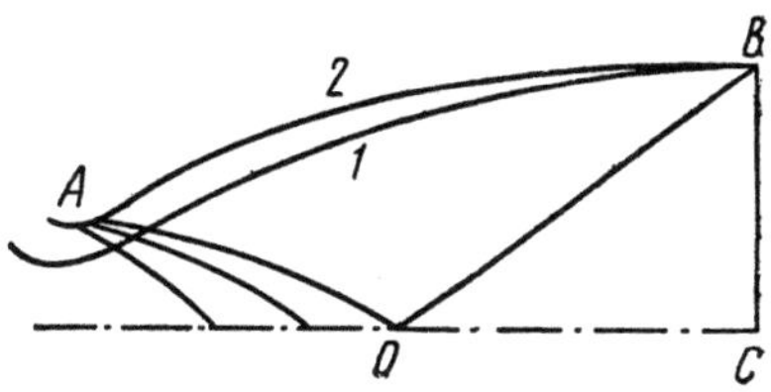

Fig.5.28. Diagram for calculating a model nozzle: (*1*) life-size nozzle, (*2*) model nozzle

done with the exponent n_1, one chooses such a characteristic AO that the mass flow across it is equal to that across BO, while the number M at the point O of the axis coincides with the value of M obtained by solving the initial-value problem on BCO. Then, solving Goursat's problem by the method of characteristics in the space between characteristics BO and AO, one finds the contour AB of the model nozzle. In Fig.5.28 the contour of the life-size nozzle (1) calculated for the case of the flow of combustion products with $n = 1.16$ is compared with that of the model nozzle (2) designed for the use of air ($n_1 = 1.4$).

Absolute values of the static pressure, temperature and efflux velocity at a nozzle's exit cross-section can be simulated by an appropriate choice of stagnation pressure and temperature values, which evidently will differ from those of a life-size nozzle due to a difference between n and n_1.

Thrust losses in jet engines are small — no more than a few percent; therefore, their experimental study requires the application of special measuring techniques. The major difficulty is to measure exactly large mass flows through a tested nozzle. To circumvent this difficulty, quite new methods of experimental study of thrust losses have been developed based on the use of differential installations for comparative tests of nozzles. The idea is to directly measure differences of thrusts developed by two nozzles under comparison, while they are oriented in opposite directions and fed strictly measured mass flows [5.6].

5.3 Main Principles of Choosing a Jet Engine Nozzle

When choosing a thrust nozzle for a jet engine taken alone or as part of a propulsion system, one has to consider the concrete tactical and technological requirements that a missile or an aircraft, for which the reactive motor is designed, must satisfy. As a rule, those who build rockets and reactive motors would like a nozzle to have extremum characteristics. Together with other components of a jet engine, a nozzle must provide a maximum thrust at the least possible weight and size of the engine as a whole. Beside these major requirements, the choice of a nozzle is often defined by requirements concerning its installation in a missile, possibilities for heat protection and many other circumstances.

An exact and mathematically strict extremum profiling of jet engine nozzles is hardly possible, for the following three main reasons. In the first place, our knowledge of the effectiveness of the transformation of thermal energy of a gas into kinetic energy of a reactive jet is only approximate. This remark refers primarily to the processes associated with viscosity and diffusion of gases resulting in boundary layer formation and mixing of flows with different parameters. Secondly, up to now, no mathematical solutions of extremum problems have been obtained for a nozzle flow as a whole, including subsonic and supersonic regions. Thirdly, the requirements of a nozzle are quite diverse and numerous. Taking into consideration structural and technological particularities of nozzle manufacturing, one can give only an approximate mathematical formulation. The thrust of a jet engine depends substantially on the environment, be it in the process of atmospheric flight or of underwater or silo launch. Interaction of outflowing exaust jets with a surrounding medium or with each other modifies the thrust of a jet engine due to modification of pressure exerted upon adjacent surfaces of a craft. Therefore, practical problems of profiling and choosing a nozzle for a jet engine can be solved only approximately [5.6] and their solution is based on an analysis of the thrust characteristics of a one-parametric family of contours. For the role of the latter a family of nozzles with a uniform flow at the exit cross-section is usually chosen. In the following examples of choosing a supersonic nozzle, we shall assume, as a rule, that the combustion chamber and the subsonic part of a nozzle are given. The pressure in the chamber, the gas composition and the flow rate at the throat will be considered known. The situation will be considered when the external pressure p_H is specified and constant.

As an example, we shall consider impulse and thrust characteristics of a family of nozzle contours with a corner point and a uniform closing characteristic. Figures 5.29,30 show nozzle contours with different numbers M_0, the lines of constant dimensionless side surface $\bar{S}_a = S_a/\pi r_*^2$ and the lines of constant impulse and thrust coefficients $K_n = \text{const}$ and $K_H = \text{const}$ (all kinds of losses are considered). Truncated contours terminating at the same line $K_n = \text{const}$ (or $K_H = \text{const}$) have equal impulses (or equal thrusts) but different lengths and expansion ratios.

If the external pressure is nonzero, then a family of nozzles contains only one maximum thrust nozzle. In the example under consideration the thrust coefficient of such a nozzle $K_{Hm} = 1.912$ (Fig.5.30). A gas in this nozzle expands on average to a pressure above atmospheric. With decreasing external pressure the point corresponding to K_{Hm} moves in the direction of larger lengths and higher expansion ratios. In the limiting case of efflux into

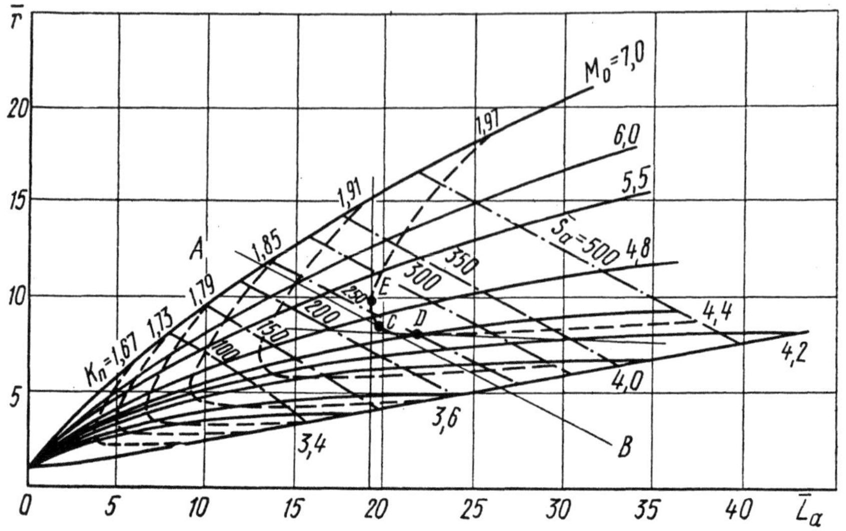

Fig.5.29. Impulse characteristics of a family of nozzles with a uniform flow at the exit cross-section

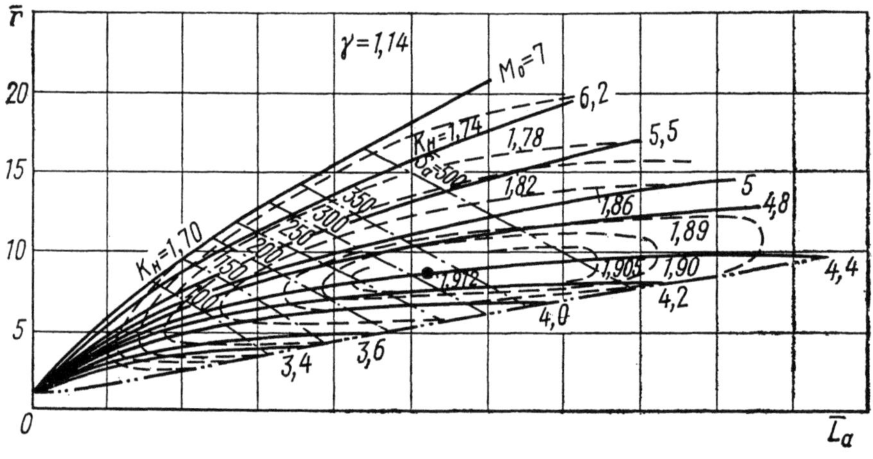

Fig.5.30. Thrust characteristics of a family of nozzles with a uniform flow at the exit cross-section

vacuum ($p_H = 0$) the expansion ratio in a real nozzle with the maximum attainable impulse $p_0/p_a \approx 10^5$-10^7.

Nozzles, whose expansion ratios exceed the limiting one, are uneconomic because the impulse increment due to larger expansion is lower than the losses due to friction and divergence. Nozzles with the maximum possible impulse and thrust should also be avoided, because moving slightly aside from the maximum, one can decrease the nozzle's weight and sizes considerably. However, this must be done carefully, taking into consideration concrete requirements for a nozzle. Usually, these requirements deal with the weight, expansion ratio and length of a nozzle.

Let the problem be the one of finding a maximum impulse nozzle with the side surface of its supersonic part constant $S_a = \text{const}$. Then, from the family of nozzle contours, truncated ones are chosen, whose end points lie at a line $S_a = \text{const}$ (e.g. AB in Fig.5.29). Evidently, the contour of a nozzle having maximum impulse at $S_a = \text{const}$ terminates at the point of contact of the lines $S_a = \text{const}$ and $K_n = \text{const}$ (point C in Fig.5.29), since displacement from the point along AB in any direction results in a decrease of the impulse. Similarly, a nozzle with maximum impulse and a fixed exit cross-section area F_a is the one whose contour terminates at the point of contact of the lines $F_a = \text{const}$ and $K_n = \text{const}$ (point D in Fig.5.29). Displacement to the left of D along the line $F_a = \text{const}$ corresponds to a decrease of friction losses with a simultaneous increase of divergence losses, while displacement in the opposite direction produces a contrary result.

The contour of a nozzle with maximum impulse and a constant length $L_a = \text{const}$ terminates at the point of contact of the lines $L_a = \text{const}$ and $K_n = \text{const}$ (point E). Displacement upwards from the point E corresponds to an increase of divergence losses which are not compensated by an increase in the impulse due to the increase of the degree of expansion. Displacement downwards from the point E corresponds to a decrease of divergence losses which, however, does not compensate the impulse drop due to the decrease of the expansion ratio.

Let us consider in greater detail the choice of a nozzle with a maximum specific impulse taking into account its weight. Choosing a nozzle for a jet engine, one must consider all the forces acting upon its walls. Usually, for an isolated jet engine these are the forces of normal pressure, the friction shear stresses at the nozzle walls and the forces of the external pressure. In the case of a missile in flight, the nozzle walls are subjected to an action of inertial and gravity forces too. A certain portion of the thrust is spent overcoming the inertial forces which are no less important than the friction forces. A nozzle chosen taking into account all the forces, including the inertial ones, may differ considerably in size from a nozzle chosen otherwise.

The nozzle weight G_a is usually proportional to its side surface. Let

$$G_a = cS_a \quad . \tag{5.50}$$

We shall be looking for a contour which, at a given side surface, possesses the maximum impulse J and thrust R. For that purpose we shall consider a family of nozzles whose truncated contours have equal side surfaces. The required properties are those of a nozzle whose contour terminates at the point of contact of the line S_a = const and the line K_n = const or the line K_H = const. Having found a number of such nozzles with different fixed side surfaces S_a, we get a subfamily of nozzles and a line of the ends of the contours corresponding to maximum impulse/thrust at S_a = const. These contours provide a solution for the problem of finding a nozzle producing maximum impulse/thrust at a fixed weight or the lightest nozzle for a fixed impulse/thrust.

The nozzle of a particular jet engine is chosen from this subfamily with the help of the weight equivalent b which defines the cost of a unit of weight expressed in the units of impulse. The weight equivalent b is determined from ballistic analysis and is equal to 0.1-0.2 m/s·N. The known thrust/impulse coefficients and side surfaces of nozzles at the line of contour ends corresponding to the maximum R_s at S_a = const are used to find an extremum nozzle using the maximum value of the effective specific thrust,

$$R_s^0 = R_s - bG_a = p_0F_*K_H/Q - bcS_a \quad . \tag{5.51}$$

Figure 5.31 presents an example of the choice of an extremum nozzle. It is seen that consideration of a nozzle's weight equivalent decreases the optimum expansion ratio. Similarly, nozzles with the maximum R_s at F_a = const, L_a = const are chosen, taking into account the weight equivalent. In this case too, one looks not for a specific impulse maximum but for the maximum of the combination $R_s - bG_a$.

It must be noted that nozzles with a maximum impulse obtained without weight consideration at a given expansion ratio cannot be recommended for use in a missile or an aircraft because the weight equivalent is not accounted for. In those cases when a nozzle with a given expansion ratio is to be chosen under size limitations, it should be done as follows.

When choosing a nozzle, it is necessary to take into account its weight with the help of, e.g. the weight equivalent b. An extremum nozzle with a fixed expansion ration F_a = const must have a maximum value not of J_s, but of the combination J_s-bG_a. The dependence of J_s on L_a has a maximum due to the contradictory behaviour of divergence and friction losses with the

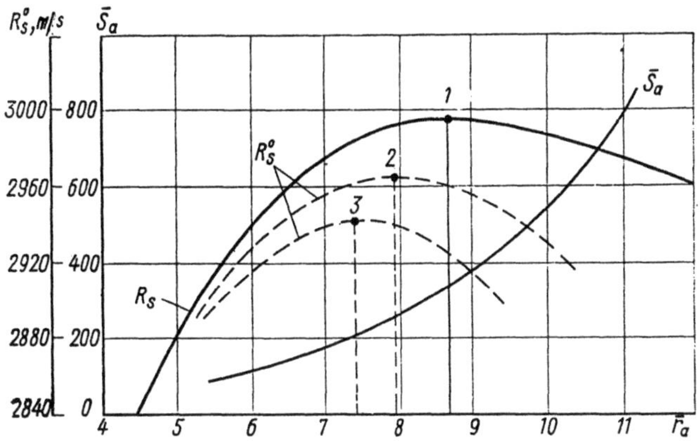

Fig.5.31. Selection of the maximum effective specific thrust nozzle with the help of the weight equivalent ($\gamma = 1.14$, $p_H/p_0 = 10^{-3}$, $\bar{T}_w = 0.9$): (1) maximum R_S nozzle at $\bar{S}_a = \text{const}$, (2) maximum R_S^0 nozzle at $b = 0.1$, (3) maximum R_S^0 at $b = 0.2$

increase of a nozzle's length. Accounting for the weight equivalent results in a decrease of the length accompanied by a slight decrease of the specific impulse. The larger the weight equivalent, the shorter a nozzle with a fixed expansion ratio compared to an original nozzle with a maximum J_s at $F_a = \text{const}$. It should be kept in mind that an extremum nozzle chosen taking into account its weight at a fixed expansion ratio differs from one chosen taking account of weight only.

For booster engines of the first rocket stages, the weight equivalent is rather small. Also, their expansion ratio is low ($\bar{F}_a < 40$). Therefore, the side surface of such nozzles is not large. That is why nozzles of the first stages very much resemble the ones with a maximum J at $F_a = \text{const}$ and $b = 0$. Due to large weight equivalents, relatively high expansion ratios and considerable side surfaces, nozzles of the upper stages tend to be shorter than those with a maximum J at $F_a = \text{const}$ and $b = 0$. If all the limitations put upon a nozzle are expressible by means of geometric characteristics (the length, the expansion ratio, side surface and so on), then the problem of choosing an extremum nozzle for a jet engine working at a constant external pressure may be solved according to the following scheme.

Let a family of contours and ballistic coefficients – specific thrust j, weight b, length ℓ, expansion ratio f and so on – be specified. The coefficients may, for instance, define the cost of a unit of specific thrust, weight, length, etc. expressed in units of the range. Then, from the entire family of nozzle contours a subfamily of nozzles with a maximum impulse at a fixed limitation—max J at $\ell L + fF + bcS = const$—may be chosen. To do this, one draws the lines of $\ell L + fF + bcS = const$ over the family of nozzle contours and finds contours of subfamilies at those points of the lines where the impulse acquires the highest value, i.e. at the points where the above lines contact the lines of constant thrust and impulse. Further, the maximum value of $jR_s + \ell L + fF + bcS$ is sought at the line formed by the ends of the nozzle contours belonging to the subfamily. The extremum nozzle for a given jet engine corresponds to the point of maximum.

The initial data array for the choice of an optimum nozzle is formed by the lines of the ends of nozzle contours with max J at $S_a = const$, $F_a = const$, $L_a = const$. Equations of these lines for nozzles with a uniform flow over the exit cross-section at $\gamma = 1.25$, $\bar{T}_w = 0.3$ may be presented in the form $\bar{r}_a = c_1 + c_2(\bar{L}_a - 10)$, where $c_1 = 5$, 4.6, 3.64 and $c_2 = 0.4$, 0.33, 0.28 at $L_a = const$, $S_a = const$, $F_a = const$, respectively. For a majority of practical problems of extremum profiling, the sizes of the supersonic parts lie between the lines connecting the ends of contours with max J at $L_a = const$ and with max J at $F_a = const$. Choosing nozzles for small expansion ratios characteristic of air-breathing reactive motors and solid-propellent rocket motors at the Mach numbers $M_0 = 1.1$-1.2, one can find that due to small angles of inclination of a nozzle generatrix, the supersonic part of a nozzle produces a negative thrust, if friction forces are taken into account. Thus, a convergent nozzle with a nonuniform velocity field at the exit cross-section may prove to be optimum. As pointed out above, the impulse of such a nozzle is smaller than that of a sonic nozzle with the same exit area, while its specific impulse is larger.

6. Flows with Physico-Chemical Transformations

The ideal gas model fails, as a rule, to describe gas flows taking place in many practical situations. Any real gas flow is accompanied by physico-chemical processes whose nature and methods of mathematical description make a problem much more complicated. The system of equations and boundary conditions presented in Chap.2 for a multivelocity, multitemperature, chemically reactive fluid, gives a general impression of the complexity of describing the flow of such a continuum in the most general case. In practice we have to cope mainly with such flows. However, though various relaxation processes take place simultaneously, they can be studied independently because their influence upon each other is, in essence, slight. In particular, nonequilibrium excitation or deactivation of vibrational degrees of freedom can be studied with the help of nonequilibrium values of concentrations of various components, obtained under the assumption that the translational and vibrational degrees of freedom are in equilibrium. The character of nonequilibrium chemical reactions in a two-phase medium depends only slightly on the dynamic and thermodynamic states of particles. Therefore, in this chapter we consider separately nonequilibrium physico-chemical processes which may take place in nozzles, among them being nonequilibrium excitation of vibrational degrees of freedom, chemical reactions and nonequilibrium two-phase flows.

6.1 Isentropic Flows and Intermolecular Interaction

We shall start the analysis with the two limiting cases of isentropic flows — the equilibrium flow and the frozen flow.

6.1.1 Isentropic Flows

Physico-chemical processes can be accounted for approximately by putting their rates equal to infinity or zero. In the first case the flow is equilibrium, in the second case it is frozen. For an equilibrium flow thermodynamic and gasdynamic parameters are defined by means of thermodynamic relationships for equilibrium processes. Concentrations of reactive components in such flows are defined by the mass action law, the energy of vibrational degrees of freedom is calculated from Einstein's formula, the partial pressure of condensing components is found from the Clausius-Clapeyron equation and velocities and temperatures of particles are assumed equal to those of the gas. It follows from the equation of entropy production (2.9) that in this case entropy remains constant for a stream tube, while in the case of reversible processes it follows from the maximum work principle that an equilibrium flow is limiting, allowing one to obtain at a nozzle's exit cross-section maximum impulse, efflux velocity, temperature and pressure compared to any other process of efflux from a fixed geometry nozzle with specified stagnation flow parameters.

Another limiting flow is a fully frozen one for which parameters characterizing a relaxation process stay constant in the flowing mixture. Such a flow is isentropic too, since the right side of the equation of entropy production (2.9) becomes zero due to zero rates of the corresponding processes. In a fully frozen flow, molar fractions of various components, the energy of vibrational degrees of freedom of molecules, velocities and temperatures of particles stay constant, while condensation and crystallization processes do not take place at all. In a frozen flow the pressure obtained at the exit cross-section of a fixed geometry nozzle at specified stagnation parameters is a minimum compared to any other efflux process, since the energy of dissipation and vibrational degrees of freedom accumulated in a gas at rest, the condensation and crystallization heats and the thermal energy of particles are not transferred to the translational and rotational degrees of freedom of molecules (the latter two always being in mutual equilibrium) and hence are not transformed into the energy of a directed gas flow.

There may exist some other models of isentropic expansion useful for different kinds of engineering estimates. For instance, one may consider a chemically frozen flow when all relaxation parameters, except the molar fractions of components, change in accordance with relationships of equilibrium thermodynamics. Sometimes it is useful to consider flows in which only phase transformations, condensation or crystallization are frozen. This allows the evaluation of the limits of influence of condensation and crystallization upon the flow parameters. Finally, one may consider a frozen two-phase flow in which all parameters, except velocity and temperature of particles, are in equilibrium while the latter stay constant in the process of efflux.

Methods of calculating equilibrium and frozen nozzle flows of complex mixtures of combustion products with all the physico-chemical transformations mentioned are presented in the first volume of the fundamental ten-volume handbook [6.1]. The other volumes contain tables of mixture parameters for various compositions, obtained in the one-dimensional approximation. Such tables, as well as h-S diagrams, allow parameters to be defined at any point of an isentropic flow, provided a thermodynamic parameter and stagnation parameters are known at that point (similar to the one-dimensional gas flow with constant specific heat ratio). In fact, the isentropic condition $S = S(p,\rho) = \mathrm{const}$, or $S = S(p,T) = \mathrm{const}$, provides a relation between the pressure and density/temperature, while the thermal and caloric equations of state together with the energy conservation equation allow temperature/density and velocity to be determined, as well as the molar fractions of various components, the weight fraction of condensate, etc.

Let us consider the main equations used while calculating the equilibrium flow of a heterogeneous mixture composed of N individual substances. We shall choose among them a minimum number m of such substances with whose help reactions of formation of the remaining $\ell = N-m$ substances can be written in a simple way. The most basic combination of such m substances consists of atoms and electrons; however, it is not obligatory to choose this combination. The m independent components and the ℓ dependent components will be denoted by subscripts k and j respectively. A reaction of formation of a dependent component (which may be an ion) M_j may be presented in the form

$$M_j = \sum_{k=1}^{m} \nu_{kj} A_k \quad , \tag{6.1}$$

where A_k is an independent component and ν_{kj} is a stoichiometric reaction coefficient. The equation of equilibrium for the reaction (the mass action law) has the form

$$\chi_j = \sum_{k=1}^{m} \nu_{kj} \chi_k \quad , \tag{6.2}$$

where χ_k, χ_j are the chemical potentials of components k and j respectively. Then, from (6.2),

$$\frac{1}{p_j} \prod_{k=1}^{m} p_k^{\nu_{kj}} = K_j(T) \quad , \tag{6.3}$$

where p_k, p_j are the partial pressures of the components k and j respectively, and K_j is an equilibrium constant,

$$K_j(T) = \exp\left\{\left[\sum_{k=1}^{m} \nu_{kj} S_k^0(T) - S_j^0(T)\right]R^{-1}\right.$$

$$\left. - \left[\sum_{k=1}^{m} \nu_{kj} h_k^0(T) - h_j^0(T)\right](RT)^{-1}\right\} \quad . \tag{6.4}$$

Here h_k^0 and S_k^0 are the enthalpy and entropy of a mole under standard conditions, and R is the universal gas constant.

If a mixture contains condensed substances, their partial pressures are assumed equal to the saturation pressures. The pressure of saturated vapours p_i^s can be determined from the condition of equality of chemical potentials of the same substance in gaseous and condensed phases (the Clausius-Clapeyron equation). It is equal to

$$p_i^s = \exp\left\{\left[S_i^0(T) - S_{is}^0(T)\right]R^{-1} - \left[I_i^0(T) - I_{is}^0(T)\right](RT)^{-1}\right\} \quad , \tag{6.5}$$

where the subscript s denotes condensed substances and I_i^0 is the total enthalpy of an individual substance.

Equations (6.3,4) are used together with conservation equations which express the condition that the number of the kth chemical element's atoms remains constant in the processes of efflux. For a heterogeneous mixture these equations are

$$\sum_{i=1}^{N} a_{ki}(n_i + n_{si}) - b_k = 0 \quad , \quad (k = 1,2,\ldots,m - 1) \quad , \tag{6.6}$$

where a_{ki} is the number of atoms of the kth type in a mole of the substance i, b_k is the number of atoms of the kth type in a mole of the initial products and n_i and n_{si} are the numbers of moles of the element i in the gaseous and condensed states respectively.

For substances present both in the gaseous and condensed states, the number of moles in the gaseous phase n_i^s is a known function of temperature determined by the relationship

$$n_i^s = \frac{p_i^s}{p} \sum_{i=1}^{N} n_i \quad . \tag{6.7}$$

The condition of electrical neutrality has the form (the possibility of the presence of charged condensed particles is ignored)

$$\sum_{i=1}^{N} a_{ii} n_i = 0 \quad , \tag{6.8}$$

where a_{li} is the ionization multiplicity. For a mixture of ideal gases, Dalton's law,

$$\sum_{i=1}^{N} p_i = p \tag{6.9}$$

is valid. Thus, at given p, T and b_k one has, for the determination of unknown n_i, n_{si} or p_i, p_{si}, the system of nonlinear equation (6.3,6-9) which has only one physically possible solution [6.2].

Some limitations are imposed by the rule of phases,

$$r = 2 + N - P - L \quad , \tag{6.10}$$

where r is the number of thermodynamic degrees of freedom, P is the number of phases and L is the number of independent chemical reactions. In the situation considered, $r = 2$, since there are two thermodynamic degrees of freedom, p and T. The number of individual substances N equals $\ell + m$ and the number of independent chemical reactions $L = \ell$; hence, $P = m$, i.e. the number of phases cannot exceed the number of elements composing a system. Since one of the phases is gaseous, the number of condensed phases P_s at $r = 2$ is equal to or less than $m - 1$.

Methods of numerical solution of the system (6.3,6-9) have been described in [6.1,3]. They are critically reviewed in [6.1]. As an example of a concrete form of this system, let us write down the system of equations for determining the products of combustion of hydrocarbons in oxygen at specified temperature and pressure. We shall limit our consideration to major individual components only: CO, CO_2, H_2, O_2, H_2O, OH, H, O. A variant of the system has the form

$$\frac{p_{CO}^2 p_{O_2}}{p_{CO_2}^2} = K_1(T) \quad , \quad \frac{p_{OH} p_H}{p_{H_2O}} = K_2(T) \quad ,$$

$$\frac{p_{H_2}^2 p_{O_2}}{p_{H_2O}^2} = K_3(T) \quad , \quad \frac{p_H^2}{p_{H_2}} = K_4(T) \quad , \quad \frac{p_O^2}{p_{O_2}} = K_5(T) \quad ,$$

$$\frac{2p_{H_2O} + 2p_{H_2} + p_{OH} + p_H}{p_{CO} + p_{CO_2}} = C_1 \quad ,$$

$$\frac{2p_{CO_2} + p_{CO} + p_{H_2O} + p_{OH} + 2p_{O_2} + p_O}{p_{CO} + p_{CO_2}} = C_2 \quad ,$$

$$p_{CO} + p_{CO_2} + p_{H_2} + p_{O_2} + p_{H_2O} + p_{OH} + p_H + p_O = p \quad .$$

Calculation of the equilibrium flow of a chemically reactive mixture in a nozzle is divided into two stages. First of all, the combustion chamber parameters are determined with the pressure specified and the temperature is determined by solving the system (6.3,6-9) under the condition of equality of enthalpy of the combustion products and the fuel. The enthalpy and entropy of a kilogram of a mixture in the combustion chamber are determined by the formulas

$$H_0 = \frac{1}{p\mu}\left[\sum_{i=1}^{N} n_i I_i^0 + \sum_{i=1}^{N} n_{si} I_{si}^0\right] \quad , \tag{6.11}$$

$$S_0 = \frac{1}{p\mu}\left[\sum_{i=1}^{N} n_i (S_i^0 - R \ln n_i) + \sum_{i=1}^{N} n_{si} S_{si}^0\right] \quad , \tag{6.12}$$

where μ is the molecular weight of the mixture. At the second stage the process of expansion is calculated with temperature specified at a cross-section and pressure determined by solving the system (6.3,6-9) under the condition $S = S_0$. Among other things, this solution gives the values of partial pressures of all the components. Then, the density is calculated from the equation of state,

$$\rho = \frac{p\mu}{RT} \quad , \tag{6.13}$$

the gas velocity from the energy equation,

$$W = \sqrt{2(H_0 - h)} \quad , \tag{6.14}$$

and the area of the cross-section (at a fixed rate) from the continuity equation

$$F = \frac{Q}{\rho W} \quad . \tag{6.15}$$

Simultaneously, the vacuum specific impulse $J_s = (\rho W^2 + p)F/Q$ and the flow rate complex $\beta = p_0 F_*/Q$ can be determined.

Figure 6.1 presents the dependence of J_s, β and T_0 on the oxidizer-to-fuel ratio α for a fuel composed of nonsymmetric dimethylhydrazine $(CH_3)_2NH_2$ — the combustible — and tetroxide nitrate N_2O_4 — the oxidizer. A set of similar graphs allows the most important characteristics of any fuel composition to be determined.

A chemically frozen outflow is calculated in a similar way under the condition $S = S_0$; however, in this case there is no need to solve the system

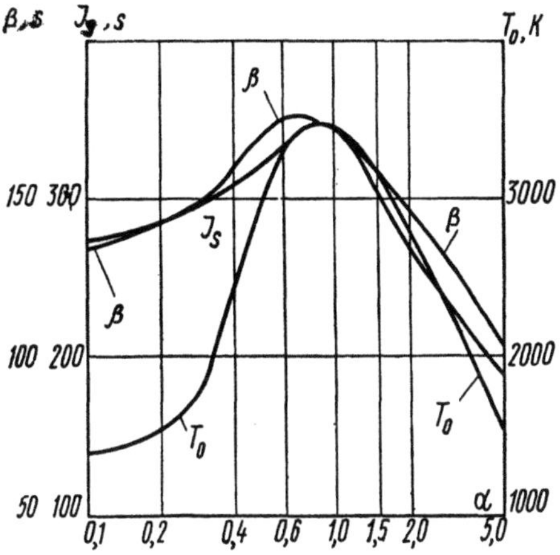

Fig.6.1. Dependence of the vacuum specific impulse J_S, the flow rate complex β and temperature T_0 on the oxidizer-to-fuel ratio at $p_0 = 150$ bar

(6.3,6-9) since the molar fractions of the components remain invariable in the process of efflux. If, besides, vibrational degrees of freedom of molecules are supposed to be frozen and the absence of phase transitions is assumed, then the process of efflux can be calculated by (2.137-139) for an ideal gas with a constant adiabatic exponent $\gamma = \gamma^0$ defined by

$$\gamma^0 = \left(\sum_{i=1}^{N} c_{pi} r_{0i}\right)\left(\sum_{i=1}^{N} c_{pi} r_{0i} - \frac{R}{\mu}\right)^{-1} ,$$

where c_{pi} is the ith component's heat capacity at a constant pressure obtained by considering the excitation of translational and rotational degrees of freedom only, and r_{0i} is the molar fraction of the ith component in the combustion chamber. Figure 6.2 presents an example of the vacuum specific impulse and the temperature dependence on the dimensionless radius $\bar{r} = \bar{F}^{\frac{1}{2}}$ for the same fuel and three limiting cases: an equilibrium flow, a chemically frozen flow and a flow with frozen chemical reactions and energy of vibrational degrees of freedom.

Sometimes, the gasdynamic parameters for equilibrium or frozen flows can be calculated approximately, if the so-called average value of the isentropic exponent n, defined by

$$n = \frac{\ln(p/p_0)}{\ln(\rho/\rho_0)} , \qquad (6.16)$$

is used. Then, to define all the necessary parameters, J_s and β among them, one can use (2.137-144) for an ideal gas with a constant adiabatic exponent $\gamma = n$. In particular, J_s and β are defined by

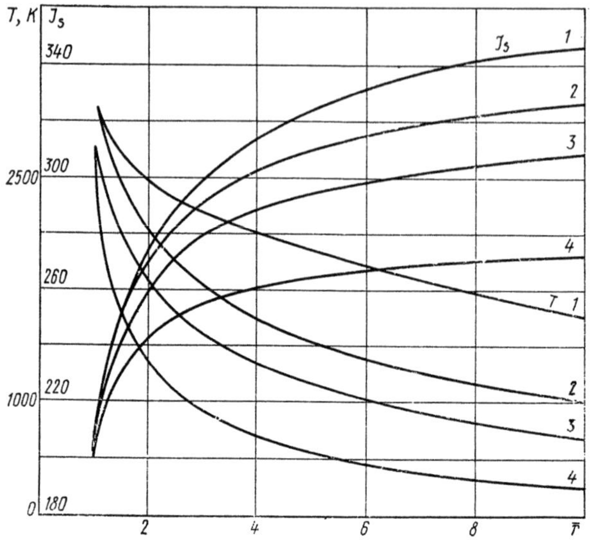

Fig.6.2. Dependence of the vacuum specific impulse J_S and temperature T on r at $\alpha = 1$, $p_0 = 150$ bar: (*1*) equilibrium flow, (*2*) chemically nonequilibrium flow, (*3*) chemically frozen flow, (*4*) flow with frozen chemical reactions and energy of vibrational degrees of freedom

$$J_s = z(\lambda)\left(\frac{n+1}{2n}\,\frac{RT_0}{\mu}\right)^{1/2} , \tag{6.17}$$

$$\beta = \frac{1}{A(n)}\left(\frac{RT_0}{\mu}\right)^{1/2} . \tag{6.18}$$

Such an approximate approach does not in practice result in errors of the flow parameters in the subsonic part of the nozzle and in the throat. For the supersonic part of a nozzle the following errors are propable: 10-15% for the temperature, 1-4% for the velocity and the specific impulse.

The same algorithms are used to calculate both two- or three-dimensional equilibrium flows and flows of an ideal gas with $\gamma = \text{const}$. However, it is necessary to have tables or analytical formulas for p/p_0, ρ/ρ_0, T/T_0, W/a_* as functions of the Mach number M which may be defined by independently calculating the equilibrium gasdynamic and thermodynamic parameters of a mixture in the one-dimensional approximation using the method described above.

6.1.2 Intermolecular Interaction

Usually, thermodynamic characteristics are defined by means of the equation of state applied to individual substances and to a mixture as a whole. Sometimes, however, intermolecular interactions of the type produced by Van der

Waals' forces may result in considerable errors in the calculated values of thermodynamic parameters. The latter can be calculated for real gases with the help of thermodynamic relationships if the equation of state is known. The most complete review of equations of state is presented in [6.4]. Note that practically all of them are empirical or semiempirical. The only equation of state obtained theoretically is the one containing virial coefficients [6.5,6], which, therefore, may be extrapolated into the region of temperatures for which experimental data are absent.

The influence of intermolecular interaction upon thermodynamic and gasdynamic parameters of products of combustion for many fuel compositions has been studied in [6.1] for a wide range of determining parameters; the handbook also contains an extensive list of publications. As a result of numerous calculations, it has been deduced that the influence of intermolecular interaction upon thermodynamic and gasdynamic parameters is insignificant (1-2% on average) since the products of combustion have a high temperature. However, in wind tunnels and shock tubes the pressure of the working media may be high (up to 1000 atm) at a rather low temperature; hence, the influence of intermolecular interaction may prove to be significant. In [6.7] isentropic expansion of air at high stagnation pressures has been studied experimentally and theoretically with the help of the equation of state [6.8], and a significant influence of the real properties of a gas upon the determining parameters has been shown.

The system of equations describing flow in the one-dimensional approximation has the form

$$h + \frac{1}{2} W^2 = H_0 \quad , \tag{6.19}$$

$$\rho W F = Q \quad , \tag{6.20}$$

$$S = S(p,T) = S_0 \quad , \tag{6.21}$$

$$\frac{p\mu}{\rho R T} = z(p,T) \quad , \tag{6.22}$$

$$h = h(p,T) \quad . \tag{6.23}$$

The function $z(p,T)$ characterizes the difference between the equation of state and that of an ideal gas, Fig.6.3. Figure 6.4 shows the dependence of the ratio of the static and stagnation pressures p/p_0 on the dimensionless area F_*/F [6.7]. Increase of the stagnation pressure p_0 at $F_*/F = \text{const}$ results in a considerable decrease of p/p_0 and a corresponding increase of

Fig.6.3. The function $z(p,T)$ in (6.22) characterizing departure from the ideal gas equation of state for air

▼

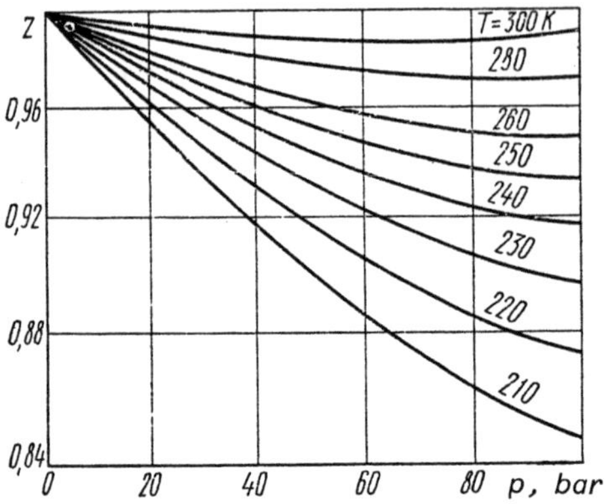

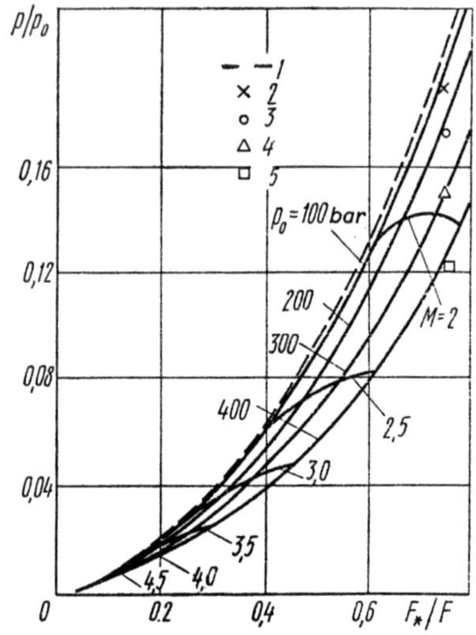

Fig.6.4. Dependence of pressure on the areas ratio at high stagnation pressures and $T_0 = 250$ K: (*1*) ideal gas at $\gamma = 1.4$, (*2*)-(*5*) experiments. Realistic properties are illustrated

the number M. Calculations show that while the total pressure increases from 100 to 700 atm, the critical pressure drop p_*/p decreases from 0.528 to 0.4. Figure 6.4 presents experimental data obtained by means of changing the static pressure in a conic nozzle with $\theta_k = 9^\circ$ at $F_*/F = 0.752$. A good agreement of the predicted and experimental data, as well as their considerable deviation from the parameters of an ideal gas with $\gamma = 1.4$, is evident. Real properties of a gas influence considerably the value of the coefficient $A(\gamma)$ in the flow rate equation (2.142): it increases by 10-15% while p_0 grows from 10 to 500 atm. Variation of stagnation temperature from 200 to 500 K does not greatly influence the dimensionless parameters, though parameters of ideal and real gases tend, evidently, to coincide as the temperature increases. Note that at large p_0 the specific heat ratio may exceed 1.4 considerably, being equal to 2-2.5.

6.2 Flows with Nonequilibrium Chemical Reactions

A relaxation process may influence the flow parameters considerably if the time of relaxation is comparable to the characteristic gasdynamic time and the change of energy associated with the process constitutes a substantial part of the total energy change.

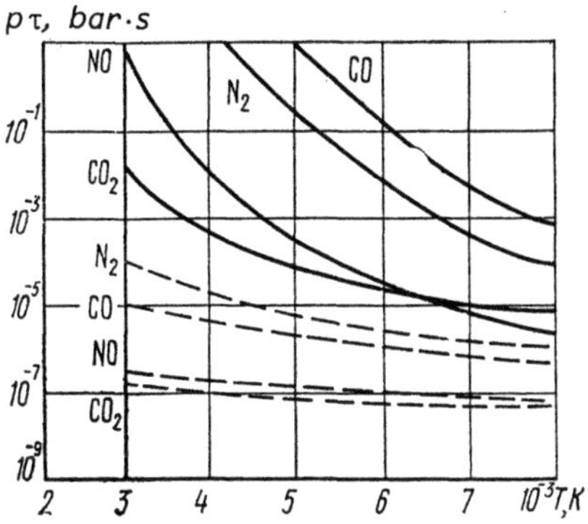

Fig.6.5. Comparison of relaxation times for dissociation (*solid curves*) and excitation (*dashed lines*) of vibrational degrees of freedom

In the case of a hot mixture with stagnation temperature $T_0 < 4500$ K flowing in a nozzle, nonequilibrium chemical reactions are most important, because their contribution to the total energy of a mixture is comparable to that of the vibrational degrees of freedom while their relaxation times exceed, as a rule, those of the vibrational degrees of freedom of molecules by one or two orders of magnitude. This is demonstrated by Fig.6.5 where the relaxation times needed to attain equilibrium between the dissociation and excitation of vibrational levels of the molecules CO_2, NO, CO, N_2 are compared at 3000-8000 K [6.9].

Nozzle gas flows are characterized by gasdynamic parameters with large gradients. Therefore, the characteristic time of a relaxation process cannot be specified in the usual sense because it depends on gasdynamic parameters and changes within broad limits. Regions may be pointed out in the flow field where the state of the gas is close to equilibrium (where the characteristic gasdynamic time exceeds the relaxation time manyfold), is nonequilibrium (where the characteristic gasdynamic time is comparable to the relaxation time) or frozen (where the characteristic gasdynamic time is much less than the relaxation time). Thus, nozzle gas flows must be studied taking into account the finite rates of chemical reactions.

6.2.1 Basic Equations, Method of Prediction and the One-Dimensional Approximation

Let us consider a chemically reactive mixture of perfect gases composed of N individual substances. We shall assume that the mixture is homogeneous, and that both the translational and the rotational degrees of freedom of the molecules are in equilibrium with the vibrational degrees of freedom. First of all, let us consider the right sides of the mass conservation equations (2.7) for individual components in the case of nonequilibrium chemical reactions. Let us write down the chemical kinetics equations. Let ℓ individual chemical reactions be going on in a mixture of gases, each reaction, say, the rth one, described by

$$\sum_{i=1}^{N} \nu'_{ir} A_i \rightleftarrows \sum_{i=1}^{N} \nu''_{ir} A_i \quad , \tag{6.24}$$

where ν'_{ir} and ν''_{ir} are stoichiometric coefficients indicating the numbers of molecules of the substance A_i taking part in the direct and reverse reactions respectively. The sum $\Sigma_{i=1}^{N} \nu_i$ is denoted by m and is called the order of a reaction. For $m = 1,2,3$, a reaction is called monomolecular, bimolecular and trimolecular respectively. At high temperatures, mostly bimolecular and trimolecular reactions go on in mixtures of gases. Reaction 2 presented in Table 6.1 is trimolecular in one direction and bimolecular in the opposite direction. In the case of a bimolecular reaction two particles collide and interact, while in the case of a trimolecular reaction there must be three particles. Trimolecular reactions in nonequilibrium nozzle flows are otherwise called recombination reactions because they result in a diminished number of gas particles; bimolecular reactions are called exchange reactions (e.g the reactions 8-10 in Table 6.1) because they are associated with the exchange of atoms and molecules between components of a mixture, while the total number of particles remains constant.

The rate of a direct chemical reaction r expressed in moles and referred to unit volume and time is equal to

$$W_+^{(r)} = K_+^{(r)}(T) \prod_{i=1}^{N} \left(\frac{\rho\alpha_i}{\mu_i}\right)^{\nu'_{ir}} \quad , \tag{6.25}$$

where α_i and μ_i are the weight fraction and the molecular weight of the ith component. The proportionality factor $K_+^{(r)}(T)$ is called the reaction rate constant. It does not depend on the concentrations of individual substances and is a function of temperature. The rate of reaction r is equal to

$$W_+^{(r)} - W_-^{(r)} = K_+^{(r)} \prod_{i=1}^{N} \left(\frac{\rho\alpha_i}{\mu_i}\right)^{\nu'_{ir}} - K_-^{(r)} \prod_{i=1}^{N} \left(\frac{\rho\alpha_i}{\mu_i}\right)^{\nu''_{ir}} \quad , \tag{6.26}$$

where $W_+^{(r)}$, $W_-^{(r)}$, $K_+^{(r)}$, $K_-^{(r)}$ are reaction rates and rate constants of the direct and the reverse reactions respectively.

According to the Arrhenius law a reaction rate constant may be presented in the form

Table 6.1. Chemical reactions and reaction rate constants for the system C-O-H-N

	Reactions	Reaction rate constants
1	$CO + O + M \rightarrow CO_2 + M$	$3.5 \cdot 10^{14}\ \exp(-2100/RT)$
2	$H + H + M \rightarrow H_2 + M$	$1.4 \cdot 10^{20} T^{-1.5}$
3	$O + O + M \rightarrow O_2 + M$	$5.5 \cdot 10^{17} T^{-0.87}$
4	$OH + H + M \rightarrow H_2O + M$	$1.2 \cdot 10^{20} T^{-1.0}$
5	$O + H + M \rightarrow OH + M$	$3.3 \cdot 10^{18} T^{-0.5}$
6	$N + N + M \rightarrow N_2 + M$	$2.7 \cdot 10^{16} T^{-0.5}$
7	$N + O + M \rightarrow NO + M$	$3.3 \cdot 10^{15}$
8	$OH + CO \rightarrow CO_2 + H$	$2.5 \cdot 10^{12}\ \exp(-5100/RT)$
9	$OH + H_2 \rightarrow H_2O + H$	$1.1 \cdot 10^{14}\ \exp(-8600/RT)$
10	$OH + OH \rightarrow H_2O + O$	$1.0 \cdot 10^{13}\ \exp(-1200/RT)$
11	$H_2 + O \rightarrow OH + H$	$1.3 \cdot 10^{13}\ \exp(-9860/RT)$
12	$O_2 + H \rightarrow OH + O$	$2.2 \cdot 10^{14}\ \exp(-16500/RT)$
13	$N_2 + O_2 \rightarrow NO + NO$	$5.2 \cdot 10^{13}\ \exp(-107000/RT)$
14	$NO + N \rightarrow N_2 + O$	$3.0 \cdot 10^{13}\ \exp(-200/RT)$
15	$NO + O \rightarrow O_2 + N$	$1.1 \cdot 10^{13}\ \exp(-41700/RT)$

$$K_+(T) = BT^{\beta} e^{-E/RT} \quad . \tag{6.27}$$

Here the factor B characterizes the number of effective collisions of reacting particles per second. The constants β and E (the latter being called the activation energy of a chemical reaction) are defined by the character of the elementary processes. As a rule, reaction rate constants are obtained experimentally. Many of them (for reactions of gaseous components) are presented in [6.10]. The characteristic reaction time or relaxation time at a known reaction rate constant is defined by the formula

$$\tau = \left[\frac{K_+^{(r)}\mu}{\rho} \prod_{i=1}^{N} \left(\frac{\rho}{\mu}\right)^{\nu_{ir}}\right]^{-1} \quad . \tag{6.28}$$

Usually, the ratio of the direct and reverse reaction rates is defined by the relation

$$K_+^{(r)}/K_-^{(r)} = K_p^{(r)}(T) \quad , \tag{6.29}$$

which, strictly speaking, is valid only under conditions close to equilibrium. [Here $K_p^{(r)}(T)$ is the equilibrium constant.]

A direct reaction r results in the formation of $\nu''_{ir} - \nu'_{ir}$ moles of the ith individual substance; hence, the mass rate of formation of a substance with a molecular weight μ_i in the process of reaction r is equal to

$$R_i^{(r)} = (\nu''_{ir} - \nu'_{ir})\mu_i W_+^{(r)} - (\nu''_{ir} - \nu'_{ir})\mu_i W_-^{(r)} \quad . \tag{6.30}$$

Thus, kinetic equations for the ith individual substance may be written:

$$\frac{d\alpha_i}{dt} = F_i(p,T,\alpha_1,\alpha_2,\ldots,\alpha_N) \quad , \quad (i = 1,2,\ldots,N) \quad , \tag{6.31}$$

where $F_i(p,\ T,\ \alpha_i) = \Sigma_{r=1}^{\ell}\ R_i^{(r)}$ is the mass rate of substance formation as a result of all ℓ reactions. The system (6.31) allows one to obtain finite relations (equations of matter conservation) similar to (6.6) whose number is equal to the number of independent components.

For the case of nozzle flows, the specific volume of a gas varies. Therefore, when calculating chemical nonequilibrium flows, one should use the mole-mass concentration equal to the number of moles of a substance referred to unit mass, $\gamma_i = \alpha_i/\mu_i = r_i/\mu$, where the r_i are the volume (mole) fractions.

Table 6.1 presents elementary chemical reactions and reaction rate constants for the system C-O-H-N, usually used in the analysis of nonequilibrium nozzle gas flows [6.1]. The reaction rate constants of trimolecular reactions have the dimension of $cm^6 \cdot mole^{-2} \cdot s^{-1}$, those of bimolecular reactions, $cm^3 \cdot mole^{-1} \cdot s^{-1}$ and the universal gas constant $R = 1.987\ cal \cdot mole^{-1} \cdot deg^{-1}$. The elementary reactions in the system H-F (without reaction 2 of Table 6.1) are presented in Table 6.2.

Let us give an example of a differential equation of chemical kinetics for H_2 in accordance with the system of reactions of Table 6.1. We have

$$\frac{d\gamma_{H_2}}{dt} = \rho\Big[K_+^{(2)}\rho\gamma_H^2\frac{1}{\mu} - K_-^{(2)}\gamma_{H_2}\frac{1}{\mu} - K_+^{(9)}\gamma_{H_2}\gamma_{OH}$$

$$+ K_-^{(9)}\gamma_{H_2O}\gamma_H - K_+^{(11)}\gamma_{H_2}\gamma_O + K_-^{(11)}\gamma_{OH}\gamma_H\Big] \quad . \tag{6.32}$$

The equation of matter conservation for H has the form

$$2\gamma_{H_2} + \gamma_{OH} + 2\gamma_{H_2O} + \gamma_H = \text{const} \quad . \tag{6.33}$$

Calculation of three-dimensional nonequilibrium flows may be divided into two stages. In the first stage, coordinates of streamlines and pressure distributions along them are determined; in the second stage, the above data

Table 6.2. Chemical reactions and reaction rate constants for the system H-F

	Reactions	Reaction rate constants
1	$H + F + M \rightarrow HF + M$	$10^{19}T^{-1}$
2	$F + F + M \rightarrow F_2 + M$	$10^{18}T^{-1}$
3	$H_2 + F \rightarrow HF + H$	$1.6 \cdot 10^{14}\exp(-1600/RT)$
4	$F_2 + H \rightarrow HF + F$	$1.2 \cdot 10^{14}\exp(-2400/RT)$
5	$H_2 + F_2 \rightarrow HF + HF$	$10^{13}\exp(-25000/RT)$

are used to calculate the rest of the parameters. Usually, the calculations of both stages are repeated to ensure a higher accuracy. The system of equations along streamlines is a system of compatibility equations along characteristics, which coincide with streamlines defined by the equations of nonequilibrium gas dynamics and, thus, contain only derivatives with respect to streamlines. The equations of this system are evidently equivalent to those in the one-dimensional approximation at a specified pressure distribution. Thus, many qualitative and quantitative characteristics of nonequilibrium nozzle flows may be studied accurately enough in the framework of the one-dimensional approximation.

A nonequilibrium nozzle flow may be described in the one-dimensional approximation by the following system of equations of motion, energy, continuity and conservation of individual components:

$$\frac{dp}{dx} + \rho W \frac{dW}{dx} = 0 \quad , \tag{6.34}$$

$$h + \frac{1}{2} W^2 = H_0 \quad , \tag{6.35}$$

$$\rho W F = Q \quad , \tag{6.36}$$

$$\frac{d\gamma_i}{dx} = F_i(p,T,\gamma_1,\gamma_2,\ldots,\gamma_N)/\mu_i W = L_i(p,T,\gamma_1,\gamma_2,\ldots,\gamma_N) - \gamma_i N_i(p,T,\gamma_1,\gamma_2,\ldots,\gamma_N) \quad , \quad (i = 1,2,\ldots,N) \quad . \tag{6.37}$$

The thermal and caloric equations of state,

$$p = R\rho T/\mu \quad , \tag{6.38}$$

$$h = \sum_{i=1}^{N} h_i(T)\gamma_i \quad , \tag{6.39}$$

are added, where the $h_i(T)$ are the enthalpies of the individual components which may be found, for instance, in [6.11].

The right side of (6.37) can be written down in two equivalent forms, the latter being convenient for numerical integration by means of implicit difference schemes (Sect.3.1). Evidently, reduction to such a form is always possible. In particular, for (6.32),

$$L = \rho\left[K_+^{(2)}\frac{\rho\gamma_H^2}{\mu} + K_-^{(11)}\gamma_{OH}\gamma_H + K_-^{(9)}\gamma_H\gamma_{H_2O}\right]/W \quad ,$$

$$N = \rho\left[K_-^{(2)}/\mu + K_+^{(11)}\gamma_O + K_+^{(9)}\gamma_{OH}\right]/W \quad .$$

The second form of the right side of (6.37) is equivalent to the form of relaxation equations used in Sect.3.1. In fact,

$$L_i - \gamma_i N_i = N_i(L_i/N_i - \gamma_i) = N_i(\gamma_i^0 - \gamma_i) \quad ,$$

where γ_i^0 is the so called conditional equilibrium value of γ_i equal to γ_i in an equilibrium flow with an infinite reaction rate ($N_i \to \infty$).

The continuity equation (6.36) may be transformed with the help of equations of motion and energy to

$$\frac{1}{W}\frac{dW}{dx} = \frac{1}{M^2 - 1}\left\{\frac{1}{F}\frac{dF}{dx} - \sum_{i=1}^{N} F_i\left[\mu - \frac{1}{T}\frac{(\partial h/\partial\gamma_i)_T}{(\partial h/\partial T)\gamma_i}\right]\right\} \quad . \tag{6.40}$$

Here the Mach number M is calculated using the frozen sound velocity. For equilibrium or frozen flows $F_i = 0$ and (6.40) transforms into the usual equation of the one-dimensional theory [e.g. (2.125)]. However, the Mach number is determined by means of the equilibrium or the frozen sound velocity respectively. It follows from (6.40) that the value $M = 1$ (as well as the velocity maximum or the pressure minimum in a subsonic flow) is attained not at the nozzle throat but downstream of it. This applies to all nonequilibrium nozzle flows and holds for two-phase nonequilibrium flows too. If the function $F = F(x)$ is specified, then, to solve a direct problem, (6.40) is used instead of (6.36).

The system (6.34-39) is composed of $N + 4$ equations and contains $N + 5$ unknowns p, ρ, T, W, F and γ_i; thus it is unclosed. It can, however, be closed by specifying any of the functions, say, F(x), or one of the functions p(x), ρ(x), W(x). In the first case, one has to solve the direct problem in the one-dimensional approximation, i.e. to determine the flow parameters for a specified stream tube $F = F(x)$. Solution of this problem in a region embracing the subsonic part of a nozzle is rather complicated, since, due to the singularity present in (6.40) at $M = 1$, one has to find the regular integral curve passing through the saddle-type singular point and having a finite derivative at it. This can be done only by means of varying the flow parameters at the initial cross-section in such a way as to avoid the solution's subsonic branch and to ensure, at the same time, regularity of the solution, i.e. to ensure that the numerator and the denominator on the right side of (6.40) tend to zero simultaneously. This results in an increase of computation time.

Therefore, an attempt to solve the direct problem in the framework of the one-dimensional approximation while studying various nonequilibrium nozzle flows, two-phase among them, seems to be wasteful. On the other hand, the advantages of solving the inverse problem with a specified longitudinal distribution of one of the gasdynamic parameters — p, ρ or W — are quite evident. In this case there is no need to use (6.40), since the configuration of the stream tube is not specified and must be found from (6.36), while the system (6.34-39) no longer contains a singular point. Also, the pressure and especially the density and velocity deviate only slightly from the corresponding equilibrium values and can be either determined by calculating, e.g. two-dimensional equilibrium flows, or measured experimentally. Finally, an algorithm for solving the system (6.34-39) with a specified distribution $p = p(x)$ may be fully used to calculate three-dimensional nonequilibrium flows. If a nonequilibrium process starts in the supersonic part of a nozzle (for instance, in the case of nonequilibrium condensation), then the necessity of passing through the singularity point is eliminated and both direct and inverse problems can be solved.

Let us describe briefly a method of solving the system (6.34-39) at a pressure distribution specified along a stream tube [6.12-15]. We shall be looking for the solution at the point x_m knowing the solution at the point x_{m-1} and the value of the pressure at the point x_m. Then the system (6.34-39) [without (6.36) and (6.38) used to determine ρ and F at the final stage of calculation] is approximated by the following system of difference equations:

$$f_i \equiv \gamma_{i(m-1)} - \gamma_{im} + \frac{h_m}{1 + (1 - s)h_m N_{im}}\{ sL_{i(m-1)} + (1 - s)L_{im} - \gamma_{i(m-1)}[sN_{i(m-1)} + (1 - s)N_{im}]\} = 0 \quad , \quad (i = 1,2,\dots,N) \ , \qquad (6.41)$$

$$f_{N+1} \equiv W_m^2 - W_{m-1}^2 + R\left[\left(\frac{T}{p\mu}\right)_m + \left(\frac{T}{p\mu}\right)_{m-1}\right](p_m - p_{m-1}) = 0 \quad , \qquad (6.42)$$

$$f_{N+2} \equiv \frac{1}{2} W_m^2 + \sum_{i=1}^{N} h_i(T_m)\gamma_{im} - H_0 = 0 \quad . \tag{6.43}$$

Here $h_m = x_m - x_{m-1}$, $0 \leqslant s \leqslant 0.5$. The system (6.41-43) may be presented in the form

$$\begin{aligned} &\mathbf{f}(\mathbf{y}_m) = 0 \quad , \\ &\mathbf{f} = (f_1, f_2, \ldots, f_{N+2}) \quad , \quad \mathbf{y}_m = (y_1, y_2, \ldots, y_{N+2})_m \quad , \\ &y_i = \gamma_i \quad , \quad (i = 1,2,\ldots,N) \quad , \quad y_{N+1} = W^2 \quad , \quad y_{N+2} = T \quad . \end{aligned} \tag{6.44}$$

At a fixed m the system (6.44) can be solved by Newton's method,

$$\mathbf{y}_m^{j+1} = \mathbf{y}_m^j - D_m^{-1}\mathbf{f}(\mathbf{y}_m^j) \quad , \tag{6.45}$$

where j is the iteration number and D is a variable Jacobi matrix. Iterations are terminated when the following conditions are fulfilled simultaneously:

$$|y_{im}^{j+1} - y_{im}^j| < \varepsilon |y_{im}^{j+1}| \quad , \quad |f_i(\mathbf{y}_m^{j+1})| < \delta_i \quad , \quad (i = 1,2,\ldots,N+2) \quad ,$$

where ε and δ_i are the allowable and closure errors respectively (Sect.3.1). Having found T, W, γ_i at the node x_m, one determines ρ from (6.38), and the area of a cross-section of the stream tube from (6.36), because the flow rate is already known. To ensure convergence of the iteration process the initial data must be specified correctly. The inverse problem for the system (6.34-38) is solved by means of formulating an initial-value problem. In the case of nonequilibrium flows the initial data must be specified in the subsonic region where the flow is close to equilibrium and where even small errors in the initial data may result in instability of a numerical solution. Therefore, the above algorithm of solution is also used to calculate the equilibrium part of flow; however, in this case equations of chemical kinetics are replaced by the mass action law (6.3), i.e. the denominator in (6.41) is substituted by $h_m(1-s)N_{im}$. The parameters thus obtained are later used as the initial data to calculate nonequilibrium flow which usually starts in the region where $p \approx (0.9\text{-}0.95)p_0$.

6.2.2 Major Characteristic Features of Chemical Nonequilibrium Flows

The major particularities of nonequilibrium flows may be studied in the one-dimensional approximation. In fact, studies of plane and axisymmetric nonequilibrium flows by means of numerical solution of nozzle gas flows [6.15]

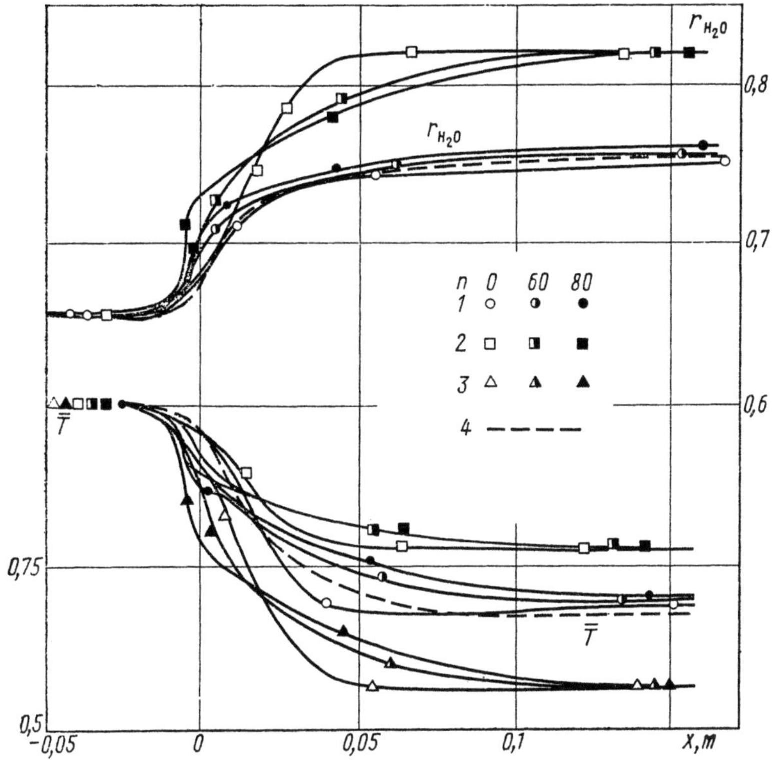

Fig.6.6. Variation of molar concentration r_{H_2O} and temperature $\bar{T}$ along streamlines at (*1*) nonequilibrium, (*2*) equilibrium and (*3*) frozen flows. Fuel composition: $H_2 + O_2$, $\alpha = 1$, $p_0 = 5$ bar, n is the layer number. (*4*) is calculated by means of one-dimensional theory

and calculations along stream tubes of an axisymmetric nozzle [6.13] show that the dependence of concentrations on the form of the stream tubes and pressure distributions along them is slight, especially in the supersonic part of a nozzle. Figure 6.6 presents the change of the molar fraction of water vapours and the dimensionless temperature $\bar{T} = T/T_0$ along streamlines. It is obvious that in equilibrium flows, concentrations differ considerably for different streamlines, while in nonequilibrium flows the difference is practically absent. Results of calculations of concentrations for a nonequilibrium flow in the one-dimensional approximation practically coincide with those of calculations accounting for the two-dimensional character of flow. At the same time, the temperature (and pressure) distributions along different streamlines differ considerably due to the two-dimensional character of flow; equilibrium, nonequilibrium and frozen flows also differ in this respect.

Before considering specific examples of calculations, let us note some general properties of nonequilibrium nozzle gas flows. Evidently, parameters of a nonequilibrium flow must, as a rule, have values intermediate between those of the two limiting flows, equilibrium and frozen (though local deviations may take place). Since in a nonequilibrium flow chemical energy is only partially released and is not fully transferred to active degrees of freedom and the energy of the directed motion of molecules, the temperature, velocity, pressure and impulse and, hence, the specific impulse in a nonequilibrium flow are below those of an equilibrium flow (though larger than in a frozen flow) (Fig.6.2). The greatest difference is observed in the values of temperature and pressure (sometimes, by tens of percent); in the case of velocity and impulse the difference is much less. The dependence of density on the flow regime is slight (though somewhat stronger than in nonequilibrium flows). Sometimes, however, local peaks of temperature and pressure may take place exceeding the corresponding equilibrium values due to an intense release of chemical energy occurring upon a sharp deviation from equilibrium, following, for instance, freezing of a flow beyond a corner point.

The impulse and specific impulse for nonequilibrium flows are always below those of equilibrium flows. The flow rate in a nonequilibrium flow at fixed stagnation parameters and throat size exceeds that of an equilibrium flow; however, the difference is small (no more than 0.5-1%). The Mach number for a nonequilibrium flow is higher than that for an equilibrium flow because of a considerable temperature drop which is not compensated by a velocity decrease.

Results of calculations of nonequilibrium flows along stream tubes may be conveniently presented in the form of the dependence of the flow parameters not on the nozzle's length, but on $\bar{r} = r/r_* = \bar{F}^{\frac{1}{2}} = (\rho_* W_*/\rho W)^{\frac{1}{2}}$, as in the case of one-dimensional equilibrium flows when the flow parameters depend not only on $\bar{r}$, but where there is also a slight dependence of the product ρW on the flow regime. Numerous fuel compositions have been studied in detail in the framework of the one-dimensional approximation; the results are presented in [6.1,16]. Graphics which allow the determination of specific impulse losses due to the nonequilibrium character of chemical reactions are presented in [6.1].

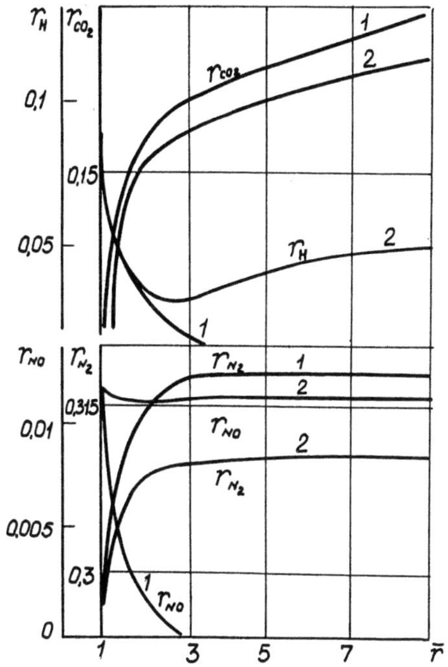

Fig.6.7. Dependence of molar concentrations r_H, r_{CO_2}, r_{NO} and r_{N_2} on $\bar{r}$ for (*1*) equilibrium and (*2*) nonequilibrium flows. Fuel composition: $(CH_3)_2NNH_2 + N_2O_4$, $\alpha = 0.9$, $p_0 = 5$ bar, $d_* = 100$ mm

Figure 6.7 shows some typical results of calculations of nonequilibrium nozzle flows. It follows from the results of calculations that composition and thermodynamic parameters of combustion products start to depart from the equilibrium values in the transonic part of a nozzle, where the gradients of gasdynamic parameters attain maximum values. In the case of low pressures and temperatures this departure for small nozzles may start in the subsonic part of a nozzle. Calculations also show that the molar fractions of CO_2, CO, N_2, O_2, H_2O and NO freeze in the vicinity of the throat and that their downstream variation does not exceed 5-10%, though for some of them (CO_2, H_2O, N_2, CO) the character of variation at a certain part of a nozzle is close to equilibrium. The molar fractions of O, H, OH and H_2 depart from the equilibrium values considerably; however, further downstream they are not frozen. In particular, the molar fraction of monatomic hydrogen has a minimum (Fig.6.7) which is absent in the case of an equilibrium reaction. Such a variation of the molar fraction of hydrogen is due to the exchange reactions being rather close to equilibrium throughout the flow, especially at high temperatures, while recombination reactions, associated with triple collisions, freeze suddenly in the transonic part of a nozzle.

The molecular weight of a mixture freezes in the vicinity of the throat and further stays constant, being 1-3% less than that in an equilibrium flow. In the case of the flow of a binary mixture with only one recombination reaction and no exchange reactions going on, the former freezes sharply in the transonic part of the flow; further downstream, the flow stays practically frozen. Such a character of reactions, when kinetics manifests itself in a narrow region only, enabled the development of an approximate method for calculating nonequilibrium flows assuming flow to be equilibrium up to a certain cross-section, and frozen further downstream.

The absolute size of a nozzle [its length or the throat diameter d_* and the form and geometrical expansion ratio $\bar{F}$ (or $\bar{r}$)], the stagnation pressure p_0 and temperature T_0 and the composition of a gas mixture at the nozzle's initial cross-section defined by the type of fuel and the oxidizer-to-fuel ratio α are the factors which influence the extent of departure from equilibrium most considerably. A change in a nozzle's absolute size is equivalent to a change in the time a gas resides in a nozzle — the shorter the time, the larger the departure of flow from equilibrium (other conditions being the same). Pressure drop also results in an increasing departure from equilibrium because the number of efficient collisions diminishes (especially in the case of trimolecular high-energy reactions), and the initial level of dissociation of combustion products increases.

Variation of temperature influences the extent of departure from equilibrium both through the rates of chemical reactions and by changing the initial level of dissociation. Recombination rate constants are practically independent of temperature; therefore, the rate of a reaction depends mainly on the presence of active radicals whose concentrations grow as the temperature increases. On the other hand, the temperature increase results in an increased extent of gas ionization at the initial cross-section of a nozzle and, hence, facilitates an increase in energy loss due to the nonequilibrium character of a process. Thus, a temperature variation may result either in an increase or a decrease of the exptent of departure from equilibrium, depending on which factor is dominant.

For C-O-H systems at $T_0 < 2300$ K a nozzle flow is frozen [6.16-19], resulting in the possibility of large specific impulse losses (up to 10%). In fact, at such temperatures the composition of combustion products may, under equilibrium conditions, sometimes be defined by means of the water vapour reaction $H_2O + CO \rightleftharpoons CO_2 + H_2$. However, the latter is not actually realized in the above form , but proceeds by means of the chain reactions presented in Table 6.1. Either the OH radical or monatomic components O and H must participate in any reaction of the chain; however, at $T_0 < 2300$ K they are practically absent. As a result, the reactions presented in Table 6.1 do not proceed and stay frozen from the combustion chamber onwards.

The composition of a gaseous mixture influences substantially the extent to which the nozzle flow parameters depart from the equilibrium values. The largest departure takes place in stoichiometric gaseous mixtures with a higher level of dissociation of combustion products at the initial cross-section. The extent of dissociation becomes smaller for enriched or lean mixtures, being accompanied by a decreasing departure from equilibrium. As the expansion ratio increases, the fraction of the energy of dissociation which can be returned to the flow in the course of equilibrium recombination grows. However, the actual fraction of the energy of dissociation which has time to transform into the kinetic energy of a flow depends slightly on the expansion ratio. Therefore, the extent of departure of a system from equilibrium increases with increasing $\bar{r}$.

An extensive series of calculations has shown that at a fixed α the molar fractions of CO_2, H_2O, CO, N_2 and H_2 (the total weight fraction of these substances in combustion products often equals 95-97%) change from 5 to 10% only, while p_0, d_*, $\bar{r}$ are varied within broad limits: 25 bar $< p_0 <$ 250 bar, 25 mm $< d_* <$ 250 mm, $1 < \bar{r} < 15$. This allows one to construct, using the results of calculations, the universal dependence of molar fractions of various components on α. Figure 6.8 presents an example of such a dependence for combustion products of the fuel $(CH_3)_2NNH_2 + N_2O_4$. The concentrations thus obtained differ from the corresponding equilibrium values by 10-20% at $\bar{r} < 5$ and by 20-60% at $5 < \bar{r} < 10$.

Nonequilibrium flow of chemical reactions results in a lower nozzle impulse compared to equilibrium flow. The difference of impulses for equilibrium

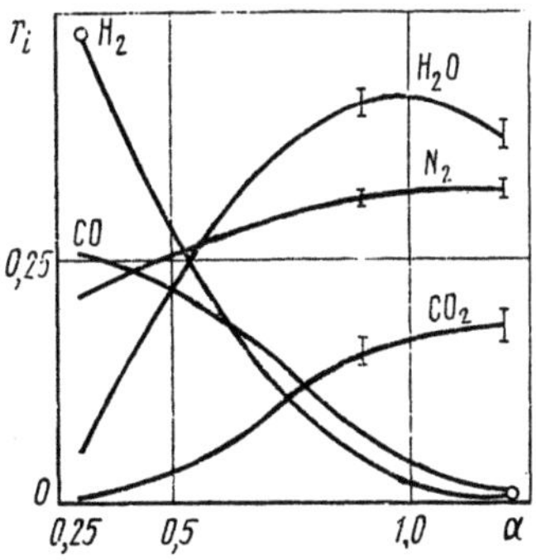

Fig.6.8. "Universal" dependence of molar concentrations of various components on α

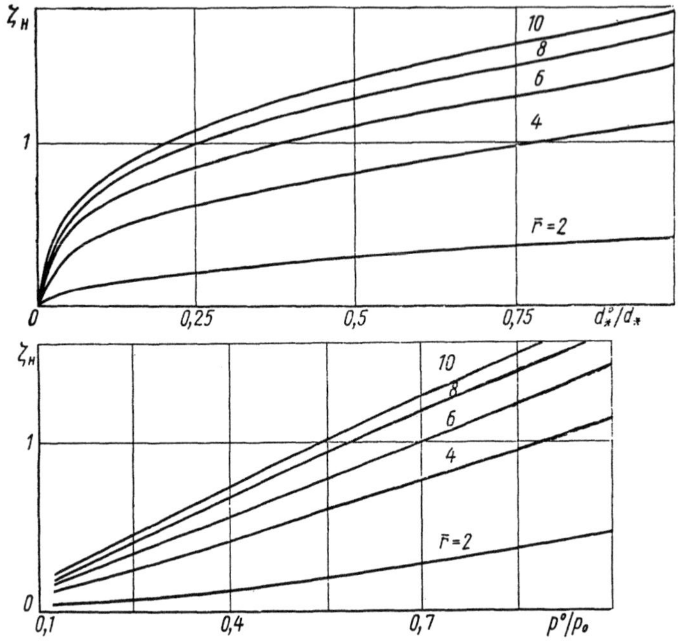

Fig.6.9. Dependence of ζ_H (%) on d_*^0/d_* ($d_*^0 = 25$ mm, $p_0 = 25$ bar) and p^0/p_0 ($p^0 = 25$ bar, $d_* = 25$ mm) for the fuel composition $H_2 + O_2$ at $\alpha = 0.8$

and nonequilibrium reaction flows, referred to the former, is called the coefficient of impulse losses due to nonequilibrium flow of chemical reactions and is denoted by ζ_H.

In [6.1], values of ζ_H are presented for a series of fuels used in liquid-propellent rocket engines, and also multiparametric graphs which allow ζ_H to be calculated. An analysis of the specific impulse losses ζ_H for various types of engines, supersonic and hypersonic ramjets among them, is presented in [6.16]. The typical dependence of ζ_H on d_*^0/d_* and p^0/p_0 is shown in Fig. 6.9, taking into consideration that $\zeta_H \to 0$ at d_*, $p_0 \to \infty$. The tendency of ζ_H to grow with the decrease of d_*, p_0 and the increase of $\bar{r}$ is evident. For the fuels O_2 + kerosene, $O_2 + (CH_3)_2NNH_2$ and $N_2O_4 + (CH_3)_2NNH_2$ at $\alpha = 0.8$, the values of ζ_H are close together, equal (at $d_* = 100$-200 mm and $\bar{r} = 5$-10) to 0.8-1.2% for $p_0 = 50$ bar and to 0.2-0.6% for $p_0 = 150$ bar (i.e. conditions characteristic for second-stage rockets). At $d_* = 10$-20 mm and the same values of p_0, $\zeta_H = 1.5$-2.5% and $\zeta_H = 0.8$-1.2%, respectively. For engines of spacecraft having lower values of p_0 ($p_0 = 10$-20 bar), the values of ζ_H increase and at d_* = 100-200 mm, equal 1.5-2.5%. It may be noted that the more nitrogen a fuel

Table 6.3. Dependence of the impulse loss coefficient ζ_H on d_* and p_0 at $\alpha = 0.8$, $\bar{r} = 5-10$

d_*, mm	p_0, bar	ζ_H, % hydrocarbon	$H_2 + O_2$
100-200	10-20	1.5-2.5	1.0-1.5
	50	0.8-1.2	0.5
	150	0.2-0.6	0.1
10-20	50	1.5-2.5	1.2-1.6
	150	0.8-1.2	0.4-0.6

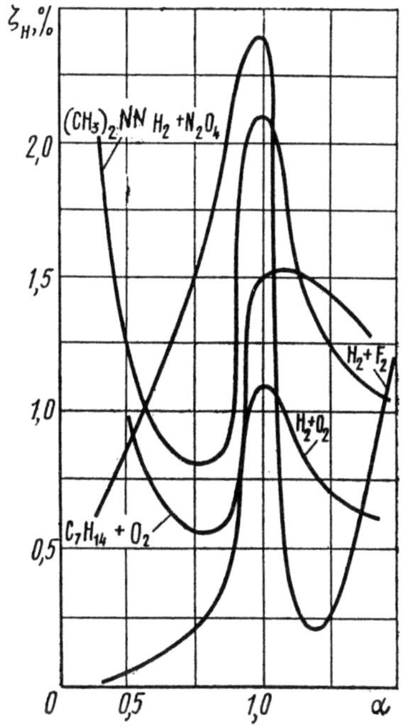

Fig.6.10. Dependence of ζ_H on α for various fuels at $p_0 = 25$ bar, $d_* = 50$ mm, $\bar{r} = 10$

contains, the less the losses due to an increase in the time of residence change. Calculations show that for the composition $H_2 + O_2$ the value of ζ_H is smaller than for hydrocarbon fuels. Some data on ζ_H are presented in Table 6.3.

The oxidizer-to-fuel ratio α has a strong influence on ζ_H (Fig.6.10). Characteristically, there is a maximum of losses near the stoichiometric value of $\alpha = 1$. A sharp increase of ζ_H at $\alpha < 0.4$ is also seen for fuels $(CH_3)_2NNH_2 + N_2O_4$ and $C_7H_{14} + O_2$. This can be explained as follows. The chemical energy accumulated in the combustion chamber transfers to active degrees of freedom of molecules in the process of recombination and exchange reactions. At high temperatures the major contribution is that of the recombination reactions, and at lower temperatures, the exchange reactions. The largest quantity of chemical energy can be accumulated in the combustion chamber at $\alpha \approx 1$, therefore, nonequilibrium flow of recombination reactions results in maximum losses at $\alpha \to 1$. At smaller and larger values of α, the contribution of chemical energy diminishes and the role of exchange reactions close to equilibrium increases as the temperature decreases, and the value of ζ_H becomes correspondingly smaller. At lower temperatures, below 2300 K ($\alpha < 0.4$), the quantity of chemical energy accumulated in the combustion chamber is smaller than at $\alpha \approx 1$; however, due to complete freezing of the exchange reactions, it is not transferred to active degrees of freedom at all, thus resulting in an increase of ζ_H.

A somewhat different situation takes place at small and large values of α for the composition $H_2 + O_2$. At low temperatures, only molecular hydrogen/oxygen and water vapours whose molar fractions stay evidently constant in the process of efflux (even in case of equilibrium) are present in the combustion chamber; thus there is no difference between equilibrium, frozen and nonequilibrium flows, and $\zeta_H = 0$. In practice, the dependence of ζ_H on α is used to choose the value of α corresponding to the maximum specific impulse. For solid-propellent rocket engines with combustion products containing condensate, the value of ζ_H is as a rule much lower than for liquid-propellent rocket engines due to a smaller fraction of chemical energy accumulated in the combustion chamber.

The entropy change in a nonequilibrium nozzle flow can be evaluated from (2.9), if the longitudinal variation of the flow parameters is known. However, it can be deduced without calculations that the change of entropy in the process of transfer from a fully equilibrium flow to a completely frozen one is nonmonotonic since under these limiting conditions entropy is constant and has the same value (at equal stagnation parameters). Therefore, in nonequilibrium flows the entropy change and, hence, the change of the total pressure do not characterize impulse losses. In fact, a frozen flow whose entropy is equal to that of an equilibrium flow, has maximum specific impulse losses. On the other hand, flows characterized by the same impulse may have different values of entropy. Thus a flow in a cylindrical tube which has deviated from equilibrium by flowing through a nozzle reaches the equilibrium state with the entropy increasing and the impulse staying constant. Calculations show that the entropy change in a chemically nonequilibrium flow is small (below 1%).

It is known that reaction rate constants differ sometimes by one or two orders of magnitude. This inspired a number of authors to study the influence of reaction rate constants upon the flow parameters (e.g. [6.16,19]). The studies have shown that the influence of reaction rate constants is most pronounced in those parts of the flow where the latter is neither equilibrium nor frozen; the influence is the slightest where the flow is close to being equilibrium or frozen. Therefore, variation of reaction rate constants produces an insignificant effect upon the flow parameters in nozzles of large liquid-propellent jet engines and solid-propellent rocket engines characterized by high pressures and temperatures in the combustion chambers, where the flow is close to equilibrium. Concentrations of components prove to be the most sensitive to variation of reaction rate constants, while the specific impulse losses are the least sensitive. For hypersonic ramjet engines ($p_0 \approx 1.6$ bar, $T_0 \approx 2700$ K) the energy-defining reaction $H + OH + M \rightleftharpoons H_2O + M$ for the H-O-N system has been found. Varying the rate constant of the reaction, one can change the extent of nonequilibrium of the flow considerably. Similar analyses which allow defining reactions to be found for a fixed system are very useful and make it possible to formulate recommendations concerning the accuracy of the reaction rate constants required to solve specific practical problems.

Numerous experimental studies of nonequilibrium nozzle flows have been carried out up to date in which the nozzle thrust, pressure distributions, temperatures and concentrations of individual components of a mixture have been measured [6.16-18]. A critical analysis of various experimental studies and their comparison with the results of calculating chemical nonequilibrium nozzle flows is presented in [6.16]. The major conclusion of the analysis is that the system of equations of chemical kinetics and the corresponding reaction rate constants presented in Tables 6.1,2, as well as numerical techniques developed in [6.13,15,19], allow one to evaluate the specific impulse losses, temperature and pressure accurately enough for practical needs. In order to predict concentrations with the same accuracy and to calculate nonequilibrium flows more accurately, further experimental studies of various reaction rate constants and of catalytic activity of various particles are needed. Figures 6.11,12 present some comparisons of numerical and experimental data.

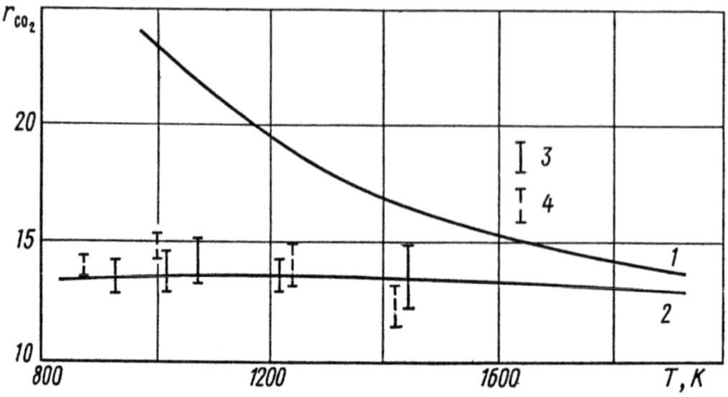

Fig.6.11

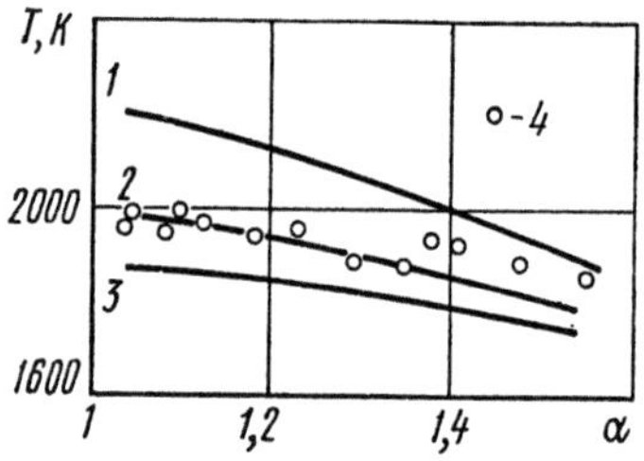

Fig.6.11. Comparison of calculated and experimental values of r_{CO_2} at $p_0 = 75$ bar, $T = 2500$ K: (*1*), (*2*) calculated equilibrium and nonequilibrium flows [6.19]; (*3*), (*4*) experiments for a conical nozzle with $\theta_k = 12.5°$ and $\theta_k = 25°$

Fig.6.12. Comparison of calculated and experimental temperature values for combustion products of hydrogen in air enriched with oxygen at $p_0/p_H = 8.3$: (*1*) calculated equilibrium, (*2*) nonequilibrium and (*3*) frozen flows, (*4*) experiment [6.16]

6.2.3 Plane and Axisymmetric Flows

Investigation of plane and axisymmetric nozzle flows is a much more complicated task than investigation of flows in the one-dimensional approximation because it is necessary to solve the system (6.34-39) repeatedly along streamlines in order to ensure convergence of iterations. The most consistent investigation of nonequilibrium flows of multicomponent mixtures is presented in [6.15], where the inverse problem of the theory of nozzles has been solved numerically. Investigation of three-dimensional nonequilibrium flows in the framework of the inverse problem of the theory of nozzles is preferable because it enables one to consider a nozzle flow as a whole and, most important, its transonic region where nonequilibrium effects are especially strong.

In Chap.3 we have described a method of numerical solution of the inverse problem of the theory of nozzles for the case of a three-dimensional nonequilibrium flow. Below, the difference scheme of [6.13] is presented for calculating plane and axisymmetric flows. In this case equations needed to determine the geometry of a streamline and pressure and velocity distributions along it are added to the system (6.34-39) which describes nonequilibrium flows in the one-dimensional approximation. Note that for a two-dimensional case W in (6.37) must be replaced by u.

We have

$$\frac{\partial r^{1+\nu}}{\partial\psi} = \frac{1+\nu}{\rho u} \quad , \tag{6.46}$$

$$\frac{\partial(pr^{1+\nu})}{\partial\psi} = \frac{(1+\nu)p}{\rho u} - r\,\frac{\partial v}{\partial x} \quad , \tag{6.47}$$

$$v = W\,\frac{\partial r}{\partial x}\left[1 + \left(\frac{\partial r}{\partial x}\right)^2\right]^{-\frac{1}{2}} \quad , \tag{6.48}$$

$$u = \sqrt{W^2 - v^2} \quad . \tag{6.49}$$

Unlike (2.107), (6.47) is presented in a form which allows the singularity at the symmetry axis to be removed. The stagnation enthalpy and the concentrations of components at the initial cross-section may generally depend on ψ and be discontinuous, i.e. flows with surfaces of tangential discontinuities can be calculated as well.

Let the solution be known at the layer ψ_n. Then, at a point of the layer ψ_{n+1} the functions $r_{mn} = r(x_m, \psi_n)$ and $p_{mn} = p(x_m, \psi_n)$ are defined by formulas

$$(r^{1+\nu})^{(k+1)}_{m(n+1)} = r^{1+\nu}_{mn} + \frac{1}{2}(1+\nu)\Delta\psi\left[\left(\frac{RT}{up\mu}\right)^{(k)}_{m(n+1)} + \left(\frac{RT}{up\mu}\right)_{mn}\right] \quad , \tag{6.50}$$

$$(pr^{1+\nu})^{(k+1)}_{m(n+1)} = (pr^{1+\nu})_{mn} + \frac{1}{2}(1+\nu)\Delta\psi\left[\left(\frac{RT}{u\mu}\right)^{(k)}_{m(n+1)} + \left(\frac{RT}{u\mu}\right)_{mn}\right] - \frac{1}{2}\Delta\psi\left[\left(r\,\frac{\partial v}{\partial x}\right)^{(k)}_{m(n+1)} + \left(r\,\frac{\partial v}{\partial x}\right)_{mn}\right] \quad , \tag{6.51}$$

where k is an iteration number. Calculations proceed as follows. At $k = 0$ all the values superscripted k are taken from the lower layer and $r^{(1)}_{m(n+1)}$, $p^{(1)}_{m(n+1)}$ are calculated by (6.50,51). The pressure distribution thus obtained is used to solve (6.34-39) along the streamline ψ_{n+1} with the initial data in the subsonic part, by means of approximating (6.41-43); as a result, $W^{(1)}_{m(n+1)}$, $T^{(1)}_{m(n+1)}$, $\gamma^{(1)}_{im(n+1)}$, $\rho^{(1)}_{m(n+1)}$ are obtained. Then, $v^{(1)}_{m(n+1)}$, $u^{(1)}_{m(n+1)}$ are calculated by (6.48,49) and the procedure is repeated using the solution of the first iteration, and so on. Derivatives with respect to x on the right sides of (6.47,48) are approximated by means of the three-point scheme (Sect.3.4).

At the initial streamline $\psi = \psi_0$, which may be the symmetry axis or a solid wall, the pressure distribution is specified; the rest of the parameters are determined by numerical integration of (6.34-39). The initial pressure distribution may be conveniently presented in the form

$$\bar{p} = \bar{p}_* + \frac{(\bar{p}_1 - \bar{p}_*)(\bar{p}_* - \bar{p}_2)(a - c)}{(\bar{p}_1 - \bar{p}_*)c + (\bar{p}_* - \bar{p}_2)a} , \qquad (6.52)$$

where $a = \exp(-\bar{x}/b)$, $c = 1$ at $\bar{x} > 0$ and $a = 1$, $c = \exp(\bar{x}/b)$ at $\bar{x} \leqslant 0$; $\bar{x} = x/r_*$, $\bar{p} = p/p_0$, $\bar{p}_1$, $\bar{p}_2$ are the values of pressure at the initial and final points of the nozzle axis, respectively, and b is a parameter characterizing the value of the pressure gradient in the transonic part of flow.

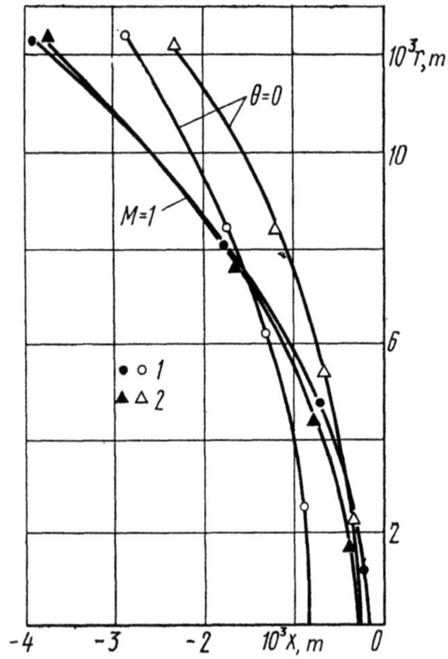

Fig.6.13. Lines of $M = 1$ and $\theta = 0$ for (*1*) nonequilibrium and (*2*) frozen flows in the throat

Figures 6.6,13 present some results of calculations of axisymmetric flows. Equilibrium, nonequilibrium and frozen flows have been calculated for the pressure distribution along the symmetry axis defined by (6.52) at $\bar{p}_1 = 0.96$, $\bar{p}_* = 0.55$, $\bar{p}_2 = 0.04$ and $b = 0.8$. The behaviour of the lines $M = 1$ and $\theta = 0$ (Fig.6.13) deserves special attention. For both frozen and equilibrium flows, these lines issue from the same point at the axis ($x = x_*$) and stretch upstream, the line $M = 1$ being always further upstream than the line $\theta = 0$ in accordance with general properties of the lines in axisymmetric and plane flows (Sect.5.1). In the case of nonequilibrium flows the lines $M=1$, $\theta=0$ stretch upstream too; however, they issue from different points of the symmetry axis, and the point on the axis where $M=1$ lies further downstream than the point corresponding to the line $\theta=0$, since in a nonequilibrium one-dimensional flow the line $M=1$ is displaced downstream of the throat [see (6.40)]. Inside the flow these lines intersect. In nonequilibrium axisymmetric and plane flows, the flow velocity, depending on the contour curvature at the throat, can be below, equal to or higher than the sound velocity; on the contrary, in equilibrium and frozen flows the flow velocity at the throat always exceeds the sound speed.

A certain increase of temperature at the streamlines with small curvature radii takes place in the transonic part of a flow. An increase of temperature beyond the corner point of a nozzle has been noted in [6.19] too. It is associated with the character of nonequilibrium processes and is absent in equilibrium and frozen flows. Actually, a large pressure gradient in the transonic part of a flow results in a sudden freezing of the chemical reactions which later start to proceed intensely in the course of further transition into the region with a smaller pressure gradient, resulting in heat release and temperature rise. Another peculiarity of temperature distribution takes place near the exit cross-section of a nozzle where the flow at small pressure gradients tends towards the local equilibrium state, resulting in an increase of nonequilibrium temperature at the exit cross-section of a nozzle and a change in the concentrations towards their equilibrium values.

The distribution of concentration of components at various streamlines of a nonequilibrium flow (Fig.6.6) is characterized by its slight dependence on the form of the streamline and the pressure distribution along it, which is in accordance with the results of calculations in the one-dimensional approximation. At the same time, concentrations of components at different streamlines of an equilibrium flow differ considerably. The dashed lines in Fig.6.6 present the results of a calculation of a nonequilibrium flow in the one-dimensional approximation, which can be used with sufficient accuracy to calculate concentrations of components; however, correct values of temperature are obtained in regions of small gradients only. Impulse losses due to a nonequilibrium flow of chemical reactions can be accurately determined in the one-dimensional approximation.

6.2.4 Approximate Methods of Calculating Nonequilibrium Flows

Approximate methods of calculating nonequilibrium nozzle flows are being developed side by side with the exact ones. One of the former—the method of sudden freezing—is used most often. It was originally proposed for a binary mixture in [6.20]; it has been generalized to the case of a multicomponent mixture in [6.21]. The main idea of the method is to replace a true nonequilibrium flow by two limiting flows, one of them equilibrium up to a certain cross-section called the section of sudden freezing, and the other frozen downstream of this section. For a binary mixture, where only one recombination reaction proceeds, such an approach is justified by the fact that in the region of high temperatures and pressures the flow is actually close to equilibrium, while in the region of large gradients of gasdynamic parameters, where pressure and temperature drop sharply, the recombination reaction slows down, since its rate is proportional to the cube of the density. Therefore, the region where the nonequilibrium reaction takes place is quite small and a true flow may be replaced by an equilibrium-frozen one without significant loss of accuracy. This is illustrated by Fig.6.14 where the variation of the molar fractions of monatomic hydrogen in the case of a nonequilibrium outflow of the binary mixture of H and H_2 is presented as a function of $\bar{r}$ at different values of d_*.

The major difficulty in using the method is associated with the determination of the cross-section, starting from which a flow can be considered frozen. According to [6.20], in an equilibrium region, $d\gamma_i/dt$ is much smaller than the dissociation and recombination rates D and R, and $D \approx R$. On the contrary, in the region where the flow is close to frozen, $D \approx \exp(-E/RT)$ decreases

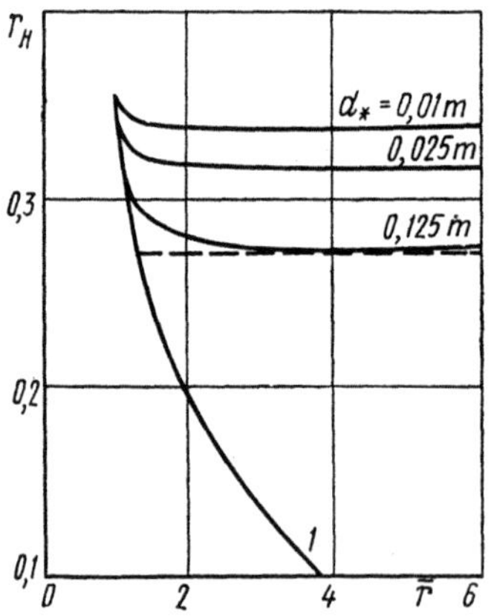

Fig.6.14. Dependence of molar atomic hydrogen concentration r_H on $\bar{r}$ at $p_0 = 10$ bar, $T_0 = 4000$ K: (*1*) equilibrium flow

so rapidly that $D \ll R$ and $-d\gamma_i/dt = R$. Therefore, one can assume that at the point of sudden freezing $-d\gamma_i/dt = AD$, while $d\gamma_i/dt$ and D may be obtained by calculating an equilibrium flow, because in front of the point of sudden freezing the flow is assumed equilibrium. The necessary value of the coefficient A is found for every particular case by means of selection with the objective of ensuring the best coincidence with the results of exact calculations. A may be put equal to or close to unity without much damage. The dashed line in Fig.6.14 presents the results of computations at such a value of A. Comparison of the exact results with those obtained by the method of sudden freezing for a binary mixture demonstrates a good agreement especially in the case where the region of transitional flow is small compared to that of the frozen flow at large expansion ratios.

Generalization of the method of sudden freezing to the case of a multicomponent mixture is quite a complicated and ambiguous procedure [6.15]. As has been pointed out above, for complex mixtures freezing of all components is not observed and one cannot always choose the single determining reaction. Therefore, the accuracy of the method of sudden freezing may be inadequate in similar cases. The above method allows one sometimes to determine the temperature of a multicomponent mixture accurately enough, but as a rule gives inaccurate values of the impulse and concentrations of individual components [6.19].

Since the computation of nonequilibrium flows is complicated, criterion dependences prove to be useful. However, the number of similarity parameters defined by the number of independent dimensionless criteria is very large (especially for complex mixtures), therefore, finding criterion dependences is a complicated procedure. Attempts have been made to determine parameters at the point of sudden freezing. It has been found that for binary mixtures and air, the dimensionless enthalpy, temperature and concentration of components at the point of sudden freezing are functions of the dimensionless stagnation temperature only. There are criterion dependences which generalize the results of computation of the impulse losses in the case where a number of determining parameters change simultaneously [6.16].

6.3 Flows with Vibrational Relaxation

As has been noted, the relaxation times of vibrational degrees of freedom are, by one or two orders of magnitude, shorter than those of chemical reactions while their energy at high temperatures is comparable to the chemical energy. Nevertheless, the process of vibrational deactivation may be non-equilibrium due to large gradients of gasdynamic parameters, especially where absolute nozzle sizes are small. Recently, interest in studying processes of vibrational deactivation in nozzles has grown in connection with the problem of the construction of gasdynamic optical quantum generators. The non-equilibrium flow of deactivation of vibrational degrees of freedom results in an inverse population of molecular vibrational levels which is necessary to initiate radiation. For practical applications it is also important to have quantitative data on the impulse losses due to relaxation of vibrational degrees of freedom. The majority of research dedicated to studying vibrational relaxation in nozzles deals with only one relaxing degree of freedom. The most complete investigation of vibrational relaxation in multicomponent mixtures has been reported in [6.22]. Below we present some results of the work.

6.3.1 Relaxation Equations and Methods of Prediction

The main molecular components of the combustion products of chemical fuels are N_2, CO_2, CO, H_2O, H_2, NO. The content of the rest of the molecular components, such as O_2 and OH, is usually insignificant and the nonequilibrium change of their vibrational energy may be ignored. The triatomic molecules CO_2 and H_2O possess three types of normal vibrations and, hence, three vibrational degrees of freedom. Multiatomic molecules with a certain vibrational degree of freedom will be considered as individual components of a mixture. Table 6.4 presents characteristic vibrational temperatures of molecules θ_V and designations of the molecules used below. Here $CO_2(\nu_1)$ and $H_2O(\nu_1)$ are the molecules which possess a valency symmetric mode of vibration, $CO_2(\nu_2)$ and $H_2O(\nu_2)$ are the molecules with a deformation mode of vibration and $CO_2(\nu_3)$ and $H_2O(\nu_3)$ are the molecules possessing a valency asymmetric mode of vibration.

In a multicomponent mixture of biatomic and multiatomic molecules, excitation and deactivation of vibrational degrees of freedom of molecules in

Table 6.4. Characteristic vibrational temperatures θ_V(K)

Molecule	N_2	$CO_2(\nu_1)$	$CO_2(\nu_2)$	$CO_2(\nu_3)$	CO	$H_2O(\nu_1)$	$H_2O(\nu_2)$	$H_2O(\nu_3)$	H_2	NO
Designation	1	2	3	4	5	6	7	8	9	10
θ_V,K	3353	1920	960	3380	3084	5260	2294	5400	6130	2700

inelastic collisions may proceed in two ways, either by means of a direct transition of kinetic energy of the colliding particles into vibrational energy of the molecule under consideration and vice versa (the process of direct excitation or deactivation, usually denoted by T-V), or by means of energy exchange between vibrational degrees of freedom of colliding molecules and inside a given molecule (the process of vibrational exchange denoted by V-V). In the mixture under consideration the following T-V processes may take place:

1) $N_2 + M \rightleftharpoons N_2^* + M - 2330\ cm^{-1}$,

2) $CO_2 + M \rightleftharpoons CO_2^*(\nu_2) + M - 667\ cm^{-1}$,

3) $CO + M \rightleftharpoons CO^* + M - 2143\ cm^{-1}$,

4) $H_2O + M \rightleftharpoons H_2O^*(\nu_2) + M - 1595\ cm^{-1}$,

5) $H_2 + M \rightleftharpoons H_2^* + M - 4260\ cm^{-1}$,

6) $NO + M \rightleftharpoons NO^* + M - 1876\ cm^{-1}$,

where M is a particle (a molecule or an atom) participating in a collision. The probability of energy transfer in a collision depends on the nature of the particle. The superscript * denotes vibrationally excited molecules and the number on the right side defines the loss of energy in the process, i.e. the decrease in kinetic energy of colliding particles in the case of excitation of the first level of vibrations of a molecule. For multiatomic molecules CO_2 and H_2O relaxation processes are written for the deformational vibration only, whose frequency ν_2 is lower than the frequencies of the other types of vibrations. The other modes of vibrations of multiatomic molecules are excited by means of intermolecular exchange of vibrational energy.

The relaxation mechanism of valency vibration modes of CO_2 and H_2O is unknown. Various ways of collisional energy relaxation of the asymmetric vibrational mode ν_3 of molecules CO_2 are possible. Relaxation of the asymmetric vibration of the CO_2 molecule may be assumed to proceed as follows:

7) $CO_2^*(\nu_3) + M \rightleftharpoons CO_2^*(\nu_1 + \nu_2) + M$,

i.e. decrease of the energy of vibrations of CO_2 by a quantum ν_3 results in the formation of a quantum ν_1 and a quantum ν_2 of vibrational energy. Since the probability of exchange of vibrational energy in a collision increases as the energy loss occurring in the process decreases, i.e. as the energy exchange between vibrational and translational degrees of freedom decreases, the following processes may be assumed to determine the relaxation of valency vibrations of $CO_2(\nu_1)$, $H_2O(\nu_1)$ and $H_2O(\nu_3)$:

8) $CO_2^*(\nu_1) + M \rightleftharpoons CO_2^*(2\nu_2) + M$,

9) $H_2O^*(\nu_1) + M \rightleftharpoons H_2O^*(2\nu_2) + M + 467\ cm^{-1}$,

10) $H_2O^*(\nu_3) + M \rightleftharpoons H_2O^*(2\nu_2) + M + 556\ cm^{-1}$.

Process 8 is a resonant one, because accidental degeneracy of the frequencies $\nu_1 = 2\nu_2$ may take place in CO_2.

Effective processes of vibrational energy exchange are possible in the mixture of gases under consideration, due to close values of vibrational frequencies of some molecules. These processes should also be included in the general system of relaxation processes. The following processes with relatively small loss of energy have been chosen as representative of similar V-V processes:

11) $N_2^* + CO_2 \rightleftharpoons N_2 + CO_2^*(\nu_3) - 19\ cm^{-1}$,

12) $N_2^* + CO \rightleftharpoons N_2 + CO^* + 187\ cm^{-1}$,

13) $CO^* + CO_2 \rightleftharpoons CO + CO_2^*(\nu_3) - 206\ cm^{-1}$,

14) $N_2^* + NO \rightleftharpoons N_2 + NO^* + 454\ cm^{-1}$,

15) $CO^* + NO \rightleftharpoons CO + NO^* + 267\ cm^{-1}$,

16) $CO_2^*(\nu_3) + NO \rightleftharpoons CO_2 + NO^* + 473\ cm^{-1}$.

Kinetic equations describing the relaxation of vibrational energy, in accordance with the above processes, may be constructed by two different methods. One can write down a system of differential equations which allow the population of every excited level of every vibrational mode to be determined by means of transition probabilities [6.23]. However, such an approach, without any simplifying hypotheses, proves to be complicated not only due to the considerable number of levels, but also due to the absence of reliable experimental and numerical data on transition probabilities for such a complicated interacting mixture of molecules. The other approach consists of formulating differential equations for the determination of the energy of every mode. Assuming a Boltzmann distribution of the energies for every mode and considering molecules to be harmonic oscillators, relaxation equations corresponding to the above processes may be written in the form:

$$\frac{d\varepsilon_1}{dt} = \varphi_1 + \varphi_{11} + \varphi_{12} + \varphi_{14} \quad , \tag{6.53}$$

$$\frac{d\varepsilon_2}{dt} = \varphi_8 - \frac{\vartheta_2}{\vartheta_4}\varphi_7 \quad , \tag{6.54}$$

$$\frac{d\varepsilon_3}{dt} = \varphi_2 - \frac{\vartheta_3}{\vartheta_4}\varphi_7 - \varphi_8 \quad , \tag{6.55}$$

$$\frac{d\varepsilon_4}{dt} = \varphi_7 - \frac{\vartheta_4}{\vartheta_1}\varphi_{11} - \frac{\vartheta_4}{\vartheta_5}\varphi_{13} + \varphi_{16} \quad , \tag{6.56}$$

$$\frac{d\varepsilon_5}{dt} = \varphi_3 - \frac{\vartheta_5}{\vartheta_1}\varphi_{12} + \varphi_{13} + \varphi_{15} \quad , \tag{6.57}$$

$$\frac{d\varepsilon_6}{dt} = \varphi_9 \quad , \tag{6.58}$$

$$\frac{d\varepsilon_7}{dt} = \varphi_4 - 2\frac{\vartheta_7}{\vartheta_6}\varphi_9 - 2\frac{\vartheta_7}{\vartheta_8}\varphi_{10} \quad , \tag{6.59}$$

$$\frac{d\varepsilon_8}{dt} = \varphi_{10} \quad , \tag{6.60}$$

$$\frac{d\varepsilon_9}{dt} = \varphi_5 \quad , \tag{6.61}$$

$$\frac{d\varepsilon_{10}}{dt} = \varphi_6 - \frac{\vartheta_{10}}{\vartheta_1}\varphi_{14} - \frac{\vartheta_{10}}{\vartheta_5}\varphi_{15} - \frac{\vartheta_{10}}{\vartheta_4}\varphi_{16} \quad , \quad \text{where} \tag{6.62}$$

$$\varphi_1 = \Phi_1(\varepsilon_1^0 - \varepsilon_1) \;, \quad \varphi_2 = \Phi_2(\varepsilon_3^0 - \varepsilon_3) \;, \quad \varphi_3 = \Phi_3(\varepsilon_5^0 - \varepsilon_5) \quad ,$$

$$\varphi_4 = \Phi_4(\varepsilon_7^0 - \varepsilon_7) \;, \quad \varphi_5 = \Phi_5(\varepsilon_9^0 - \varepsilon_9) \;, \quad \varphi_6 = \Phi_6(\varepsilon_{10}^0 - \varepsilon_{10}) \quad ,$$

$$\varphi_7 = \Phi_7\left(r_3\frac{R}{\mu}\right)^{-2}(\vartheta_2\vartheta_3)^{-1}\left\{\frac{1}{2}\varepsilon_2\varepsilon_3\left(\varepsilon_4 + r_4\frac{R}{\mu}\vartheta_4\right)\exp\left(\frac{\vartheta_2 + \vartheta_3 - \vartheta_4}{T}\right)\right.$$
$$\left. - \varepsilon_4\left[r_3\frac{R}{\mu}\left(\varepsilon_2\vartheta_3 + \frac{1}{2}\varepsilon_3\vartheta_2\right) + \frac{1}{2}\varepsilon_2\varepsilon_3 + \left(r_3\frac{R}{\mu}\right)^2\vartheta_2\vartheta_3\right]\right\} \quad ,$$

$$\varphi_8 = \Phi_8\left(r_3 \frac{R}{\mu} \vartheta_3\right)^{-1}\left[\frac{1}{2}\, \varepsilon_3^2 - \varepsilon_2\left(\varepsilon_3 + r_3 \frac{R}{\mu} \vartheta_3\right)\right] ,$$

$$\varphi_9 = \Phi_9\left(r_7 \frac{R}{\mu} \vartheta_7\right)^{-2}\left[\varepsilon_7^2\left(\varepsilon_6 + r_6 \frac{R}{\mu} \vartheta_6\right)\exp\left(\frac{2\vartheta_7 - \vartheta_6}{T}\right) - \varepsilon_6\left(\varepsilon_7 + r_7 \frac{R}{\mu} \vartheta_7\right)^2\right] ,$$

$$\varphi_{10} = \Phi_{10}\left(r_7 \frac{R}{\mu} \vartheta_7\right)^{-2}\left[\varepsilon_7^2\left(\varepsilon_8 + r_8 \frac{R}{\mu} \vartheta_5\right)\exp\left(\frac{2\vartheta_7 - \vartheta_8}{T}\right) - \varepsilon_8\left(\varepsilon_7 + r_7 \frac{R}{\mu} \vartheta_7\right)^2\right] ,$$

$$\varphi_{11} = \Phi_{11}\left(r_4 \frac{R}{\mu} \vartheta_4\right)^{-1}\left[\varepsilon_4\left(\varepsilon_1 + r_1 \frac{R}{\mu} \vartheta_1\right)\exp\left(\frac{\vartheta_4 - \vartheta_1}{T}\right) - \varepsilon_1\left(\varepsilon_4 + r_4 \frac{R}{\mu} \vartheta_4\right)\right] ,$$

$$\varphi_{12} = \Phi_{12}\left(r_5 \frac{R}{\mu} \vartheta_5\right)^{-1}\left[\varepsilon_5\left(\varepsilon_1 + r_1 \frac{R}{\mu} \vartheta_1\right)\exp\left(\frac{\vartheta_5 - \vartheta_1}{T}\right) - \varepsilon_1\left(\varepsilon_5 + r_5 \frac{R}{\mu} \vartheta_5\right)\right] ,$$

$$\varphi_{13} = \Phi_{13}\left(r_4 \frac{R}{\mu} \vartheta_4\right)^{-1}\left[\varepsilon_4\left(\varepsilon_5 + r_5 \frac{R}{\mu} \theta_5\right)\exp\left(\frac{\vartheta_4 - \vartheta_5}{T}\right) - \varepsilon_5\left(\varepsilon_4 + r_4 \frac{R}{\mu} \vartheta_4\right)\right] ,$$

$$\varphi_{14} = \Phi_{14}\left(r_{10} \frac{R}{\mu} \vartheta_{10}\right)^{-1}\left[\varepsilon_{10}\left(\varepsilon_1 + r_1 \frac{R}{\mu} \vartheta_1\right)\exp\left(\frac{\vartheta_{10} - \vartheta_1}{T}\right) - \varepsilon_1\left(\varepsilon_{10} + r_{10} \frac{R}{\mu} \vartheta_{10}\right)\right] ,$$

$$\varphi_{15} = \Phi_{15}\left(r_{10} \frac{R}{\mu} \vartheta_{10}\right)^{-1}\left[\varepsilon_{10}\left(\varepsilon_5 + r_5 \frac{R}{\mu} \vartheta_5\right)\exp\left(\frac{\vartheta_{10} - \vartheta_5}{T}\right) - \varepsilon_5\left(\varepsilon_{10} + r_{10} \frac{R}{\mu} \vartheta_{10}\right)\right] ,$$

$$\varphi_{16} = \Phi_{16}\left(r_{10}\frac{R}{\mu}\vartheta_{10}\right)^{-1}\left[\varepsilon_{10}\left(\varepsilon_4 + r_4\frac{R}{\mu}\vartheta_4\right)\exp\left(\frac{\vartheta_{10} - \vartheta_4}{T}\right)\right.$$

$$\left. - \varepsilon_4\left(\varepsilon_{10} + r_{10}\frac{R}{\mu}\vartheta_{10}\right)\right] \ .$$

Here ε_i is the ith component's vibrational energy per unit mass of mixture and ε_i^0 is the equilibrium value of the vibrational energy at the temperature of the gas. By Einstein's formula,

$$\varepsilon_i^0 = Rr_i\gamma_i\vartheta_i\left[\exp\left(\frac{\vartheta_i}{T}\right) - 1\right]^{-1} \quad ,$$

where the degree of degeneration $\gamma_i = 1$ at $i \neq 3$ and $\gamma_3 = 2$, r_i is the molar concentration of the ith component and Φ_i is the rate of the ith relaxation process which depends on the temperature and pressure of the gas.

The rates of relaxation processes depend on the character of the energy transfer [6.23]. In the case where vibrational energy is transferred into translational energy,

$$\Phi_i = p\sum_j r_j/\tau_{ij} \quad , \quad (i = 1,2,\ldots,6 \ , \ j = 1,3,4,5,7,9,10) \quad , \tag{6.63}$$

where τ_{ij} is the relaxation time of the ith process for the case of collisions with molecules of the jth kind at $p = 1$ atm and where p is the pressure. In the case of intermolecular transfer of vibrational energy, Φ_i is also determined by (6.63) at $i = 7,8,9,10$. In the case of intermolecular exchange of vibrational energies,

$$\Phi_i = p\frac{r_j}{\tau_i} \quad , \quad (i = 11,12,13,14,15,16 \ , \ j = 4,5,4,10,10,10) \quad , \tag{6.64}$$

where τ_i is the relaxation time of the ith process at $p = 1$ atm. Usually, the dependence of the relaxation times of elementary processes on temperature is determined by experimental data and whenever possible correlated with the help of the Landau-Teller theory, which gives a linear dependence of $\ln\tau$ on $T^{-1/3}$ [6.24]. Reference [6.22] recommends the following formulas for the computation of relaxation times based on a detailed analysis of experimental data:

Process 1

$$
\begin{aligned}
\ln\tau_{1,1} &= 220T^{-1/3} - 24.81 \quad , \\
\ln\tau_{1,3} &= 240T^{-1/3} - 25.76 \quad , \\
\ln\tau_{1,5} &= 220T^{-1/3} - 24.81 \quad , \\
\ln\tau_{1,7} &= 33.5T^{-1/3} - 16.55 \quad , \\
\ln\tau_{1,9} &= 86.9T^{-1/3} - 19.94 \quad , \\
\ln\tau_{1,10} &= 223{,}9T^{-1/3} - 24.96 \quad .
\end{aligned}
\tag{6.65}
$$

Process 2

$$
\begin{aligned}
\ln\tau_{2,1} &= 40.1T^{-1/3} - 17.51 \quad , \\
\ln\tau_{2,j} &= 40.1T^{-1/3} - 17.91 \quad , \quad (j = 3,5,10) \quad , \\
\ln\tau_{2,7} &= -48.3T^{-1/3} - 12.55 \quad , \\
\ln\tau_{2,9} &= -16.8T^{-1/3} - 15.3 \quad .
\end{aligned}
\tag{6.66}
$$

Process 3

$$
\begin{aligned}
\ln\tau_{3,1} &= 174T^{-1/3} - 23.3 \quad , \\
\ln\tau_{3,3} &= 191T^{-1/3} - 24.23 \quad , \\
\ln\tau_{3,5} &= 174T^{-1/3} -23.3 \quad , \\
\ln\tau_{3,7} &= 72T^{-1/3} -21.68 \\
\ln\tau_{3,9} &= 67T^{-1/3} - 19.5 \quad , \\
\ln\tau_{3,10} &= 177T^{-1/3} - 23.57 \quad .
\end{aligned}
\tag{6.67}
$$

Process 4

$$
\begin{aligned}
\ln\tau_{4,j} &= -36.8T^{-1/3} - 10.85 \quad , \quad (j \neq 7) \quad , \\
\ln\tau_{4,7} &= -36.8T^{-1/3} - 13.15 \quad .
\end{aligned}
\tag{6.68}
$$

Process 5

$$
\begin{aligned}
\ln\tau_{5,j} &= 100T^{-1/3} - 20.3 \quad , \quad (j = 1,3,5,10) \quad , \\
\ln\tau_{5,j} &= 100T^{-1/3} - 21.67 \quad , \quad (j = 7,9) \quad ,
\end{aligned}
\tag{6.69}
$$

Process 6

$$
\begin{aligned}
\ln\tau_{6,j} &= 100T^{-1/3} - 18.3 \quad , \quad (j \neq 10) \quad , \\
\ln\tau_{6,10} &= -14.9 \quad .
\end{aligned}
\tag{6.70}
$$

Process 7

$$
\begin{aligned}
\ln\tau_{7,1} &= -29.4 + 255T^{-1/3} - 1000T^{-2/3} \quad , \\
\ln\tau_{7,3} &= -22.5 + 101T^{-1/3} - 223T^{-2/3} \quad , \\
\ln\tau_{7,5} &= -19.6 + 74.2T^{-1/3} - 160T^{-2/3} \quad , \\
\ln\tau_{7,7} &= -14.7 - 13.8T^{-1/3} + \ln f(T) \quad , \\
\ln\tau_{7,9} &= -16.3 + 10.1T^{-1/3} + \ln f(T) \quad , \\
\ln\tau_{7,10} &= -18.6 + 67.27T^{-1/3} - 204T^{-2/3} \quad ,
\end{aligned}
\tag{6.71}
$$

where

$$
f(T) = \left[1 - \exp\left(-\frac{\vartheta_4}{T}\right)\right]\left[1 - \exp\left(-\frac{\vartheta_2}{T}\right)\right]^{-1}\left[1 - \exp\left(-\frac{\vartheta_3}{T}\right)\right]^{-1} \quad .
$$

Processes 11-15

$$
\begin{aligned}
\ln\tau_{11} &= -17.1 + 0.38 \cdot 10^{-2}T - 0.13 \cdot 10^{-5}T^2 \\
\ln\tau_{12} &= -0.4 \cdot 10^{-3}T - 13.15 \quad , \\
\ln\tau_{13} &= -0.8 \cdot 10^{-3}T - 14 \quad , \\
\ln\tau_{14} &= 48.5T^{-1/3} - 17.3 \quad , \\
\ln\tau_{15} &= 20.7T^{-1/3} - 16.4 \quad .
\end{aligned}
\tag{6.72}
$$

The following remarks may be made regarding Processes 8, 9, 10 and 16. The energy relaxation for the symmetric vibration mode of CO_2 molecules (Process 8) has been subjected to superficial experimental study only. Due to Fermi resonance, CO_2 displays a strong interaction between the symmetric and deformational modes of vibration. Thus, the process of exchange of vibrational energy may be expected to proceed very rapidly. The experimental data obtained up to date do not contradict this conclusion and allow one to assume that the symmetric and deformational modes of vibration stay in mutual equilibrium during the process of relaxation. Similarly, due to a total absence of experimental data, the valency vibrational modes of H_2O molecules during relaxation (Processes 9 and 10) are usually assumed to be in equilibrium with the deformational mode of vibration. Note that in the temperature interval of practical importance, the contribution of these modes to the vibrational energy of H_2O molecules is, due to high characteristic temperatures, small compared to that of the deformational mode.

In order to satisfy the condition of local equilibrium of vibrational modes, the relaxation times of the Processes 8-10 are formally assumed to be two orders of magnitude smaller than the relaxation time of the corresponding process for the deformational mode. There are no experimental data for the Process 16. Since the Processes 14 and 16 have practically equal energy losses, it is usually assumed that their relaxation times coincide. A nozzle flow with relaxation of vibrational degrees of freedom is described in the one-dimensional approximation (as in the case of flows with nonequilibrium chemical reactions) by the system (6.34-36,38,39), in which the relaxation equations (6.53-62) for the energies of various vibrational modes of molecules should be used instead of the relaxation equations (6.37) for concentrations of chemical components. Consideration of (6.53-62) shows that any of them can be presented in the form characteristic of relaxation equations, similar to the chemical kinetics equations (6.37), i.e. in the form

$$W \frac{d\varepsilon_i}{dx} = L_i(p,T,\varepsilon_1,\varepsilon_2,\ldots,\varepsilon_N) - \varepsilon_i N_i(p,T,\varepsilon_1,\varepsilon_2,\ldots,\varepsilon_N) \quad .$$

As before, one should solve the inverse problem specifying pressure distribution along a nozzle. The difference method of solving the system of equations is based, similarly to the case of chemical equilibrium flows, upon the use of implicit difference schemes for the numerical solution of relaxation equations. Therefore, replacing γ_i in (6.43) by ε_i, the same system (6.41-43) may be used.

In the case where flows are nonequilibrium the values sought for the energies of vibrational degrees of freedom enter expressions for the enthalpy of individual components $h_i(T)$ in linear combination. When solving the resulting system of difference equations of the type (6.41-43), it is assumed that the γ_i in (6.42) are known as a result of calculating chemical nonequilibrium flow. The method of solution is similar to the one used for chemical nonequilibrium flow and has been described in detail in the preceding section. After determining the energy of the vibrational mode ε_i, the vibrational temperature T_i is found from Einstein's formula under the assumption of a Boltzmann distribution of the energy among the levels of a mode. Also, knowing the values of the energies ε_i, one can determine the populations of various levels [6.23] if necessary. A calculation for a nozzle flow with relaxation of vibrational degrees of freedom proceeds as follows. First of all, a chemical nonequilibrium flow for equilibrium excitation of the vibrational degrees of freedom of molecules is calculated by the method described in Sect.6.2. Computation of a chemical nonequilibrium flow is necessary to determine the molar concentrations of components used in the system of equations for a vibrational nonequilibrium flow. The distributions of the molar concentrations thus obtained are further used to calculate the flow along a stream tube with vibrational relaxation and to determine all the flow parameters, including the area of the stream tube (using the continuity equation). This algorithm does not take into account the mutual influence of chemical and vibrational relaxations. Computations show that in the expansion of combustion products of modern chemical fuels in nozzles, the processes may be separated in the first approximation, since vibrational nonequilibrium usually becomes noticeable at such a distance from the nozzle exit cross-section that the chemical composition is already frozen.

Since the vibrational state of a molecule changes in binary collisions, the rate of the vibrational energy change is directly proportional to the gas pressure. Therefore, it can easily be deduced from the relaxation equations that for a vibrational nonequilibrium nozzle gas flow the parameter $\Psi = p_0 L_0$, where p_0 and L_0 are characteristic values of pressure and of a linear dimension respectively, is a similarity parameter.

6.3.2 Results of Calculations in the One-Dimensional Approximation

Let us consider the effect of vibrational nonequilibrium upon the nozzle flow parameters and specific impulse losses for the example of the expansion of combustion products of the fuel $(CH_3)_2NNH_2 + N_2O_4$ [6.22]. Figure 6.15 presents the dependence of dimensionless vibrational temperatures T_v of various components on the dimensionless nozzle radius $\bar{r} = r/r_*$ at different values of Ψ and for oxidizer-to-fuel ratio $\alpha = 1.2$. We see that as the parameter Ψ decreases the vibrational temperatures of components deviate progressively from the equilibrium ones and may even freeze at sufficiently small values of Ψ and large expansion ratios. The smaller the value of Ψ, the closer to the throat the deviation of vibrational temperatures from their equilibrium values occurs. When Ψ decreases, the vibrational temperatures of H_2, NO, N_2 and CO exhibit the biggest deviation from the equilibrium values. For instance, if, at $\Psi = 625$ bar·mm, the vibrational tempera-

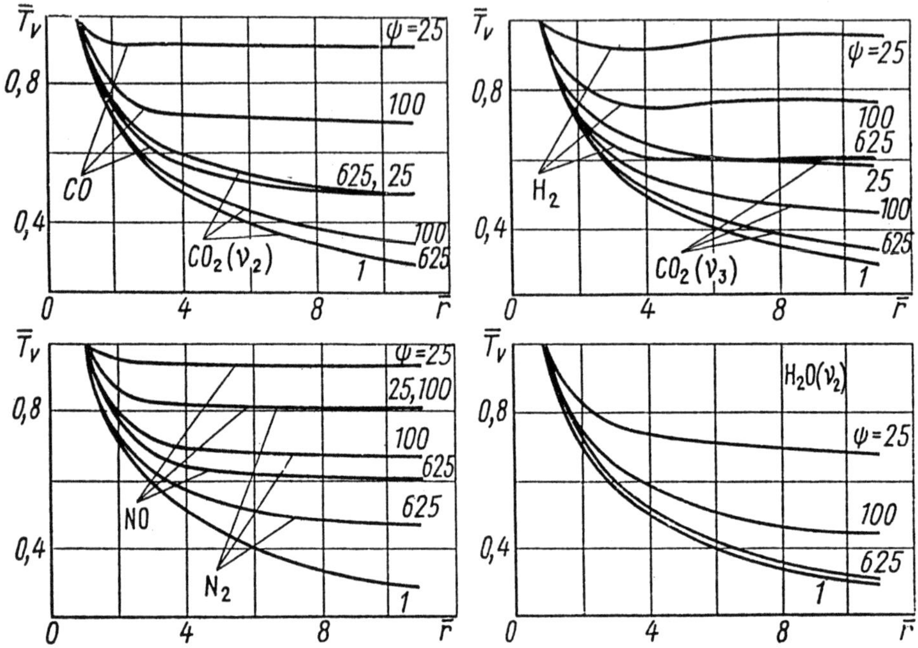

Fig.6.15. Dependence of vibrational temperature $\bar{T}_V = T_V/T_*$ on $\bar{r}$ for various Ψ (bar·mm): (*1*) equilibrium temperature

tures of CO_2 and H_2O are close to equilibrium, the vibrational temperature of CO at $\bar{r} = 10$ exceeds the equilibrium value more than 1.5 times.

Deviation of vibrational temperatures and, hence, of vibrational energies of components from their equilibrium values results in a change of the gas-dynamic parameters in a nozzle flow. The general character of the variation of parameters in a nonequilibrium flow with vibrational relaxation is similar to that in the case of a chemical nonequilibrium flow. The gas temperature, pressure, velocity and impulse in a nonequilibrium flow are below the corresponding values in an equilibrium flow, but exceed those of a frozen flow. As before, the gas density depends on the process character only slightly. Calculations show that at small values of Ψ (< 100 bar·mm) the gas pressure and temperature deviate from the corresponding equilibrium values by 10-20%. In the limiting case of a vibrationally frozen flow starting from the throat (Ψ = 0) these deviations may reach 40-50% (Fig.6.2).

Let us consider impulse losses ζ_{KH} due to vibrational nonequilibrium. Figure 6.16 shows the dependence of impulse losses ζ_{KH} on $\bar{r}$. We shall com-

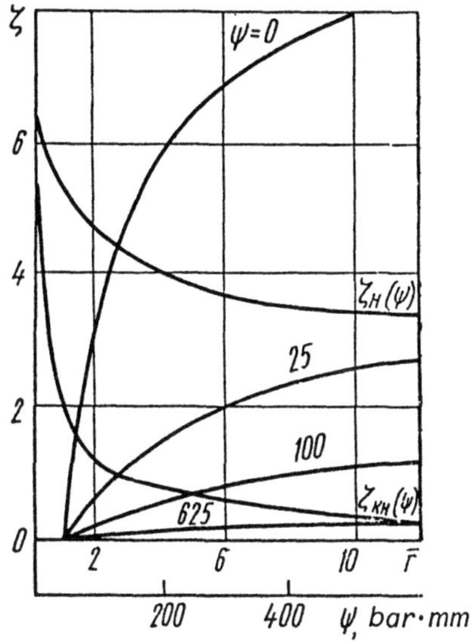

Fig.6.16. Dependence of impulse losses due to vibrational relaxation ζ_{KH} and due to nonequilibrium flow of chemical reactions ζ_H

pare (under identical conditions) specific impulse losses due to the nonequilibrium flow of chemical reactions and vibrational nonequilibrium. Since recombination takes place at triple collisions, impulse losses due to chemical nonequilibrium cannot be characterized by the parameter Ψ only; they also depend on p_0 and d_*. The values of dimensionless impulse losses (Fig. 6.16) due to the nonequilibrium flow of chemical reactions correspond to $p_0 = 25$ bar. In the limiting cases of frozen flow the maximum specific impulse losses due to chemical and vibrational nonequilibrium have close values. In a vibrationally frozen flow the main contribution to impulse losses belongs not to the components N_2, CO, $CO_2(\nu_3)$ whose characteristic vibrational temperatures exceed 3000 K, but to $H_2O(\nu_2)$, $CO_2(\nu_2)$ and $CO_2(\nu_1)$ whose characteristic vibrational temperatures are much lower and, hence, contribution of vibrational energy into enthalpy is much greater. However, since the relaxation times of recombination processes are usually much longer than those of vibrational relaxation, at fixed values of p_0 and d_* the degree of completion of relaxation processes, and, hence, the losses caused, are much greater than for the processes of vibrational deactivation. The losses due to vibrational nonequilibrium decrease sharply as the parameter Ψ increases and at $\Psi > 100$ bar·mm are some times smaller than the losses due to chemical nonequilibrium.

In frozen flows impulse losses decrease sharply at values of the oxidizer-to-fuel ratio $\alpha < 0.9$, because at moderate α the combustor temperature decreases and causes a decrease of the fraction of total vibrational energy in the gas enthalpy; simultaneously, the molar concentrations of CO_2 and H_2O, whose vibrational energy constitutes a considerable part of the total vibrational energy of the combustion products, also decrease. At the same time, at $\Psi = 100$ bar · mm, losses do not practically depend on the oxidizer-to-fuel ratio. This can be explained by the fact that in the latter case, unlike frozen flows, a decrease of the fraction of vibrational energy is accompanied by an increasing degree of completion of the process of vibrational deactivation, due to a decrease in the temperature of a nozzle flow. As stated above, for chemical nonequilibrium flows, concentrations of components depend only slightly on the character of the pressure distribution along a stream tube (Fig.6.6). On the contrary, in the case of flows with vibrational relaxation, the vibrational temperature, as well as the gas temperature, displays a considerable dependence on the character of the pressure distribution. Since pressure gradients at a nozzle contour are smaller than at the axis, then at equal $\bar{r}$, vibrational temperatures at the axis exceed those at the contour. Note that a two-dimensional nozzle flow with vibrational relaxation has been studied in [6.25] whose authors have numerically solved the inverse problem of the theory of nozzles by the method developed in [6.14] (see Chap.3).

The results of [6.22] allow us to conclude that when analyzing a vibrational nonequilibrium nozzle flow in detail, one must take into account the variation of pressure distributions at different streamlines, while the results obtained in the one-dimensional approximation may be used to estimate approximately the effect of nonequilibrium upon the flow parameters and impulse losses.

6.3.3 Plane and Axisymmetric Flows

Studies of plane and axisymmetric flows in nozzles have been reported in [6.25-28]. The inverse problem of the theory of nozzles is dealt with in [6.25-27]. Let us consider in detail [6.26,27]. The system of equations of gas dynamics is written in the form of (6.34,35,38,39,46-48). It is supplemented with equations of vibrational relaxation (6.53-62). The method of numerical solution is analogous to the one described in Sect.6.2.3. The numerical method of solving the inverse problem includes calculation, with the help of (6.51,52), of distributions of pressure and velocity components and determination of the pattern of streamlines at an unknown layer ψ_{n+1}.

Using the pressure distribution along the streamline ψ_{n+1} thus obtained, one solves a system of equations similar to (6.35-39) [where γ_i in (6.37) should be replaced by ε_i in accordance with the system (6.53-62)] with the help of approximating equations of the type (6.41-43) and with initial data specified in the subsonic part of the flow. At the initial streamline symmetry axis for the sub-, trans- and supersonic parts of flow, the pressure distribution is given in the form of (6.52), while the other parameters are defined by integration of the system (6.35-39). It is supposed that in the subsonic part of the flow, equilibrium excitation of the vibrational degrees of freedom takes place.

In [6.27] the problem of calculating flow in a plane nozzle with nonequilibrium relaxation of vibrational degrees of freedom is solved numerically in the framework of the inverse problem. Vibrational temperatures for different modes are obtained, the inverse population of levels is found and the value of the gain factor is calculated. The influence of geometrical parameters of a nozzle upon the inverse population of levels and the gain factor K_u are investigated and the importance of taking into consideration two-dimensional effects is demonstrated. Figure 6.17 shows lines of constant Mach number for a plane nozzle with a corner point taking into consideration (solid lines) and ignoring (dashed lines) vibrational relaxation. Considerable changes of the flow field due to vibrational relaxation are evident, especially in the purely supersonic region. The same figure shows (in its lower part) distributions of the Mach number, gain factor K, temperature of translational degrees of freedom of the mixture T and vibrational temperatures T_1 for the symmetric and deformation modes of CO_2, T_2 for asymmetric vibrations of CO_2 and T_3 for the vibrational mode of nitrogen. Distributions along the nozzle's contour are shown by solid lines and those along the symmetry axis by dashed ones. It is evident that in the subsonic part of the nozzle the real flow differs very little from the equilibrium one. In the transonic part of the flow, the exchange between vibrational and translational degrees of freedom (V-T exchange) of the nitrogen molecules stops because the gradients of the gasdynamic parameters become large as does the relaxation time. Then the V-T exchange between CO_2 and H_2O slows down. Rapid and continuous exchange between levels of asymmetric oscillations and vibrationally excited nitrogen ensures transmission of the vibrational energy from nitrogen to the third type of CO_2 oscillations. This is why T_2, after achieving its minimum value, starts to approach T_3. As long as the relaxation time of V-T exchange for CO_2 molecules with symmetric oscillations is small, the value of T_1 is al-

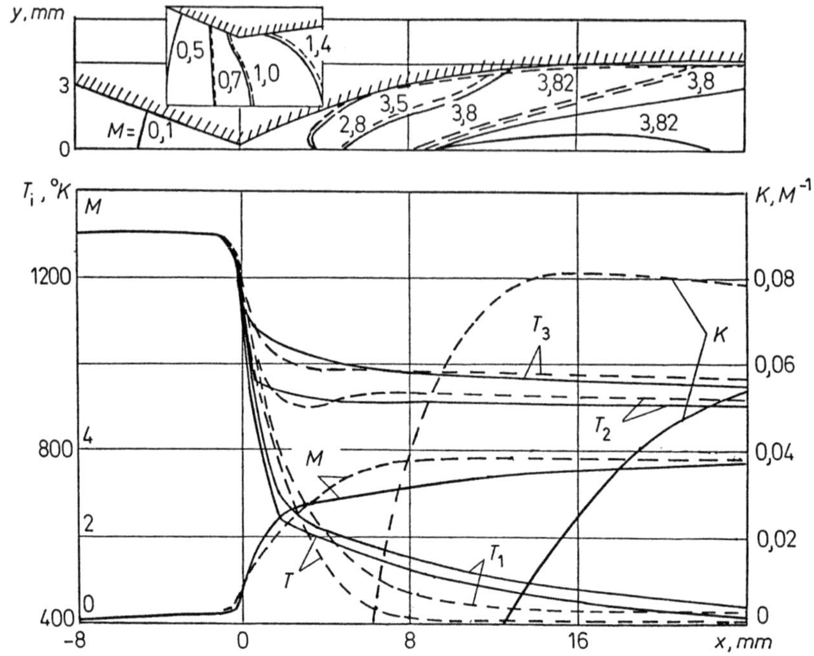

Fig.6.17 (lower part). Distributions of Mach number M, vibrational temperatures and the gain factor K along the symmetry axis (---) and along the nozzle's contour (——) for a plane nozzle flow with vibrational relaxation. T is the temperature of translational degrees of freedom, T_1 is the temperature of symmetric and deformational modes for CO_2, T_3 is the vibrational temperature of nitrogen (upper part). Lines of M = const taking into account (——) and ignoring (---) vibrational relaxation

most equal to T. Figure 6.17 shows a marked difference between distributions of parameters along the nozzle's axis and its contour, the difference being especially pronounced for the gain factor. It is worthwhile pointing out that it is possible to calculate contours of nozzles with a corner point in the framework of the inverse problem.

In [6.27] results of numerical profiling are presented and the influence of a nozzle's configuration and gasdynamic parameters upon the characteristics of gasdynamic lasers is analyzed.

When two-dimensional nonequilibrium flows are investigated for a family of similar nozzles and in nozzles with varying angle of inclination of the converging part, it is shown that the subsonic part of a nozzle should be profiled so that the sonic line is straight. The geometrical parameters of the sub- and supersonic parts of a nozzle must be so related as to provide attachment of flow and a sonic surface close to plane. A parametric study of supersonic parts of two classes of plane nozzles for gasdynamic lasers designed on the basis of uniform and symmetric characteristics at the exit section has been carried out. Parametric investigation of the influence of the degree of expansion upon total pressure, temperature and gas composition has shown that the minimum losses of useful vibrational energy in the resonator cavity correspond to nozzles profiled on the basis of uniform characteristics.

The efficiency with which thermal energy is transformed into energy of coherent electromagnetic radiation depends essentially on the geometry of a nozzle which defines properties of vibrationally nonequilibrium flows in the working part of a gasdynamic laser. The supersonic part of a nozzle has a strong influence upon the character of nonequilibrium processes. At present, gasdynamic lasers are being designed using a family of nozzles with a corner point profiled on the basis of a uniform closing characteristic. However, this family of nozzles does not contain the shortest nozzles. A higher cooling rate of gases can be achieved in shorter nozzles with a nonuniform distribution of parameters at the exit section. Therefore, in this work, parametric profiling of plane and axisymmetric nozzles providing prescribed nonequilibrium gasdynamic parameters at the exit section is done on the basis of solving the inverse problem using the classical method of characteristics. In particular, a family designed on the basis of a symmetric closing characteristic, providing parallel flow at the exit section and an absence of shock waves throughout the flow field is considered.

A parametric study of the characteristics of two classes of nozzles, the first one designed on the basis of a uniform closing characteristic and the second one designed on the basis of a symmetric closing characteristic, has been made. Nozzles profiled on the basis of a symmetric characteristic possess higher values of the angle of contour inclination at the corner point, thus providing more intense flow acceleration both beyond the corner point and throughout the whole of the supersonic region.

For gasdynamic lasers working on a mixture of gases ($CO_2 + N_2 + H_2O$) the influence of the geometrical parameters upon the nozzle efficiency coefficient η has been analyzed [6.29]. It has been found that for nozzles belonging to the second class, the value of η, averaged over a cross-section, is higher at the exit by 0.04 than the corresponding value for nozzles of the first class. However, calculations have shown that at the exit sections of nozzles belonging to the first class, gasdynamic parameters are far from being uniform. The longitudinal velocity component and the Mach number decrease in the direction from the axis to the nozzle wall, while pressure and density increase in the same direction. Thus, the Mach number nonuniformity constitutes about 60%. Such a strong nonuniformity of gasdynamic parameters at the exit section of nozzles of the second class may result in additional gasdynamic perturbations, which, in their turn, may cause part of the vibrational energy to be lost in the vibrationally relaxing flow beyond the exit section of a nozzle array. Therefore, in order to study the quality of laser flows correctly for both classes of nozzles, relaxing flows of gaseous mixtures in the resonator cavity beyond the nozzles' exit section are analyzed numerically. In order to simulate interacting flows for the purpose of calculation, a region is selected, at the upper and lower boundaries of which the planes of interaction are substituted with equivalent (with respect to boundary conditions) plane walls of a rectilinear channel.

The mixed problem of calculating isentropic flow in this region with the help of the data obtained at exit sections of the nozzles is solved by means of the layerwise net method of characteristics which is "smearing" with respect to weak shock waves. Calculations which ignore the influence of radiation generation show that nonuniformity of gasdynamic parameters at exit sections of short nozzles strongly affects gas flow in the resonator cavity. Nonuniform profiles of gasdynamic parameters at the downstream sections start to straighten, and the peripheral part of flow, where pressure is large, compresses its central part, thus generating a suspended shock whose intensity grows when x becomes larger.

As a result, the presence of a shock wave beyond the exit section of nozzle 2 results in a considerable decrease of η for nozzle 2 and lower mean values of the gain factor K. Hence, the flow of a mixture of gases $CO_2 + N_2 + H_2O$, discharged through a nozzle with uniform parameters at the exit section, stores more vibrational energy.

The above numerical analysis allows one to conclude that of the two classes of nozzles considered for gasdynamic lasers using CO_2 as the working medium, nozzles designed on the basis of a uniform characteristic are more efficient from the point of view of minimum vibrational energy losses in the resonator cavity. The latter nozzles have a sharp edge at the exit section of the nozzle array and provide a high degree of uniformity for all gasdynamic parameters.

6.4 Two-Phase Flows

Recently considerable attention has been paid to the study of nozzle flows of dusted gases, the research being stimulated mainly by the necessity of predicting the characteristics of solid-propellent rocket engines. The presence of solid and liquid particles of various sizes results in a considerable complication of the physical picture of dusted gas flow compared to pure gas flow and, hence, in a complication of the mathematical description of the phenomena and the solution methods. Terms accounting for mass, momentum and energy exchange between particles and gas must be added to the equations of motion and new equations describing the motion of particles and phase transformations must be introduced. The general system of equations must be closed by means of phenomenological relations and equations concretizing expressions for mass, momentum and energy fluxes due to interaction of phases.

A very general system of equations describing the flow of a multivelocity fluid composed of a number of continua has been presented in Chap.2. Below we shall concretize these equations for some special, but practically important, flows and consider physical particulars of two-phase flows and methods of calculation.

6.4.1 Basic Concepts: One-Dimensional Approximation, Equilibrium and Frozen Flows

When studying the interaction between two phases, one has to know the physical properties of both. For the gas phase, physico-chemical transformations are usually assumed to be equilibrium; also, for the sake of simplicity, the gas is supposed to be ideal with a constant heat capacity. Combustion products of solid-propellent rocket engines often contain a considerable amount of metal oxides whose physical properties are presented in Table 6.5, where η and σ are coefficients of viscosity and surface tension respectively and c_s is the heat capacity.

Usually, the combustion temperature exceeds the melting points of oxides and liquid particles subjected to surface tension acquire a spherical shape. The particles' size is an important characteristic of a two-phase medium. A system may be monodisperse or polydisperse, the former containing particles of equal sizes. In a polydisperse mixture, containing particles of different sizes, the size distribution of particles may be expressed by normalized functions of density of number $f(r)$ or weight $g(r)$ distributions, such that

Table 6.5. Physical properties of oxides (for solids at the normal temperature, for liquids at the melting point)

Oxide	T_m K	μ	$10^{-3}\rho$, kg/m^3 Solid	Liquid	Coefficients η, N·s/m	σ, N/m	$10^{-3} c_s$ J/kg·deg Solid	Liquid	Melting heat, kJ/kg
Al_2O_3	2302	102	3.96	3.06	0.06	0.7	0.9	1.4	1149.7
BeO	2820	25	3.01	2.56	-	0.3	-	2.7	2845.3
MgO	3075	40	3.58	-	-	-	0.85	1.7	1920.8
B_2O_3	723	70	1.82	1.70	0.50 (2100 K)	0.11 (2100 K)	-	1.8	330.6
Li_2O	1700	30	2.01	-	-	-	1.8	3.4	1639.3

$df = f(r)dr$ is the fraction of particles whose radii belong to the interval $r + dr$, while $dg = g(r)dr$ is the weight fraction of particles in the same interval [6.30]. Evidently, $g(r) = f(r)m_r/m$, where m_r is the mass of a particle with radius r and m is the arithmetic mean mass of particles in the mixture. The function f(r) allows one to introduce the integral function of distribution

$$J(r) = \int_0^r f(r)dr \quad ,$$

which defines the fraction of particles with dimensions from 0 to r. The functions f(r) and g(r) are used to introduce some average particle radii by means of the formula

$$r_{nk} = \left\{\left[\int_0^\infty r^n f(r)dr\right]\left[\int_0^\infty r^k f(r)dr\right]^{-1}\right\}^{1/(n-k)} \quad , \tag{6.73}$$

where $n = 1$, $k = 0$ corresponds to the arithmetic-mean radius, $n = 2$, $k = 0$, to the root-mean-square radius, $n = 3$, $k = 0$, to the root-mean-cube radius, $n = 3$, $k = 2$, to the mean-volume-surface radius and $n = 4$, $k = 3$, to the mean-mass radius. Evidently, the mean radii differ considerably and for an equiprobable distribution with $f = \text{const}$ at $0 \leqslant r \leqslant r^0$, one has $r_{10} = 0.5\ r^0$, $r_{20} \approx 0.58\ r^0$, $r_{30} \approx 0.63\ r^0$, $r_{32} \approx 0.75\ r^0$, $r_{43} \approx 0.8\ r^0$.

Usually, in practice, the following normal-logarithmic distribution is used·

$$f(r) = \frac{\lg e}{\sqrt{2\pi}\, r \lg\sigma} \exp\left[-\left(\frac{\lg r - \lg r_0}{\sqrt{2}\ \lg\sigma}\right)^2\right] \quad , \tag{6.74}$$

where r_0 is the expectation value and $\lg\sigma$ is the standard deviation of the logarithm of particle radii. For the normal-logarithmic distribution, just as for the equiprobable one, $r_{10} < r_{20} < r_{32} < r_{43}$ and the mean dimension increases substantially with increase in σ.

Let us write down in the one-dimensional approximation equations describing the stationary flow of a multivelocity and multitemperature continuum without phase transformations. The heat capacities of the gas c_p and the particles c_s will be assumed constant. The particles will also be assumed to be noninteracting uniformly heated spheres. The pressure produced by the particles can be ignored, but their volume will be taken into consideration. Weight fractions of particles of different sizes will be considered constant. The possibility of the variation of weight fractions due to the interaction between particles and condensation will be considered later. The above assumptions allow one to write the general system (2.1-6) for a special case without mass exchange between a mixture's components, i.e. $F_i \equiv 0$ and when the momentum exchange between the gas and the particles is due to aerodynamic forces exerted upon the latter by the gas flow. The force of interaction is assumed to be a sum of pressure and friction forces, and energy exchange is caused by the work done by aerodynamic forces and by heat transfer. Parameters of particles of the ith fraction, $(i = 1,2,\dots,N)$, will be denoted by the subscript is. The following equations for the continuity and motion of gas and particles, the equation of the mixture's energy, that of the particles' energy (or the equation of convective heat exchange between gas and particles) and the equation of state for the gas may then be obtained:

$$\rho W F = (1 - \alpha_s) Q \quad , \tag{6.75}$$

$$\rho_{is} W_{is} F = \alpha_{is} Q_i \quad , \quad \alpha_{is} = g_{is} \alpha_s \quad , \quad (i = 1,2,\dots,N) \quad , \tag{6.76}$$

$$W \frac{dW}{dx} + \frac{1}{\gamma\rho^0} \frac{dp}{dx} + \frac{\alpha_s}{1 - \alpha_s} \frac{\rho W}{\rho^0} \sum_{i=1}^{N} g_{is} \frac{dW_{is}}{dx} = 0 \quad , \tag{6.77}$$

$$W_{is} \frac{dW_{is}}{dx} + \frac{1}{\gamma\rho_s^0} \frac{dp}{dx} = \frac{3C_{xi}\rho^0 r_*}{8\rho_s^0 r_{is}} (W - W_{is})|W - W_{is}|$$

$$= \frac{\bar{C}_{xi}}{St_{is}} (W - W_{is}) \quad , \quad (i = 1,2,\dots,N) \quad , \tag{6.78}$$

$$(1 - \alpha_s)\left(c_p T + \frac{1}{2} W^2\right) + \alpha_s \sum_{i=1}^{N} g_{is}\left(c_s T_{is} + \frac{1}{2} W_{is}^2\right)$$

$$= H_0 = [(1 - \alpha_s) c_p + \alpha_s c_s] T_0 \quad , \tag{6.79}$$

$$\frac{dT_{is}}{dx} = \frac{3\alpha^0(T - T_{is})r_*}{r_{is}\rho_s^0 c_s W_{is} a_*}$$

$$= \frac{c_p Nu_{is}}{3c_s Pr W_{is} St_{is}} (T - T_{is}) \quad , \quad (i = 1,2,\ldots,N) \quad , \tag{6.80}$$

$$p = \rho^0 T \quad , \quad \frac{\rho}{\rho^0} + \sum_{i=1}^{N} \rho_{is}/\rho_s^0 = 1 \quad . \tag{6.81}$$

Here p,ρ,T,W are referred to their respective critical values p_*,ρ_*,T_*,a_*; heat capacities are referred to γ, R, x is referred to r_*; $\gamma = a_*^2\rho_*/p_*$ and $p_* = R\rho_* T_*$. The term on the right side of (6.78), characterizing momentum exchange between a gas and particles, is equal to the force a gas exerts upon the particles, and according to the law of momentum conservation equals the force which the particles exert upon a gas. Similarly, the term on the right side of (6.80) characterizes energy exchange between a gas and particles.

In (6.75-81) α_s is the total fraction of particles, r_{is} and α_{is} are the radius and the weight fraction of the ith particles, ρ and ρ_{is} are the densities of the gas and particles, i.e. their masses referred to the occupied volume, ρ^0 and ρ_s^0 are the true densities of the gas and particles, i.e. the masses of the gas and particles referred to the volumes occupied by the gas only and by the particles only, R is the gas constant of the gaseous phase, $\bar{C}_{xi}$ is the drag coefficient C_{xi} referred to the drag coefficient C_{xi}^0 in Stokes' flow ($C_{xi}^0 = 24/Re_{is}$), α^0 is the heat transfer coefficient and Re_{is}, St_{is}, Nu_{is} and Pr are the Reynolds, Stokes, Nusselt and Prandtl numbers respectively defined as

$$Re_{is} = \frac{2\rho^0|W - W_{is}|r_{is}}{\eta} \quad , \quad St_{is} = \frac{2\rho_s^0 r_{is}^2 a_*}{9\eta r_*} \quad , \tag{6.82}$$

$$Nu_{is} = \frac{2r_{is}\alpha^0}{\lambda} \quad , \quad Pr = \frac{c_p\eta}{\lambda} \quad ,$$

where η and λ are coefficients of viscosity and heat conduction. Let us also write down the equation of entropy production for a mixture, which is [6.31] the sum of the products of thermodynamic forces $[\rho_{is}(W - W_{is})/T,\ \rho_{is}(T - T_{is})/T_{is}]$ and the fluxes $[f,q]$, and equals

$$W\frac{dS_\Sigma}{dx} = \frac{1}{T}\sum_{i=1}^{N}\frac{\rho_{is}}{\rho St_{is}}\left[\bar{C}_{xi}(W - W_{is})^2 + \frac{Nu_{is}c_p(T - T_{is})^2}{3Prc_s T_{is}}\right] \quad , \tag{6.83}$$

where S_Σ is the entropy of a mixture,

$$S_\Sigma = (1 - \alpha_s)S + \sum_{i=1}^{N} \alpha_{is} S_{is} \quad .$$

S_{is} is the entropy of the particles belonging to the ith fraction and S is the entropy of the gas. With the help of equations of motion and energy, (6.75) for $\rho = \rho^0$ may be transformed into

$$\frac{M^2 - 1}{W} \frac{dW}{dx} = \frac{1}{F}\frac{dF}{dx} + \frac{\alpha_s}{1 - \alpha_s}\left[- \gamma \frac{\rho W}{p} \sum_{i=1}^{N} g_{is} \frac{dW_{is}}{dx} + \frac{1}{c_p T} \sum_{i=1}^{N} g_{is} \frac{d}{dx}\left(c_s T_{is} + \frac{W_{is}^2}{2}\right)\right] \quad . \tag{6.84}$$

Here the Mach number M has been determined by means of the frozen sonic speed which in a nonequilibrium two-phase flow is simply equal to the sound velocity in a gas. According to the general property of nonequilibrium flows, it follows from (6.84) that the condition $M = 1$ (as well as the maximum velocity or the minimum pressure for a completely subsonic flow) is attained not at the throat, but downstream of it. In the case of nonequilibrium two-phase flows the displacement may be significant, especially at large weight fractions of particles, unlike one-phase nonequilibrium flows for which the displacement, though present, is small. For equilibrium or frozen flows (6.84) transforms into the usual equation of one-dimensional theory, but the Mach number is defined by the equilibrium or frozen sonic speeds, respectively.

The system (6.75-81) corresponds to the case of discrete size distribution of particles. For a continuous distribution the sums in (6.75-81) must be replaced by integrals. Before starting to analyze the system of equations, let us present semiempirical formulas used to calculate the drag coefficient and the Nusselt number. The former depends on the numbers Re_{is} and $M_{is} = |W - W_{is}|/a$, where $a = \sqrt{\gamma RT'}$ is the sonic speed in the gas. The Nusselt number also depends on Re_{is} and M_{is} and also on the number Pr. At small Reynolds numbers ($Re < 0.1$) the drag coefficient is defined by the classical Stokes formula, and the Nusselt number equals 2. At larger values of Re and M one has to take into account the effects of inertia, compressibility and rarefaction of flow past a particle. At $Re = 0.1\text{-}10^3$ the standard drag curve for a sphere in incompressible fluid is approximated by [6.32]

$$C_{x0} = \frac{21.1}{Re} + \frac{6.3}{\sqrt{Re}} + 0.25 \quad , \tag{6.85}$$

which takes into account the forces of inertia.

Reference [6.33] presents experimental values of the drag coefficient for a sphere at Re = 200-1000 and M = 0.2-1.0, approximated by

$$C_x = C_{x0}(1 + 1.2MC_{x0})^{-1}(1 - 0.445M + 4.84M^2 - 9.73M^3 + 6.94M^4) \quad , \qquad (6.86)$$

where rarefaction is accounted for by the second multiplier. The Nusselt number Nu^0 for incompressible fluid may be determined from [6.34]

$$Nu^0 = 2 + 0.459Re^{0.55}Pr^{0.33} \quad . \qquad (6.87)$$

Compressibility and rarefaction at Re = 2-100 and M = 0-0.7 may be taken into account by means of the following semiempirical dependence:

$$Nu = Nu^0(1 + 3.42Nu^0MRe^{-1}Pr^{-1})^{-1} \quad . \qquad (6.88)$$

For nozzle gas flows the numbers Re_{is} and M_{is} may change from 1 to 200 and from 0 to 1 respectively. Calculating C_x and Nu by the Stokes formulas results in considerable errors; thus, (6.85-88) should be used.

An analysis of the system (6.75-81) shows that in two-phase flows, besides the usual gasdynamic criteria, some new ones are encountered, namely the numbers M_{is}, Re_{is}, St_{is} and c_p/c_v. For a Stokes flow regime, which is approximately realized only in the subsonic part of a nozzle, there are two similarity criteria: the Stokes number St_{is} and c_p/c_v. In this case the Stokes number characterizes the ratio of the inertial path of a particle to the system's characteristic size. Evidently, the value of the Stokes number defines various flow regimes which may take place. At $St_{is} = 0$ the flow is equilibrium, since the inertial path is zero, and a particle acquires the velocity and temperature of the gas instantaneously. At $St_{is} = \infty$ the flow is frozen, while at intermediate values of the Stokes number it is nonequilibrium with dissipation and entropy increase occurring during a finite time of interphase momentum and energy exchange.

Let us start by considering the limiting flow regimes, ignoring the volume occupied by particles. In an equilibrium flow ($St_{is} = 0$) the gas velocity and temperature equal those of the particles and the momentum is a maximum as in all equilibrium flows. By means of a simple transformation, introducing the gas constant, density, specific heat capacity and adiabatic exponent for a certain fictitious gas with the help of the formulas

$$R^0 = R(1 - \alpha_s) \quad , \qquad \rho^0 = \rho(1 - \alpha_s)^{-1} \quad ,$$

$$c_p^0 = (1 - \alpha_s)c_p + \alpha_s c_s \quad ,$$

$$\gamma^0 = \gamma\left(1 + \frac{\alpha_s}{1 - \alpha_s}\frac{c_s}{c_p}\right)\left(1 + \gamma\frac{\alpha_s}{1 - \alpha_s}\frac{c_s}{c_p}\right)^{-1} \qquad (6.89)$$

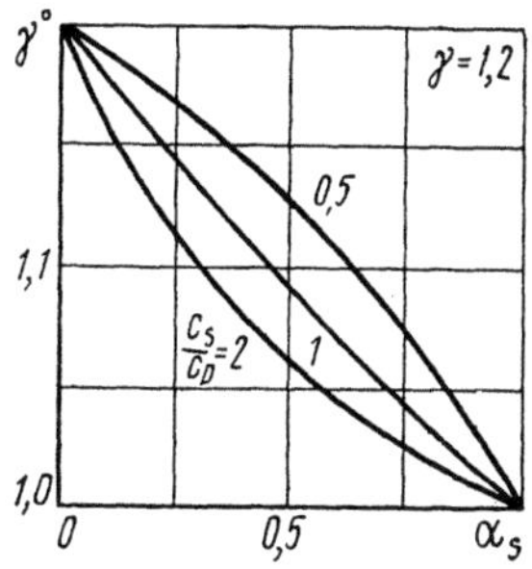

Fig.6.18. Dependence of effective adiabatic exponent γ^0 on weight fraction α_s

and assuming the fictitious gas to have the same pressure, temperature and velocity as the mixture, one can present an equilibrium flow of the mixture as a flow of the fictitious gas with adiabatic exponent γ^0, gas constant R^0 and heat capacity c_p^0 and use gasdynamic functions (2.137-139) to calculate the equilibrium flow. Figure 6.18 shows the dependence of γ^0 on α_s (note that $\gamma^0 < \gamma$).

In the other limiting case of frozen flow ($St_{is} = \infty$) the parameters of a gas and particles are determined independently. Gas flow is described by gasdynamic functions of an ideal gas with an adiabatic exponent γ, while for particles one has $T_{is} = \text{const}$, $W_{is} = \text{const}$. The minimum impulse corresponds to a frozen flow in which particles have almost zero velocities and their temperature is close to the stagnation temperature of the gas.

Thus, for nonequilibrium two-phase flows, the flow of the gaseous phase is the limiting frozen one, and according to the general property of nonequilibrium flows, increasing deviation from equilibrium results in an increase of the adiabatic exponent and in a corresponding decrease of the pressure, temperature and specific impulse. In accordance with (6.83), both limiting flows are isentropic and possess equal entropies of the mixture. However, the entropies of the gas and the particles both change in such a way that the total entropy stays constant. Two more limiting cases of isentropic two-phase flows are possible: in one of them the gas and particles possess equal velocities, but the latter have a constant temperature; in the other, the gas and particles possess equal temperatures, but the particles have equal velocities.

Let us compare parameters of a mixture in the limiting cases. We shall introduce the mean-mass velocity of a mixture, which characterizes the specific impulse, by the formula

$$W_s = \sum_{i=1}^{N} \alpha_{is} W_{is} + (1 - \alpha_s) W \quad . \tag{6.90}$$

Parameters of an equilibrium flow will be denoted by the subscript 1, those of a flow with equal velocities of the gas and particles, but with the particle temperature equal to the gas stagnation temperature, by 2, parameters of a flow with equal gas and particle temperatures but with almost zoro particle velocity, by 3, and parameters of a frozen flow, in which the velocity of the particles is close to zero, while their temperature equals the gas stagnation temperature, by 4. For the sake of simplicity, we shall consider outflow into vacuum. Then, from the energy equation (6.79) one easily obtains

$$
\begin{aligned}
W_{s1} &= W_1 = \left\{2c_pT_0\left[1 + \alpha_s\left(\frac{c_s}{c_p} - 1\right)\right]\right\}^{\frac{1}{2}} \;, \\
W_{s2} &= W_2 = [2c_pT_0(1 - \alpha_s)]^{\frac{1}{2}} \;, \\
W_{s3} &= \left\{2c_pT_0(1 - \alpha_s)\left[1 + \alpha_s\left(\frac{c_s}{c_p} - 1\right)\right]\right\}^{\frac{1}{2}} \;, \\
W_3 &= \left\{2c_pT_0\left[1 + \alpha_s\left(\frac{c_s}{c_p} - 1\right)\right]\right\}^{\frac{1}{2}}(1 - \alpha_s)^{-\frac{1}{2}} \;, \\
W_{s4} &= (1 - \alpha_s)\sqrt{2c_pT_0} \;, \\
W_4 &= \sqrt{2c_pT_0} \;.
\end{aligned}
\tag{6.91}
$$

It follows from the above formulas that the equilibrium flow velocity W_{s1} is always the largest of the mean-mass velocities, while the fully frozen flow velocity W_{s4} is always the smallest. Evidently, $W_{s2} > W_{s3}$ at $c_s/c_p < 1$, $W_{s2} = W_{s3}$ at $c_s/c_p = 1$ and $W_{s2} < W_{s3}$ at $c_s/c_p > 1$. Thus, only when $c_s/c_p < 1$, do the decrease in velocity and the impulse losses due to the total velocity lag exceed those due to the total temperature lag. Note that just such a situation takes place in cases of practical interest. Thus, for a particle of Al_2O_3, $c_s = 1420\ \mathrm{J\cdot kg^{-1}\cdot deg^{-1}}$, while for a typical composite solid propellent $c_p \approx 3000\text{-}3500\ \mathrm{J\cdot kg^{-1}\cdot deg^{-1}}$. W_3 is the maximum velocity of the gaseous phase, since in this case the thermal energy content fully transforms into the kinetic energy of directed motion of a pure gas (without particles). W_2 is the minimum velocity. The gas velocity in a frozen flow W_4 exceeds the mean-mass equilibrium flow velocity at $c_s/c_p < 1$, is equal to it at $c_s/c_p = 1$ and is below it at $c_s/c_p > 1$.

The above relationships between gas velocities and mean-mass velocities are determined by the contribution of particles to the mixture's stagnation enthalpy, since, depending on the relationship between c_p and c_s, the presence of particles may decrease, increase or keep constant the mixture's stagnation enthalpy compared to that of the gaseous phase.

Using (6.89) and the continuity and energy equations, one can deduce that the mixture flow rate,

$$Q_1 = \frac{A(\gamma^0)}{1 - \alpha_s} \frac{p_0 F_*}{\sqrt{RT_0}} \quad , \tag{6.92}$$

for an equilibrium flow with fixed pressure, stagnation temperature and throat area, is approximately $(1 - \alpha_s)^{\frac{1}{2}}$ times less than the mixture's flow rate,

$$Q_1 = \frac{A(\gamma)}{\sqrt{1 - \alpha_s}} \frac{p_0 F_*}{\sqrt{RT_0}} \quad , \tag{6.93}$$

for a frozen flow. In an equilibrium flow, the mixture's flow rate exceeds that of the gaseous phase in a frozen flow approximately by the factor $(1 - \alpha_s)^{\frac{1}{2}}$ due to the decrease of the gas constant. However, since the flow rate of the gaseous phase is $(1 - \alpha_s)$ times less than that of the mixture, the total flow rate of the mixture in a frozen flow is $(1 - \alpha_s)^{\frac{1}{2}}$ times more than the flow rate in an equilibrium flow.

The increase of flow rate in nonequilibrium flows compared to equilibrium ones is, at equal stagnation parameters and area F_*, a general property of nonequilibrium flows. This fact has been noted above while considering flows with nonequilibrium excitation of vibrational degrees of freedom. The only difference is that in two-phase flows the flow rate increases considerably – up to tens of percents – especially if the weight fraction of particles is large, while in one-phase nonequilibrium flows the flow rate increase does not exceed 1%.

It is interesting to compare the numbers M^0 and M at the nozzle throat, calculated with the help of the equilibrium a^0 and frozen a sonic speeds. We have

$$M^0 = \frac{W}{a^0} = W[\gamma^0 RT(1 - \alpha_s)]^{-\frac{1}{2}} \quad , \quad M = \frac{W}{a} = W(\gamma RT)^{-\frac{1}{2}} \quad , \tag{6.94}$$

or

$$M^0 = M\sqrt{\frac{\gamma}{\gamma^0(1 - \alpha_s)}}$$

and $M^0 > M$. Figure 6.19 shows the dependence of the number M^0 on M. Thus, if a flow is close to equilibrium, then $M^0 \approx 1$ at the throat, while the Mach number determined by the frozen sonic speed at this section is less than 1.

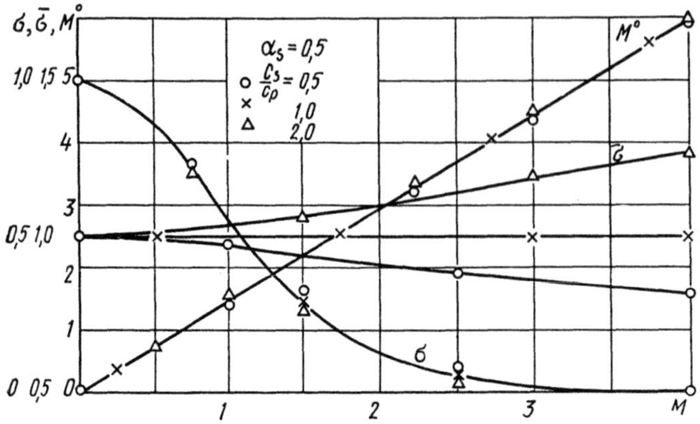

Fig.6.19. Dependence of M^0, σ and $\bar{\sigma}$ on the number M

For instance, at $\alpha_s \approx 0.9$ and $\gamma = 1.4$, $M = 0.37$ at the throat, while $M = 1$ is attained in the divergent section of a nozzle at $\bar{F} \approx 7$. The Mach number calculated by the frozen sonic speed determines the direction of the characteristics in two-phase flows, and if $M < 1$ in the divergent section of a nozzle, then small disturbances propagate upstream of the region where $M < 1$. This does not take place in pure gas flows for which the flow in the divergent section of a nozzle is supersonic.

Let us use the limiting flows to consider variation of two-phase mixture parameters at the exit cross-section of a nozzle as a function of the weight fraction of particles. Of all possible methods of comparison we shall consider the two most usable ones. In the first case, it is assumed that variation of the weight fraction of particles is not accompanied by change of the nozzle's geometry, i.e. $\bar{F}$, stagnation pressure and temperature, the gas constant, c_p and c_s, but that the flow rate of the mixture does change due to the variation of the weight fraction of particles. In the second case, the flow rate, stagnation pressure and temperature, the gas constant, c_p, c_s and the area of the nozzle's exit section stay constant, but the change of the condensate weight is accompanied by a variation of the area of the throat. The first method of comparison in the case of equilibrium flow of a mixture, shows that an increase of the weight fraction is equivalent to decreasing γ^0 (Fig.6.18) and Fig.2.6 can be used to compare parameters at $\bar{F} = \text{const}$. At $q(\lambda) = \text{const}$, in an equilibrium flow, an increase of the weight fraction of the condensate (i.e. decrease of γ^0) results in an increase of the pressure, temperature, Mach number and impulse (though not the specific impulse) compared to those in a pure gas flow, i.e. in a fully frozen flow.

The second method of comparison leaves the flow rate and mass flux Q/F unchanged. To compare the flow parameters Fig.2.6 may also be used, but one has to keep in mind that the change of the weight fraction of the particles proportional to $(1-\alpha_s)$ results in changing F_*. This method of comparison shows that an increase of the weight fraction of the condensate decreases the nozzle pressure and impulse, but increases the temperature, Mach number and specific impulse. Using these methods of comparison, one has to keep in mind that in practice, an increase of the weight fraction of the condensate may be accompanied by a simultaneous change of the stagnation pressure and temperature, gas constant, c_p and c_s.

Let us now consider variation of the stagnation pressure and temperature in a two-phase flow. Let the nozzle flow be equilibrium up to a certain point. Then, at a point of the flow field with a specified static pressure, velocity and temperature, the stagnation pressure is determined by means of gasdynamic formulas for a fictitious gas, i.e. $p_0^0 = p/\pi\ (M^0)$, where p_0^0 is the reservoir stagnation pressure. If now we realize the reverse process, decelerating the mixture in such a way that the kinetic energy of the gas and particles are transformed into thermal energy, then at zero velocity the stagnation pressure will be equal to p_0^0. However, if only the gaseous phase is decelerated, while the velocity and temperature of the particles stay unchanged and equal to their values at the point under consideration, i.e. an isentropic but "frozen" deceleration is realized, then the stagnation pressure p_0 at the point with zero velocity will be below p_0^0. The total pressure losses may be defined by the formula

$$\sigma = p_0/p_0^0 = \pi(M^0)/\pi(M) \quad . \tag{6.95}$$

The change of the stagnation temperature may be determined in a similar way. The value of $\bar{\sigma} = T_0/T_0^0$ is determined by

$$\bar{\sigma} = \frac{T_0}{T_0^0} = \left(1 + \frac{\gamma - 1}{2} M^2\right)\left[1 + \frac{\gamma^0 - 1}{2} (M^0)^2\right]^{-1} \quad . \tag{6.96}$$

An analysis of this formula and the energy equation show that $T_0^0 > T_0$ if $c_s < c_p$, $T_0^0 = T_0$ if $c_s = c_p$ and $T_0^0 < T_0$ if $c_s > c_p$. Figure 6.19 presents the dependence of σ and $\bar{\sigma}$ on M. If the particles have velocity and temperature lags relative to the gas, then T_0 may exceed T_0^0 at any c_s/c_p. The fact that the stagnation temperature of a two-phase jet T_0^0 is higher than the temperature at the initial cross-section of a nozzle has been observed experimentally and reported in [6.35].

The above estimates of the stagnation pressure and temperature change in two-phase flows are naturally the limiting ones since they have been made for the limiting flows. However, they do correctly reflect qualitative tendencies and remain valid in the case of nonequilibrium flows.

One can obtain a closed solution of the system (6.75-81) by putting $W_{is} = KW$, $T_{is} = LT + (1-L)T_0$ [6.36], where K is a constant and

$$L = \left[1 - \frac{6(K - 1)\Pr\bar{C}_x c_s}{K\,\mathrm{Nu}\,c_p}\right]^{-1} .$$

Such a flow, which is for a mixture with constantly lagging particles, may be replaced by a fictitious gas flow for which

$$c_{pk} = \frac{(1 - \alpha_s)c_p + \alpha_s c_s L}{(1 - \alpha_s) + \alpha_s K^2} , \qquad R_k = \frac{(1 - \alpha_s)R}{(1 - \alpha_s) + \alpha_s K} ,$$

$$\gamma_k = \frac{\gamma\left(1 + \frac{\alpha_s}{1 - \alpha_s} K\right)\left(1 + \frac{\alpha_s L}{1 - \alpha_s} \frac{c_s}{c_p}\right)}{\gamma\left(1 + \frac{\alpha_s}{1 - \alpha_s} K\right)\left(1 + \frac{\alpha_s L}{1 - \alpha_s} \frac{c_s}{c_p}\right) - (\gamma - 1)\left(1 + \frac{\alpha_s}{1 - \alpha_s} K^2\right)} , \qquad (6.97)$$

$$\rho_k = \rho\left(1 + \frac{\alpha_s}{1 - \alpha_s} K\right) , \quad W_k = W , \quad T_k = T , \quad p_k = p .$$

Flow of such a gas is calculated by means of gasdynamic functions for an ideal gas, similarly to the calculation for an equilibrium flow. Equilibrium and frozen flows may be obtained by putting $K = 1$, $L = 1$ and $K = 0$, $L = 0$ respectively.

According to the solution (6.97), the velocity of the gas, and, hence, that of the particles changes along a nozzle in a linear manner and their difference also increases linearly. Therefore, it has a limited application, since in real flows $W - W_{is}$ has a maximum.

6.4.2 One-Dimensional Nonequilibrium Flows Without Phase Transitions; Impulse Losses

The limiting situations considered in the previous section are never realized in practice. Particles can be accelerated by aerodynamic forces only, generated by gas flow past them, and to be accelerated, they must be moving slower then the gas. Similarly, particles transfer their heat to the medium if their temperature is higher than that of the surrounding gas. Differences of velocities and temperatures between the gas and particles result in a nonequilibrium motion of the mixture. It follows from (6.78,80) that, under otherwise equal conditions, the difference of velocities and temperatures in geometrically similar nozzles will be larger, the larger the Stokes number (i.e. the larger the particles and the density of the substance which constitutes them, and the smaller the absolute dimensions of a nozzle and viscosity of the gas). An exact determination of the parameters of the gas and particles in the framework of the one-dimensional approximation may be obtained only by numerical integration of the system (6.75-80). This is done using the same methods used to solve the system describing nonequilibrium flows with physicochemical transformations presented in Sect.6.2. As before, the inverse problem with a given pressure $p = p(x)$ or density $\rho = \rho(x)$ distribution should be solved; when solving the relaxation equations (6.78,80), one should use implicit difference schemes. Therefore, directing the reader to Sect.6.2, we shall write down only the formulas analogous to (6.41-43) when the function $p = p(x)$ is specified. Ignoring the volume occupied by the particles, we have

$$W_m^2 - W_{m-1}^2 + \frac{R}{\gamma}\left[\left(\frac{T}{p}\right)_m + \left(\frac{T}{p}\right)_{m-1}\right](p_m - p_{m-1}) + h_m\left\{\left[\sum_{i=1}^{N} \frac{\rho_{is}}{\rho} St_s^{-1}\bar{C}_{xi}(W - W_{is})\right]_m + \left[\sum_{i=1}^{N} \frac{\rho_{is}}{\rho} St_{is}^{-1}\bar{C}_{xi}(W - W_{is})\right]_{m-1}\right\} = 0 \quad , \tag{6.98}$$

$$(1 - \alpha_s)\left(c_pT + \frac{1}{2}W^2\right)_m + \sum_{i=1}^{N} \alpha_{is}\left(c_sT_{is} + \frac{1}{2}W_{is}^2\right)_m - H_0 = 0 \quad , \tag{6.99}$$

$$W_{is(m-1)} - W_{ism} + \frac{h_m}{1 + h_m(1 - s)N_{Wim}}\{sL_{wi(m-1)} + (1 - s)L_{Wim} - W_{is(m-1)}[sN_{Wi(m-1)} + (1 - s)N_{Wim}]\} = 0 \quad , \tag{6.100}$$

$$T_{is(m-1)} - T_{ism} + \frac{h_m}{1 + h_m(1 - s)N_{Tim}}\{sL_{Ti(m-1)} + (1 - s)L_{Tim} - T_{i(m-1)}[sN_{Ti(m-1)} + (1 - s)N_{Tim}]\} = 0 \quad , \tag{6.101}$$

where

$$L_{Wi} = N_{Wi} W \quad , \quad N_{Wi} = \frac{\bar{C}_{xi}}{W_{is} St_{is}} \quad ,$$

$$L_{Ti} = N_{Ti} T \quad , \quad N_{Ti} = \frac{Nu_{is} c_p}{3 Pr St_{is} W_{is} c_s} \quad .$$

The values of W, W_{is}, T and T_{is} are determined at the point x_m, and one then calculates ρ from (6.81), the stream tube area F from (6.75) and the condensate density ρ_{is} from (6.76). The initial data for integrating (6.75-81) are usually fixed in the subsonic region assuming that at the initial section the gas and particles possess equal velocities and temperatures.

Let us point out some general regularities of parameter variation in the case of nonequilibrium two-phase nozzle flows. Evidently, the velocity of the particles is below both the equilibrium velocity and the gas velocity in a nonequilibrium flow, while their temperature is higher. The gas velocity and temperature may be equal to, higher than or lower than the corresponding parameters for an equilibrium flow, depending on c_s/c_p. The mean-mass velocity of the mixture in a nonequilibrium flow is lower than in an equilibrium one. The pressure and specific impulse of nonequilibrium flow at a fixed area ratio are lower than in an equilibrium flow, but higher than in the frozen one. The Mach number in a nonequilibrium flow calculated from the frozen sonic speed is lower than the equilibrium flow Mach number determined by the sonic speed of a fictitious gas with γ^0 and R^0, due to a decrease of the gas constant in the latter. (Note that for a chemically equilibrium flow it is vice versa).

Figure 6.20 presents some results of calculations. The depicted differences of velocities and temperatures of the gas and the particles are typical for nonequilibrium two-phase nozzle flows. They increase in the subsonic and supersonic sections of a nozzle due to the increase of the velocity gradient, and for tiny particles they achieve maximum in the supersonic vicinity of the throat (where the velocity gradient is a maximum).

For an approximate estimation of the gas parameters in nonequilibrium two-phase flows one can sometimes use gasdynamic functions. If deviation from equilibrium is known to be slight and the velocity, the temperature and the stagnation parameters are given, then, having determined the number M^0 using the equilibrium sonic speed and using gasdynamic functions, one can find all the remaining parameters, corresponding approximately to the nonequilibrium flow parameters. In this case the use of the frozen Mach number may result in large errors in the determination of the pressure and density.

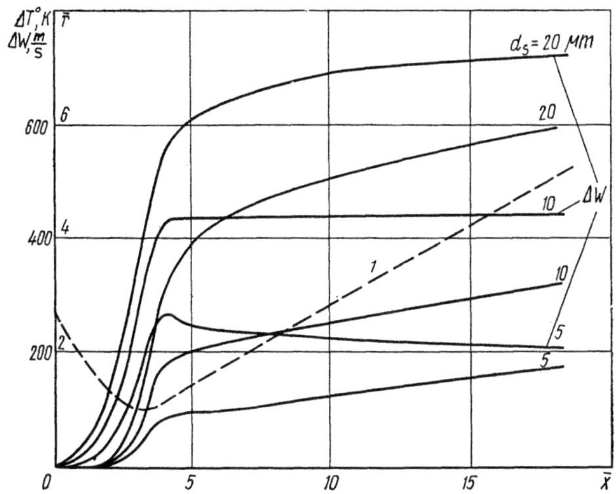

Fig.6.20. Dependence of gas and particle velocities and temperatures on $\bar{x}$ at $\alpha_s = 0.3$, $d_* = 100$ mm, $p_0 = 40$ bar, $T_0 = 3200$ K: (*1*) nozzle contour

Analogously, using the value of p/p_0 at a point, one may find all the remaining parameters. The results of calculations show that even for particles as large as 20μm and $\alpha_s < 0.6$ the gas parameters in a nonequilibrium flow differ from those for an equilibrium flow calculated with the help of the adiabatic exponent γ^0 by no more than 5-10%.

The value of the loss coefficient ζ_s associated with the velocity and temperature lags of particles is of considerable practical significance. Usually, ζ_s is defined as the difference of impulses for equilibrium and non-equilibrium flows referred to the impulse of an equilibrium flow at equal $\bar{r} = r/r_*$ and flow rates. In order to determine ζ_s, a real polydispersive mixture may be substituted by an equivalent monodispersive mixture with mean-mass particle radius r_{43} [6.32]. It has also been found that the one-dimensional approximation allows the losses of this kind to be determined accurately enough for practical needs. The value of ζ_s increases as the condensate weight fraction and size of the particles increases and decreases with increasing p_0 and d_*. Results of calculating ζ_s in a wide interval of the determining parameters are presented in [6.32] together with formulas for calculating ζ_s taking into account the shape of a nozzle. Thus, for a conical nozzle with $\theta_k = 15^\circ$ at $p_0 = 40$ atm, $\bar{R}_2 = 2$, $\bar{r} = 2.5$, the value of ζ_s may be found from the following approximate formula:

$$\zeta_s = 20\alpha_s d_s^{0.75} d_*^{-0.5} \quad , \qquad (6.102)$$

valid for $\alpha_s < 0.4$ (here d_s is measured in micrometers and d_* in millimetres). Calculations show that the effect of the other parameters (p_0, $\bar{R}_2$, $\bar{r}$, θ_k) is not so great and may be accounted for with the help of coefficients close to unity.

The decrease of β in a nonequilibrium flow may be determined by the approximate formula:

$$\Delta\beta/\beta = 24\alpha_s d_s^{0.75} d_*^{-0.50} \quad . \tag{6.103}$$

The question of the impulse loss dependence on α_s for large weight fractions of particles has been studied in [6.37]. It has been found that impulse losses attain a maximum value in the interval of α_s from 0 to 0.9. This can easily be confirmed analytically for the case of outflow into vacuum if one assumes that $W_s = KW$, $K < 1$. Then, we have

$$\zeta_s = 1 - [1 + \alpha_s(K - 1)][1 + \alpha_s(K^2 - 1)]^{-\frac{1}{2}} \quad , \tag{6.104}$$

and it follows that ζ_s is maximum at $\alpha_s = (1 + K)^{-1}$, i.e. at $\alpha_s > 0.50$. Figure 6.21 shows the dependence of ζ_s on α_s for various values of K and presents experimental data [6.37]. Evidently, with decreasing K, i.e. with increasing particle lag, the maximum of ζ_s is displaced towards larger α_s. However, in the case of a frozen flow ($K = 0$), ζ_s increases with increasing α_s. As noted above, at $\alpha_s < 0.4$ the dependence of ζ_s on α_3 is closer to linear. The existence of the loss maximum is a consequence of the contrary effect which α_s and the difference of the gas and particle velocities (which decreases with increasing α_s) have upon ζ_s.

The above computations have been carried through ignoring the volume occupied by the particles; this usually results in insignificant errors only. In fact, even in the case of large condensate weight fractions (up to 95%), $\rho_s^0 \gg \rho_{is}$ and, hence, from (6.81) one obtains $\rho \approx \rho^0$, while the term $(1/\gamma\rho_s^0)$ dp/dx on the right side of (6.78) may be ignored. Evidently, the influence of the number of particles on the drag coefficient may prove to be more important.

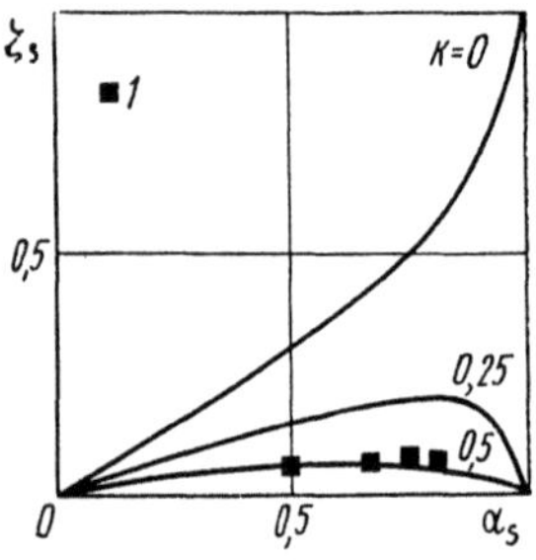

Fig.6.21. Dependence of two-phase losses ζ_s (%) on the weight fraction α_s: (*1*) experiment

6.4.3 One-Dimensional Flows with Interacting Particles

In a polydispersive mixture flowing in a nozzle, particles of different sizes attain different velocities. Thus, relative velocities of particles 1 and 5 μm in diameter near the throat may reach 250 ms^{-1}; the same quantity for particles 1 and 20 μm in diameter may reach 700 ms^{-1}. Relative velocities of particles and gas are of the same order of magnitude. This provides conditions for the interaction of particles with each other, leading to coagulation and fragmentation accompanied by mass, momentum and energy exchange. Aerodynamic forces may also cause deformation and fragmentation of particles. Investigations of these quite complicated physical processes have resulted in the development of methods of calculating nozzle flows of a polydispersive condensate taking into account the inertial coagulation and fragmentation of particles. The most conclusive studies of these problems have been reported in [6.32,38,39]. Below we present the main results of the investigations.

Coagulation in a polydispersive mixture can be studied in two ways. The first is based upon studying the evolution of particles belonging to various size fractions and is called Lagrange's method. The second one is, on the contrary, associated with monitoring the change in the number of particles of fixed dimensions and is called Euler's method. Both methods give close results. Below we describe Lagrange's method which is simpler and more convenient for computer manipulation.

Let us write down, in the one-dimensional approximation, the system of equations describing a polydispersive mixture flow with coagulation. We shall introduce the function n defining the number of particles in a unit volume. Further, only those size distributions of particles which can be readily compared to experimental data will be considered. We have

$$g_{is} = \left(\frac{\alpha_s}{1-\alpha_s}\frac{\rho W}{\rho_{is}W_{is}}\right)^{-1} \quad , \quad \rho_{is} = m_{is}n_{is} \quad , \tag{6.105}$$

or

$$n_{is} = \frac{\alpha_s}{1-\alpha_s}\frac{\rho W}{m_{is}W_{is}}\, g_{is} \quad .$$

In the case of a one-dimensional motion the particles' trajectories are straight lines and if coagulation is assumed to take place every time they collide, it can be easily deduced that an increase of the mass of the ith particle due to the absorption of a smaller j^{th} particle ($j = 1,2,\ldots,i-1$) is

$$\frac{dm_{is}}{dx} = \frac{1}{W_{is}} \sum_{j=1}^{i-1} K_{ij} n_{js} m_{js} \quad .$$

If the real character of the interaction is taken into account, then

$$\frac{dm_{is}}{dx} = \frac{1}{W_{is}} \sum_{j=1}^{i-1} K_{ij} n_{js} m_{js} \ni_{ij} \Phi_{ij} \quad , \tag{6.106}$$

where $\ni_{ij}$ is the capture coefficient, Φ_{ij} is the collision efficiency coefficient and $K_{ij} = \pi(r_{is} + r_{js})^2 |W_{js} - W_{is}|$ is the coagulation constant for interacting particles.

The decrease in the number of ith particles due to their capture by the larger j^{th} ones ($j = i+1, i+2, \ldots, N$) is determined by the relationships

$$\frac{dg_{is}}{dx} = -\frac{n_{is}}{W_{is}} \sum_{j=i+1}^{N} K_{ij} n_{js} \ni_{ij} \Phi_{ij} \quad , \tag{6.107}$$

$$\frac{dg_{is}}{dx} = \frac{1-\alpha_s}{\alpha_s \rho W} \left(n_{is} \frac{dm_{is}}{dx} + m_{is} \frac{dn_{is}}{dx} \right) \quad , \tag{6.108}$$

where $i = 1, 2, \ldots, N$. These formulas and the expression for the coagulation constant are derived using classical methods from the kinetic theory of gases.

Coagulation of smaller particles with the larger ones results in the variation of their velocity and temperature in accordance with the law of momentum and energy conservation at collisions. The following relationships are true for the velocities and temperatures of particles formed by coagulation:

$$\frac{\partial W_{is}}{\partial x} = \frac{1}{m_{is} W_{is}} \left[\sum_{j=1}^{i-1} K_{ij} n_{js} m_{js} \ni_{ij} (W_{js} - W_{is}) + m_{is} \sum_{j=i+1}^{N} K_{ij} n_{js} \ni_{ij} (1 - \Phi_{ij})(W_{js} - W_{is}) \right] \quad , \tag{6.109}$$

$$\frac{\partial T_{is}}{\partial x} = \frac{1}{c_s m_{is} W_{is}} \left[\sum_{j=1}^{i-1} K_{ij} n_{js} m_{js} \ni_{ij} (E_{js} - E_{is}) + m_{is} \sum_{j=i+1}^{N} K_{ij} n_{js} \ni_{ij} (1 - \Phi_{ij})(E_{js} - E_{is}) \right] \quad , \tag{6.110}$$

where $E_{is} = c_s T_{is} + W_{is}^2/2$. The signs of the partial derivatives are used to denote the change of parameters taking place as a result of interaction between particles.

Adding the right sides of (6.109,110) to the corresponding right sides of (6.78,80) and putting $\ni_{ij} = \Phi_{ij} = 1$, we see that the system (6.75-81) together with (6.106-110) describes a nonequilibrium two-phase nozzle flow with coagulation, ignoring the real details of the particles' interaction. The latter can be accounted for by means of introducing the capture, fragmentation and collision efficiency coefficients.

Let us study each of the processes in detail. The capture coefficient characterizes the ratio of the number of particles of a certain size which have collided with a larger particle, to the number of particles which would have collided in the case of rectilinear relative motion. To determine the capture coefficient the following formulas may be used [6.30]:

$$\ni_{ij}^{0} = [St_{ij}(St_{ij} + 0.125)^{-1}]^2 \tag{6.111}$$

for potential flow past a sphere and

$$\bar{\ni}_{ij} = \left[1 + \frac{0.75\ \ln(4St_{ij})}{2St_{ij} - 1.2}\right]^{-2} \tag{6.112}$$

for viscous flow. Here,

$$St_{ij} = \frac{(W_i - W_j) r_i^2 \rho_s}{9 r_j \eta}$$

is the Stokes number and $r_i < r_j$. The critical Stokes number, below which the capture coefficient is zero, is equal to 0.607 and 0.0417 for viscous and potential flows, respectively.

If there is a considerable difference of velocities between particles and the gas the former may be deformed and broken by aerodynamic forces. The flow regime is defined by the Reynolds number Re_{is}, while the stability of the particles is governed by the Weber number,

$$We = \rho\, d_{is} (W - W_{is})^2 \sigma^{-1} \quad .$$

When the latter attains the critical value, the liquid particles disintegrate [6.40]. The Weber number attains its maximum value near the throat and changes from 5 to 30 while a particle's size changes from 5 to 10 μm. Calculations and experiments show that the critical Weber numbers are equal to 10-20. Therefore, fragmentation may be significant, especially for large weight fractions of particles and high stagnation pressures. The spectrum of particles upon fragmentation plays an important role in the determination of particle sizes. Since the problem is extremely complicated, it is assumed that a fragmenting drop is divided into two halves.

The collision efficiency coefficient, equal to the ratio of the change of a particle's mass to the total mass of particles colliding with it, $\Phi_{ij} = \Delta m_i/\Sigma_{j=1}^{i} m_j$, is another important parameter characterizing the interaction of particles. There are empirical formulas for calculating the coefficients Φ_{ij} (see [6.38]) obtained by reducing experimental data under conditions which only partially correspond to nozzle flows:

$$\Phi_{ij} = 1 - 0.115 Re_s^{0.94} \Gamma^{0.36} (r_i/r_j)^{0.88} \quad , \tag{6.113}$$

$$(2.5 \leqslant r_j/r_i \leqslant 12 \quad , \quad 15 \leqslant Re_s \Gamma^{0.4} \leqslant 70)$$

and

$$\Phi_{ij} = 1 - 0.247 Re_s^{0.434} \Gamma^{0.133} (r_i/r_j)^{0.273} \quad , \qquad \text{where}$$

$$Re_s = d_{is} \rho_s \eta_s^{-1} (W_{is} - W_{js}) \quad , \quad \Gamma = \eta_s^2/(2 r_j \sigma_s \rho_s)$$

is the so-called stability criterium. Since both experimental and theoretical solutions of the problem are extremely complicated, one has usually to use these formulas even though they are so different.

The above relationships are valid for the interaction of liquid particles. In the case of solid particles, interaction processes may prove to be no less complicated for mathematical description due to the possibility of fragmentation of particles, their reflection and so on. However, if one assumes that no such processes take place and that impacts are inelastic and kinetic energy transform into heat which is equally divided between particles, while the size distribution of particles and the sizes themselves remain invariable (m_i = const, n_i = const, g_i = const), then the following equations for determining the velocity and energy change in the process of interaction may be obtained:

$$\frac{\partial W_{is}}{\partial x} = \frac{1}{W_{is}} \sum_{j=1}^{N} K_{ij} m_{js} n_{js} \ni_{ij} \frac{W_{js} - W_{is}}{m_{is} + m_{js}} \quad , \tag{6.114}$$

$$\frac{\partial T_{is}}{\partial x} = \frac{1}{4 c_s W_{is}} \sum_{i=1}^{N} K_{ij} m_{js} n_{js} \ni_{ij} \frac{(W_{js} - W_{is})^2 (m_{is} + 3 m_{js})}{(m_{is} + m_{js})^2} \quad .$$

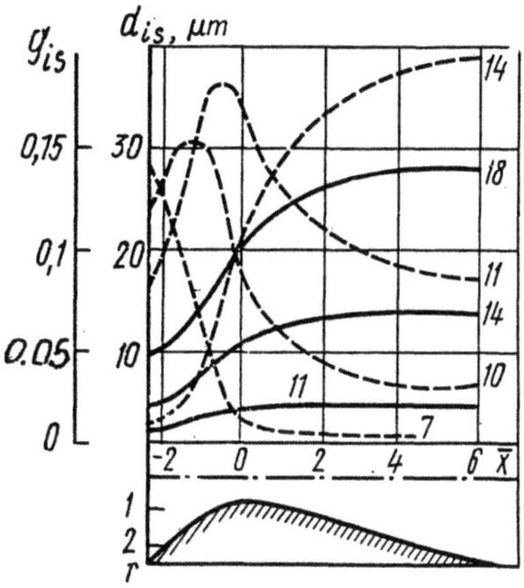

Fig.6.22. Variation of particles' sizes and relative weight fractions for different species in the process of coagulation in a nozzle at $\alpha_s = 0.28$, $d_* = 100$ mm, $p_0 = 40$ bar, $T_0 = 3200$ K, $\theta_0 = 45°$, $\theta_k = 15°$, $R_2 = 2$ (numbers denote species)

Let us present some results of calculations for nonequilibrium two-phase flows with coagulation made by the Lagrange method. At the nozzle's initial cross-section particles are distributed according to the normal-logarithmic law with $\sigma = 1.5$, $r_0 = 1.1$ and $d_{43} = 2$ μm. The initial function of distribution density is represented by 20 fractions, logarithmically equidistant in the interval from 0.5 to 15 μm. Figure 6.22 shows the variation of diameters of various fractions (the solid lines) and of their relative weights (the dashed lines). The nozzle contour is shown in the same figure. It can be seen that the weight fraction of the smaller particles decreases (curve 7) due to absorption by the larger ones and the fractions of the latter grow (dashed curve 14). The weight fraction of intermediate particles (dashed curves 10,11) at first increases; later, however, when the number of smaller particles diminishes, the fraction of intermediate particles decreases because the number of particles in these fractions is not compensated by their growth.

Computations show that the growth of particles is considerably more intense for larger absolute dimensions of a nozzle. Despite the fact that an increase of absolute dimensions results in a decrease of the lag and the coagulation constant, the time of residence of the particles increases proportionally to the nozzle diameter and the latter effect proves to be more important. The average dimensions of the particles increase also as their concentration and the chamber pressure increase. Figure 6.23 illustrates some of the tendencies mentioned above. The effect of coagulation upon impulse losses is shown in Fig.6.24. The data presented in Figs.6.22-24 show the importance of coagulation in the case of a polydispersive mixture with liquid particles. At the same time, one has to keep in mind that elementary processes

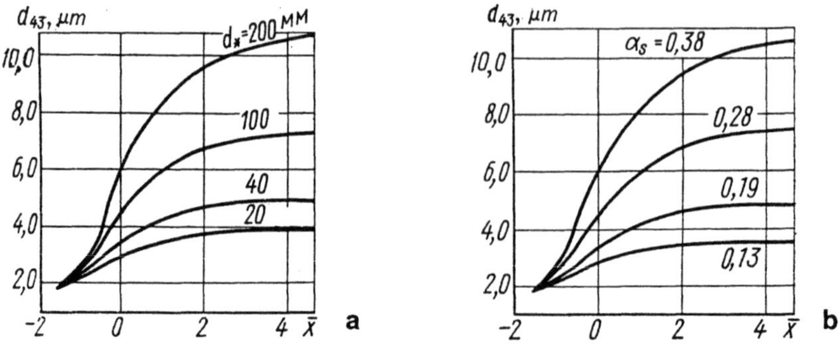

Fig.6.23a,b. Variation of average particle sizes as a function of the throat diameter at $\alpha_s = 0.28$ and the weight fraction at $d_* = 100$ mm

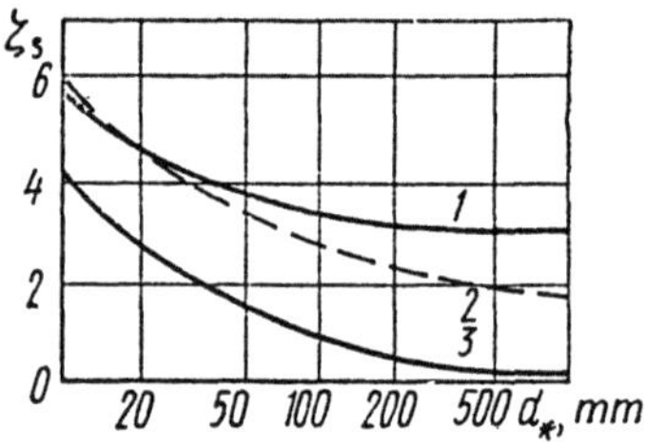

Fig.6.24. Variation of two-phase losses ζ_s (%) as a function of the throat diameter ($\alpha_s = 0.3$, $\bar{r}_a = 2.5$): (*1*) taking coagulation into account, (*2*) taking into account coagulation, fragmentation, capture and collision efficiency coefficients, (*3*) ignoring coagulation

of particle interactions have been inadequately studied and both qualitative and quantitative data on the effect of coagulation may change substantially with further accumulation of knowledge about the processes.

6.4.4 One-Dimensional Flows with Phase Transitions

The above results refer to two-phase nonequilibrium flows without phase transitions. However, flows in aerodynamic nozzles and nozzles of rocket engines may be accompanied by the processes of condensation and crystallization which are nonequilibrium due to the finite residence time of a gas in a nozzle.

Let us start by considering condensation. This may take place if the partial pressure of a mixture component exceeds its saturation pressure in the process of expansion. Generally speaking, at large expansion ratios all mixture components condense in turn because the Poisson adiabatic $d \ln p/dT = C_p/RT$ in the p-T diagram necessarily crosses the curve (described by the Clausius-Clapeyron equation) defining the saturated vapour pressure over a plane surface. This equation is

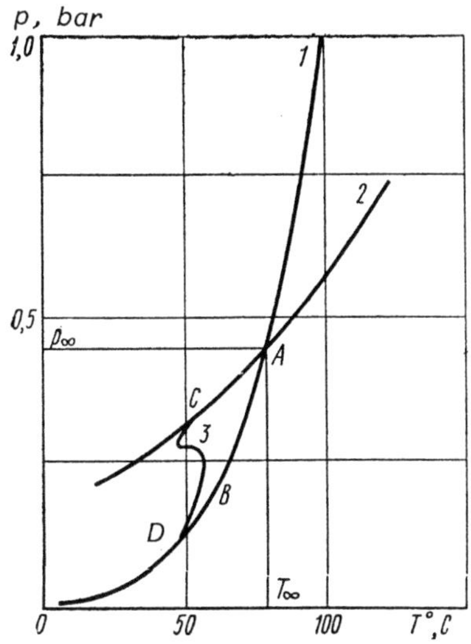

Fig.6.25. (p-T)-diagram for water vapours: (1) saturation curve (equilibrium flow), (2) Poisson's adiabatic (frozen flow), (3) nonequilibrium flow with nonequilibrium condensation

$$\frac{d \ln p_{i\infty}}{dT} = \frac{L}{R_s T^2} \quad , \tag{6.115}$$

where $p_{i\infty}$ is the saturation pressure of the condensing i^{th} vapour component over a plane surface, L is the condensation heat and R_s is the gas constant of the condensing component. Figure 6.25 presents the p-T diagram for water vapour. If, after reaching the saturation point A, further expansion proceeds in thermodynamic equilibrium, then the vapour pressure and temperature are related by (6.115) and the process of expansion of the corresponding mixture component will proceed not along the Poisson adiabatic AC, but along the curve AB (Fig.6.25) and will be accompanied by condensation of the gaseous phase of the component. However, according to numerous experimental data, actual condensation does not take place in the section corresponding to the point A, or for a certain distance downstream, and the whole process proceeds along a curve ACD (nonequilibrium condensation). The nonequilibrium flow parameters possess, as always, values intermediate between those of equilibrium expansion, corresponding initially to expansion along the Poisson adiabatic and the saturation curve, and the parameters of frozen expansion, when the time is insufficient for phase transitions to take place, and expansion proceeds strictly along the Poisson adiabatic. In the process of equilibrium expansion the condensation heat is released gradually and the pressure, temperature and impulse are higher than for a nonequilibrium flow. On the other hand, the same parameters of a nonequilibrium flow exceed those of a frozen flow.

In equilibrium and frozen flows with condensation, the mixture's entropy remains constant and equal to that at the nozzle's initial cross-section; however, entropies of the gaseous and liquid phases taken alone do change in an equilibrium flow.

The variation of the state of the vapour along AC is accompanied by a more intense temperature drop than in the case of expansion along the saturation curve. Thus, a condensation lag in the process of expansion is accompanied by a more intense drop of the vapour temperature which turns out to be below the saturation temperature at a fixed pressure. The condensation lag in high-gradient flows is due to a lack or total absence of condensation nuclei. However, the state of a supercooled gas is unstable since small disturbances may result in supersaturation which is absolutely stable. The state of supersaturation may be changed either by heterogeneous condensation at foreign particles or by homogeneous condensation at the nuclei created by the condensing component. In the case of the heterogeneous condensation, the main component may condense at nuclei formed by foreign particles. If an admixture, whose quantity is insignificant, has a higher temperature and saturation pressure compared to the main component, then, though its condensation does not change the flow parameters significantly, the nuclei thus formed may become the centres of condensation of the main component. Therefore, the latter may start to condense at the slightest supersaturation. Thus, according to [6.41], if nitrogen contains only 0.05% of water vapour, its supersaturation does not take place in the process of expansion. According to [6.42], condensation of nitrogen in air is an equilibrium process.

Numerous investigations of air condensation in wind tunnels show that as the initial pressure and, hence, the saturation temperature, decreases, supercooling increases, since a lower pressure results in the reduction of the number of molecular collisions and, hence, in a decrease of the number of condensation nuclei formed by foreign components. Therefore, at low saturation temperatures and pressures, vapour condensation proceeds homogeneously. Below we consider homogeneous condensation which usually takes place in wind tunnel nozzles and in nozzles (jets) of rocket engines. Heterogeneous condensation in nozzles has been considered in [6.43,44].

The problem of spontaneous generation of condensation nuclei is the most important problem of the theory of homogeneous condensation [6.45,46]. The usual theory of phase transitions does not consider the time history of such transitions; it considers equilibrium between the initial and new phases under the assumption that the latter is fully developed and that the interface of the phases is plane. The phases may stay in equilibrium with each other infinitely.

Evidently, growth of a new phase at a finite rate, after it has appeared and grown sufficiently, is possible only in the case of a certain departure from equilibrium. If the formation rate of the new phase is small, then the initial phase stays rather long in the so-called "metastable" state. Supercooled vapours and superheated liquids present examples of the state. A new phase appears in the original metastable phase in the form of nuclei. One may extend the concept of equilibrium of phases A and B in such a way that the nuclei B of specified sizes and shape stay in equilibrium with the medium A, which preserves its metastable state, i.e. is stable with respect to the already formed phase B, separated from the phase A by a plane surface. Thomson was the first to consider the question for the example of superheated vapour condensation. He has shown that the pressure of vapour in equilibrium with a liquid drop at a fixed temperature is larger, the smaller the drop's radius. Thus, vapour supersaturated in the usual sense, with respect to a large drop, may be nonsaturated with respect to a tiny droplet.

Let us write down some relationships necessary to understand the physical particulars of the processes of nonequilibrium condensation in a nozzle. According to Gibbs' distribution, the number of complexes containing g molecules each, for instance, condensation nuclei formed spontaneously, for an equilibrium system is equal to

$$N_g = N_1 \exp[-\Delta\Phi/(kT)] \quad , \tag{6.116}$$

where N_1 is the number of vapour molecules, $\Delta\Phi$ is the energy of formation of a nucleus containing g molecules, and k is Boltzmann's constant.

The formation energy of a nucleus of radius r is equal to

$$\Delta\Phi = (\varphi_B - \varphi_A)g + 4\pi r^2\sigma \quad , \tag{6.117}$$

where φ_B and φ_A are the thermodynamic potentials per molecule of gas and liquid respectively, and σ is the surface tension. Since $gV_s = (4/3)\pi r^3$, where V_s is the volume occupied by a molecule of liquid,

$$\Delta\Phi = g(\varphi_B - \varphi_A) + 4\pi\sigma g^{2/3}\left(\frac{3}{4\pi}V_s\right)^{2/3} \quad . \tag{6.118}$$

The equilibrium state of a system is characterized by the relationship

$$\frac{d\Delta\Phi}{dg} = \varphi_B - \varphi_A + 4\pi\sigma\,\frac{dr^2_{cr}}{dg} = 0 \quad , \tag{6.119}$$

from which Thomson's formula may be obtained:

$$\ln \frac{p_i}{p_{i\infty}} = \frac{2\sigma_i V_{is}}{r_{icr} kT} , \tag{6.120}$$

$$r_{icr} = \frac{2\sigma_i V_{is}}{kT \ln (p_i/p_{i\infty})} , \tag{6.121}$$

where the subscript i refers to the ith condensing component and p_i is the partial pressure of the component. The formula (6.120) allows the saturation pressure to be determined for a drop whose radius is equal to r.

It can easily be seen that vapour which is saturated with respect to large drops and supersaturated with respect to a plane surface is nonsaturated with respect to small droplets. Drops whose radius is equal to r_{cr} defined by (6.121) are called critical drops or nuclei. At fixed p and T, subcritical drops ($r < r_{cr}$) are surrounded by nonsaturated vapour and, therefore, evaporate; at the same time, supercritical drops ($r > r_{cr}$) are immersed in supersaturated vapour and, hence, grow. The ratio $s = p/p_\infty$ is called the supersaturation, and the value $\Delta T = T_\infty - T$ is called the supercooling. For drops whose radii are of the order of a micron, the saturation pressure exceeds that at a plane surface by 0.1%, while at $r = 6 \cdot 10^{-4}$ μm the supersaturation is approximately equal to 6.

With the help of (6.120) the formula for $\Delta\Phi$ may be rewritten in the form

$$\frac{\Delta\Phi}{kT} = \frac{\sigma}{kT} \left(\frac{4}{3} \pi V_s^2\right)^{1/3} g_{cr}^{2/3} \left[3 \left(\frac{g}{g_{cr}}\right)^{2/3} - 2 \frac{g}{g_{cr}}\right] . \tag{6.122}$$

It follows from (6.118) and (6.122) that in the state of nonsaturation (with respect to a plane surface), when $\varphi_B > \varphi_A$, $\Delta\Phi$ is a positive monotonically increasing function of the argument g/g_{cr}, the vapour is stable, since drops of critical dimensions are absent and accidentally formed nuclei disappear, their number decreasing with the increase in $\Delta\Phi$ defined by (6.116).

For drops with a critical radius equal to infinity the state of saturation ($\varphi_A = \varphi_B$, $s = 1$) is stable. In supersaturated vapours ($\varphi_A > \Phi_B$, $s > 1$) $\Delta\Phi$ is no longer a monotonic function of the variable g/g_{cr}: it grows at $r < r_{cr}$, attains a maximum value at $r = r_{cr}$ and decreases at $r > r_{cr}$. Hence, according to Thomson's formula, the nuclei whose radii are smaller than the critical one evaporate, and their number diminishes with increasing r. On the contrary, at $r > r_{cr}$ the nuclei grow and their number increases with increasing r. Thus, at $\varphi_A > \varphi_B$ formation of a new phase may start; however, the rate of the process is limited by the necessity of crossing the barrier of $\Delta\Phi_{cr}$, similar to the activation energy in chemical reactions. In this case a vapour stays in the metastable state corresponding to the maximum of the potential $\Delta\Phi$. However, the state of the new phase in the form of critical size nuclei at the point of the maximum is unstable. The probability of the formation of a new phase in a unit volume is defined by

$$P = A \exp\left(-\frac{B\sigma^3}{T(\Delta T)^2}\right) \quad . \tag{6.123}$$

These considerations allow one to construct the following procedure for nonequilibrium condensation in nozzle flows. At the initial stage of the process, near the saturation point A the degree of vapour supersaturation increases, since the drops of critical size must be large, and the probability of their formation is small. Due to increasing supersaturation, the size of critical nuclei decreases, and the probability of their formation grows. Because of a rapid decrease in fluctuation probability as their sizes increase, the beginning of the phase transition is defined by the probability of formation of critical nuclei; therefore, only the formation of such nuclei is taken into consideration while calculating nonequilibrium condensation. Furthermore, due to an increase in the degree of supersaturation at the initial stage of the condensation process, those nuclei, which at the initial moment were of critical size, will soon become supercritical and will continue growing. The rate of formation of the critical nuclei is large and equals about 10^{19} particles per unit volume; therefore, despite their tiny sizes ($\approx 10^{-7}$cm), the total surface at which condensation takes place is rather large. Rapid formation of these nuclei and their further growth result in an intense increase of the liquid-phase mass and heat release. Usually, L/c_pT is of the order of unity and, therefore, even a small quantity of the liquid phase may influence the flow parameters considerably. Heat release not only prevents further increase of supersaturation but also results in a decrease of the degree of supersaturation. Formation of new nuclei, which largely depends on supersaturation, stops immediately and further condensation proceeds at newly formed nuclei.

Thus, all condensation centres form, as a rule, at the very beginning of condensation, once sufficient supersaturation is attained. The transition from the metastable state to the state of saturation which follows, may proceed rather rapidly and be shock-like (the so-called condensation shocks). However, under certain conditions a rather smooth transition to the state of saturation may be observed. Figures 6.26,27 show typical pressure and temperature distributions in a nozzle where nonequilibrium condensation

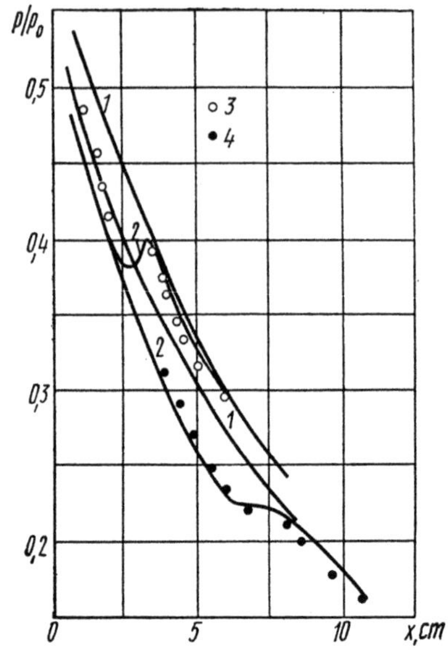

Fig.6.26

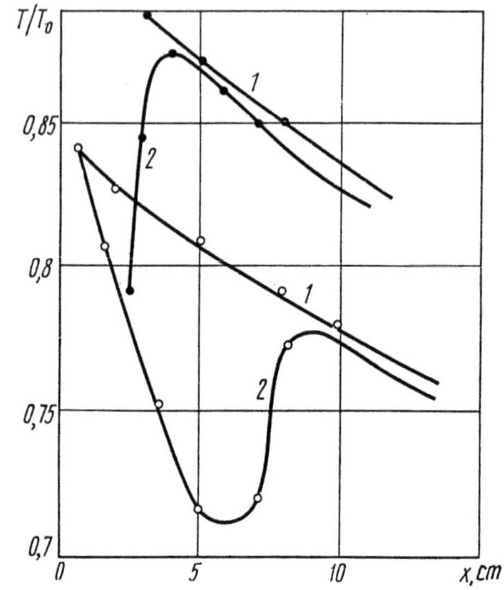

Fig.6.27

Fig.6.26. Pressure distribution along a nozzle in the case of nonequilibrium condensation. Calculation under conditions of experiments (*3*), (*4*): (*1*) equilibrium flow, (*2*) nonequilibrium flow. Experiment for a conical nozzle at $r_* = 1$ cm [6.47]: (*3*) $p_0 = 4.9$ bar, $T_0 = 442$ K, $T_\infty = 415$ K, (*4*) $p_0 = 2$ bar, $T_0 = 445$ K, $T_\infty = 370$ K

Fig.6.27. Pressure distribution along a nozzle's length in the case of nonequilibrium condensation (see legend to Fig.6.26)

takes place. We see that at higher pressures transition to saturation is almost shock-like, while at lower pressures it is smooth. At the same time, large gradients may prevent saturation being reached. Thus, the process of condensation in nozzle flows may be divided into two stages: at the first stage, nuclei formation plays the most significant role; at the second stage, nuclei growth is most important, and nuclei formation may be ignored. Figure 6.28 presents variations of the liquid-phase mass fraction (water vapour), the number of particles per unit mass N and the rate of particle formation J^0 along the wall and axis of a conical nozzle [6.48]. We see that the process of nuclei formation terminates at a very short segment, where the rate of their formation has a sharp maximum; after that the number of nuclei remains constant, and they grow.

The classical theory of condensation nuclei formation has been developed in [6.45,46], where an approximate expression for the stationary nonequilibrium function of the nuclei's size distribution and the formula for the rate

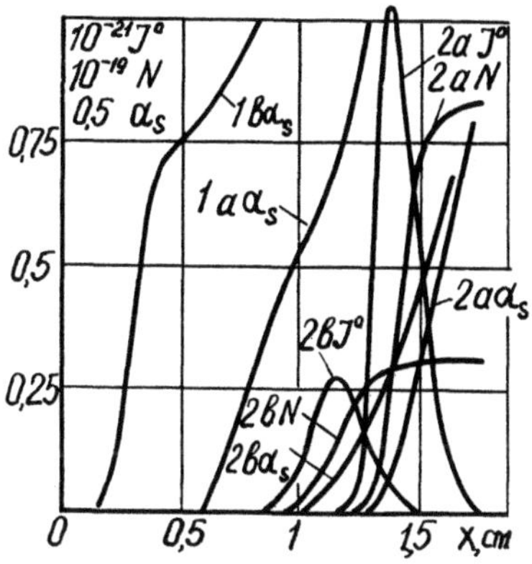

Fig.6.28. Variation of the nuclei formation rate J^0, the number of particles per unit mass N and the mass fraction of liquid phase α_s along axis (*a*) and wall (*b*) of a conical nozzle at $r^*=1$ cm, $\theta_k=12^\circ$, $\bar{R}_2=2$: (*1*) equilibrium flow, (*2*) nonequilibrium flow

of nuclei formation have been obtained. Despite some attempts to revise the formulas in [6.46], numerous calculations and experiments show that nowadays they are adequate for practical needs. Let us write down, without stating the assumptions made in the course of derivation, the equation for the nonequilibrium function of the nuclei size distribution f(g,t) [6.46]. We have

$$\frac{\partial f}{\partial t}=\frac{\partial}{\partial g}\left(D\,\frac{\partial f}{\partial g}\right)+\frac{1}{kT}\,\frac{\partial}{\partial g}\left(fD\,\frac{\partial\Delta\Phi}{\partial g}\right)\quad, \tag{6.124}$$

where $D=\alpha_{ik}S_g\beta$, α_{ik} is the condensation coefficient, S_g is the area of a nucleus surface and β is the flux of vapour molecules referred to unit surface and time. If the function f(g,t) is known, then the rate of formation of complexes containing g molecules J(g,t) is defined by

$$J(g,t)=-\,DN_g\,\frac{\partial}{\partial g}\left(\frac{f}{N_g}\right)\quad, \tag{6.125}$$

where N_g is the Gibbs distribution function [see (6.116)]. Assuming that the process of condensation nuclei formation is stationary ($df/dt=0$) and $f=N_g$ at small g and $f=0$ at $g\gg g_{cr}$, one can deduce that the following approximate expression for the rate of critical nuclei formation is valid:

$$J_i^0=\frac{\alpha_{ik}}{\rho_{is}}\left(\frac{p_{i\infty}}{kT}\right)^2\left(\frac{2\sigma_i k}{\pi R_{is}}\right)^{\frac{1}{2}}\exp\left(-\,\frac{4\pi r_{icr}^2\sigma_i}{3kT}\right)\quad. \tag{6.126}$$

If expansion proceeds at large gasdynamic parameter gradients, the stationary condition may not be fulfilled. To estimate the effect of a nonstationary condition upon the rate of formation, (6.124) has been solved numerically [6.49]. As an initial condition, it has been assumed that at the point of saturation at $t=0$, the distribution of nuclei of arbitrary sizes coincides with the Gibbs distribution. The boundary conditions are similar to

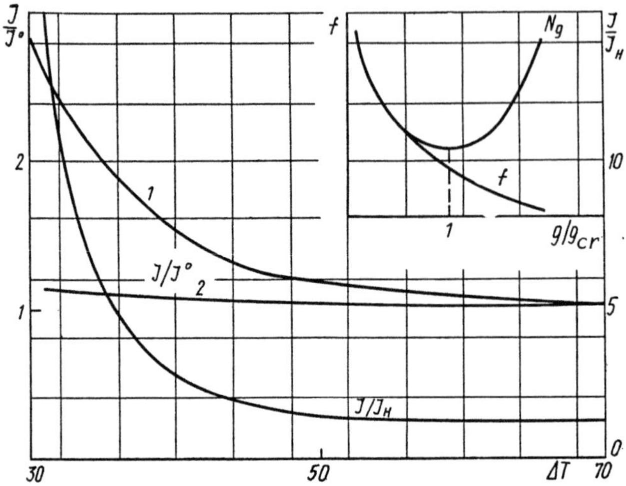

Fig.6.29. Dependence of the dimensionless nuclei formation rate J/J_H on $\Delta T = T - T_\infty$ ($T_\infty = 323$ K) at $dT/dt = 10^7 K\cdot s^{-1}$, (1) J/J_0 at $dT/dt = 10^7 K\cdot s^{-1}$ and (2) $dT/dt = 10^3 K\cdot s^{-1}$. Comparison of the nonequilibrium $f(g,t)$ and equilibrium N_g distribution functions around the point $g = g_{cr}$

those used while deriving (6.126), i.e. for all t, $f(g,t) = N_g$ at small g and $f(g,t) \to 0$ at $g \to \infty$. Equation (6.124) has been integrated numerically starting from the saturation point; the value of $\partial T/\partial t$ has been specified. Qualitative comparison of the nonequilibrium $f(g,t)$ and the equilibrium N_g distribution functions is presented in the upper right part of Fig.6.29. The figure shows that the boundary conditions have been chosen correctly and that in the vicinity of the point $g = g_{cr}$ the equilibrium distribution gives an unrealistically large number of condensation nuclei. The numerical solution allows one to evaluate the validity of the stationary condition hypothesis and various approximate assumptions used to derive (6.126). Evidently, the discrepancy between the rates of stationary J and nonstationary J_H nuclei formations must depend on the physical characteristics of the condensing component, the extent of supercooling, the temperature gradient $\partial T/\partial t$ and the saturation temperature. Figure 6.29 presents the dependence of J/J_H and J/J^0 on $\Delta T = T_\infty - T$ for water vapours. The dependence of J and J_H has been obtained by solving (6.124) numerically under the assumption that in the stationary case $\partial f/\partial t = 0$ and that J^0 denotes the rate of stationary nuclei formation calculated from (6.126). The computations show that the rate of stationary nuclei formation exceeds the nonstationary one, the difference increasing with decreasing ΔT. However, at large ΔT, $J \approx J^0 \approx J_H$. The difference increases also as the saturation temperature and the condensation coefficient decrease since these are accompanied by a decrease in the number of collisions per second and longer time periods are required to reach a stationary distribution. Sometimes, J^0 and J_H may differ by an order of magnitude, as in the case of vapours of B_2O_3. Nevertheless, (6.126) can evidently be used in many practically important cases, since the maximum difference between J^0 and J_H corresponds to small ΔT, when the rate of nuclei formation and the liquid-phase mass are small.

At the second stage of condensation, the determining role belongs to the growth of condensation nuclei formed earlier. If the mean free path of molecules exceeds the mean drop size considerably (the situation characteristic of nozzle flows), then the Knudsen formula should be used,

$$\frac{dr_i}{dt} = \frac{\alpha_{ik}}{\rho_{is}}\left[\frac{p_i}{\sqrt{2\pi R_{is}T}} - \frac{p_{i\infty}(T_{is})}{\sqrt{2\pi R_{is}T_{is}}}\right] \quad , \tag{6.127}$$

obtained under the condition that the number of molecules colliding with a drop and those flying away from its surface are equal. In a continuous flow, where transfer of vapour molecules to a drop is defined by diffusion, the Maxwell law,

$$\frac{dr_i}{dt} = \frac{D_i[p_i - p_{i\infty}(T_{is})]}{\rho_{is}R_{is}Tr_{is}} \quad , \tag{6.128}$$

should be used, where D_i is the diffusion coefficient of a condensing component. A drop's temperature may be found from the energy balance differential equation which for tiny droplets can be reduced to

$$\begin{aligned} &\alpha_{ik}\left[1 - \left(\frac{T_{is}}{T}\right)^{\frac{1}{2}} \frac{p_{i\infty}(T_{is})}{p_i}\right] - (1 - \alpha_{ik})\beta_i\left(\frac{T_{is}}{T} - 1\right) + \frac{2\gamma_i\alpha_{ik}}{\gamma_i + 1} \\ &\times \left[1 - \left(\frac{T}{T_{is}}\right)^{\frac{1}{2}} \frac{p_{i\infty}(T_{is})}{p_i}\right]\left(\frac{L_i}{c_pT} - 1\right) - \frac{\beta_i^0 p}{p_i}\left(\frac{R_i}{R}\right)^{\frac{1}{2}}\left(\frac{T_{is}}{T} - 1\right) = 0 \quad , \end{aligned} \tag{6.129}$$

where α_{ik} is the condensation coefficient, β_i and β_i^0 are the thermal accommodation coefficients of a condensing component and an inert gas and γ_i is the adiabatic exponent of a condensing component. The last term in (6.129) refers to an inert noncondensing gas and is zero if the latter is absent. It follows from (6.127,129) that the increment of a drop's radius and its temperature do not depend on the absolute value of the radius.

In order to describe the condensation process quantitatively, it is necessary to know the mass fraction of precipitated liquid, which can be determined, for instance, from the mass function of the particle (not only the critical ones) distribution in units of f_{im}. The equation for the distribution function f_{im} (derived similarly to Boltzmann's equation, see [6.50]) has the form

$$\frac{df_{im}}{dt} + \frac{\partial}{\partial r_i}\left(f_{im}\frac{dr_i}{dt}\right) = \frac{J_i^0}{\rho}\,\delta(r_i - r_{icr}) \quad , \tag{6.130}$$

where δ is the Dirac function (since only those new nuclei are considered which have critical dimensions) and ρ is the mixture density. This equation must be solved at every point of a flow simultaneously with the system of gasdynamic equations. A rather simple method of determining the mass fraction of the liquid phase and the distribution function has been proposed in [6.48] for the case when the increments of a drop's radius and temperature do not depend on the absolute value of the radius, as for the majority of actual nozzle flows.

The essence of the method is as follows. Let us divide a streamline into small segments and determine the increment of the liquid-phase mass fraction at the jth segment. Parameters referring to the segment k, at which formation of drops takes place, will be denoted by the subscript k, and parameters referring to the segment under consideration, by the subscript j. Let $N_k = J_k^0 \Delta l_k / \rho_k W_k$ be the number of critical nuclei formed per unit mass during the time needed to pass the segment k whose length is Δl_k. Then, an increment of the liquid-phase mass fraction per unit mass during the time needed to pass the jth segment is equal to

$$\Delta\alpha_j = 4\pi\rho_s\left(\sum_{k=1}^{j-1} N_k r_{kj}^2 \Delta r_j + \frac{1}{3} N_j r_{jcr}^3\right) \quad , \tag{6.131}$$

where Δr_j is the increment of a drop's radius at the segment Δl_j.

Taking into consideration that $r_{kj} = r_{k,j-1} + \Delta r_{j-1}$, $r_{(j-1)(j-1)} = r^*_{j-1}$, the latter expression may be reduced to

$$\Delta\alpha_j = 4\pi\rho_s\left(A_j \Delta r_j + \frac{1}{3} N_j r_{jcr}^3\right) \quad , \tag{6.132}$$

where $A_j = \Sigma_{k=1}^{j-1} N_k r_{kj}^2$. Putting $D_j = \Sigma_{k=1}^{j-1} N_k$ and $B_j = \Sigma_{k=1}^{j-1} N_k r_{k(j-1)}$, one can easily obtain the following recurrence relations,

$$D_j = D_{j-1} + N_{j-1}, \quad B_j = B_{j-1} + D_{j-1}\Delta r_{j-2} + N_{j-1} r_{(j-1)cr} \quad , \tag{6.133}$$

$$A_j = A_{j-1} + N_{j-1} r_{(j-1)(j-1)}^2 + D_j \Delta r_j^2 + 2B_j \Delta r_j \quad ,$$

which facilitate the calculation of $\Delta\alpha_j$ substantially. The value of Δr_j is obtained from (6.127) with the help of (6.129). Since the particles formed upon condensation are small and usually do not exceed some Ångstroms, the gas and particles may be assumed to have equal velocities and the particles' temperature must be calculated using (6.129), instead of (6.80). In this connection, coagulation of particles due to velocity differences may be ignored.

Generally, a number of components may be condensing in a flow simultaneously; however, according to the rule of phases (6.10), the number of condensing components cannot exceed that of the independent elements of which the mixture's components are formed. Condensation may be accompanied by chemical reactions, and the weight/molar fractions of noncondensing components should be obtained either from the equations of chemical kinetics (if reactions are nonequilibrium), or from the law of mass action (if reactions proceed in equilibrium). The molar fractions of condensing components should be found from the algebraic equations of material balance whose number, in accordance with the rule of phases, is equal to the number of condensing components.

Let the number of independent elements in a mixture be equal to m and the number of components in the gaseous phase, to N, of which the first m components are condensing. The system of equations describing a flow with nonequilibrium chemical reactions and nonequilibrium condensation in the equilibrium approximation has the form

$$\rho WF = Q \quad , \tag{6.134}$$

$$W \frac{dW}{dx} + \frac{1}{\rho}\frac{dp}{dx} = 0 \quad , \tag{6.135}$$

$$\frac{1}{2} W^2 + h = H_0 \quad , \tag{6.136}$$

$$\frac{d}{dx}\left(\frac{\alpha_i}{\mu_i}\right) = F_i \quad , \quad (i = m+1,\ldots,N) \quad , \tag{6.137}$$

$$\sum_{i=1}^{N} a_i \frac{\alpha_i}{\mu_i} + \sum_{i=1}^{m} a_{is} \frac{\alpha_{is}}{\mu_i} = b_k \quad , \quad (k = 1,2,\ldots,m) \quad , \tag{6.138}$$

$$\frac{d\alpha_{is}}{dx} = f^0(p_i, p, T, T_{is}) \quad , \tag{6.139}$$

$$T_{is} = f^1(p_i, T, p_{i\infty}) \quad , \tag{6.140}$$

$$p = R\rho T \quad , \tag{6.141}$$

$$p_i = p\alpha_i\mu/\mu_i \quad , \tag{6.142}$$

$$h = \sum_{i=1}^{m} \alpha_{is} h_{is} + \sum_{i=1}^{N} \alpha_i h_i \quad , \tag{6.143}$$

where $h_{is} = h_i - L_i$ and the functions f^0 and f^1 are defined with the help of (6.126,127,129,130). Usually, nonequilibrium condensation begins in the supersonic section of a nozzle. Therefore, one can solve numerically both the direct problem with a specified variation of the stream tube area $F = F(x)$ and the inverse problem. In the case of the inverse problem it is preferable to specify the density distribution of the mixture $\rho = \rho(x)$ since it depends only slightly on whether the flow is equilibrium or not. The following iterative procedure may be used to solve (6.134-143) numerically. Let the solution be sought at the node x_n if the solution at the node x_{n-1} is known and ρ_n is specified. Then, putting in the first iteration $T_n^{(1)} = T_{n-1}$, one obtains $p_n^{(1)}$ from (6.141), and $T_{isn}^{(1)}$ and $\alpha_{isn}^{(1)}$ from (6.140) and (6.139) respectively, putting on the right sides of the equations in the first approximation the values of functions which are equal to their values at a known point $(n-1)$. Further, $\alpha_{in}^{(1)}$, $W_{in}^{(1)}$ and a correct value of $T_n^{(2)}$ are obtained from (6.135-137) by the method presented in Sect.6.2 [see (6.41-43)]. Then, the entire process is repeated until convergence is attained using, whenever necessary in consequent iterations, the values of functions obtained in the previous iteration.

The same procedure is used in solving the direct problem, except that originally ρ_n is obtained from (6.134) under the condition that $W_n^1 = W_{n-1}$. If chemical reactions are absent, $\alpha_i = \text{const}$ $(i = m+1,\dots,N)$, and there is no need for (6.137). Otherwise, the procedure of calculation remains the same. Note that the method of calculation presented takes into account both the critical nuclei formation and their growth.

Let us consider some results of computations. We shall start by presenting a number of characteristic constants. Thus, for water we have $L = 2700$ kJ/kg, $c_p = 1870$ J/kg·deg, $\sigma = (128 - 0.19\ T)\cdot 10^{-3}$ N/m, $\rho_s = (1.11 - 4\cdot 10^{-4} T)$ g/cm^3. The values of supersaturation s or supercooling ΔT are the main parameters which define the beginning of nonequilibrium condensation. The following elementary considerations allow one to estimate the minimum possible value of ΔT.

Let only some tens of critical size condensation nuclei be formed in a unit volume of a mixture. The moment this occurs may be considered as the beginning of condensation. Then from (6.116) one can determine the value of the critical radius at $\Delta\Phi = \Delta\Phi_{cr}$ and a fixed number of condensation nuclei [remember that the equilibrium distribution (6.116) gives an excessive number of condensation nuclei compared to the nonequilibrium one]; then, the value of supercooling may be obtained from (6.120), which is the minimum possible one denoted by (ΔT_{min}). For the case of formation of a single critical nucleus, (6.120) gives for water vapour $\Delta T_{min} \approx 25\text{-}30^{\circ}$. Note that the known experimental supercooling values always exceed T_{min} obtained in such a way.

It follows from the above analysis of the process of nonequilibrium condensation, as well as from (6.120,126,128,129), that the value of supercooling is determined by the local conditions in the vicinity of the point of intersection of Poisson's adiabatic and the saturation curve, i.e. it must depend on the saturation temperature T_{∞}, the temperature gradient dT/dt, the mixture isentropic exponent γ^{0} and the noncondensing component concentration c^{0}. The determining parameters of the process of nonequilibrium homogeneous condensation which define the degree of supercooling, sizes of condensation nuclei and their number at Wilson's point have been introduced at a phenomenological level in [6.51,52], where results of parametric studies of condensation in nozzles are also presented. It is shown that in the absence of a noncondensing component, the main determining parameters include the temperature at the dew point T_{∞} and temperature gradient dT/dt. In [6.53] these determining parameters have been obtained analytically and by means of comparison of calculated and experimental data in a wide range of T_{∞} (280-520 K) and dT/dt ($1\text{-}10^{8}$K/s); T_{∞} was shown to be the major determining parameters.

The value of ΔT, as well as the degree of departure from equilibrium, must increase with increasing dT/dt and decreasing γ^{0}. However, the influence of γ^{0} on ΔT, drop size and the total number of drops is unimportant. The concentration of noncondensing components c^{0} has a slight effect upon the value of supercooling; however, it increases the rate of drop growth considerably, especially at low T_{∞} [see (6.127,129)]. Actually, the noncondensing components, contributing nothing to the liquid-phase mass fraction, take back some heat from a drop whose temperature tends to the gas temperature and, thus, help to intensify the drop's growth.

The effect of the saturation temperature T_{∞} upon ΔT is more complicated. Evidently, at very small T_{∞} (and, hence, ρ_{∞}) ΔT may be quite large, especially at large dT/dt, since the liquid-phase formation rate and the rate of condensation nuclei growth decrease with decreasing T_{∞}. As T increases from small values to moderate ones (e.g. from -70 to 0°C for water vapour), the value of ΔT decreases to a certain value, and then, depending on the values of c^{0}, γ^{0} and dT/dt, it may continue decreasing, remain constant or increase. Elaborate experimental studies of water vapour expansion in plane nozzles

[6.47] have demonstrated that ΔT stays practically constant and equal to 50-60 K while the temperature T_∞ changes from 50 to 150°C. On the other hand, with the increase of T_∞, the process of condensation has a growing effect upon the mixture pressure and temperature (see Figs.6.26,27). We see that in the presence of nonequilibrium condensation, a decrease of T_∞ changes the character of the pressure distribution considerably. If an almost shock-like pressure increase (the so-called condensation shock) is characteristic at large T_∞, and is followed by a gradual approach of the equilibrium value, at small T_∞ a smooth pressure increase takes place. It should be noted at the same time that a sharp pressure increase may result in the local pressure (but not momentum) being higher than the equilibrium one. Calculations show that even spontaneous condensation does not exterminated supercooling completely, and a decrease of T results in an increase of the supercooling of the remainder. Therefore, a sharp drop in supercooling in the first region of spontaneous condensation at large dT/dt and low T_∞ may be followed by a new build-up of supercooling and formation of a new region of spontaneous condensation.

Let us finally consider the influence of the number M upon ΔT. On the one hand, it is evident that the rates of condensation nuclei formation and growth are defined by T_∞ and p_∞ and do not depend on the absolute flow velocity and, hence, on the number M. However, the influence of the same mass fraction of condensed liquid phase upon the flow parameters will be different at different Mach numbers. In fact, heat input in the transonic region at $M \approx 0.8$-1.5 may change the flow parameters considerably, while in the supersonic region the same amount of heat has a much smaller effect on the flow parameters. According to [6.48] (see also Table 6.6) the number M at $M_\infty > 1.5$ does not practically influence ΔT, the drop sizes and the rates of their formation and growth. Analysis of experimental and calculated data shows that at $M > 1.5$ radii of critical nuclei at the points where condensation of water vapour, nitrogen and oxygen begins, are equal to 6-7 Å and each such condensation nucleus contains 8-10 molecules (compare with the water molecule radius equal to 2.3 Å and those of nitrogen and oxygen equal to 1.75 Å). The data allows one to define the value of ΔT from (6.120). The ratio T_k/T_∞ proves to be constant and, in a certain interval of T_∞, equal to 0.875 and 0.7 for water vapour and nitrogen respectively. The dependence of M_k on M_∞ (M_k and T_k are the Mach number and the temperature at the point where condensation begins, and M_∞ is the Mach number at the saturation point) has the form

$$M_k = \left[\frac{2}{\gamma - 1}\frac{T_\infty}{T_k}\left(1 + \frac{\gamma - 1}{2}M_\infty^2 - \frac{T_k}{T_\infty}\right)\right]^{-\frac{1}{2}} \quad . \tag{6.144}$$

This dependence agrees well with experimental data at the above values of T_k/T_∞ and may be used to evaluate condensation lag in wind tunnels [6.42,47, 54].

The Mach number of a nonequilibrium flow calculated using the frozen sonic speed exceeds that of a frozen flow, since the temperature of a nonequilibrium flow is higher than that of a frozen flow. At the same time, the Mach number of an equilibrium flow calculated using the equilibrium sonic speed exceeds the Mach number of a nonequilibrium flow. Note also that in an equilibrium flow the sonic speed at the saturation point and, hence, the Mach number, are discontinuous at the point of transition from Poisson's adiabatic to the saturation curve, which is absent, of course, in equilibrium flows. The results of calculations show that ρW and the mixture density depend only slightly on the character of the process. Figure 6.30 shows the effects of T_∞ and dT/dt, average particle sizes and the number of particles per unit volume upon ΔT on the basis of data obtained by numerical solution of (6.134-143) [6.55].

Let us compare results of experiments and calculations. In a number of papers attempts have been undertaken to attain agreement of experimental and numerical results by varying the condensation and thermal accommodation coefficients. Since data for the first coefficient are quite contradictory (the values of α_k for water vapours vary from paper to paper from 0.02 to 1),

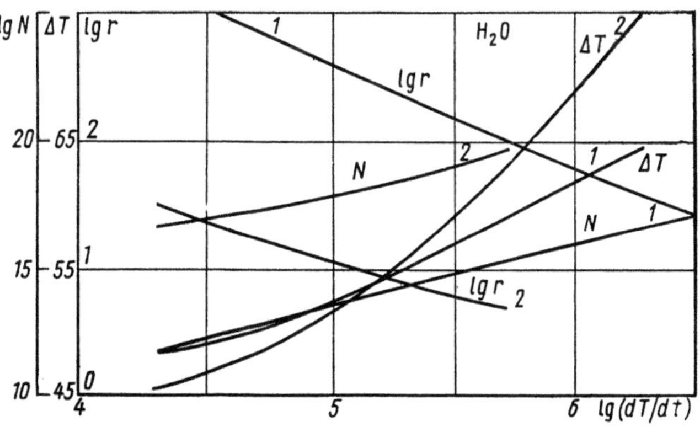

Fig.6.30. Dependence of supercooling ΔT(K), particle radius r (Å) and the number of particles per unit volume N on temperature gradient dT/dt($K\cdot s^{-1}$) at (*1*) $T_\infty = 373$ K and (*2*) $T_\infty = 273$ K

it seems to be expedient to choose a dependence of α_k on temperature which would make calculated results and reference experimental data agree both in pressure and particles size distribution, in the best possible way. However, one dependence of this kind is insufficient; in addition, the formation energy in (6.126) should be taken with a certain coefficient ζ. Such an approximation gives for water vapour $\alpha_k = \exp(-3.22 + 875 T^{-1})$, $\zeta = -1.18 + 0.00795 T$ at $T > 273$ K and $\alpha_k = \zeta = 1$ at $T < 273$ K. The use of these approximations for numerical integration of the system (6.134-143) gave a good agreement between calculated and experimental data (Fig.6.26) [6.55].

Nonequilibrium condensation of water vapour, nitrogen and oxygen may take place in wind tunnel nozzles. In nozzles of rocket engines these components do not condense due to high stagnation temperatures; however, they do condense in the exhaust jets. However, for combustion products of fuels containing boron, nonequilibrium condensation of B_2O_3 vapours may take place in the process of efflux. Calculations of nonequilibrium condensation for the fuel composition $B_5H_9 + H_2O_2$ have shown that impulse losses do not exceed 1% even at the condensation coefficient $\alpha_k = 10^{-3}$ [6.32]. Generally, the same laws are true for flows in exhaust jets of rocket engines and nozzle flows in the presence of nonequilibrium condensation. However, the particle radii in jets are smaller than in nozzle flows because of the smaller T_∞. Thus, for the fuel composition $N_2O_4 + (CH_3)_2NNH_2$ at $\alpha = 0.9$ and $p_0 = 100$ bar, $\alpha_{sH_2O} = 0.3$, $T_\infty = 223$ K and condensed particles have radii from 5 to 6 Å.

Let us now consider crystallization of liquid particles. The process may start if the expansion in a nozzle is accompanied by the gas temperature dropping below the fluid particle melting temperature. Condensation of certain components in nozzles with large expansion ratios may be followed by their crystallization. In practice, the phenomenon of crystallization is encountered, as a rule, in the case of fuels with Al and Be additives whose combustion products contain Al_2O_3 and BeO oxides with melting points of 2303 and 2830 K, respectively, at a combustor temperature in excess of 3000 K. The particles of these oxides are in a liquid state in the combustion chamber and upon reaching the melting point in the process of expansion, must start crystallizing in accordance with the condition of equilibrium. Equilibrium thermodynamics requires that the process of crystallization proceeds at a constant temperature corresponding to the melting point, and this region of flow may be described by gasdynamic functions for an isothermal flow. Along the part of a nozzle where crystallization takes place, crystallization heat transfers to the gas gradually until the whole of the liquid phase transforms into the solid state.

However, the process is not in equilibrium. The rate of heat transfer from particles to a gas as well as the rate of formation of nucleation centres are finite, and the above processes do not have time for completion during the short period of residence in a nozzle. The actual amount of crystallization heat transferred to a gas is much smaller than that corresponding to an equilibrium flow of the process. Crystallization kinetics is somewhat similar to condensation kinetics. The mass fraction of the new phase, as before, is defined by the rate of formation of the latter's nuclei (this time tiny crystals) and the rate of their further growth. The probability of their formation may be defined by (6.123) (as in the case of a liquid phase). However, this time σ is the interphase surface tension and ΔT is the supercooling of the liquid (for water $\sigma \approx 10$ dyne/cm and for aluminium oxide $\sigma \approx 200$ dyne/cm). The probability of formation of at least one nucleation centre is lower, the smaller the mass of the liquid. It is known that tiny droplets of water may be supercooled by tens of degrees below zero. The time needed to freeze micron-size droplets is measured in seconds or minutes; for submicron-size droplets it is measured in hours. However, the length of time a particle resides in the supersonic section of a nozzle where crystallization usually takes place, is of the order of 10^{-4}-10^{-5}s. Thus, due to the small dimensions of the particles, the short residence time and a rather large value of σ, equilibrium crystallization does not take place and, most probably, particles of metal oxides, upon attaining the melting point, stay in the liquid supercooled state. On the other hand, the efficiency of crystallization heat transfer to a gas in a real process of particle flow in a nozzle is substantially lower than in an equilibrium flow. In fact, the temperature of particles in a nonequilibrium two-phase flow is higher than the gas temperature and the difference is greater, the larger the particles. Therefore, crystallization heat is released in a nonequilibrium two-phase flow at a lower pressure and a lower gas temperature compared to an equilibrium flow and, thus, is less efficient than in the latter case.

Figure 6.31 presents the temperature dependence for a gas and particles of different sizes on the dimensionless radius of a nozzle. We see that the larger the particles, the lower the gas temperature from which crystallization begins. The process is accompanied by an increasing difference in the gas and particle temperatures since the latter remains constant and equal to the melting temperature while the former continues to decrease.

Calculations of the vacuum impulse increment due to crystallization for particles of different sizes, taking into consideration the finite rate of

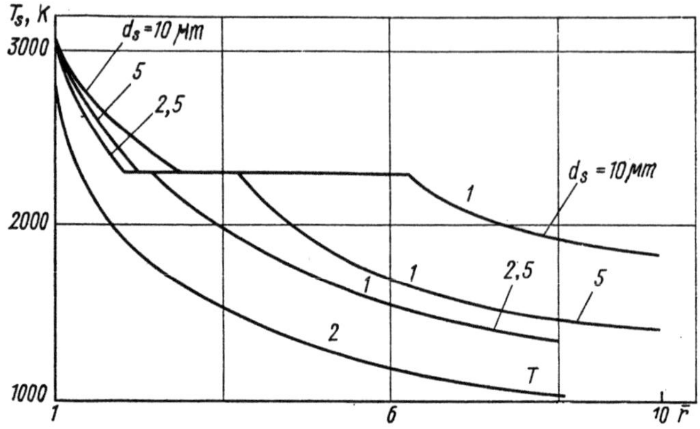

Fig.6.31. Dependence of gas and particle temperature on $\bar{r}$: (*1*) nonequilibrium flow with crystallization, (*2*) frozen flow without crystallization

heat transfer from particles, show according to (6.80) that the greater part of the impulse increment is due to particles smaller than 3-5 μm in diameter [6.32]. However, these are the particles for which the crystallization lag probability is the greatest. For particles 10 to 15 μm in diameter, the impulse increment due to a finite rate of heat transfer is negligible. Thus, the finiteness of the rate of heat transfer from particles to a gas and the crystallization nuclei formation lag make it possible to ignore crystallization in the case of a two-phase mixture expansion in a nozzle.

Let us estimate impulse losses due to the absence of crystallization. For the sake of simplicity, the case of outflow into vacuum will be considered. From the energy equation one gets

$$
\begin{aligned}
W_1 &= \{2T_0[(1-\alpha_s)c_p + \alpha_s c_s]\}^{\frac{1}{2}} , \\
W_2 &= \{2T_0[(1-\alpha_s)c_p + \alpha_s(c_s - Q^0/T_0)]\}^{\frac{1}{2}} , \qquad (6.145) \\
\zeta_s^0 &= 1 - \frac{W_2}{W_1} \approx \frac{\alpha_s Q^0}{2T_0[(1-\alpha_s)c_p + \alpha_s c_s]} ,
\end{aligned}
$$

where W_1 is the velocity of outflow into vacuum in the case of an equilibrium flow with crystallization, W_2 is the velocity of outflow into vacuum in the case of a frozen flow in the absence of crystallization and Q^0 is the melting heat. Then, the maximum impulse losses due to the absence of crystallization, ζ_s^0, for combustion products containing Al_2O_3 with $\alpha_s = 0.325$, $T_0 = 3250$ K and $c_s/c_p = 0.35$ are equal to about 2%. It follows from (6.145) that ζ_s increases with increasing Q^0 and α_s and decreasing T_0.

Concluding this section, we would like to point out that the monograph [6.56] discusses systematically the theory of condensation and considers various aspects of practical applications. It also gives a detailed account of modern numerical methods of solving equations describing nonequilibrium flows with phase transformations and presents results of parametric numerical studies of jet, nozzle and other kinds of flows. The monograph pays considerable attention to creating a single system of determining parameters describing the process of nonequilibrium condensation.

6.4.5 Flows in Axisymmetric and Plane Nozzles

Examination of two-phase flows in the one-dimensional approximation produces a number of qualitative details. However, two-dimensional effects play an important role in dusted gas flows both due to the irregular distribution of particles over the cross-sections of a nozzle and due to the possibility of particles hitting the walls of the sub- and supersonic sections of a nozzle because of the different (both in value and sign) action the gas exerts upon particles at different points of a nozzle. As a result, the particle trajectories do not coincide with streamlines, while the velocity vector and temperature of particles in the trans- and supersonic regions exhibit a rather strong dependence on the subsonic region flow parameters. Therefore, to describe a two-phase nozzle flow correctly, one has to calculate the sub-, trans- and supersonic regions simultaneously.

Let us start by presenting equations describing plane ($\nu = 0$) and axisymmetric ($\nu = 1$) two-phase flows of a polydisperse matrix of a gas and particles without phase transitions:

$$\frac{\partial}{\partial x}(\rho u y^{\nu}) + \frac{\partial}{\partial y}(\rho v y^{\nu}) = 0 \quad , \tag{6.146}$$

$$\frac{\partial}{\partial x}(\rho_{is} u_{is} y^{\nu}) + \frac{\partial}{\partial y}(\rho_{is} v_{is} y^{\nu}) = 0 \quad , \tag{6.147}$$

$$u\frac{\partial u}{\partial x} + v\frac{\partial u}{\partial y} + \frac{1}{\gamma\rho^0}\frac{\partial p}{\partial x} = \sum_{i=1}^{N}\frac{\rho_{is}}{\rho} f_{ix} \quad , \tag{6.148}$$

$$u\frac{\partial v}{\partial x} + v\frac{\partial v}{\partial y} + \frac{1}{\gamma\rho^0}\frac{\partial p}{\partial y} = \sum_{i=1}^{N}\frac{\rho_{is}}{\rho} f_{iy} \quad , \tag{6.149}$$

$$u\frac{\partial}{\partial x}\left(c_pT + \frac{1}{2}W^2\right) + v\frac{\partial}{\partial y}\left(c_pT + \frac{1}{2}W^2\right) = \sum_{i=1}^{N}\frac{\rho_{is}}{\rho}(\mathbf{f}_i\mathbf{W}_{is} + q_{is}) \quad , \tag{6.150}$$

$$u_{is}\frac{\partial u_{is}}{\partial x} + v_{is}\frac{\partial u_{is}}{\partial y} + \frac{1}{\gamma\rho_s^0}\frac{\partial p}{\partial x} = f_{ix} \quad , \tag{6.151}$$

$$u_{is}\frac{\partial v_{is}}{\partial x} + v_{is}\frac{\partial v_{is}}{\partial y} + \frac{1}{\gamma\rho_s^0}\frac{\partial p}{\partial y} = f_{iy} \quad , \tag{6.152}$$

$$u_{is}c_s\frac{\partial T_{is}}{\partial x} + v_{is}c_s\frac{\partial T_{is}}{\partial y} = q_{is} \quad , \tag{6.153}$$

where $i = 1,2,\dots,N$ and u, v, u_{is}, v_{is} are the projections of velocity vectors of a gas and particles upon the axes x and y of a Cartesian coordinate system. The projection of the force f_i exerted by the gas upon the particles and the value of the heat flux q_{is} [see (6.78,80,86,88)] are determined from

$$f_{ix} = St_{is}^{-1}\bar{C}_{xi}(u - u_{is}) \quad , \quad f_{iy} = St_{is}^{-1}\bar{C}_{xi}(v - v_{is}) \quad , \tag{6.154}$$

$$q_{is} = \frac{Nu_{is}c_p}{3PrSt_{is}}(T - T_{is}) \quad .$$

Since the trajectories of particles are characteristics of the system (6.146-153), its last three equations at $\rho_s^0 \gg \rho^0$ may be presented in the form

$$u_{is}\frac{du_{is}}{dx} = f_{ix} \; , \quad v_{is}\frac{dv_{is}}{dx} = f_{iy} \; , \quad u_{is}\frac{dT_{is}}{dx} = q_{is} \; , \; i = 1,2,\dots,N \tag{6.155}$$

where differentiation is performed with respect to particle trajectories at which the following relation holds:

$$\frac{dr_{is}}{dx} = \frac{v_{is}}{u_{is}} \quad . \tag{6.156}$$

Calculation of plane and axisymmetric two-phase flows in nozzles may be carried out in two different ways. One of them (which gives higher precision) involves calculating flows by means of numerical integration of the system (6.146-153) and allows the mutual influence of the gas and the particles to be taken into account. Since the system of equations for the supersonic part of a nozzle is hyperbolic, various types of the method of characteristics are used in its integration [6.57-60].

Flows in the sub- and transonic parts of a nozzle are calculated either by the method of stabilization, or by using a numerical algorithm to solve the inverse problem of the theory of nozzles [6.61].

In [6.61] the inverse method is used to solve the problem of dusted gas flow in plane and axisymmetric nozzles. The basic ideas of this method and its algorithm have been discussed in Chap.3 and Sects.6.3.2 and 6.3.3. The main advantages of such an approach are its high precision and the small amount of computer time needed, characteristic of the inverse method. It also allows flows in sub-, trans- and supersonic regions to be calculated with the help of the same algorithm, which is designed in such a way that both mono- and polydisperse flows can be treated. In [6.61], the distribution of the gas flow velocity is given at the nozzle axis and the parameters of the gas and the particles are fixed at the entrance section. The inverse problem consists in determining the flow field in the vicinity of the nozzle's symmetry axis using the initial data and, thus, is reduced to solving an initial-value problem. Unlike the case of purely gaseous flow, two initial-value problems must be solved in the case of a two-phase dusted flow. One of them is solved in the transverse direction in order to obtain some gasdynamic parameters, and the other one is solved in the longitudinal direction, providing the particles' parameters and the remaining gasdynamic parameters.

Solution of the inverse problem results in the determination of parameters of the gas and the particles and gives the gas streamlines and the trajectories of the particles. Evidently, any streamline of the gas flow may be assumed to be a rigid nozzle wall. However, unlike the pure gas case, the shape of the contour of a nozzle corresponding to a certain gas flow rate will be different at different values of the particles' parameters; in particular, it will differ for particles of different sizes. The inverse problem may be conveniently solved with the help of independent variables ψ and x, where ψ is the streamline function and x is the longitudinal Cartesian coordinate.

The system of equations describing gas flow with particles of a condensed phase differs from the usual gasdynamic equations, for the right-hand sides of the equations of motion and energy contain in the former case additional members, which account for the action of the particles upon the gas; also, equations are added which describe the motion of and the heat exchange between the particles and the gas. In ψ, x coordinates this system of equations coincides with the one formed by (2.111-113).

The method of solving the system (2.111-113) in the framework of the inverse problem is similar to the one used to solve a corresponding system of equations in the case of nonequilibrium flows (Sects.6.3.2,3.3). The only difference is a somewhat different way of determining initial data at the symmetry axis. In the case of two-phase flows an asymptotic expansion needs to be constructed in the vicinity of the symmetry axis (both for plane and axisymmetric flow) enabling all the flow parameters to be found at the symmetry axis (Sect.4.2.3). As usual, the velocity distribution is given at the symmetry axis and values of u_{is}, T_{is}, v_{is}, ρ_{is} are fixed at the plane $x = x_0$ for all ψ, as in the case of nonequilibrium flows. One may also use another approximate method of calculation, based on separate solution of equations for the gaseous phase and the particles [6.62,63]. In the process of such calculations, the entire field of gasdynamic parameters, particle parameters, their trajectories and the zones occupied only by gas are obtained by taking into account the mutual influence of the gas and the particles. An approximate approach is possible based on the separate solution of equations for the gaseous phase and the particles. It is assumed that the gas parameters do not change under the influence of the particles and may be obtained as a result of independent calculations for a gas with a fictitious adiabatic exponent γ^0, i.e. the gas parameters correspond to an equilibrium flow. At the same time, the particle parameters are determined by numerical integration assuming the gas parameters are constant. In this case the system (6.146-153) separates into two independent systems (6.146, 148-150) for a fictitious gas with $\gamma = \gamma^0$, in which the sums on the right sides of these equations are absent, and the system (6.147,151-153 or 155) for the particles. The former system is solved either by considering the inverse problem of the theory of nozzles or by the method of characteristics in the supersonic region. Such a calculation allows one to obtain the gas parameters as functions of coordinates at all points of the flow field. In order to obtain the particle parameters u_{is}, v_{is}, T_{is}, the system (6.155) is integrated under the assumption that the gas parameters are known and constant. The validity of such an approximate approach is confirmed by the results of calculations in the one-dimensional approximation presented above, according to which the gas parameters of nonequilibrium and equilibrium two-phase flows practically coincide. Evidently, the second approximate method is much simpler. Calculations performed by the two methods show that the values of the gas and the particle parameters obtained in both cases are quite close in the subsonic part of flow. In its transonic part, the parameters of the gas and the particles differ by 10-15%, which may result in different trajectories in the supersonic part of flow and at the points in the supersonic part where sedimentation of the particles on the nozzle's contour occurs [6.62]. This difference is essential if the specific impulse losses are to be determined precisely. However, rather good qualitative behaviour of the gas and particles' parameters and their trajectories may be obtained by means of the second approximate method of calculation. Figures 6.32,33 present results of such calculations.

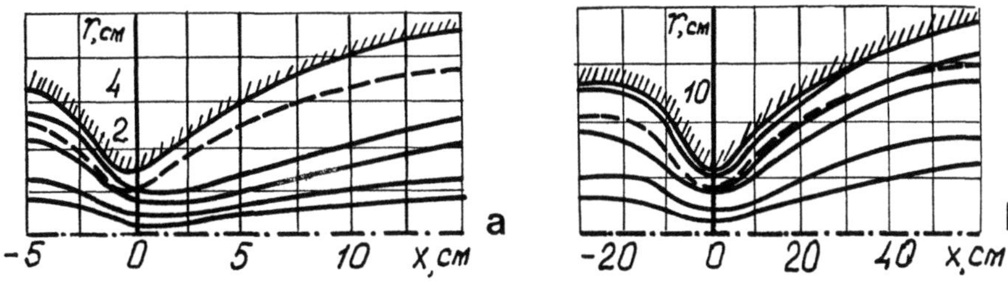

Fig.6.32a,b. Trajectories of particles (——) and gas streamlines (---) in Laval nozzles at (**a**) $d_s = 10$ µm, $d_* = 25$ mm and (**b**) $d_s = 2.5$ µm, $d_* = 100$ mm

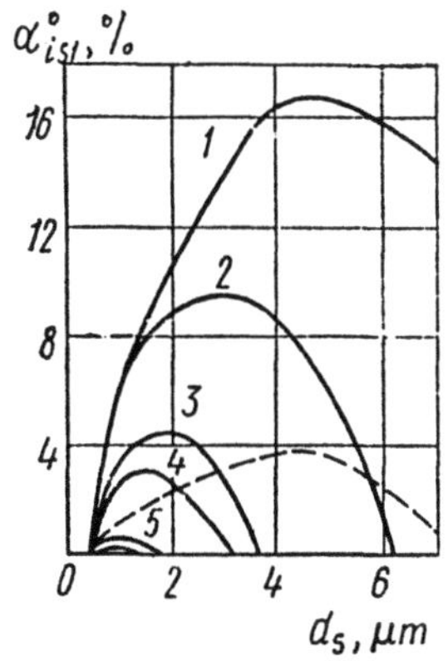

Fig.6.33. Weight fraction of particles precipitated upon a nozzle contour in the region of supersonic flow at $\gamma^0 = 1.14$, $d_* = 200$ mm, $\bar{r}_a = 5.5$ (——) and at $\bar{r}_a = 3.6$, $\bar{L}_a = 7.5$ (---): (*1*) $\bar{L}_a = 16$, (*2*) $\bar{L}_a = 12$, (*3*) $\bar{L}_a = 10$, (*4*) $\bar{L}_a = 9$, (*5*) $\bar{L}_a = 8$

Figure 6.32 shows trajectories of particles for a number of typical situations [6.62]. For large particles ($d_s > 10$ µm) and narrow throats ($d_* < 25$ mm), trajectories in the supersonic section of a nozzle are close to straight lines. Trajectories of small particles are curved and at large nozzle diameters they practically coincide with streamlines. Such behaviour of the particles is explained by an increase of their velocity and temperature lags with the increase of their dimensions and the gradient of the gasdynamic parameters accompanied by the decrease of the gas density. The gas streamlines and the praticle trajectories are rather close in the subsonic section of a nozzle; however, they differ considerably in its trans- and supersonic sections, especially for large particles and small throat diameters.

For all particles there exists a trajectory touching a nozzle wall in the subsonic section. These limiting trajectories divide the flow of particles in a nozzle. Those particles whose trajectories lie above the limiting one and intersect a nozzle contour in the subsonic section, strike the nozzle wall. Their number, naturally, increases as their diameter increases and that of the throat decreases. The weight fractions of particles α^0_{is} which hit the nozzle wall in its subsonic section in the intervals of $d_s = 1\text{-}20$ μm and $d_* = 10\text{-}200$ mm may be obtained from the relationship $\alpha^0_{is} = 8\cdot 10^6 d^2_{is} d^{-1}_*$ (α^0_{is} is expressed in percent if d_{is} and d_* are taken in millimetres).

Downstream of the point of contact the limiting trajectory becomes the line of tangential discontinuity, since above it the particle density is zero, and the space between the nozzle wall and the limiting trajectory is filled with pure gas. For particles with $d_s > 10$ μm and nozzles whose $d_* < 50$ mm, the pure gas zone occupies a considerable part of the cross-section. Thus, in a nozzle with $d_* = 25$ mm pure gas occupies about 70% of the exit section area, i.e. the separation of large particles takes place, which concentrates them near the axis, because both components of their velocities are much smaller than those of the gas in the transonic region of flow.

With increasing throat area, the pure gas zone at a nozzle's exit cross-section diminishes and at a certain $d_* = d_{*1}$ the limiting trajectory hits the wall at the exit cross-section. With further increase in d_* the nozzle wall will also be intersected by some trajectories lying below the limiting one. The particles moving along these trajectories will fall on the supersonic part of a nozzle contour. This is accompanied at first by an increase of the particle weight fraction α^0_{is1} with increasing d_*; at the same time the point of intersection of the limiting trajectory with the nozzle wall moves upstream. However, as d_* further increases, the gas and particle velocities become quite close, and starting with a certain value of $d_* = d_{*2}$, α^0_{is1} begins to diminish. Thus, the dependence of α_{is1} on the throat diameter for a fixed particle size is nonmonotonic. The same refers to the dependence of α^0_{is1} on the size of particles at a fixed throat diameter.

Figure 6.33 shows the dependence of α_{is1} on d_s for a number of nozzles. With decreasing length, the weight fraction of particles striking a nozzle wall decreases as well as the size of particles whose limiting trajectories intersect the supersonic part of a nozzle contour. Note that for nozzle contours providing a maximum thrust at a fixed exit cross-section area in the absence of heat exchange between a gas and a nozzle wall, the number of particles falling on the walls is rather large. This effect should be, naturally, taken into consideration when choosing an optimum nozzle profile for two-dimensional flows. In particular, the optimum length of a nozzle in the case of a two-phase flow should be chosen taking into account not only the losses due to friction, scattering and the velocity and temperature lags of particles, but also considering the losses due to precipitation of particles in the supersonic part of a nozzle. Sometimes, the end segment of a nozzle contour should be changed in order to prevent precipitation of particles [6.62,64].

Let us assume that the particles striking the wall in the sub- or supersonic part of a nozzle take no further part in the thrust generation. Then, besides the impulse losses due to velocity and temperature lags of particles, losses due to precipitation of particles appear. Evidently, one can assume that the latter impulse losses are equal approximately to the precipitated particles' momentum referred to the nozzle impulse. In the case of a monodisperse dusted gas these losses can be calculated from

$$\zeta_w^0 = \alpha_{is}(\alpha_{is}^0 + \alpha_{is1}^0)u_{is}/J^0 \quad . \tag{6.157}$$

Estimated impulse losses ζ_w^0 due to precipitation of particles 2.5 μm in diameter in the subsonic part of a nozzle with $d_* = 25$ mm for the particle weight fraction $\alpha_{is} \approx 0.3$ are equal approximately to 0.3%. With an increase in the diameter d_*, the impulse losses ζ_w^0 due to particle precipitation in the subsonic part of a nozzle will decrease approximately proportionally to d_*.

If the size distribution of particles is known, then (6.157) can be generalized to the case of a polydisperse dusted gas flow. In this case, we have

$$\zeta_w^0 = \alpha_s \sum_{i=1}^{N} (g_i\alpha_{is}^0 + g_{i1}\alpha_{is1}^0)\frac{u_{is}}{J^0} \quad , \tag{6.158}$$

where g_i, g_{i1} and u_{is} are the weight fraction of the ith particles and their velocity at the section where precipitation takes place. For a supersonic contour providing the maximum thrust at a fixed exit section area, these losses are equal approximately to 0.5% at $\bar{r}_a = 5.25$ and $d_* = 200$ mm.

The monodisperse flow of a dusted gas in the supersonic part of an axisymmetric nozzle has been calculated exactly, taking into account interaction between gas and particles, in [6.58,59].

As was pointed out above, the particles' paths and their parameters are accounted for accurately enough if calculations are performed using the approximate method. This conclusion is confirmed by calculations made in [6.58] where it has been shown that the positions of the limiting paths are practically independent of the weight fraction of the particles, just as in the approximate method, where such a dependence is absent completely. The largest difference in the gas parameter values obtained by the approximate method and by solving the full equations takes place in the pure gas zone in the supersonic region of flow. Here the gas velocity is higher and the temperature lower than in the central part of the flow (Fig.6.34). The particle

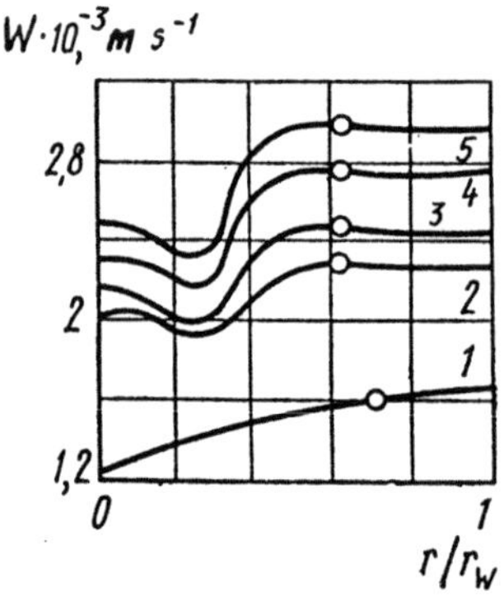

Fig.6.34. Gas velocity distributions over various cross-sections of an axisymmetric nozzle for a nonequilibrium two-phase flow at $d_s = 5$ μm, $\alpha_s = 0.25$: (1) $\bar{x} = 0.34$, $\bar{r} = 1.18$, (2) $\bar{x} = 2.12$, $\bar{r} = 1.81$, (3) $\bar{x} = 3.0$, $\bar{r} = 2.15$, (4) $\bar{x} = 3.64$, $\bar{r} = 3.16$, (5) $\bar{x} = 9.23$, $\bar{r} = 4.5$. Points at the curves correspond to the limiting line, r_w is the value of r at the nozzle wall

parameters, including the density ρ_{is}, are discontinuous at the limiting path, the latter being the line of tangential discontinuity. The gas parameters at $\rho_{is} \ll \rho_s^0$ stay continuous, though their derivatives may be discontinuous. The presence of the tangential discontinuity line, naturally, intensifies the nonuniformity of parameters at a nozzle cross-section in the case of two-phase flows compared to pure gas flows. The condensate density increases in the transonic region in the direction from the axis to the limiting trajectory, staying practically uniform at the sections of the region of developed supersonic flow.

Simultaneously with the calculation of the flow parameters, the momentum of a two-dimensional two-phase nonequilibrium flow may be determined at any cross-section of a nozzle. The difference between the momentum here and that of an equilibrium two-dimensional two-phase flow at the same section includes the losses due to precipitation of particles in the sub- and supersonic regions ζ_w^0, the losses due to velocity and temperature lag of particles ζ_s, the losses due to absence of crystallization ζ_s^0 and, finally, the losses due to additional flow nonuniformity contributed by particles. The analysis of numerical results shows that values of ζ_s and ζ_s^0 may be calculated accurately enough in the one-dimensional approximation, ζ_w^0 may be obtained from (6.158) and the losses due to additional flow nonuniformity contributed by particles may be neglected compared to the losses associated with the nonuniformity of a two-dimensional two-phase equilibrium flow.

Flows with nonequilibrium condensation in axisymmetric and plane nozzles are described by a system of equations similar to (6.146-153). The lag of liquid particles may be ignored because of their small sizes; however, equations must be introduced describing the condensation kinetics. Therefore, a two-dimensional nonequilibrium flow with condensation must be calculated with the help of the system (6.146,148,149) and (6.136-143) keeping in mind

Table 6.6. Parameters of water vapour flow in a conical nozzle

N^0N^0	p_0,bar	T_0K	T_∞K	M_∞	ΔT Axis	ΔT Contour	M_k Axis	M_k Contour
1	2.6	467	370	1.26	57.0	58.4	1.75	1.76
2	4.3	941	370	3.06	57.2	56.7	3.51	3.50
3	18.9	757	370	2.52	57.6	57.5	2.54	2.54
4	18.0	653	392	2.00	52.1	51.8	2.39	2.37
5	18.0	900	348	3.12	63.0	63.2	3.64	3.64

that in this case, ρ in (6.146,148,149) denotes the mixture density, and the sums on the right sides of (6.148) and (6.149) must be omitted.

A water vapour flow in a conical nozzle has been calculated in [6.48] using the method of characteristics. Some calculated results are presented in Table 6.6 and Fig.6.28. They show that the lines of M = const coincide with the lines of the start of nonequilibrium condensation. The difference of velocity gradients at the axis and walls of a nozzle results in differences in the values of J^0, N and α_s; however, it does not influence the values of M_k and ΔT. The data presented in Table 6.6 demonstrate the above-stated fact that there is a slight dependence of ΔT on the number M at a constant T_∞.

Calculations show that the geometry of streamlines and the mixture density are the same for equilibrium and nonequilibrium flows. Therefore an approximate method of calculation can be proposed for two-dimensional nonequilibrium flows with condensation and with nonequilibrium chemical reactions and vibrational relaxation. First of all, a two-dimensional equilibrium flow in a nozzle is calculated and the positions of the streamlines are determined together with the density distributions along them. Then, the system (6.134-143) is solved in the one-dimensional approximation and the flow parameters are obtained taking into account nonequilibrium condensation.

Calculations of flows in axisymmetric nozzles with large expansion ratios have shown that two condensation zones may exit because after the first zone a secondary increase of supersaturation may take place at low pressures and temperatures [6.65]. Experimental and numerical studies of condensation in a flow past a convex plane angle were performed in [6.66]. This investigation included taking shadow photographs of the flow and measuring pressure distributions. It showed that calculated and experimental data coincide at the condensation coefficient α_k=0.04 in the presence of condensation characteristics of the centred rarefaction wave bend, while the Mach number becomes smaller along them. The bending of characteristics may result in their intersection, i.e. in the formation of shock waves generated by spontaneous condensation (by condensation "shocks"). Such shock waves may appear in nozzles in the vicinity of zones of spontaneous condensation. However, their intensity in the region of generation is, as a rule, low.

6.5 Conductive Gas Nozzle Flows in the Presence of an Electromagnetic Field

The interest in magnetic hydrodynamics has been to a large extent inspired by the development of a new method of thermal energy conversion into electric power with the help of magnetogasdynamic energy transformers — MGD generators. The basic idea of the method is as follows. It is well known that the thermal energy of any fuel may be transformed into the kinetic energy of combustion gases by attaching a nozzle to the combustor. If the gas flows between two capacitor plates placed in a transverse magnetic field, then a difference of potential will arise at the plates and closure of the circuit will cause an electric current to flow. However, attempts to realize the idea came across a number of difficulties associated with both the creation of electromagnetic circuits and inadequate knowledge in the field of gasdynamics of electroconductive media. In the general case of an arbitrarily oriented magnetic field, a conductive gas flow has a three-dimensional structure even in axisymmetric nozzles and its study is quite a complicated task. Let us analyze in some detail electroconductive gas flows in circular nozzles in the presence of an axial magnetic field. Neglecting the Hall effect and supposing that the magnetic Reynolds number is small, i.e. ignoring the induced magnetic field compared to the applied one, we see that an axisymmetric Laval nozzle flow preserves axial symmetry. Consideration of even such a simple conductive gas flow allows us to find some qualitative particularities associated with the magnetic field effect. Under the above assumptions, the system of gasdynamics equations together with the Maxwell equations acquires the following vector form [6.67]:

$$\nabla(\rho \mathbf{W}) = 0 \quad ,$$

$$\rho(\mathbf{W}\nabla)\mathbf{W} + \nabla p = f \quad , \tag{6.159}$$

$$(\mathbf{W}\nabla)S = -\frac{\gamma - 1}{\rho\gamma}(\mathbf{f}\mathbf{W}) \quad , \tag{6.160}$$

$$\mathrm{div}\,\mathbf{B} = 0 \quad , \quad \mathrm{rot}\,B = 0 \quad , \quad \mathrm{rot}\,\mathbf{E} = 0 \quad .$$

where $\mathbf{B}$ and $\mathbf{E}$ are the magnetic field induction and electric field strength respectively, $\mathbf{f} = \rho_e \mathbf{E} + \mathbf{j} \times \mathbf{B}$ is the Lorentz force, $\mathbf{j} = \sigma(\mathbf{E} + \mathbf{W} \times \mathbf{B})$ is the current density, σ is the conductance and ρ_e is the charge density. In the presence of an ionized additive with ionization potential J, the latter's ionization is insignificant and free electrons stay in equilibrium with the gas. The concentration of electrons is related to the temperature and pressure by the Saha formula:

$$\sigma = \sigma_{01}\left(\frac{T}{T_{01}}\right)^{3/4}\left(\frac{p_{01}}{p}\right)^{\frac{1}{2}}\exp\left[-\frac{1}{2}J\left(\frac{1}{T}-\frac{1}{T_{01}}\right)\right] \quad . \tag{6.161}$$

Since the field $\mathbf{B}$ is solenoidal and irrotational, a potential ϕ may be introduced such that $\mathbf{B} = \mathrm{grad}\phi$, where ϕ satisfies the Laplace equation $\Delta\phi = 0$. Similarly, for stationary problems the Maxwell equations allow one to introduce a scalar electric potential φ such that $\mathbf{E} = -\mathrm{grad}\varphi$, where φ for isotropically conductive media satisfies the Poisson equation,

$$\Delta\varphi = \mathrm{div}(\mathbf{W} \times \mathbf{B}) \quad . \tag{6.162}$$

If the distribution of a magnetic field $B_0(x)$ along the symmetry axis is known, then the components B_x and B_r are defined by

$$B_x = \frac{\partial\Phi}{\partial x} = \frac{1}{\pi}\int_0^{\pi} B_0(x + ir\cos\eta)d\eta \quad ,$$
$$B_r = \frac{\partial\Phi}{\partial r} = \frac{i}{\pi}\int_0^{\pi} B_0(x + ir\cos\eta)\cos\eta\, d\eta \quad . \tag{6.163}$$

In a plane case we have

$$B_x = B_0(x + iy) \quad , \quad B_y = iB_0(x + iy) \quad . \tag{6.164}$$

This expression is analogous to the solution of the inverse problem for an incompressible fluid (Sect.4.1).

Let us further consider the boundary conditions. If the nozzle walls are supposed to be isolators, then the normal component of electric current is zero at the boundary streamlines. Similarly, it is assumed that the normal components of electric current are zero at the inlet and exit cross-sections of a nozzle. Then, if the applied magnetic field is axial, it follows from (6.162) and the boundary conditions that $E_x = E_r = 0$, and if the nozzle is axially symmetrical, the circumferential component of the electric field strength E_ϑ is zero too. Then,

$$j_x = j_r = 0 \quad , \quad j_\vartheta = \sigma(uB_r - vB_x) \quad ,$$

i.e. the flow field contains only circumferential short-circuit currents. Hence, both components of the Lorentz force lie in the axial plane. In this case, a gas flow in an axisymmetric nozzle is axially symmetric and in a cylindrical coordinate system the flow parameters depend on x and r only, while the Maxwell equations are not used to solve the gasdynamics system (6.159).

For curvilinear channels, the system (6.159,160) may be conveniently presented in the curvilinear orthogonal coordinates system ψ, σ. [See (2.117-120); below, the variable σ is denoted by s.] We have

$$\frac{\partial r^2}{\partial \psi} = \frac{2u}{\rho W^2} \quad , \tag{6.165}$$

$$\frac{\partial}{\partial \psi}(r^2 x) = -\frac{rv}{\rho W^2} + \frac{2xu}{\rho W^2} \quad , \tag{6.166}$$

$$\frac{\partial}{\partial \psi}(r^2 p) = \frac{\gamma r N f_\psi}{\rho W} - \gamma \frac{rg}{W}\left(v\frac{\partial u}{\partial s} - u\frac{\partial v}{\partial s}\right) + \frac{2pu}{\rho W^2} \quad , \tag{6.167}$$

$$\frac{\partial p_0}{\partial s} = \frac{N p_0 f_s}{gp} \quad , \tag{6.168}$$

$$v = -Wg\frac{\partial r}{\partial s} \quad , \quad u = -Wg\frac{\partial x}{\partial s} \quad ,$$

$$g = \left[\left(\frac{\partial x}{\partial s}\right)^2 + \left(\frac{\partial r}{\partial s}\right)^2\right]^{-\frac{1}{2}} \quad , \tag{6.169}$$

$$W = \left[\frac{2}{\gamma - 1}\left(1 - \frac{p}{\rho}\right)\right]^{\frac{1}{2}} \quad , \tag{6.170}$$

$$\rho_0 = p_0 \quad , \tag{6.171}$$

$$p = p_0(\rho/\rho_0)^\gamma \quad . \tag{6.172}$$

The system (6.165-172) is made nondimensional by referring the pressure and density to the corresponding stagnation parameters p_{01}, ρ_{01} at the inlet section, the velocities to $a_{01} = \sqrt{\gamma p_{01}/\rho_{01}}$, the linear dimensions to a characteristic length L, the stream functions to $\rho_{01}a_{01}L^2$, the conductance to σ_{01}, the induction to its value B_{00} at the centre of the cylindrical coordinates system and the electromagnetic force to $\sigma_{01}a_{01}B_{00}^2$. The MGD-interaction parameter N used in the system of equations is defined by

$$N = \frac{\sigma_{01}LB_{00}^2}{\rho_{01}a_{01}} \quad .$$

The projections of the Lorentz force upon the axes s and ψ are

$$f_s = -\frac{\sigma}{W}(u^2B_r^2 + 2uvB_xB_r + v^2B_x^2) \ ,$$
$$f_\psi = \frac{\sigma}{W}(uvB_r^2 + v^2B_xB_r + u^2B_xB_r - uvB_x^2) \ . \tag{6.173}$$

Let us comment on the above system of equations. Equation (6.168) has been obtained from the entropy production equation (6.160) and allows the total pressure variation due to Joule heating to be determined. It has the following general form:

$$W\frac{\partial p_0}{\partial s} = \frac{\rho_0 T_0}{\rho T}\left[\left(\frac{T}{T_0} - 1\right)N_V + \frac{W}{g}f_s\right] \ , \tag{6.174}$$

where $N_V = \mathbf{jE}$ is the electric power input per unit volume. Since, due to the boundary conditions and the axial character of the magnetic field, $E = 0$ and, hence, $N_V = 0$, then (6.174) converts into (6.168). Similarly, (6.170) at $N_V = 0$ is a consequence of the equation of the total enthalpy H_0 conservation along a streamline which in the general case has the form

$$\rho W\frac{\partial H_0}{\partial s} = N_V \ , \quad H_0 = \frac{\gamma}{\gamma - 1}RT_0 \ .$$

The remaining equations of the system are the conventional equations (2.117-121) of motion, continuity, direction and state presented in a typical form in the variables s,ψ.

A number of investigations has been performed recently dedicated to studying conductive gas flows in axisymmetric Laval nozzles in the presence of an axial magnetic field and under the assumption that the induced magnetic fields may be neglected, i.e. that the magnetic Reynolds number is small, which is true for the majority of practically important flows in MGD-generators. Moreover, results of special calculations show that omitting the corresponding terms in the equations of magnetic gas dynamics does not affect the flow parameters significantly at $Re_m < 20$. In [6.68] the method of characteristics has been used to calculate flows in the supersonic parts of plane and axisymmetric nozzles with a corner point in the presence of a constant magnetic field parallel to the symmetry axis. In [6.69] the method of stabilization has been applied to solve the direct problem of the theory of nozzles under the same assumption about the magnetic field. Finally, in [6.70] the inverse problem of the theory of nozzles in the presence of a nonuniform axial magnetic field has been solved numerically and flows have been calculated both in an axisymmetric Laval nozzle and in annular radial nozzles.

In order to solve the inverse problem, the pressure distribution, the streamline equations $r = r_0(s)$, $x = x_0(s)$ and the distribution of the magnetic field induction $B = B_0(x)$ must all be specified at the initial streamline $\psi = \psi_0$. As in the case of three-dimensional and nonequilibrium flows, the inverse problem is split into two initial-value problems. For (6.165-167) it is solved in the direction of ψ, while for (6.168) it is solved in the direction of s. The initial condition for (6.168) must be put at the inlet nozzle section $s = s_0$. The sequence of solution is as follows. Knowing the initial streamline and pressure distribution along it, one finds the total pressure p_0 from (6.168). This immediately allows the total pressure losses due to Joule heating of the gas to be evaluated. We have

$$p_0 = \exp\left(N \int \frac{f_s}{gp}\, ds\right) = \exp\left[-N \int \frac{\sigma}{gpW} (uB_r + vB_x)^2 ds\right] \quad . \tag{6.175}$$

All the functions under the integration sign, due to the boundary conditions, are known functions of s. Then, ρ_0, ρ, W, u, v are determined at the initial streamline ψ_0. In order to obtain the values of the variables at the next layer, $r = r(s)$, $x = x(s)$ and $p = p(s)$ are first found from (6.165-167), for instance, by means of the difference scheme described in Sect.3.4; then, all the rest of the unknown parameters are determined in the same sequence as at the initial streamline. Thus, the entire flow field is calculated in the vicinity of the initial streamline and the geometry of other streamlines, each of which may be considered to be a nozzle wall, is determined.

Some qualitative features of conductive gas flows in nozzles should be noted. From (6.175) we see that an increase of the interaction parameter is accompanied by an exponential decrease of the total pressure associated with Joule energy dissipation. The latter results also in an increase of the gas temperature and a decrease of the density and velocity (compared to $N = 0$) if the static gas pressure stays constant. Thus, application of a magnetic field decelerates the gas and results in total pressure losses. Evidently, the effects become stronger with increasing N, and in particular, a supersonic flow, upon application of an intense magnetic field, may become subsonic.

Figures 6.35-37 demonstrate the above effects. Distributions of the Mach number M along the axis and the wall of the supersonic part of an axisymmetric nozzle are presented in Fig.6.35 for various values of the interaction parameters N [6.68]. As was pointed out above, the magnetic field decreases the number M compared to the value of M for a nonconductive gas. In the case of a constant longitudinal magnetic field (for which the cal-

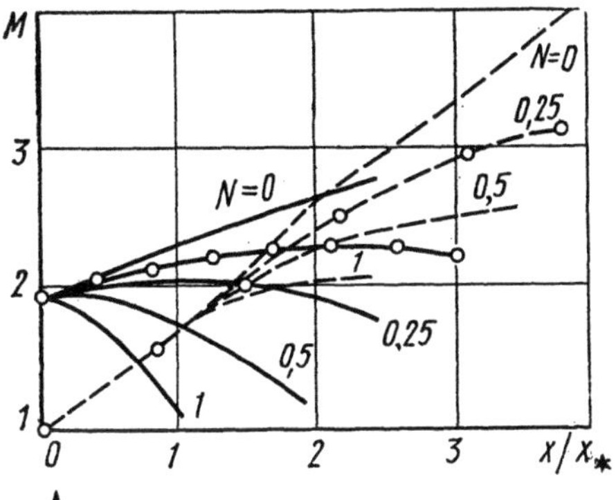

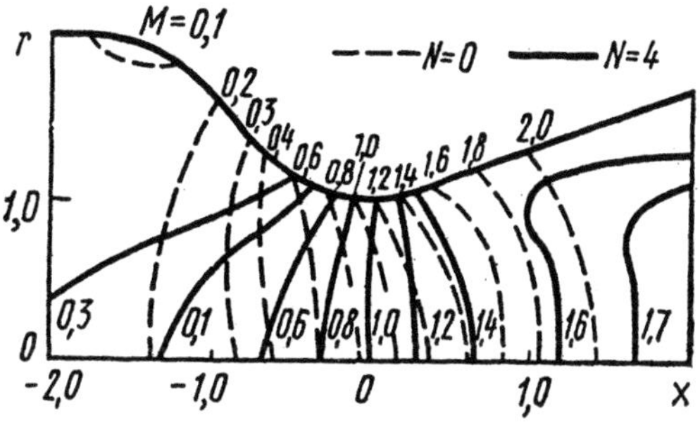

Fig.6.35. Variation of the number M along the axis (---) and the wall (——) of an axisymmetric nozzle with a corner point in the presence of a magnetic field, for various values of the interaction parameters

Fig.6.36. Lines of M = const in an axisymmetric nozzle [6.69]

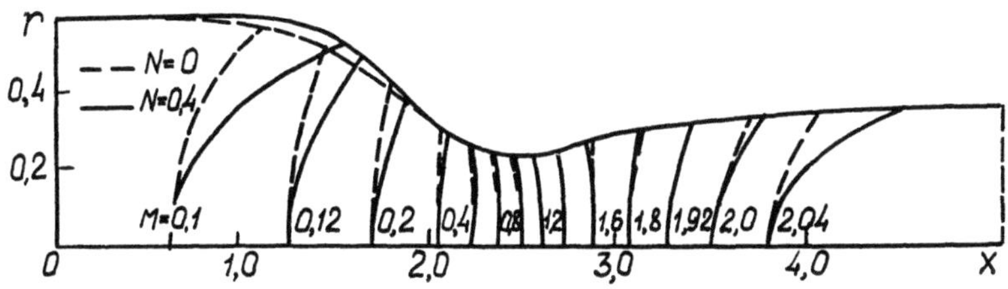

Fig.6.37. Lines of M = const in an axisymmetric nozzle [6.70] in the presence of magnetic field

culations have been done) the effect is noticed most of all at a nozzle wall, where the angle between the magnetic field direction and a streamline, and, hence, the Lorentz force, have their maximum values. Here, at large N, the flow decelerates so intensely that it becomes subsonic. At the nozzle axis the Lorentz force is zero and, hence, the effect of the magnetic field is weaker. Small circles in Fig.6.35 denote results of calculations carried out with the assumption that conductance does not stay constant but depends on the temperature raised to the fifth power. Since in this case, due to the temperature decrease, the conductance also decreases downstream, the effect of the magnetic field weakens compared to the case of constant conductance [see also (6.161)].

Figures 6.36,37 present the lines M = const in the sub-, trans- and supersonic parts of a nozzle at various values of the interaction parameter in the presence of a constant magnetic field. We see that an increase in N results in a considerable increase of the flow nonuniformity. The pressure nonuniformity due to the magnetic field may become 3 to 5 times more intense. Note that the Lorentz force is directed to the nozzle axis in the subsonic part and to the wall in the supersonic part. It follows from the figures and analysis of the equations that near the walls the gas density decreases more intensely than near the symmetry axis and that the effect increases as the interaction parameter increases. As a result, a gas is displaced from the nozzle walls to the axis and a fictitious channel with smoother (than the actual nozzle) walls forms, through which the major portion of the gas flows. This results in a certain straightening of the lines $p = \text{const}$ and $M = \text{const}$ in the vicinity of the throat. However, the flow rate coefficient diminishes due to the action of the magnetic field because the total pressure decreases as a consequence of Joule dissipation. At the axis of a nozzle the Lorentz force is zero and the total pressure remains unchanged. The losses of the total pressure do not exceed 5%. Application of a magnetic field results in a decrease of the nozzle impulse compared to that in the absence of a magnetic field. However, this decrease does not exceed $\sim$0.5% at realistic values of the interaction parameter.

6.6 Multilayer Nozzle Flows

Sometimes gas flows in the nozzles of modern jet engines in such a way that the flow consists of a number of layers with different physical properties. Such flows develop on film-cooled liquid-propellent rocket motors, in solid-propellent rocket motors with pure (free of particles) gas flow in the pre-wall layer and in some types of low-thrust jet engines. Evidently, such flows are accompanied by interlayer mixing and diffusion of various components. Studying such flows, taking into account viscous dissipation, mixing and diffusion, is quite a complicated task for both experimental and theoretical investigation. However, in many practically important nozzle flows, mixing does not influence the flow parameters significantly and may be ignored. We shall deal with such flows below.

6.6.1 One-Dimensional Approximation

Let us start by considering the theory of one-dimensional multilayer nozzle flows [6.71]. In accordance with the hypotheses of the one-dimensional theory, it will be assumed that the static pressure across a nozzle is constant and equal for all the layers while the remaining parameters may vary from layer to layer (unlike a single-layer flow). The temperature, density, velocity, adiabatic exponent, pressure and the stagnation temperature are among these parameters. Let there be n layers in a nozzle with a stationary adiabatic and isentropic flow of a perfect gas with constant thermodynamic properties in each layer. (Note that the assumptions exclude the possibility of flow mixing.) Then the following relationships are valid at any point of a nozzle:

$$F = \sum_{i=1}^{n} F_i \ , \qquad \frac{dF}{dx} = \sum_{i=1}^{n} \frac{dF_i}{dx} \ , \tag{6.176}$$

where F_i is the cross-sectional area of the ith layer. The flow may be calculated in each layer separately as a one-dimensional single-layer gas flow. Then, according to (2.125), we have

$$\frac{dF_i}{dx} = \frac{F_i}{\gamma_i}\left(\frac{1}{M_i^2} - 1\right)\frac{d}{dx}\ln p \ . \tag{6.177}$$

From (6.176) and (6.177) it follows that

$$\frac{d}{dx}\ln p = \frac{1}{\beta}\frac{dF}{dx} \ , \qquad \beta = \sum_{i=1}^{n} \frac{F_i}{\gamma_i}\left(\frac{1}{M^2} - 1\right) \ . \tag{6.178}$$

Later we shall see that β is an important parameter of a multilayer flow allowing sub- and supersonic flows to be distinguished.

A multilayer nozzle flow may be conveniently studied under the assumption that the pressure at the initial cross-section of a nozzle is an independent variable. If the stagnation pressure p_{0i} and temperature T_{0i} are specified for all layers, as well as γ_i and R_i, then the values of the mass flow parameter Q_i depend on the local pressure and the local cross-sectional area only and according to (2.127) and (2.142) are defined by the relationship

$$Q_i = \frac{F_i p_{0i}}{\sqrt{T_{0i}}}\left(\frac{p}{p_{0i}}\right)^{1/\gamma_i}\left\{\frac{2\gamma_i}{R_i(\gamma_i - 1)}\left[1 - \left(\frac{p}{p_{0i}}\right)^{\gamma_i/(\gamma_i-1)}\right]\right\}^{\frac{1}{2}} \ . \tag{6.179}$$

At the initial cross-section of a nozzle the F_i are known and the pressure is given and therefore, (6.179) allows Q_i to be obtained. Then, from (6.179) it follows that the local values of F_i depend only on the local pressure, the known stagnation parameters and the values of γ_i and R_i in a layer. Local values of the Mach number M_i also depend on the local pressure only:

$$M_i^2 = \frac{2}{\gamma_i - 1}\left[\left(\frac{p_{0i}}{p}\right)^{(\gamma_i-1)/\gamma_i} - 1\right] \quad . \tag{6.180}$$

Thus, the local value of β defined by the relationship (6.178) is a function of the pressure at the initial cross-section of a nozzle, the local pressure and the given stagnation parameters, γ_i and R_i, in every layer. Hence, (6.178) at a given pressure at the initial cross-section may be integrated from the initial cross-section up to the nozzle exit and the whole flow pattern may be obtained for various values of the pressure at the initial cross-section (Fig.6.38).

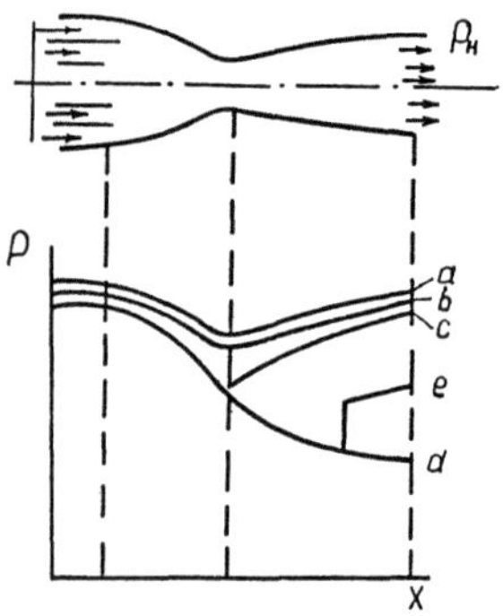

Fig.6.38. Multilayer nozzle flow

If the latter pressure is high enough, then it follows from (6.180) that the M_i are small and β is positive throughout the nozzle. Therefore, the pressure p, in accordance with (6.178) will vary similarly to F (the curves a and b in Fig.6.38); in particular, the minimum value of p will be attained at the throat. Integration of (6.178) also shows what external pressure is needed at fixed p_{0i}, T_{0i}, γ_i and R_i in order to preserve the flow regime. Since

$$\frac{d\beta}{dp} = \sum_{i=1}^{n} \frac{F_i}{\gamma_i^2 p M_i^4}\left[(1 - M_i^2) + 2\left(1 + \frac{1+\gamma_i}{2} M_i^2\right)\right] > 0 \quad , \tag{6.181}$$

β changes similarly to the pressure and, hence, for the regimes a and b the minimum value of β is attained at the throat.

If the pressure at the initial cross-section is decreasing, β decreases too and may become zero. Then, evaluating the indeterminate form in (6.178), one gets

$$\frac{d}{dx}\ln p = \pm\left(\frac{d^2F}{dx^2}\left\{\sum_{i=1}^{n}\frac{F_i}{\gamma_i^2 M^2}\left[(1 - M_i^2)^2 + 2\left(1 + \frac{1+\gamma_i}{2}M^2\right)\right]\right\}^{-1}\right)^{½} . \qquad (6.182)$$

Since $d^2F/dx^2 > 0$, the derivative $d\ln p/dx$ is positive or negative at the throat depending on the sign before the radical. In Fig.6.38 the curve c corresponds to the positive radical and the curve d to the negative one. As is similar in the case of a single-layer flow, the throat is a saddle-type singularity and two types of isentropic flow are possible in the divergent part of a nozzle. In the interval $p_d < p_H < p_c$ no value of the external pressure corresponds to an isentropic flow. One may expect that (similar to one-dimensional single-layer flows) such values of the external pressure may be attained by means of a shock wave (the curve e) forming at a point of the curve d.

Qualitatively the flow parameters change along the curve c in the same way as along the curves a and b; p and β stay minimum at the throat. Note that the plus sign in (6.182) is chosen only if the external pressure is rigorously equal to the value corresponding to the curve c. According to (6.181) $d\beta/dx < 0$ along the curve d; therefore, β reverses its sign at the throat and it follows from (6.178,181) that p and β must decrease in the divergent part of a nozzle and the flow must be continuously accelerating throughout the nozzle. Integrating (6.178), one can find the value of the external pressure at which such a flow regime takes place. If the pressure at the initial cross-section is below the one corresponding to the curve c, then no isentropic flow may exist, since β becomes zero in front of the throat and the derivative dp/dx becomes infinite.

Now let us consider the effect of the external pressure p_H. If $p_H > p_c$, then it evidently affects the pressure at the initial cross-section and, hence, the flow rate in each layer. If $p_H < p_c$, then it has no influence upon these parameters. Such a flow will be called a "choked" multilayer flow. The phenomenon of "choking", which in a single-layer flow corresponds to the sonic line formation at the throat (and here only), may be explained in the case of a multilayer flow by considering the propagation of small disturbances. In the framework of the one-dimensional theory the motion of a wave is considered under the assumption that the cross-sectional area stays constant. A small planar pressure disturbance cannot propagate in different flows with a different absolute velocity in each of them, since it must satisfy the condition that the static pressure is constant at the interface. In this connection, a wave must be continuous and must propagate as a single complex wave, since a pressure increment cannot vary within it from layer to layer. If we denote the velocity of a complex wave propagating upstream by α, then at $\alpha > 0$ the flow, evidently, is multilayer and subsonic, at $\alpha = 0$ it is multilayer and sonic and at $\alpha < 0$ it is multilayer and supersonic.

An analytical expression for α may be obtained by considering each flow separately as a flexible stream tube and writing down relationships for the laws of mass, momentum and energy conservation across a complex wave. We have

$$\frac{\Delta F_i}{F_i} = \frac{\Delta p_i}{\gamma_i p}\left(\frac{1}{[\alpha(\gamma_i R_i T_i)^{-1/2} + M_i]^2} - 1\right) = 0 \tag{6.183}$$

where ΔF_i and Δp_i are the area and pressure variations in layers across the wave. Since $\Sigma\, \Delta F_i = \Delta F = 0$, $p_i = p$, $\Delta p_i = \Delta p$, then one gets from (6.183),

$$\sum_{i=1}^{N} \frac{F_i}{\gamma_i}\left(\frac{1}{[\alpha(\gamma_i R_i T_i)^{-1/2} + M_i]^2} - 1\right) = 0 \quad . \tag{6.184}$$

Using the definition of β and (6.184), we obtain

$$\beta = \sum_{i=1}^{n} \frac{F_i}{\gamma_i}\left(\frac{1}{M_i^2} - \frac{1}{[\alpha(\gamma_i R_i T_i)^{-1/2} + M_i]^2}\right) \quad . \tag{6.185}$$

Hence, β and α have always the same signs, i.e. at $\beta > 0$ a flow is multilayer and subsonic, at $\beta = 0$ it is sonic and at $\beta < 0$ it is supersonic, similar to a single-layer flow. "Choking" of a multilayer flow takes place at the throat, since β may become zero only here. The condition of "choking", $\beta = 0$, for a multilayer flow is a generalization of the condition $M = 1$ at the throat, which defines "choking" in a one-dimensional single-layer flow.

The above analysis shows that a multilayer flow cannot be considered "choked" if the Mach number in one of the layers equals 1. The only necessary condition for "choking" is that β must be zero, while at the throat the Mach numbers for all the layers need not be equal to 1. At the same time, the condition of all M_i being less/more than 1 is not sufficient for the entire multilayer flow to be subsonic/supersonic in the above sense. In essence, it follows from the definition of β that individual flows affect the value of β in proportion to their cross-sectional areas. However, the subsonic and supersonic regimes of multilayer flows, from the point of view of the pressure behaviour, are analogous to those of a uniform nozzle flow.

The theory developed above allows one to obtain, for a two-layer flow, the following formula relating the parameters of the two flows at a point of a nozzle:

$$\frac{Q_2\sqrt{T_{02}}}{Q_1\sqrt{T_{01}}} = \left\{\frac{F}{F_{*1}}\left(\frac{2}{\gamma_1 - 1}\right)^{\frac{1}{2}}\left(\frac{\gamma_1 + 1}{2}\right)^{(\gamma_1+1)/2(\gamma_1-1)}\right.$$

$$\left. - \left(\frac{p_{01}}{p}\right)^{1/\gamma_1}\left[1 - \left(\frac{p}{p_{01}}\right)^{(\gamma_1-1)/\gamma_1}\right]^{\frac{1}{2}}\right\} \qquad (6.186)$$

$$\times\left\{\frac{\gamma_2 R_1(\gamma_1 - 1)}{\gamma_1 R_2(\gamma_2 - 1)}\left[1 - \left(\frac{p}{p_{02}}\right)^{(\gamma_2-1)/\gamma_2}\right]^{\frac{1}{2}}\right\}\frac{p_{02}}{p_{01}}\left(\frac{p}{p_{02}}\right)^{\frac{1}{2}} ,$$

where

$$F_{*1} = \frac{Q_1}{p_{01}}\left(\frac{T_{01}R_1}{\gamma_1}\right)\left(\frac{\gamma_1 + 1}{2}\right)^{(\gamma_1+1)/2(\gamma_1-1)} .$$

In the case of an "unchoked" flow, this relation allows p_{02}/p_{01} and p_H/p_{01} (or p_H/p_{02}) to be related at fixed $Q_2\sqrt{T_{02}}/Q_1\sqrt{T_{01}}$, γ_1, γ_2, R_1 and R_2. Some typical results are presented in Fig.6.39. Intersections of the curves presented in the figure and corresponding to various values of F_{*1}/F_a and F_{*1}/F_* (F_* and F_a denote, as earlier, areas of the throat and the exit cross-section of a nozzle, respectively) for the same nozzle indicate the moment when the "choking" condition is fulfilled (the curve c in Fig.6.38). At values of p_{01}/p_H below the values corresponding to the point of intersection, p_{02}/p_{01} is defined by the curve for an "unchoked" flow; at larger p_{01}/p_H the value of p_{02}/p_{01} is independent of p_{01}/p_H.

The character of a "choked" flow is determined by the conditions at the throat where $\beta = 0$. Thus, we have

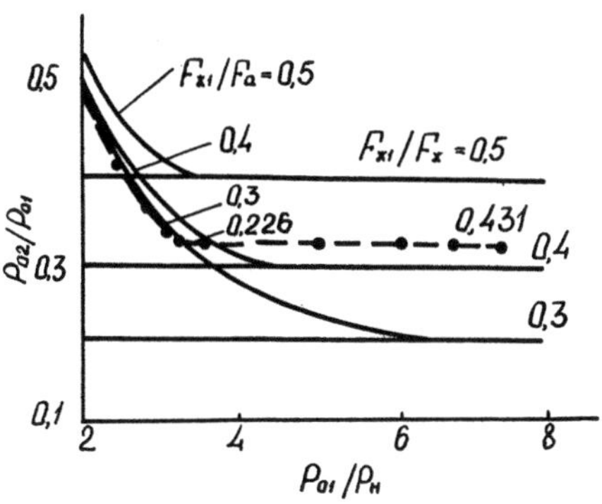

Fig.6.39. Typical relationships between the flow parameters at $R_1 = R_2 = 287$ m/K, $\gamma_1 = \gamma_2 = 1.4$, $Q_2\sqrt{T_{02}}/Q_1\sqrt{T_{01}} = 0.315$. The points show experimental data [6.71]

$$\frac{Q_2\sqrt{T_{02}}}{Q_1\sqrt{T_{01}}} = \frac{\gamma_2 p_{02}}{\gamma_1 p_{01}}\left\{\frac{\gamma_1 - 1}{2}\left[\left(\frac{p}{p_{01}}\right)^{(1-\gamma_1)/\gamma_1} - 1\right]^{-1} - 1\right\}$$

$$\times \left(\frac{p}{p_{02}}\right)^{1/\gamma_1}\left\{\frac{2\gamma_2}{R_2(\gamma_2 - 1)}\left[1 - \left(\frac{p}{p_{02}}\right)^{(\gamma_2-1)/\gamma_2}\right]\right\}^{\frac{1}{2}}$$

$$\times \left\{1 - \frac{\gamma_2 - 1}{2}\left[\left(\frac{p}{p_{02}}\right)^{(\gamma-\gamma_2)/\gamma_2} - 1\right]^{-1}\right\} \qquad (6.187)$$

$$\times \left(\frac{p}{p_{01}}\right)^{1/\gamma_1}\left\{\frac{2\gamma_1}{R_2(\gamma_1 - 1)}\left[1 - \left(\frac{p}{p_{01}}\right)^{(\gamma_1-1)/\gamma_1}\right]\right\}^{\frac{1}{2}} .$$

If the "choking" condition is satisfied at the throat, then (6.186) and (6.187) allow the flow parameters to be determined uniquely at any cross-section of a nozzle with fixed F/F_* similarly to a single-layer "choked" flow for which the area ratio defines all the flow parameters uniquely.

Some calculated results are presented in Fig.6.40 which allow one to obtain p_{02}/p_{01} for the "choked" flow, if F_{x1}/F_* and $Q_2\sqrt{T_{02}}/Q_1\sqrt{T_{01}}$ are given. The same figure shows the dependence of M_1 and M_2 on $Q_2\sqrt{T_{02}}/Q_1\sqrt{T_{01}}$ for "choking" at the throat. We see that velocities of all layers are equal to 1 at the nozzle throat. In [6.71] a nozzle has been tested at $F_{*1}/F_a = 0.226$, $F_{*1}/F_* = 0.431$.

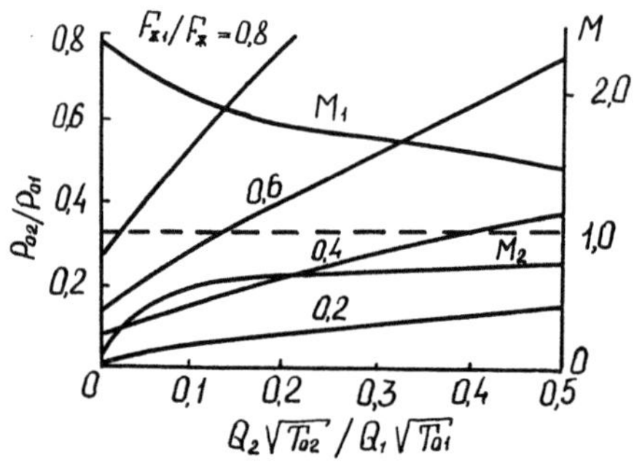

Fig.6.40. Relationships between the flow parameters in a nozzle under "choking" conditions at $F_1^*/F^* = 0.431$. The Mach number M is taken for the throat

Figure 6.39 demonstrates a good agreement between experimental data and the results of calculating a multilayer flow by means of the one-dimensional theory. This means that for the flow under consideration, mixing and two-dimensional effects may be ignored.

Sometimes it is necessary to compare the multilayer flow impulse J and the specific impulse J_s with those of a single-layer flow with the same mass and energy fluxes. Such a comparison is based on the use of the above "choking" conditions and relations similar to (6.186) and (6.187). Let us make such a comparison in the one-dimensional approximation using the example of a two-layer flow with nonuniform total pressure and temperature. The flow parameters of the first layer will be denoted by the subscript 1, those of the second layer by the subscript 2 and the flow parameters of the two-layer flow by the superscript 0; the flow parameters of the single-layer flow with the same mass and energy fluxes will be left without indices. For the sake of simplicity, it will be assumed that the layers consist of moving gases with constant heat capacities. Then, in accordance with (2.142) and (2.143) and the conditions of comparison, we have

$$Q = Q_1 + Q_2 = Q^0 \ ,$$

$$QT_0 = Q_1T_{01} + Q_2T_{02} \quad \text{or} \quad T_0 = g_1T_{01} + (1 - g_1)T_{02} \ , \tag{6.188}$$

$$g_i = Q_i/Q \ ,$$

$$J^0 = Q_1J_{s1} + Q_2J_{s2} \ , \quad J = QJ_s \ ,$$

$$J_{si} = z(\lambda_i)\sqrt{\frac{\gamma_i + 1}{2\gamma_i} R_iT_{0i}} \ ,$$

$$J_s^0 = g_1J_{s1} + (1 - g_1)J_{s2} \ .$$

Here mixing is ignored, as earlier, and the static pressure is considered equal for the layers.

Let us start by considering a flow with a nonuniform distribution of the total pressure, and put $p_{02} < p_{01}$. Then, according to the "choking" condition, $\lambda_1 > 1$ and $\lambda_2 < 1$. Since at the throat the function $z(\lambda)$ has a minimum, it is evident that $J_s^0 > J_s$ and $J^0 > J$, and

$$\frac{J_s^0}{J_s} = \frac{1}{z(\lambda)}\left[g_1z(\lambda_1) + (1 - g_1)z(\lambda_2)\right] \ . \tag{6.189}$$

From Fig.6.40 it follows that at $p_{02}/p_{01} = 0.3$ and $g_2/g_1 = 0.5$, $J_s^0/J_s = 1.09$ at the throat. However, an increment in the specific impulse of the nonuniform flow decreases as the expansion ratio increases and $J_s^0/J_s \to 1$.

Let us next consider a flow with a nonuniform distribution of total temperature. Due to the equality of the static and total pressures in the layers, the velocity coefficients λ_1 and λ_2 are equal and coincide with those of a uniform flow with the same static pressure. However, at the sections with equal static pressures, the cross-sectional area of a single-layer flow exceeds that of a two-layer flow under the condition of equality of the mass and energy fluxes. Actually, it may be easily shown that

$$\frac{F}{F^0} = \frac{\sqrt{g_1 + (1 - g_1)T_{02}/T_{01}}}{g_1 + (1 - g_1)\sqrt{T_{02}/T_{01}}} > 1 \quad . \tag{6.190}$$

This relationship is valid independent of the expansion ratio. It is valid at the throat too. Thus, due to $\lambda_1 = \lambda_2 = \lambda$ we have $J/J_0 = J_s/J_s^0 = F/F^0 > 1$, i.e. in the case of a nonuniform distribution of the total temperature the impulse and specific impulse of a uniform flow exceed those of a nonuniform one. Thus, at $T_{02}/T_{01} = 4$, $g_1 = 0.5$ we have $J = 1.05\ J^0$. Similar conclusions have been drawn by comparing impulse of a two-layer flow in [6.72], which contains an analysis and numerical data for the general case of a two-layer flow with different total pressures and temperatures.

Let us consider an equilibrium two-layer flow using the condition that the total pressures in the layers are equal but have different oxidizer-to-fuel ratios α — the situation characteristic for nozzles of liquid-propellent jet engines with film cooling. Comparing two-layer flows with a single-layer one, it is convenient to choose as an independent variable not α, but the dimensionless oxidizer flow rate $\bar{Q}_0 = Q_0/Q$, $\bar{Q}_{0i} = Q_{0i}/Q_i$, where Q_0 is the total oxidizer flow rate and Q_{0i} and Q_i are the oxidizer flow rate and the total flow rate through a layer respectively. Such a choice of the independent variable is due to $\bar{Q}_0$ being a linear function of g_1, $\bar{Q}_0 = g_1 Q_{01} + (1-g_i)\bar{Q}_{02}$ while α is a nonlinear function of g_1. In the case of a two-layer flow the specific impulse of a nozzle is a linear function of g_1 too; $J_s^0 = g_1 J_{s1} + (1-g_1)J_{s2}$ for fixed J_{s1} and J_{s2}. A typical dependence of J_s on Q_0 is presented in Fig.6.41. From the geometrical construction presented in the figure it follows that if $\bar{Q}_{01}$ and $\bar{Q}_{02}$ exceed $\bar{Q}_{02}^k$, the latter being the value of $\bar{Q}_{02}$ at the point of inflection of the function $J_s(\bar{Q}_0)$, then $J_s^{0I} < J_s^I$ for all g_1, and the straight line I passes through the points with coordinates $\bar{Q}_{02}^I$, J_{s2}^I, and $\bar{Q}_{01}$, J_{s1}. Thus, in this case, the nonuniform distribution of α (or $\bar{Q}_{0i}$) over a cross-section results in decreasing the specific impulse of a two-layer flow compared to the specific impulse of a uniform flow with the same mass and energy fluxes. If $\bar{Q}_{02} < \bar{Q}_{02}^k$, then $J_s^{0II} > J_s^{II}$ at $g_1 < g^0$ and $J_s^{0II} < J_s^{II}$ at $g_1 > g_0$, i.e. in the former case the two-layer flow has a larger specific impulse while in the latter it is smaller than in a single-layer

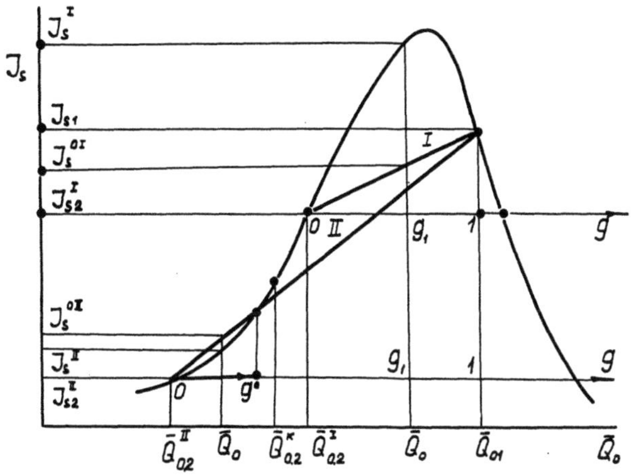

Fig.6.41. Graphical determination of J_s^0 and J_s

flow. Note that the inequality (6.190) may be proved by means of an analogous geometrical construction.

In solid-propellent rocket motors a nonuniform field of the condensed phase concentrations may form, for some reason, at the inlet of a nozzle. Complicated flows exist in the region which, due to the separation of particles, may result in a redistribution of particles in the flow. The initial concentration nonuniformity may depend on the grain configuration, the technology of grain manufacturing and the particulars of the combustion process. Finally, formation of pure gas zones due to inertial effects is also a source of such nonuniformities. A nonuniform distribution of particle concentrations results in the specific impulse of a nozzle decreasing compared to the case of a uniform distribution.

Let us evaluate the specific impulse losses due to such nonuniformities for some limiting cases. Let the weight fraction of particles in a uniform flow be equal to α. We shall assume that at the initial cross-section of a nozzle particles have been redistributed in such a way that all of them have concentrated in one layer containing, naturally, the gas phase too, while the other layer consists of a pure gas. It will also be assumed that the total pressures and temperatures of the layers stay equal to the initial values, since particles have been redistributed due to mechanical mixing only. In addition, we shall assume that the gas is perfect and the flowing gas-particle mixture stays in equilibrium, i.e. no velocity and temperature lags take place. The nonuniformity degree is characterized by $\eta = Q_g/Q$ where Q is the total flow rate of the gaseous phase and Q_g is the gas flow rate in the pure gas layer. It is evident that $0 \leqslant \eta \leqslant 1$; at $\eta = 0$ the condensate is distributed uniformly, while at $\eta = 1$ all the gas stays in the pure gas layer. The flow parameters of the layer containing particles will be denoted by the subscript 2, while those of the pure gas flow are denoted by 1. Knowing the values of c_p, c_s and γ_1 and taking into account that the mixture flows in equilibrium, one may, by specifying η, obtain from (6.89) γ_2, R_2 and α_2 for the layer containing particles. We have

$$\gamma_2^0 = \gamma_1 \frac{1 - \alpha_2(1 - c_s/c_p)}{1 - \alpha_2(1 - \gamma_1 c_s/c_p)} \ , \tag{6.191}$$

$$R_2 = (1 - \alpha_2)R_1 \ , \qquad \alpha_2 = \frac{\alpha_0}{1 - \eta(1 - \alpha_0)} \ .$$

The following formula is valid for the specific impulse of a two-layer flow under the assumption that there is an absence of mixing and an equality of the static pressure:

$$J_s = J_{s1}\eta(1 - \alpha_0) + J_{s2}[1 - \eta(1 - \alpha_0)] \ . \tag{6.192}$$

The specific impulse losses due to flow nonuniformity are determined from

$$\xi_\eta = 1 - \frac{J_s}{J_s|_{\eta=0}} \ . \tag{6.193}$$

Evidently, ξ_η increases with increasing η. Calculations show that at $\eta = 0.4$, $\bar{r}_a = 5$ and $c_s/c_p = 0.5$, $\xi_\eta \approx 2\%$.

6.6.2 Axisymmetric Flows

Multilayer plane and axisymmetric nozzle flows are calculated either by means of the numerical solution of the inverse problem of the theory of nozzles or by the method of characteristics applied to the supersonic region. The coordinates s, ψ used to solve the inverse problem (Sect.2.1) facilitate the calculation of flows with tangential discontinuity surfaces to which belong stream surfaces dividing the flow into regions with different physical properties. Such calculations are usually done ignoring the mixing of layers though total enthalpies and pressures may vary.

Let us consider a method of solving the inverse problem for a two-layer flow of a perfect gas. We shall assume that in the region limited by the streamlines $\psi = 0$, $\psi = \psi_0$ the gas has the adiabatic exponent $\gamma = \gamma_1$, the total pressure p_{01}, the total temperature T_{01}, while above the streamline $\psi = \psi^0$

the adiabatic exponent is $\gamma = \gamma_2$, the total pressure is equal to p_{02} and the total temperature is T_{02}. Up to the streamline $\psi = \psi^0$ a flow with a given velocity distribution at the initial surface is calculated by means of (3.92-95) and all the flow parameters are determined at the line. Since at the streamline $\psi = \psi^0$ both flows must have equal angles of velocity inclination and equal static pressures, then for the flow parameters of the second layer the following relationships must hold:

$$p_2 = \sigma\left(\frac{\gamma_2 + 1}{2}\right)^{\gamma_2/(\gamma_2-1)}\left[1 - \frac{\gamma_1 - 1}{\gamma_1 + 1}\,(u_1^2 + v_1^2)\right]^{\gamma_1/(\gamma_1-1)} \quad , \quad \rho_2 = p_2^{1/\gamma_2} \quad ,$$

$$v_2 = \frac{v_1}{\sqrt{u_1^2 + v_1^2}}\left[\frac{\gamma_2 + 1}{\gamma_2 - 1} - \frac{2}{\gamma_2 - 1}\,p_2^{(\gamma_2-1)/\gamma_2}\right]^{\frac{1}{2}} \quad , \quad v_1 u_2 = u_1 v_2 \quad , \quad \sigma = \frac{p_{01}}{p_{02}} \quad ,$$

where all the functions are referred to their critical values. After determining the parameter denoted by the subscript 2 at the stream surface $\psi = \psi^0$, an analogous calculation is done for the flow field at $\psi > \psi^0$. The above method has been used for a series of calculations of two-layer axisymmetric nozzle flows [6.73]. Some of the results are presented in Fig.6.42. We see that the sonic line in the second layer is situated further downstream than in the first one, while the values of W at both sides of the tangential

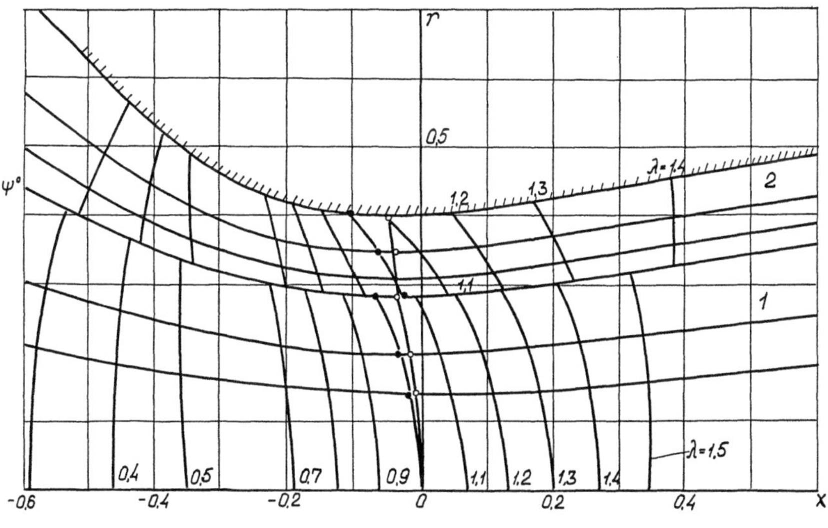

Fig.6.42. Two-layer nozzle flow at $\gamma_1 = 1.14$, $\gamma_2 = 1.4$, $p_{02}/p_{01} = 1$. (•) and (○) show the sonic line and the line $\theta = 0$ respectively

discontinuity surface are almost equal in the subsonic region though they differ substantially starting from the transonic region.

In [6.73] a method of characteristics has been developed for calculating the flow in the supersonic region with a tangential discontinuity surface. The initial data obtained by calculating the transonic region of the two-layer flow were used to calculate a flow in a nozzle with a corner point whose contour had been obtained by solving Goursat's problem for a single-layer flow at $\gamma = 1.14$ and $M_0 = 4.6$. This time it proved worthwhile to use the compatibility equation in the form of (2.63), since this allows one to use the same procedure for calculating parameters at the inner points and at the points on the line of the tangential discontinuity.

Figure 6.43 shows the pressure distribution along the contour of a nozzle with a corner point for two-layer flows with $\gamma_1 = 1.14$ and $\gamma_2 = 1.4$ at various values of $\bar{r}^0$ (i.e. at various flow rates in the prewall layer) and the pressure distribution at the same contour in a single-layer flow with $\gamma = 1.14$. We see that at $\bar{r}^0 > 0.95$ the pressure at the contour increases starting from the point C where the first characteristic starts to reflect from the surface of the tangential discontinuity. To the right of the point C the prewall layer flow starts to "feel" the presence of the inner layer. The pressure increases because flows with different γ turning through the same angle in the Prandtl-Meyer flow develop different pressures: the larger γ, the more intensely the pressure drops. At relatively large prewall layer

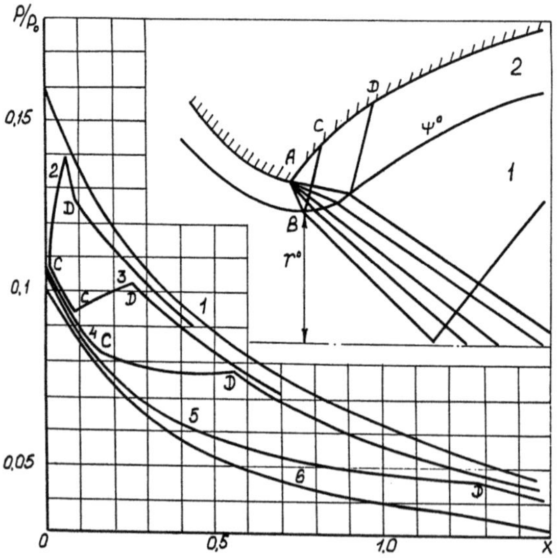

Fig.6.43. Pressure distribution at a nozzle wall for a two-layer flow. (*1*) Single-layer flow $\gamma = 1.14$. Two-layer flow with $\gamma_1 = 1.14$, $\gamma_2 = 1.4$: (*2*) $\bar{r}^0 = 0.99$, (*3*) $\bar{r}^0 = 0.95$, (*4*) $\bar{r}^0 = 0.9$, (*5*) $\bar{r}^0 = 0.8$, (*6*) $\bar{r}^0 = 0.7$

thicknesses the influence of the inner layer starts to be felt mainly to the right of the point D where the pressure derivative is discontinuous and starts varying more intensely while the pressure approaches that of a single-layer flow. Thus, starting from the point D the prewall flow is defined mainly by the inner layer. Up to the point D, the disturbances produced by the inner layer are weakened by the rarefaction wave issuing from the point A. In thin prewall layers the influence of the inner layer is "felt" in the immediate vicinity of the corner point. The pressure in the prewall layer tends to become equal to that in the inner layer and, as long as the latter is larger (though the flows turn through the same angle), the pressure increases. If a layer is thinning, the difference between the static pressures in single-layer and two-layer flows at a nozzle wall naturally decreases; however, the Mach numbers of the flows may differ considerably due to the different values of the adiabatic exponent. Note that the pressure increases in a flow past a corner point only if the adiabatic exponent in the prewall layer is larger than that in the flow's central part. Calculations show that no pressure increase is observed if a nozzle contour in the vicinity of the corner point is rounded off by a circular arc with $R_2^0 \approx 0.3$-$0.5\ r_*$.

7. Special Nozzles, Three-Dimensional Flows, Viscosity Effect

This chapter deals with nozzles of special configurations, such as various annular nozzles, aerodynamic and conical nozzles. Some results of up-to-date rare investigations of three-dimensional nozzle gas flows are presented. In particular, swirling flows are considered. Much attention is paid to slightly perturbed flows and the corresponding perturbations of the side forces and moments. The last section presents calculated and experimental results of studying viscous nozzle flows at low Reynolds numbers when the influence of viscosity is not restricted by the thin boundary layer near the nozzle walls but embraces the whole of the flow.

7.1 Annular Nozzles and Wind Tunnel Nozzles

Annular nozzles may be used in jet engines and wind tunnels. In the first case their important advantage—a shorter length compared to that of axisymmetric circular ones—is used; also, they allow large thrusts to be obtained under off-design conditions.

7.1.1 Some Schemes of Annular Nozzles and Methods of Calculation

Various schemes of axisymmetric nozzles with annular throats (the so-called "disc-type" nozzles, nozzles with a central body and nozzles with a rectilinear upper wall) may be obtained by modifying elements of an annular nozzle. Annular nozzles differ in throat inclination which may be zero, positive or negative (Fig.7.1a-c). Those with zero inclination are subdivided into three groups—with external, internal and double expansion (Fig.7.1d-f).

The design of annular nozzles that provide maximum thrust for minimum length is based upon the same ideas as designing circular ones. In both

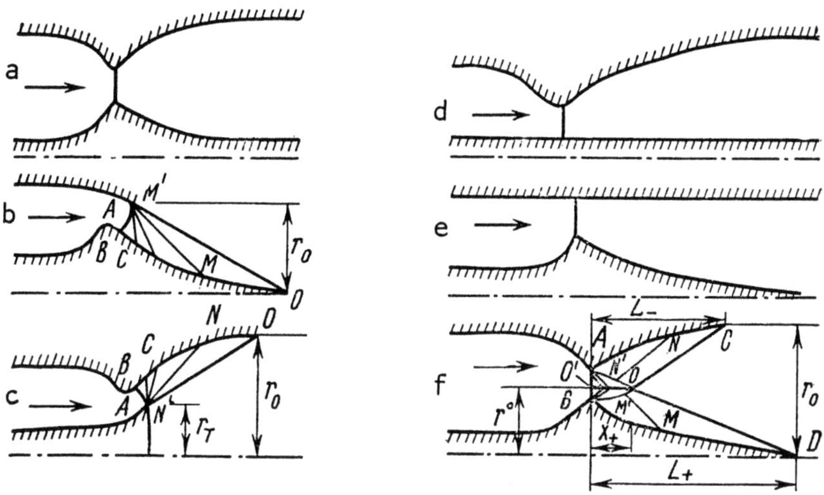

Fig.7.1a-f. Diagrams of annular nozzles

cases a gas is accelerated while flowing past corner points or highly curved segments of the transonic contour of a nozzle. The contour of the straightening part of a nozzle is designed with the help of either variational or uniform closing characteristics. However, truncated annular nozzles constructed on the basis of uniform or variational characteristics differ even less than the axisymmetric ones, since flows in annular nozzles are very much like the plane ones, for which the above families of contours are identical.

Let us consider a method of calculating annular nozzles with double expansion and corner points (Fig.7.1f). Firstly, the method of characteristics is used to calculate the axisymmetric rarefaction flow. The calculation proceeds along characteristics of the first family in the region of the wave O'BO and along the second family characteristics in the region of the wave O'AO. Each characteristic terminates at the point where the angle of the velocity vector inclination is zero. The calculation gives coordinates of the streamline and the line at which $\theta = 0$ and the gasdynamic parameters along the lines. Characteristics AO and BO are boundary characteristics of the rarefaction waves. The point O lying at the line $\theta = 0$ is such that the uniform characteristic OD terminates exactly at the symmetry axis and the gas flow rate through OD is equal to the flow rate through the characteristic OB (at all points of the uniform characteristic OD, $\theta = 0$ and $M = M_0$ = const). Then, another uniform characteristic OC is drawn from the point O, also satisfying the condition of equality of flow rates through OC and

AO. Next, the Goursat problem is solved in the space between characteristics OD, OB and OC, AO and the family of streamlines is obtained, in particular the upper and the lower contours of the nozzle. Since the line $\theta = 0$ practically coincides with the middle line of the throat O'O and may be considered to be a streamline, nozzles d and e (Fig.7.1) are special cases of a double expansion nozzle (type f) in which the line OO' is assumed to be a solid wall. The above property is a consequence of the fact that the gas flow in the vicinity of the annular orifice is close to plane, since according to the compatibility equation the axisymmetric nature of a flow is practically not felt in the regions far from the symmetry axis where dr/r is small.

Exact claculations of flows in annular nozzles of various schemes confirm the plane character of the flow in them [7.1-3]. In particular, the Mach number distribution along the line $\theta = 0$ corresponds to a similar distribution for outflow from a plane orifice with a rectilinear sonic line (see Fig.5.17, curve 3). In order to calculate this distribution approximately, one may use (5.24) assuming $\bar{x}$ to be the ratio of the length of an annular orifice to its semiwidth. In the plane case the Prandtl-Meyer flow takes place in the regions AOC and BOD. Evidently, the flow in these regions of an annular nozzle is close to the Prandtl-Meyer flow. The latter property is confirmed also by results of exact calculations. Characteristics NN' and MM' are almost straight lines and the gasdynamic parameters change along them only slightly. Flows in the b and c types of nozzles possess analogous properties. They may be approximately considered to be the Prandtl-Meyer centred plane flows. The error due to this assumption evidently increases near the symmetry axis and is smaller, the further from the axis. Therefore, the contours of nozzles c and d, as well as the upper contour of nozzle f, and the gas parameters at them are calculated exactly enough under the assumption that the flow is plane.

The plane character of flow in an annular nozzle allowed the development of the approximate method of rectilinear characteristics (Sect.2.2) according to which the flow parameters at the characteristics belonging to both families are assumed to be constant while the axisymmetric nature of the flow is accounted for by means of the flow rate equation. In particular, coordinates of the upper contour of the nozzle f (e.g. those of the point N) are defined by

$$r_N^2 = r_{N'}^2 + \frac{(\psi_* - \psi_{N'})\sin(\alpha_{N'} + \theta_{N'})}{q(M_{N'})\sin\alpha_{N'}} \quad , \tag{7.1}$$

$$x_N = x_{N'} + (r_N - r_{N'})\cotan(\alpha_{N'} + \theta_{N'}) \quad , \tag{7.2}$$

while coordinates of the lower contour (e.g. of the point M) are defined by

$$r_M^2 = r_{M'}^2 - \frac{\psi_{M'}\sin(\alpha_{M'} - \theta_{M'})}{q(M_{M'})\sin\alpha_{M'}} \quad , \tag{7.3}$$

$$x_M = x_{M'} - (r_M - r_{M'})\operatorname{cotan}(\alpha_{M'} - \theta_{M'}) \quad . \tag{7.4}$$

Here ψ is the gas flow rate through the throat, and $\psi_{N'}$, $\psi_{M'}$ are the gas flow rates through the segments of characteristics AN' and BM'. The values of α and θ at the points M and N are assumed to be equal to the corresponding values at the points M' and N'. The flow in the region of rarefaction waves ABO differs insignificantly from a plane flow. Therefore, the values of $\psi_{N'}$ and $\psi_{M'}$ may be approximately determined if the flow parameters at characteristics of plane rarefaction waves and the value of the flow rate for a plane flow at the points N' and M' are known. The flow rate ψ^a in an axisymmetric flow is determined by the flow rate ψ^p in the corresponding plane flow with the help of the formula

$$\psi^a = 2r^0\left(\frac{1}{2}\,h \pm \psi^p\right) \quad , \tag{7.5}$$

where the minus-sign corresponds to characteristics of the second family, the plus-sign to those of the first family, and h is the orifice width.

Specification of the number M_0 and radii r_0, r_T facilitates the determination of the coordinates for contours of nozzles b and c. The flow is calculated upstream of the characteristic AO until it attains sonic speed at the corner point, and the nozzle contour is determined as a streamline with the flow rate equal to that through the characteristic AO. Examples of calculation and some tables of contours are presented in [7.2]. In order to determine the coordinates of a contour approximately, one can use (7.1-4) as before, assuming that points N',M' correspond to the point A, at which the values of α and θ are related by (2.150). The value of $\psi_* - \psi_{N'}$ in (7.1) must be substituted by $(r_0^2 - r_T^2)q(M_0)$, and the value of $\psi_{M'}$ in (7.3) by $r_0^2 q(M_0)$. By changing α at the corner point in the limits from $\alpha_0 = \arcsin M_0^{-1}$ to $\alpha = \pi/2$, one can determine the coordinates of the contour and the values of parameters at it. Figure 7.2 compares results obtained by the exact method of characteristics and by the approximate method of rectilinear charateristics [7.1].

Let us consider the dependence of the length of the upper contour L_- of a circular nozzle and that of the lower contour L_+ on the number M_0 (Fig. 7.1f). From the flow rate equation for the lower half of an annular nozzle we get

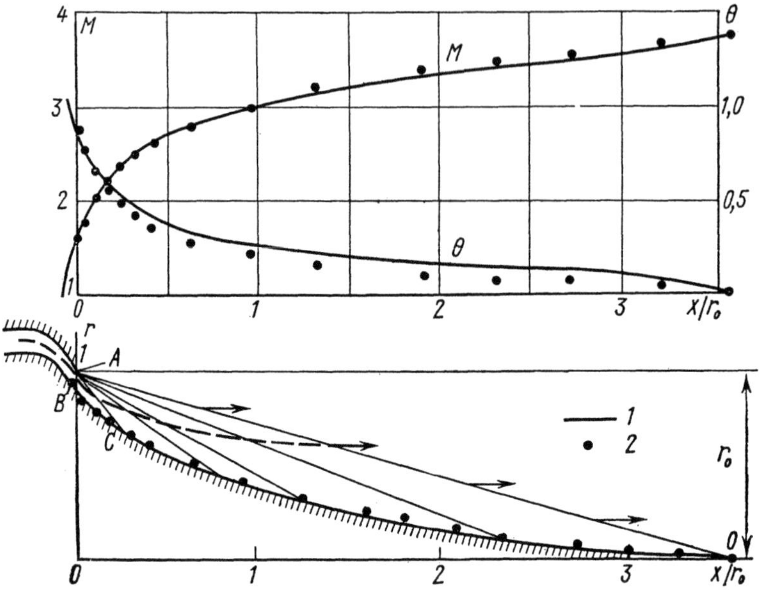

Fig.7.2. Contour of a nozzle's central body and distribution of M and θ along the contour: (*1*) method of characteristics [7.1], (*2*) method of rectilinear characteristics

$$\frac{h}{2r^0} = 1 - \sqrt{1 - q(M_0)} \quad . \tag{7.6}$$

The formula allows one to calculate the circular gap at a fixed r^0 as a function of the Mach number M_0. At a fixed h/r^0 it also defines the number M_0. Naturally, at a fixed value of r^0, the size of the circular gap decreases as M_0 increases. It follows from (7.6), that at a fixed throat size for a circular nozzle with corner points situated on the same plane, there is only one value of M_0 at which the uniform characteristic issuing from the point 0 reaches the nozzle axis. Using (7.6), we get

$$L_+ = r^0(x_+/r^0 + \sqrt{M_0^2 - 1}) \quad ,$$

$$L_- = r^0[x_+/r^0 + (r_0/r^0 - 1)\sqrt{M_0^2 - 1}] \quad . \tag{7.7}$$

Then, according to (5.24,7.6) and assuming the flow is plane, we have

$$\frac{x_+}{r^0} = \frac{1 - \sqrt{1 - q(M_0)}}{q(M_0)} \quad , \tag{7.8}$$

$$L_+ = r^0\left[\frac{1 - \sqrt{1 - q(M_0)}}{q(M_0)} + \sqrt{M_0^2 - 1}\right] \quad , \tag{7.9}$$

$$L_- = r^0\left[\frac{1 - \sqrt{1 - q(M_0)}}{q(M_0)} + \left(\frac{r_0}{r^0} - 1\right)\sqrt{M_0^2 - 1}\right] \tag{7.10}$$

Let us next calculate r_0/r^0 using the flow rate equation for the upper half of a circular nozzle. We get

$$\frac{r_0}{r^0} = \left\{1 + \frac{2}{q(M_0)}\left[1 - \sqrt{1 - q(M_0)}\right]\left[1 + \frac{1}{2}\left[1 - \sqrt{1 - q(M_0)}\right]\right]\right\}^{\frac{1}{2}} \quad . \tag{7.11}$$

Then, independent of the value of the adiabatic exponent, $r_0/r^0 \to \sqrt{2}$ at $M_0 \to \infty$. The exact value of r_0/r^0 is equal to 1.33 at $M_0 = 3.25$ and 1.38 at $M_0 = 4$ [7.3].

Let us compare the lengths of the upper and lower contours of an axisymmetric annular nozzle with that of a conventional axisymmetric nozzle with a corner point which has the same throat area and Mach number M_0.

The length of the supersonic part of a conventional axisymmetric nozzle with a corner point and a uniform and parallel flow at the exit cross-section is equal to

$$L^0 = r_* \left[\frac{\ell^0}{r_*} + \sqrt{\frac{M_0^2 - 1}{q(M_0)}}\right]$$

or, according to (5.24) at $\gamma = 1.4$, to

$$L^0 = \frac{r_*}{\sqrt{q(M_0)}}\left(1.17 + \sqrt{M_0^2 - 1}\right) \quad . \tag{7.12}$$

The condition of equality of the critical section areas of annular and conventional nozzles gives

$$r_* = r^0\sqrt{2h/r^0} = 2r^0\left[1 - \sqrt{1 - q(M_0)}\right]^{\frac{1}{2}}$$

and, hence,

$$\frac{L^0}{L_+} = 2\left[\frac{1 - \sqrt{1 - q(M_0)}}{q(M_0)}\right]^{\frac{1}{2}} \frac{1.17 + \sqrt{M_0^2 - 1}}{\left\{\left[1 - \sqrt{1 - q(M_0)}\right]/q(M_0)\right\} + \sqrt{M_0^2 - 1}} \quad , \tag{7.13}$$

$$\frac{L^0}{L_-} = 2\left[\frac{1 - \sqrt{1 - q(M_0)}}{q(M_0)}\right]^{\frac{1}{2}} \frac{1.17 + \sqrt{M_0^2 - 1}}{\left\{\left[1 - \sqrt{1 - q(M_0)}\right]/q(M_0)\right\} + (r_0/r^0 - 1)\sqrt{M_0^2 - 1}} \quad . \tag{7.14}$$

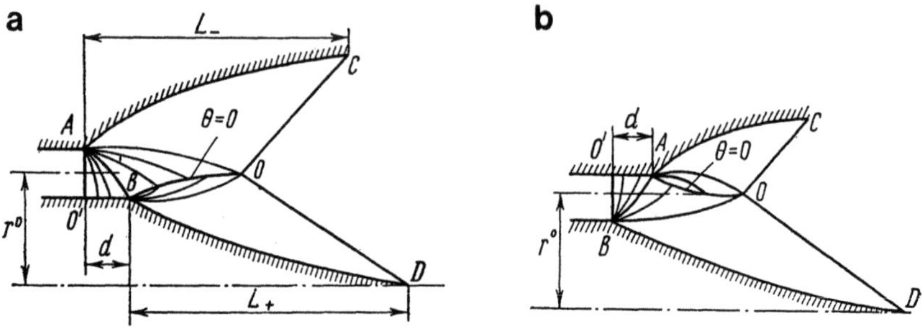

Fig.7.3a,b. Annular nozzles with displaced corner points

Thus, $L^0/L_+ \to \sqrt{2}$ and $L^0/L_- \to 3.4$ at $M_0 \to \infty$. Hence, the use of an annular nozzle with two corner points lying in the same plane and providing uniform parallel flow at the exit allows a central nozzle to be shortened compared to a conventional axisymmetric one with a corner point, by at least $\sqrt{2}$ times. At the same time, the length of the upper contour of an annular nozzle shortens compared to that of a conventional axisymmetric nozzle by 3 to 3.5 times, approximately.

Introduction of a new parameter d – the displacement of the corner points – allows different values of M_0 to be obtained at a fixed size of the circular gap. Figure 7.3 presents schematically the results of calculations for circular nozzles with displaced lower and upper corner points. The method of calculation is similar to the one used in calculating flows in annular nozzles at $d = 0$. First, the rarefaction wave originating from the point A (or B) and finishing at the rectilinear wall 0'B (or 0'A) is calculated and its length defines the value of the corner points' displacement. This wave interacts with the rarefaction wave issuing from the point B (or A). Since, as the flow turns, the velocity vector inclinations at points A and B have opposite signs, then the rarefaction waves' interaction results in the formation of a line issuing from B (or A), at every point of which the flow velocity is parallel to the axis (the line $\theta = 0$). There is a single point 0 at the line, such that the flow rate through the characteristic B0 is equal exactly to the flow rate through the uniform characteristic 0D issuing from 0 and reaching the nozzle axis. From the point 0 a uniform characteristic 0C is drawn so that the flow rates through characteristic A0 and 0C are equal. The Goursat problem is solved in the space between characteristics 0C, 0A and 0B, 0D, and the family of streamlines is obtained, the upper and the lower nozzle contours among them.

Only one number M_0 exists for nozzles with displaced corner points at fixed h/r^0 and d, at which the condition of gas flow rate equality through characteristics BO and OD is fulfilled.

Note that the ordinate of the point O at the line $\theta = 0$ differs from the average radius of the critical section r^0 by no more than 2%, while the distribution of M with respect to x along the line $\theta = 0$ is close to the distribution along the axis of a plane jet.

Since point O lies practically at the middle line, the number M_0 for an annular nozzle with a displaced lower corner point exceeds that for a similar nozzle with the same critical section dimensions but with corner points lying in the same plane. An annular nozzle with a displaced upper corner point has a smaller number M_0 compared to that of the corresponding annular nozzle with corner points lying in the same plane.

It follows from simple geometrical considerations that the largest decrease of the length of an annular nozzle L compared to that of a conventional axisymmetric nozzle is attained at $L_+ = L_-$. In this case the length of an annular nozzle is approximately twice as short as that of the corresponding conventional axisymmetric nozzle, and the average radius of the critical section is two times smaller than the exit section radius. A nozzle length may be decreased and the number M_0 increased (at constant dimensions of the critical section) by displacing the lower corner point.

The method of characteristics allows the flow parameters to be determined in the supersonic part of a nozzle only. Solving the inverse problem, one can calculate the entire flow field. This is done by means of specifying an analytical curve ℓ which is a streamline of the flow sought, at which an analytical distribution of pressure/velocity in the sub-, trans- and supersonic regions is specified and the initial-value problem is solved for the gasdynamics equations. One of the nozzle walls may be assumed to be the curve ℓ and the pressure/velocity distribution may be chosen to obtain the required flow (the other wall coincides with an appropriate streamline). However, most often the pressure/velocity distribution is unknown and to evaluate it in the first approximation, the one-dimensional theory is used. For annular nozzles of the types d, e, f (Fig.7.1) the curve ℓ is specified as $r = \text{const}$. Some results of calculations are presented in Figs.7.4,5. Here any streamline may be assumed to be a nozzle wall. At $r > r^0$ the flow corresponds to the nozzle d, and at $r < r^0$, to the nozzle e. The figures show pressure $\bar{p} = p/p_*$ distributions along the streamline $\psi = 0.06$ (curve 1), the line ℓ (curve 2), the streamline forming the central body contour (curve 3)

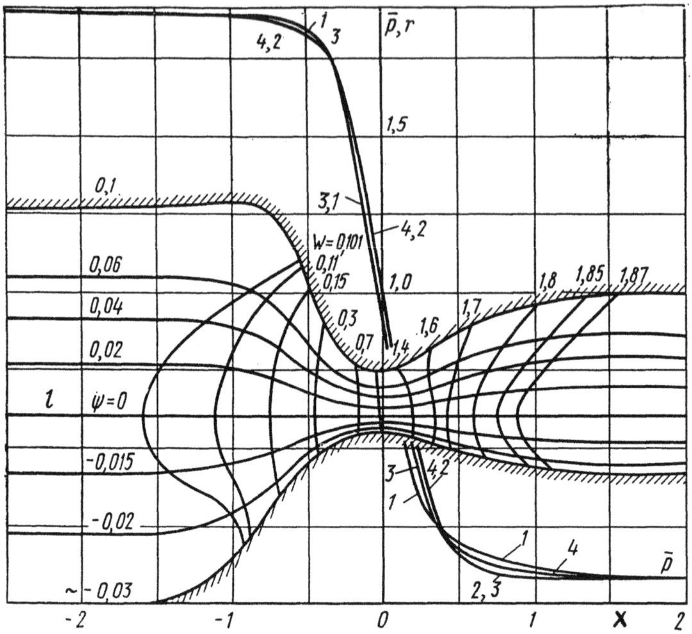

Fig.7.4

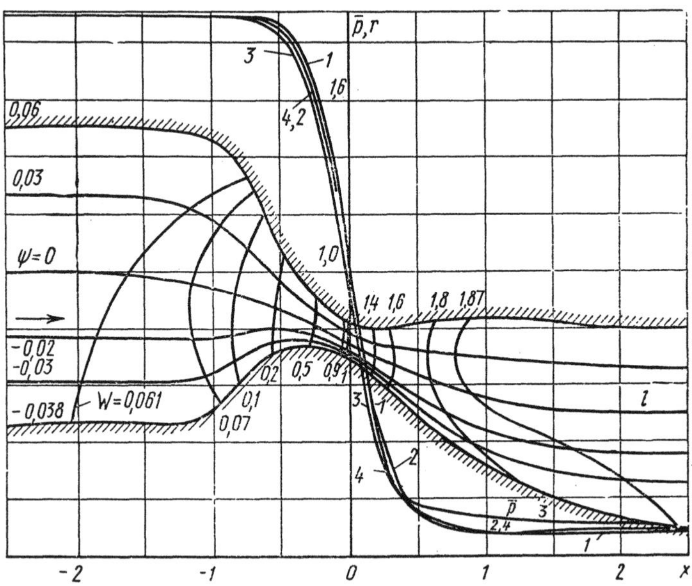

Fig.7.5

Fig.7.4. Flow in an annular axisymmetric double-expansion nozzle

Fig.7.5. Flow in an annular axisymmetric double-expansion nozzle with an inclined throat

and the distribution obtained from the one-dimensional theory (curve 4). The pressure distribution along 1 is practically one-dimensional.

Comparison of contours obtained by solving the inverse problem for axisymmetric, plane and one-dimensional cases shows that near the throat their coordinates differ by 1% only. However, as a whole, the flow in an annular nozzle differs substantially from the one-dimensional one, especially at large x in the vicinity of the symmetry axis, where large positive pressure gradients are observed. The flow may be considered plane only in zones small compared to the distance from the symmetry axis. The best coincidence with a plane flow takes place in the transonic region.

Let us compare experimental and calculated data (Fig.7.6). In [7.4] the inverse method of comparison was applied, with a measured pressure distribution used as initial data while solving the initial-value problem. Then, the calculated central body contour was compared to that of the tested nozzle. This method of comparison requires highly accurate measurements of pressure. The pressure was measured at the upper rectilinear wall of an annular nozzle with a central body whose contour contained a corner point. About 200 measurements were made and approximately 1000 manometer readings were reduced. Then the experimental data (Table 7.1) were smoothed by means of the following normalized convolution operator:

$$p(x) = \frac{1}{\Omega}\int_{-\delta}^{\delta} p(x - \zeta)\exp\left(-\frac{\delta^2 t}{\delta^2 - \zeta^2}\right)d\zeta \quad , \tag{7.15}$$

where

$$\Omega = \int_{-\delta}^{\delta} \exp\left(-\frac{\delta^2 t}{\delta^2 - \zeta^2}\right)d\zeta \quad ,$$

and t is the smoothing parameter. The experimental data were reduced at $t = 23$, $\delta = 3$. The pressure measurement error and the error due to smoothing did not exceed 1%.

Figure 7.6 presents the results of the comparison. Here curve 1 is the central body contour of the nozzle used in the tests, 2 denotes the calculated streamline, 3 is the central body contour corrected by the boundary layers at the central body and at the rectilinear wall, and 4 is the smoothed pressure distribution at the rectilinear wall: little circles denote experimental data.

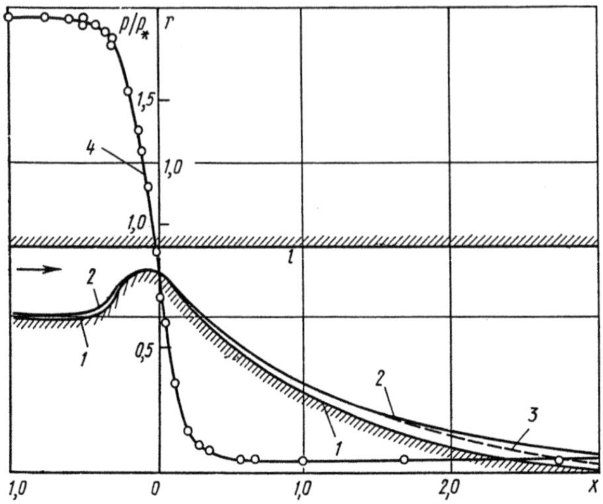

Fig.7.6. Comparison of calculated and experimental data for pressure distribution and a nozzle contour shape at $\gamma = 1.4$

Table 7.1. Experimental values of pressure for an annular nozzle

x	p/p_*	x	p/p_*	x	p/p_*
3.00	0.0383	0.25	0.1246	-0.10	1.2983
2.50	0.0383	0.20	0.1679	-0.12	1.3597
2.00	0.0383	0.16	0.2129	-0.14	1.4186
1.70	0.0383	0.14	0.2434	-0.16	1.4738
1.40	0.0384	0.12	0.2835	-0.20	1.5668
1.15	0.0384	0.10	0.3385	-0.25	1.6597
1.00	0.0385	0.08	0.4138	-0.30	1.7300
0.88	0.0396	0.06	0.5114	-0.35	1.7786
0.76	0.0410	0.04	0.6234	-0.42	1.8198
0.67	0.0420	0.02	0.7426	-0.50	1.8467
0.58	0.0438	-0.00	0.8615	-0.60	1.8641
0.50	0.0496	-0.02	0.9748	-0.75	1.8750
0.42	0.0603	-0.04	1.0773	-1.00	1.8803
0.35	0.0751	-0.06	1.1604	-1.50	1.8818
0.30	0.0938	-0.08	1.2335	-2.00	1.8819

7.1.2 Off-Design Flow Regimes in Annular Nozzles

It is a characteristic feature of some annular nozzles (types a, c, e) (Fig. 7.1) that at off-design regimes, when the external pressure p_H exceeds p^0 defined in the one-dimensional approximation by the ratio of the nozzle's exit section area to the throat area, their thrust coefficients are considerably higher than those of conventional Laval nozzles. This is due to the high atmospheric pressure being transmitted to the central body contour along the characteristics of the first and second families upon their reflection from the jet boundary. In a conventional Laval nozzle at $p_H > p^0$ the static pressure

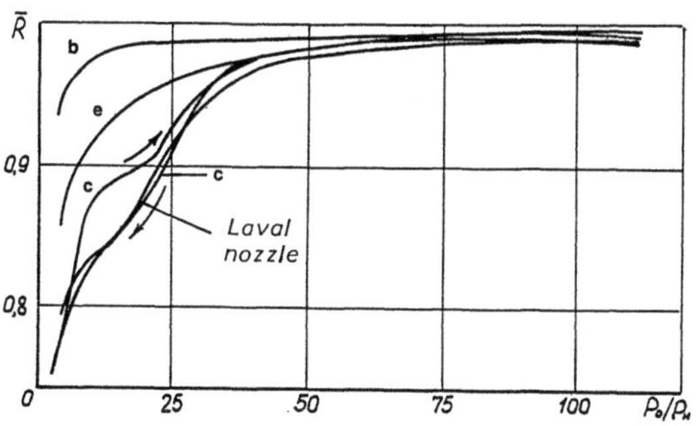

Fig.7.7. Comparison of thrust coefficients for a number of annular nozzles (see Fig.7.1) for off-design regime

at a large part of its surface is considerably lower than the external pressure; only at the end part of a nozzle is it close to the latter. Thrust coefficients of various nozzles are presented in Fig.7.7, which demonstrates clearly the advantages of annular nozzles at off-design regimes. Here the thrust coefficient is equal to the actual thrust of a nozzle referred to the thrust of an ideal nozzle with the flow at the exit section parallel to the axis, $p_H=p^0$ and the flow rate equal is to that of the actual nozzle. The thrust coefficient acquires maximum values for annular nozzles of the types b and e, since in these cases the external pressure acts upon almost the entire contour. Below we consider characteristic features of off-design flows in such nozzles [7.1,5]. In off-design regimes calculation of practically all types of nozzles is quite a complicated task, first of all due to viscous effects which, together with separated flows, play a significant role, but which cannot be described strictly enough at present. To deal with the flows, semi-empirical methods are being developed. Off-design flow regimes in nozzles b and c are among the rare ones which can be calculated exactly enough, since no intense shock waves form in them. At $p_H > p^0$ the flows can be numerically calculated by the method of characteristics. In the case of nozzle b, calculation of flow in the annular region between the jet boundary and the body's contour starts from the characteristic AC of the rarefaction wave, at whose point A the pressure equals p_H (Fig.7.8). Evidently, upstream of the characteristic, the flow parameters are the same as for the design regime at $p_H = p^0$.

Figures 7.8-10 present typical geometries of the jet boundary and characteristics and distributions of the static pressure p referred to the stagnation pressure. Experimental data are shown by various symbols and calculated ones by solid lines. The wavy structure of the jet boundary and the pressure distribution at the central body is produced by reflection of alter-

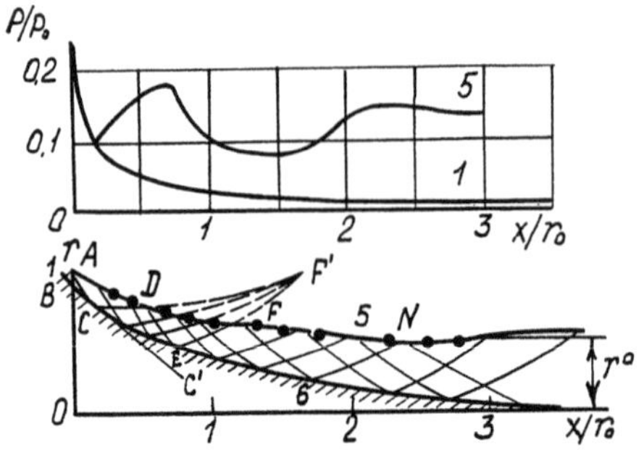

◄ Fig.7.8. Flow field in an annular nozzle and pressure distribution at the central body at $M_0 = 3.7$, $\gamma = 1.4$: (*1*) $p^0/p_H = 1$, (*5*) $p^0/p_H = 0.11$

Fig.7.9. Pressure distribution at the central body and the jet boundary at $M_0 = 3.7$, $\gamma = 1.4$ for various values of p^0/p_H: (*1*) $p^0/p_H = 1$, (*2*) $p^0/p_H = 0.67$, (*3*) $p^0/p_H = 0.38$, (*4*) $p^0/p_H = 0.15$, (*6*) $p^0/p_H = 0.09$

▼

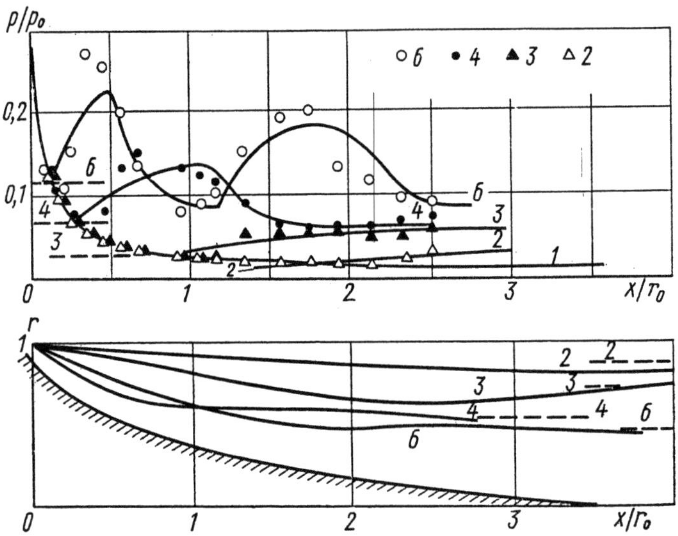

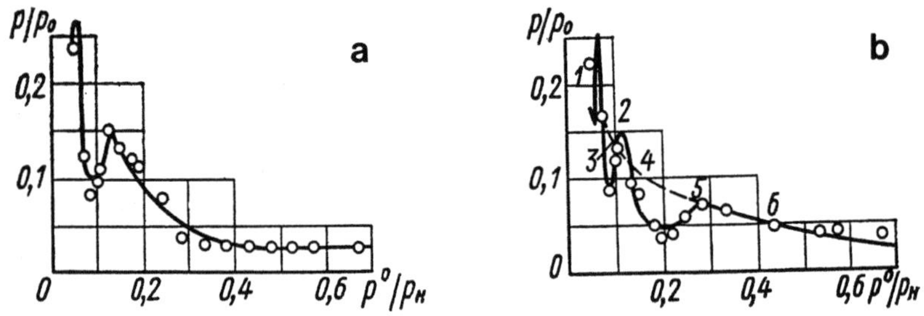

Fig.7.10. Dependence of pressure on p^0/p_H at the points of the central body with coordinates $\bar{x} = 0.93$ (**a**) and $\bar{x} = 2.5$ (**b**) at $M_0 = 3.7$, $\gamma = 1.4$

nating rarefaction and compression waves from the boundaries of the jet and the body. In fact, characteristics AC, CD, DE, EF, FG, GN, etc. separate different analytical regions of flow (Fig.7.8). At these characteristics, derivatives of the gasdynamic parameters and curvature of streamlines, the jet boundary among the latter, are discontinuous. Thus, the characteristic AC separates the analytical rarefaction flow ABC and the flow in triangle ACD defined by the values of parameters at the characteristic AC and at the segment AD of the jet boundary; the characteristic CD divides the compression flow in the region CDE defined by the body's contour and the values of parameters at the characteristic CD from the flow inside triangle ACD. Derivatives of the pressure distribution at the body are correspondingly discontinuous at the points C, E and G.

It is most important for the comprehension of the flow pattern that triangle CDE is occupied by a compression flow. To explain this, we shall assume that the flow inside CDE is plane. Then characteristics AC, CD and the jet boundary AD are rectilinear and if, starting from the point C, the body's contour CC' were rectilinear, then the flow inside CDE would be uniform with constant parameters. However, since the wall CE is curved, the region is occupied by a compression flow analogous to one taking place in a uniform supersonic flow past a concave wall. Such a flow is known to be closed by a suspended shock originating at the point F' where characteristics intersect. Dashed lines in Fig.7.8 show characteristics of an imaginary compression flow which would take place if a uniform flow with $p = p^0$ in a certain region above the line AD as well as between the characteristics AC and CD existed. Generally speaking, the point F' may lie both inside and outside the jet. However, calculations show that it always stays outside the jet. Compression waves originating in triangle CDE reflect from the jet boundary as rarefaction waves. The latter, upon reflecting from a body's surface, remain rarefaction waves, while from a jet boundary they reflect as compression waves. The further flow structure is defined by an alternating system of rarefaction and compression waves which reflect from the wall and the jet boundary. Upon reflecting from a solid wall, the wave intensity conserves its value and sign, while upon reflecting from a jet boundary it conserves its value but changes sign.

Far enough from a body the flow parameters become pratically constant, the pressure at a section becomes equal to the atmospheric one, and the velocity vector is parallel to the axis. Using the continuity equation at fixed values of p^0/p_H and M_0, one can find the radius r^0 of an isobaric section of a jet whose pressure is equal to atmospheric and whose velocity vector is parallel to the axis. We have

$$p_H/p_0 = \pi(M^0) \quad , \quad r^0 = [q(M_0)/q(M^0)]^{1/2} \quad , \qquad (7.16)$$

where M^0 is the Mach number at the isobaric section found from the first relation. The values of p_H/p_0 and r^0 are shown in Figs.7.8 and 7.9 by dashed lines. It is seen that already in the vicinity of the end part of the central body, the jet radius is practically equal to r^0, i.e. the flow straightens rapidly enough. The flow isentropicity results in the thrust coefficient of a nozzle with a full central body being equal to unity for off-design regimes in the framework of ideal liquid. In other words, the nozzle under consideration is ideally self-regulating. Naturally, viscosity effects and the truncation of the central body (in order to minimize the total losses taking into account friction, the weight equivalent and so on), result in the thrust coefficients of real nozzles (of the annular scheme under consideration) being somewhat less than unity at off-design regimes. However, this difference is insignificant and such a nozzle scheme has the best thrust characteristics compared to other schemes (Fig.7.7).

Let us consider the dependence of the static pressure at a fixed point of a body on the ratio p^0/p_H (Fig.7.10), whose nonmonotonic variation is also associated with the alternation of rarefaction and compression waves. Even a small change of p^0/p_H at $0.05 < p^0/p_H < 0.2$ results in a considerable pressure variation at a point. This is because even a small change of p^0/p_H results, for points in the vicinity of the boundary characteristics AC, EF, FG, etc., in their removal from the region of rarefaction/compression into the region of compression/rarefaction. Since at small p^0/p_H such regions are much narrower than at large p^0/p_H, the dependence of p/p_0 on p^0/p_H has multiple maxima and minima. When with increasing p^0/p_H, a fixed point happens to lie at the segment CE, the pressure starts to decrease monotonically. Upon the transition of a fixed point onto the segment BC, the pressure stops changing, since here it does not depend on p_H.

An analogous flow pattern takes place at off-design regimes for annular nozzles of the type c, which sometimes are called disc-type nozzles. If the base region is not closed, then p_H equals 60 to 90% of the atmospheric pressure value. The base pressure propagates to the nozzle contour along characteristics and an alternating system of rarefaction and compression waves appears resulting in nonmonotonic pressure variation at the contour.

The presence of external flow past the base may result in a certain change for the worse in annular nozzles' thrust characteristics at off-design regimes, because a zone of separated flow may form next to the base region with

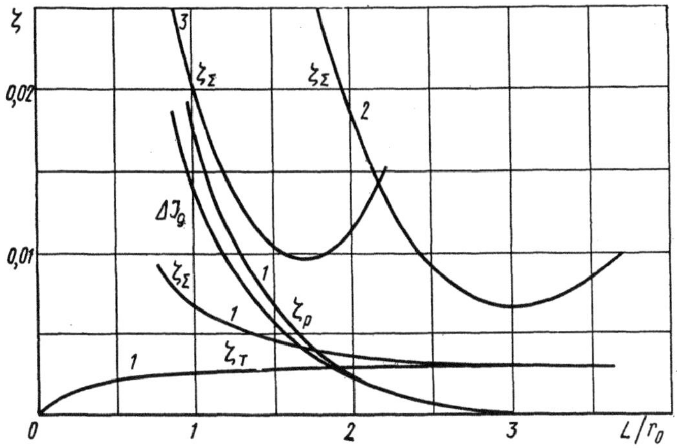

Fig.7.11. Specific impulse losses in annular nozzles at $M_0 = 3.7$, $\gamma = 1.4$: (*1*) nozzle b, (*2*) Laval nozzle, (*3*) nozzle c

pressure below atmospheric, which results in pressure at the central body dropping below atmospheric too.

Let us compare the total impulse losses in an annular nozzle and in a conventional Laval nozzle at design regimes. Figure 7.11 presents experimental and calculated values of impulse loss coefficients as functions of the central body length. For an annular nozzle, the total losses were calculated taking into account experimental values of the base pressure with the central body differently truncated. The losses were obtained from

$$\zeta_\Sigma = \zeta_T + \zeta_p - \Delta J_g \quad , \tag{7.17}$$

where $\Delta J_g = p_g r_T^2 [p_0 \varepsilon(1) q(\lambda^0) z(\lambda^0)]^{-1}$ and r_T is the relative radius of the central body's end face. It follows from Fig.7.11 that the dependence of ζ_Σ on the length (ignoring the base pressure effect) for type b nozzles does not have the typical minimum characteristic of annular nozzles (due to ζ_T and ζ_p having opposite tendencies with increasing nozzle length). Therefore, even a threefold truncation of the central body practically does not change its impulse. At equal total losses an annular nozzle is three times shorter than a conventional Laval nozzle. Evidently, truncation of the central body results in a certain decrease of the thrust coefficient; however, it stays high enough. An analogous conclusion may be made with respect to disc-type nozzles, though their advantage compared to circular nozzles is less considerable. The total losses for such a nozzle are also shown in Fig.7.11.

A disc-type nozzle has an ambiguous dependence of the thrust coefficient on the ratio p_0/p_H (Fig.7.7). This dependence, obtained with increasing pressure ahead of the nozzle, differs from a similar dependence obtained with decreasing pressure, the latter effect being caused by different transformations of the base flow in these two cases.

In order to be able to predict characteristics of an annular nozzle, one has to have data on the base pressure. Some experimental data and a method of calculating the type b nozzles are presented in [7.5]. It is shown that in the case of a closed base zone (i.e. if a sonic line exists in the base zone and the end face pressure does not depend on the external pressure), the base pressure decreases as the central body length increases. However, if the base zone is unclosed, then the base pressure is almost equal to the external pressure (which takes place for disc-type nozzles too). A typical dependence of the base pressure on the central body length is presented in Fig.7.12.

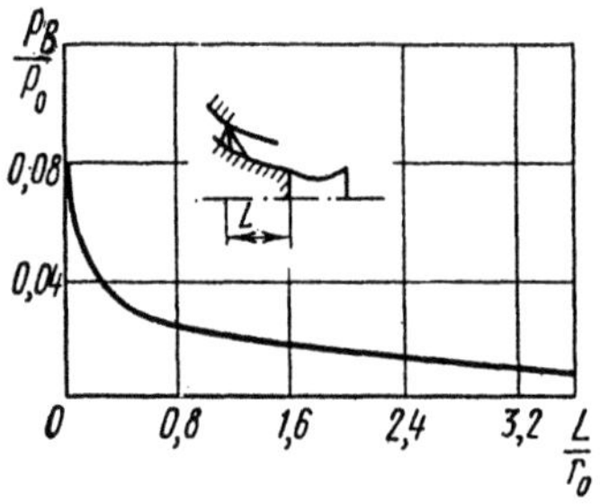

Fig.7.12. Dependence of the base pressure at the central body (nozzle b) on the length for the closed base region regime ($M_0 = 3.7$, $\gamma = 1.4$)

7.1.3 Wind Tunnel Nozzles

Wind tunnel nozzles are designed to provide uniform flow at the exit cross-section. Therefore, their contours are calculated by solving the inverse problem of the theory of nozzles, since it is quite a complicated task to specify a nozzle contour providing a given degree of flow uniformity while solving the direct problem.

Solution of the problem of nozzle profiling is divided into two stages: firstly, some method is used to calculate the nozzle contour which provides a uniform flow at the exit cross-section in the framework of the ideal gas model, then a semiempirical correction is introduced taking into account viscosity.

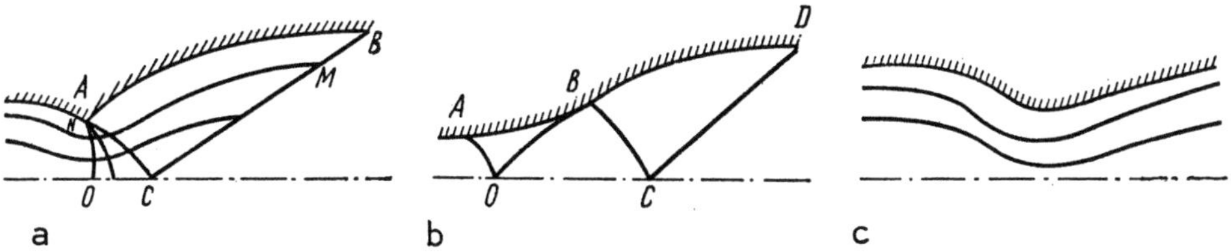

Fig.7.13a-c. Profiling of aerodynamic nozzles

A nozzle contour providing uniform flow at the exit cross-section may be designed in a number of ways. One of these assumes that a nozzle's generatrix has an inflection at a point A (Fig.7.13a). A certain initial characteristic AO is specified and flow in the rarefaction wave OAC is calculated. Then Goursat's problem is solved in the space between the characteristic AC and the uniform characteristic BC, the flow rate through which is equal to the flow rate through AC, and a family of streamlines is obtained, each of which may be assumed to be a contour of a wind tunnel nozzle. Evidently, the contour AB has the shortest relative length. The drawback of the method is the necessity of choosing the nozzle contour, especially in the transonic part upstream of the point A, corresponding to the given initial characteristic AO. If this condition is not satisfied, then a descrepancy between the true flow in the transonic region and the one assumed in the calculations will result in a nonuniformity of the flow parameters at the exit cross-section.

Another method (Fig.7.13b) used, in particular, in [7.6,7] is as follows. A certain experimental or calculated velocity distribution is given at the segment OC of the nozzle axis and the method of characteristics is used to calculate flow in the characteristic triangle OBC. Then, the contour of the required nozzle AD is determined by solving Goursat's problem in the space between characteristics CB, CD and OA, OB, the characteristic CD being uniform, while the characteristic AO is obtained somehow by calculating the transonic region. The second method has the same drawback as the first one, and besides, gives nozzles which are too long.

The third approach (Fig.7.13c) proposed in [7.8] uses a method of solving the inverse problem of the theory of nozzles, which allows the sub-, trans- and supersonic regions of flow to be calculated using a single scheme. Contours of plane, axisymmetric and annular nozzles of various types are designed in a similar way; it is necessary only to choose the velocity or pressure distribution at the initial streamline (where the initial data are specified), which provides uniform flow at the exit of the shortest nozzle. Since nozzles with a corner point possess the shortest supersonic part possible, the corresponding experimental or calculated velocity/pressure distribution may be taken as the initial one. In particular, the experimental pressure distribution presented in Table 7.1 may be used.

Starting from a certain basic pressure distribution p(x) at the initial line, one can obtain an array of distributions for various values of M_0. Let a distribution $p_0(x)$ and two values of pressure $p_- = p_0(-\infty)$ and $p_+ = p_0(+\infty)$ be given for infinite points of the subsonic and supersonic parts of flow respectively. Then, the following family of pressure distributions may be obtained:

$$p(x) = p'_+ + \frac{p'_- - p_+}{p_- - p_+}[p_0(x) - p_+] \quad , \tag{7.18}$$

where p'_- and p'_+ are new values of pressure at infinite points of the sub- and supersonic parts of flow respectively.

If $p_0(x)$ is a smoothed distribution obtained experimentally and corresponding to a flow in a nozzle with a corner point, then the function p(x)

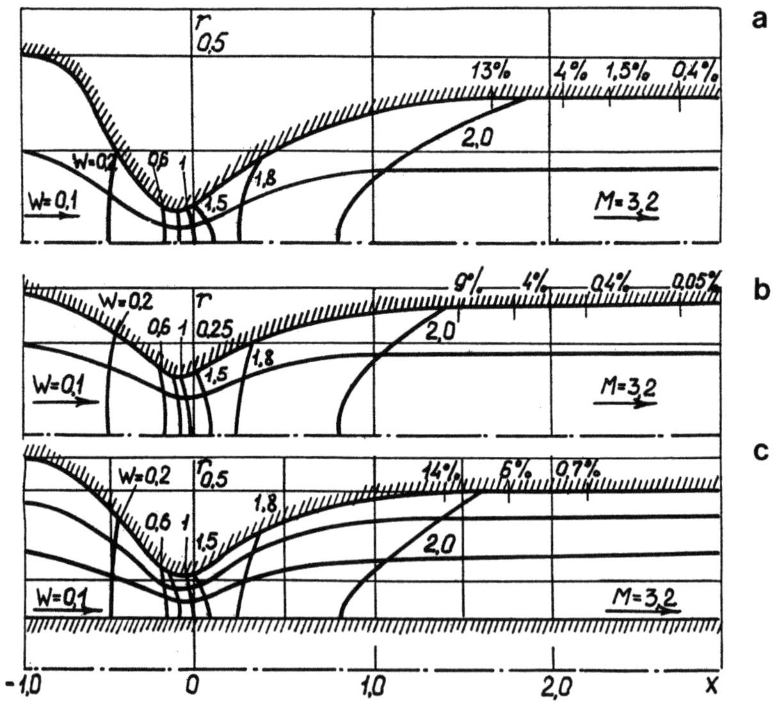

Fig.7.14a-c. Contours of plane (a), annular (b) and axisymmetric (c) aerodynamic nozzles

generates a flow with streamlines approaching the corner point closely enough. Figure 7.14 presents contours of plane, axisymmetric and annular wind tunnel nozzles with $M_0 = 3.2$ obtained by solving numerically the inverse problem using the experimental pressure distribution of Table 7.1 as the initial data. For some sections pressure nonuniformity is presented (in percent).

7.2 Conical Nozzles

Conical nozzles are widely applied for solving many scientific and engineering problems. They are used as thrust nozzles, wind tunnel nozzles and for studying various nonequilibrium problems. Many studies of gas flows in conical nozzles have been reported [7.9-13]. The most detailed analysis of transonic and supersonic perfect gas flows in conical nozzles is presented in [7.9,12]. This section deals with properties of gas flows in supersonic parts of conical nozzles obtained by means of the numerical method presented in Sect.3.2.2 (see [7.9]).

The assumption often used for the flow in the supersonic part of a nozzle, i.e. that it is close to that from a point source, is not always justified. In fact, the structure of flow in a conic nozzle is much more complicated. Under certain conditions flow may decelerate in the vicinity of the conjunction of the radius and conical parts of a nozzle (point C in Fig.7.15). The compression wave generates a shock which, after multiple reflections from the symmetry axis and the nozzle contour, can influence the flow parameter distributions considerably.

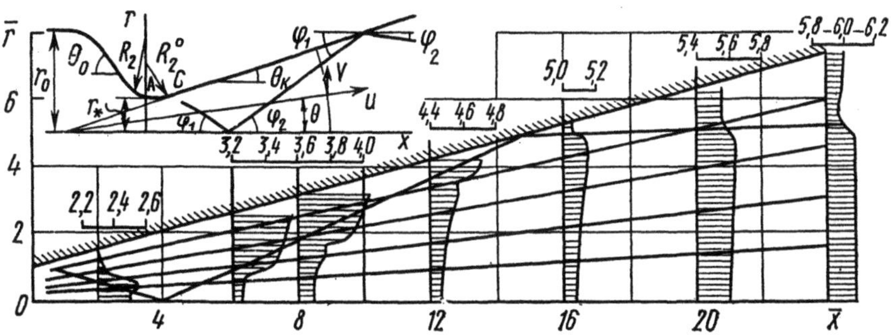

Fig.7.15. Flow in a conical nozzle; streamlines and distribution of Mach number in the cross-sections

Table 7.2. Conical nozzle flow parameters

Variant number	$\bar{R}_2$	$\bar{R}_2^0$	θ_k deg.	γ	Axis				Contour				
					$\bar{x}$	φ_1 deg	φ_2 deg	p_2/p_1	$\bar{x}$	$\bar{r}$	φ_1 deg	φ_2 deg	p_2/p_1
1	2.00	0.0	15	1.40	3.65	20	21	4.36	14.00	4.74	12	10	1.04
2		0.5			3.75	20	21	3.99	14.60	4.89	12	10	1.03
3		1.0			3.90	20	21	2.62	15.20	5.08	12	10	1.01
4		2.0			4.30	20	21	1.77	16.90	5.52	12	10	1.00
5	0.25	0.5	15	1.40	4.15	20	21	14.58	15.20	5.04	12	10	1.12
6	0.50				4.05	20	21	9.45	15.00	5.03	12	10	1.07
7	1.00				3.85	20	21	5.38	14.70	4.96	12	10	1.06
8	2.00	0.5	5	1.40	1.40	35	37	1.74	3.00	1.26	32	27	1.09
					5.60	32	28	1.97	8.90	1.78	23	21	1.05
					13.80	24	19	1.37	22.20	2.92	15	12	1.01
9			10		2.30	24	28	2.25	6.10	2.07	18	18	1.09
					15.50	18	16	1.98	-	-	-	-	-
10			10		6.50	20	14	5.56	-	-	-	-	-
11	2.00	0.0	5	1.40	1.35	35	37	1.78	3.00	1.26	32	27	1.13
					5.55	32	28	1.88	8.75	1.76	24	22	1.06
					13.80	26	20	1.50	21.60	2.88	16	13	1.01
12			10		2.25	24	28	2.60	6.00	2.05	18	18	1.14
					15.30	19	17	2.01	-	-	-	-	-
13			20		6.30	20	14	8.55	-	-	-	-	-
14	2.00	0.5	15	1.14	3.00	23	26	2.46	3.70	3.29	21	17	1.05
15				1.25	3.30	21	23	2.93	10.40	3.78	17	13	1.05
16			20	1.14	4.50	26	19	2.35	18.00	7.48	14	11	1.01
17				1.25	5.10	23	17	2.78	-	-	-	-	-
18			10	1.14	2.00	27	33	1.89	4.90	1.86	23	23	1.08
					10.60	23	22	1.69	21.00	4.68	15	14	1.01
19				1.25	2.15	26	31	1.94	5.35	1.92	21	21	1.08
					12.20	21	20	1.65	-	-	-	-	-

Let us consider the unfluence of the geometrical parameters of a nozzle ($\bar{R}_2$, $\bar{R}_2^0$, θ_k) and the adiabatic exponent γ upon the structure of a flow using the results of [7.9], where studies have been done for a wide interval of determining parameters (the bar denotes values referred to r_*). Table 7.2 presents the variants considered. Here are presented the values of $\bar{R}_2$, $\bar{R}_2^0$, θ_k, γ, coordinates of points at the axis and the contour where the shock wave falls, the angles of inclination of the falling and reflected shock waves φ_1, φ_2 and the value of pressure p_2 behind the reflected shock wave referred to p_1 in front of the falling shock. The numbers of the variants presented in the table will later be used when analyzing the calculated results.

A typical gas flow pattern in a conical nozzle is shown in Fig.7.15 (Table 7.2, variant 2), where streamlines, the shock wave position and the distributions of the Mach number for a number of sections $\bar{x}$ = const are presented.

We see that a suspended shock wave forms in the vicinity of the point C where the radius and conical parts of the nozzle conjugate, and its intensity is greater the closer it is to the symmetry axis. In fact, the flow in the vicinity of the point C is directed from the symmetry axis and, in the case of an analytic contour containing no points of curvature discontinuity, it would continue accelerating along a circular arc. However, a conical part is a kind of obstacle which results in the discontinuity of derivatives in the flow parameter distributions at the point of conjugation at the wall. The intensity of acceleration decreases sharply, and under certain conditions the flow may decelerate at the wall. The decelerating flow thus formed in the vicinity of the point C is accompanied by the intersection of characteristics belonging to the same family, resulting in the formation of a suspended shock wave.

The Mach number derivative discontinuity at the wall near the point of conjugation may be obtained by means of (5.12), which for the case under consideration acquires the form

$$\left[\frac{dM}{ds}\right] = M\left(1 + \frac{\gamma - 1}{2}M^2\right)\left[(M^2 - 1)^{\frac{1}{2}}\bar{R}_2^0\right]^{-1} \quad .$$

This formula gives results which agree well with those obtained by numerical methods. The shock wave generated at the point of conjugation can reflect repeatedly from the axis and walls of a nozzle. While approaching the symmetry axis, its intensity increases, and further downstream decreases due to the influence of rarefaction waves. As a rule, the intensity of the wave is

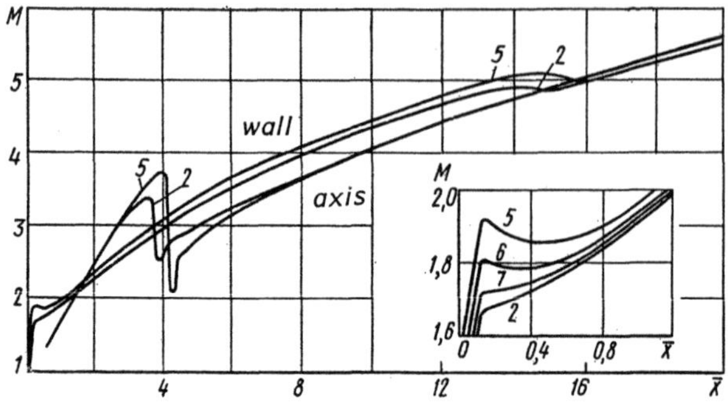

Fig.7.16. Distribution of the Mach number along the nozzle axis and contour (variants 2, 5, 6, 7)

not large. The presence of the shock wave results in a nonmonotonic distribution of the flow parameters over the cross-sections.

Figure 7.16 presents the Mach number distribution along the axis and contour of conical nozzles. We see that $\bar{R}_2$ affects the Mach number distributions and the shock wave intensity in the immediate supersonic vicinity of the point of conjugation, and that a decrease of $\bar{R}_2$ is accompanied by an increase of the shock wave intensity. Coordinates of the points where the shock wave intersects the symmetry axis and the contour of a nozzle exhibit only a slight dependence on $\bar{R}_2$, the points moving downstream as the latter decreases. At small $\bar{R}_2$, a segment appears at a nozzle contour downstream of the conjugation point, where the number M is practically constant or even has a minimum corresponding to a pressure increase. The effect of $\bar{R}_2^0$ upon the flow parameters is also slight, though more significant than the influence of $\bar{R}_2$. With increasing $\bar{R}_2^0$ the number M gradients along the curvilinear part of a nozzle contour and along the symmetry axis decrease. The shock wave intensity somewhat increases as $\bar{R}_2^0$ decreases and attains its maximum value for a nozzle with a corner point. With increasing $\bar{R}_2^0$ the shock wave moves downstream. The above-calculated Mach number axis and contour distributions agree very well – for position and intensity of shock waves – with the experimental data contained in [7.10,11,13]. Numerical and experimental data from [7.11] are compared in Fig.7.17.

In order to explore the effect of the angle θ_k upon the flow in the supersonic region, calculations were carried out for nozzles with small rounding-off radii, $\bar{R}_2^0 = 0.5$, and with a corner point, $\bar{R}_2^0 = 0$. The shock wave positions and their intensities exhibit a strong dependence on the angle θ_k (Table 7.2)

As expected, calculations show that a gas with a larger γ, at a fixed nozzle length, expands to a larger number M. The shock wave becomes more

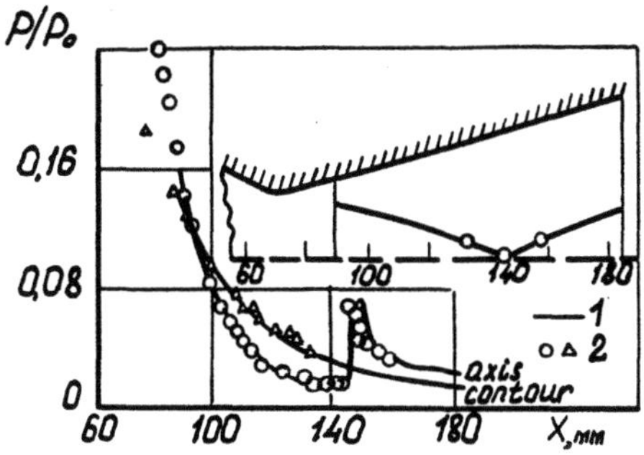

Fig.7.17. Position of the shock and static pressure distribution (p_0 is the total pressure: (*1*) calculated, (*2*) experimental [7.11]

intense too, and contacts the axis and the nozzle wall further downstream of the throat. Though the influence of γ in the transonic part of a nozzle is small, it may be considerable in the supersonic part, especially at large values of θ_k. This refers both to shock wave positions and to the flow as a whole. Decreasing γ results in a quicker smoothing of the flow parameters over the cross-sections of a nozzle.

For practical purposes it is most important to know the values of coefficients characterizing impulse losses due to divergence, i.e. the difference between the impulses of actual and ideal nozzles, which are assumed to have equal total pressures, flow rates and exit cross-section areas. In addition, an ideal nozzle has a uniform flow at the exit parallel to the axis. Let us consider two kinds of loss coefficients, one corresponding to the real flow in a nozze (ζ_p), and the other calculated under the assumption that the nozzle flow is produced by a source (ζ_{pk}). The first coefficient is calculated using (5.39), and the second one with the help of [7.9]:

$$\zeta_{pk} = 1 - \frac{J_k}{J^0} = 1 - \cos^2\frac{\vartheta_k}{2}\,\lambda_k\left(1 + \frac{1}{\gamma M_k^2}\right)\left[\lambda^0\left(1 + \frac{1}{\gamma M^{02}}\right)\right]^{-1} \quad .$$

Velocity coefficients λ^0, λ_k are obtained from the relationships $q(\lambda^0) = \mu r_k^{-2}$, $q(\lambda_k) = \mu r_k^{-2}\cos^2(\theta/2)$ where r_k is the radius of a nozzle's exit cross-section, the superscript 0, as usual, denotes parameters of an ideal nozzle and the subscript k denotes parameters of a source flow. The following asymptotic relation is true:

$$\lim_{r_k \to \infty} \zeta_{pk} = \sin^2\frac{\theta_k}{2} \quad . \qquad (7.19)$$

The formula is often used to calculate losses due to divergence in conical nozzles; however, it may result in an error up to 0.5%.

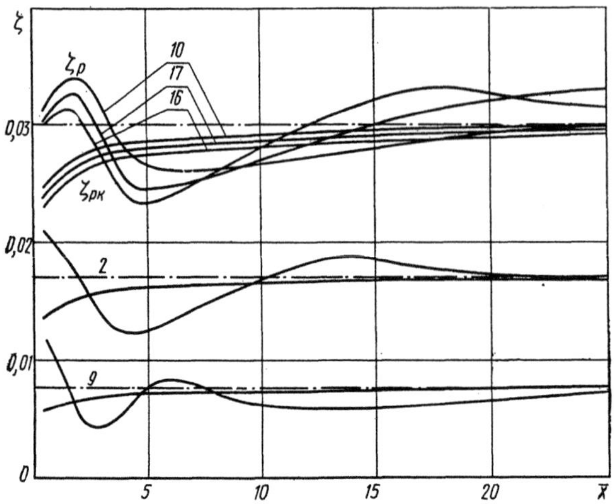

Fig.7.18. Dependence of divergence losses ζ_p and ζ_{pk} on the length of a conical nozzle for various θ_k and γ

Some results of calculating the values of ζ_p and ζ_{pk} are presented in Fig. 7.18, where the asymptotic values of ζ_{pk} are shown by dash-dot lines. The effect of the radii $\bar{R}_2$ and $\bar{R}_2^0$ upon the coefficients ζ_p and ζ_{pk} (as well as on the entire nozzle flow) is small (0.1% or less). The adiabatic exponent γ plays an important role which increases as the angle θ_k increases. With decreasing γ, the values of ζ_p and ζ_{pk} become somewhat smaller. The most important role belongs to the angle θ_k. Its increase is accompanied by increasing losses due to divergence, since nonuniformity of the flow becomes more pronounced. The amplitude of coefficients' variation increases too, evidently, due to an increasing difference between the actual nozzle flow and the one generated by a source in the transonic region. The values of ζ_p oscillate about the asymptotic value given by (7.19), i.e. initial perturbations of the source flow may result both in a decrease and an increase of losses compared to the source flow.

Conical nozzles are widely used as an object of study, not only because their manufacture is so simple, but also due to a widely spread opinion that flows in such nozzles are close to source flows. However, realization of a potential source flow requires special profiling (Sect.2.2), since the presence of a sonic line in the form of a circular arc results in the appearance of limiting lines [7.14] and, as the above-calculated results show, formation of shock waves. However, in source flows all gasdynamic parameters are constant at circular arcs drawn from the source centre, and the velocity vector is directed along radii. The greatest difference is observed near the throat. Here flow at first accelerates intensely, then compression waves form in the vicinity of the conjugation point and the flow velocity decreases. Farther along the contour the flow accelerates again, and slows down at the location where the shock wave crosses the nozzle contour.

The difference of the pressures in actual and source flows decreases downstream, but stays rather large for a considerable part of a nozzle length reaching 10 to 15%. This is to be held in mind when experimentally studying nonequilibrium flows in which the effects result in pressure variations of the same order of magnitude. The differences between parameters of actual and source flows depend mainly on the angle θ_k and γ and increase as the latter increases. At small θ_k and γ a nozzle flow generally differs little from a source flow; however, even at $\theta_k = 5^\circ$ the pressure difference near a shock wave reaches 10%.

7.3 Swirling Nozzle Flows

Swirling gas flows have been under study for some time already and have found wide application for controlling thrust. Swirl mainly affects the flow rate through a nozzle, whose thrust changes accordingly. At the same time, the specific impulse stays practically unchanged. In other words, swirl may be used to control the mass flow rate. If a nozzle's length is restricted and prohibits a uniform flow from being obtained at the exit cross-section (which, as we know, provides a maximum thrust), then the use of swirl may result in increasing the thrust [7.15].

While developing rotating solid-propellent rocket motors, it has been found that their characteristics depend largely on rotation. This is due to forces intensifying combustion and the influence of rotation on the combustor and nozzle flow fields. By intensifying combustion, swirl affects the entire gasdynamic flow pattern and may throttle a nozzle's throat. Both effects result in increasing pressure in the combustion chamber which in its turn accelerates the propellent burning rate.

Flow swirl is used also for reversing thrust of air-breathing jet engines. It also helps to prolong the residence of radioactive fuels in nuclear rocket engines and stabilizes electric arcs in electric heaters. Swirl helps to perfect the performance of combustion chambers too. It intensifies mixing and combustion in general, making the latter more stable. (The rate of turbulent burning increases about three-fold.) Swirl suppresses flame pulsations and jet noise and makes combustion more complete, thus decreasing pollution of the environment. It also allows a combustor to be made smaller and the weight of a jet engine to be reduced. In the case of air-breathing jet engines, swirl applied throughout a channel helps to reduce the length by more than 10%.

In MGD generator channels flow is swirled by the Lorentz force. However, this kind of swirl may be eliminated by swirling a flow appropriately at the initial cross-section of an MGD channel. This is the aim, for instance, of swirling flows in radial nozzles of MGD generators.

7.3.1 Introductory Notes

Originally flow swirl was considered in [7.16] for inviscid incompressible flows which were assumed to be potential flow, thus having an empty core. However, further experiments have shown that in slightly swirled flows no empty cores form and intense pulsations arise. Only artificial air injection along the symmetry axis allowed an empty core to be formed and stabilize the flow. Despite such an artificial way of core formation, theoretical and experimental values of the flow rate coefficients agree quite well. A further decrease of flow swirl eliminates pulsations; however, air injection no longer facilitates the formation of an empty core.

It has been shown in [7.17] that when the ratio of the initial tangential velocity at a nozzle wall to the initial axial velocity (the Frenkel number) exceeds 1.92, a liquid starts rotating like a solid body. Such a situation takes place in the case of slowly rotating rocket motors with a cigarette-burning solid-propellent grain. However, the hypothesis of solid body rotation overdefines the problem.

Various experiments with swirling flows made in ducts and nozzles show that the solid body rotation takes place in the central part of the flow only, i.e. in the immediate vicinity of the symmetry axis, outside of which the flow stays practically potential and rotates as a free vortex (Fig.7.19a). This is accompanied by the formation of an empty core along the axis (actually a reverse flow region is formed).

If the outflow is subcritical, then the recirculation zone stretches throughout the nozzle and out into the environment (Fig.7.19b,c). This results in suction of the external air into the nozzle through the zone. If the outflow is supercritical, then the recirculation zone breaks into two zones, internal and external (Fig.7.19c). In the latter case there is no recirculating flow at the throat. This corresponds to the so called "choked" nozzle flow when the supersonic part of the flow does not affect its subsonic part.

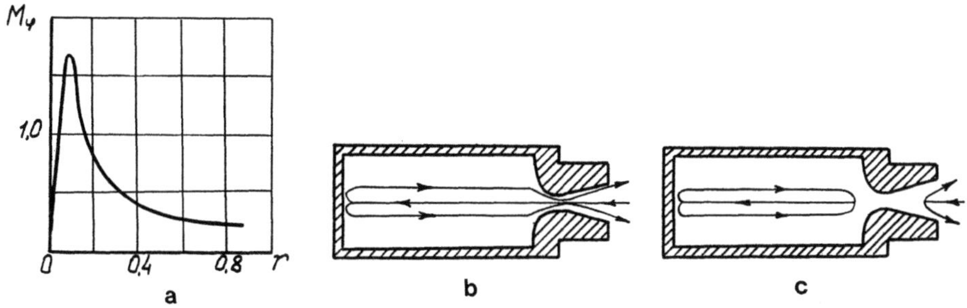

Fig.7.19a-c. Dependence of the Mach number (obtained with the help of the circumferential velocity component) on r (**a**) calculated by pressure measured experimentally in a vortex chamber (**b** and **c**)

Experiments have shown that in rapidly rotating solid-propellent rocket motors, toroidal recirculation zones form in the subsonic part of a nozzle and near its walls [7.18]. In [7.19] an analytical flow model has been proposed using both solid body rotation near the symmetry axis and potential vortex flow in the peripheral zone.

Two "conjugate" flow regimes may exist in swirling flows, one at low swirl intensities when finite length standing waves are absent and another at high swirl intensities when standing waves, recirculation and cavitation zones may appear. At first, increasing swirl or a nozzle diameter has practically no effect on the flow. However, when a certain critical value of swirl intensity or nozzle diameter is attained, the entire flow pattern changes abruptly. Formation of a reverse flow is defined by the ratio of the angular momentum to the axial momentum. The critical value of the ratio corresponds to the Frenkel number 1.92. At subcritical flow regimes (flows with "filled" cores), the cores where liquid rotates as a solid body grow with increasing distance downstream of the initial cross-section, and may occupy the whole of a channel's cross-section. At supercritical flow swirls, a zone of reverse axial flow appears near the walls. Due to energy exchange, vortex swirl at the initial cross-section of a channel transits to solid body rotation at the exit cross-section. Thus, the swirl of the central flow tubes decreases while that of the peripheral ones increases.

Formation of reverse axial flows is observed in intensely swirled jets too. The effect of rotation upon the wave structure of an axisymmetric jet is qualitatively analogous to that of off-design conditions. Recirculating flows are unstable and accompanied by the formation of large vortices whose characteristic dimensions are proportional to the width of the zone.

Many swirl studies have been dedicated to flow stability. Vortex destruction proceeds as abrupt structural changes of the central part of a swirling flow. Rapid deceleration, deformation and expansion of the core take place and recirculation zones form. Velocity and pressure distributions change in the surrounding flow. These changes, in their turn, result in an alteration of the general gasdynamic pattern of a nozzle flow, in particular, heat exchange. For instance, in the case of a laminar boundary layer in a convergent conical nozzle, flow swirl results in an increasing heat flux; however, the increase is insignificant near the throat where the heat flux is maximum.

Vortex destruction may occur in swirling duct flows too, e.g. it takes place in suction channels of turbines and pumps with air intakes. In ducts, a vortex core, upon destruction, usually becomes axisymmetric and almost closed, and the flow downstream returns to the original state. However, a different type of destruction is observed in divergent channels where the core acquires the shape of a double spiral (the so-called helicoidal destruction). Still another type of destruction, when the core has the shape of a conventional spiral, is observed, usually behind wings, though it may also take place in ducts. Similar vortex destruction modes are characteristic of nozzle flows as well as of flows in cyclone separators and other devices using flow swirl and in jets.

It should be noted that vortex destruction is a sudden transition process from one dynamic state of an axisymmtric flow to another, analogous to a hydraulic jump in open channels. Vortex destruction and boundary layer separation have much in common. For intensely swirled flows, as well as for flows with positive pressure gradients, large axial gradients of parameters are characteristic. This corresponds to the deceleration of a flow in the longitudinal direction followed by vortex destruction, while the main flow stays stable with respect to axisymmetric disturbances and unstable with respect to nonaxisymmetric ones.

In the core of a swirling flow an important dissipative role belongs to viscous forces, especially at large velocity gradients. Results of theoretical investigations compared to experimental data have shown that viscous forces must be taken into account mostly while calculating intensely swirled flows.

The majority of theoretical investigations, excluding the most recent ones, were based on the hypothesis of a radially balanced flow, according to which the normal velocity component may be neglected. The assumption is valid for calculating flows in slender nozzles. Besides the above-mentioned works, of considerable interest is [7.20], where the integral similarity criterion for swirling flows has been derived, and [7.21] where an analytical solution for helicoidal flows has been proposed. Recently, swirling flows in circular and annular nozzles have been investigated in the framework of the direct and inverse problems, taking into account the two-dimensional character of flow [7.22-24].

7.3.2 Radially Balanced Flows

As a rule, rotation of a nozzle flow produces a complex three-dimensional flow pattern including separation zones. Therefore, theoretical and experimental studies of such flows often prove to be impossible. Thus, it is of interest to consider the simplest types of swirling flows in order to understand their quantitative features.

Let us assume that swirling of a flow does not result in its becoming nonaxisymmetric, i.e. the flow parameters are independent of the circumferential coordinate. Then, it follows from the equation of motion, e.g. from (2.98), that for straight-axis channels,

$$wr = \Gamma(\psi) \quad , \tag{7.20}$$

where w is the circumferential velocity component. Thus, in an axisymmetric swirling flow the product wr stays constant along a streamline together with the entropy S and the total enthalpy H_0.

From the equations of motion written in Crocco's form, see (4.65-67),

$$\operatorname{rot} \mathbf{W} \times \mathbf{W} = T\nabla S - \nabla H_0 \quad . \tag{7.21}$$

From the formula for $\operatorname{rot} \mathbf{W}$ in cylindrical coordinates,

$$\operatorname{rot} \mathbf{W} = \frac{1}{r}\left(\frac{\partial rw}{\partial r} - \frac{\partial v}{\partial \varphi}\right)\mathbf{e}_x + \frac{1}{r}\left(\frac{\partial u}{\partial \varphi} - \frac{\partial rw}{\partial x}\right)\mathbf{e}_r + \left(\frac{\partial v}{\partial x} - \frac{\partial u}{\partial r}\right)\mathbf{e}_\varphi \quad , \tag{7.22}$$

it follows that at $\Gamma(\psi) = \text{const}$, $S(\psi) = \text{const}$, $H_0(\psi) = \text{const}$, an axisymmetric swirling flow is potential, i.e. an isentropic and isoenergetic flow at $\Gamma(\psi) = \text{const}$ is in this case irrotational. If $\Gamma(\psi) \neq \text{const}$, then the conditions $S(\psi) = \text{const}$ and $H_0(\psi) = \text{const}$ do not ensure potentiality of a flow, and in accordance with (7.21) this means that the vectors $\operatorname{rot} \mathbf{W}$ and $\mathbf{W}$ are collinear:

$$\operatorname{rot} \mathbf{W} = \lambda \mathbf{W} \tag{7.23}$$

where λ is an arbitrary function. Such flows are helicoidal [7.25]. The form of the function $\Gamma(\psi)$ is specified by initial conditions and defines the dependence of the circumferential velocity component on r at any cross-section of a nozzle. For example, if $\Gamma = \mathrm{const}$, then the circumferential velocity component changes in a vortex-like manner. If $\Gamma(\psi) \sim \psi$, then $w \sim r$ and the circumferential velocity component changes as in the case of a solid body rotation. As stated above, the latter situation takes place in the vicinity of the symmetry axis while the former is characteristic of the peripheral parts of flow (Fig.7.19a). $\Gamma(\psi)$ may vary in other ways too. If one uses, for instance, the formula

$$\Gamma(\psi) = \Gamma_w(1 - e^{-\delta\psi})(1 - e^{-\delta})^{-1} \quad , \tag{7.24}$$

where Γ_w is the wall value of Γ at $\psi = 1$ and δ is a certain parameter, then at small δ and ψ solid body swirling is obtained, while at large δ, vortex-like swirling is obtained. At other values of ψ the velocity profile is similar to the one shown in Fig.7.19a.

Let us consider radially balanced flows, i.e. flows in which the normal velocity components may be ignored. Such flows are analogous to the usual one-dimensional flows without swirl because they are also realized in slender nozzles. A radially balanced flow takes place also near the nozzle axis. In cylindrical coordinates an axisymmetric radially balanced flow of an ideal gas with $\gamma = \mathrm{const}$ is described by the following system of equations:

$$\begin{aligned} &\frac{\partial p}{\partial r} = \gamma\rho\,\frac{w^2}{r} \quad , \\ &wr = \Gamma(\psi) \quad , \quad p/\rho^\gamma = S(\psi) \quad , \\ &\frac{1}{2}\,(u^2 + w^2) + \frac{1}{\gamma - 1}\,\frac{p}{\rho} = H_0(\psi) \quad , \\ &d\psi = \rho u\, dr^2 \quad , \end{aligned} \tag{7.25}$$

where velocity components, pressure and density are referred to their critical values. The system describes a flow completely if the channel form is specified and the functions $\Gamma(\psi)$, $S(\psi)$ and $H_0(\psi)$ are known. It follows from the first of the equations that the derivative of pressure with respect to radius in a radially balanced flow is positive, i.e. the pressure increases with increasing distance from the symmetry axis. Since the pressure derivative at the axis must be finite, the velocity component w in real flows must be zero at the axis. It follows from Bernoulli's equation that for intense swirls, i.e. at large values of w, an empty core may form near the symmetry axis. In particular, in a potential swirling flow at $\Gamma(\psi) = \text{const}$ we get, for an empty core, the radius

$$r_0 = \Gamma[(\gamma + 1)/(\gamma - 1) - u^2]^{-\frac{1}{2}} \quad . \tag{7.26}$$

An isentropic potential swirling flow with $\Gamma(\psi) = \text{const}$ is the simplest example of radially balanced flows. From (7.22) it can easily be deduced that, unlike the pressure and the circumferential velocity, the velocity component u is constant over a nozzle's cross-section in such a flow. The pressure may be obtained from the first of equations (7.25) to an accuracy of a constant. The unknown constant is expressed by means of the velocity component u. In order to obtain the velocity u, it is necessary to know the regime of nozzle flow, similarly to the usual one-dimensional flows. If a "choked" flow takes place when the external pressure does not influence the nozzle flow, then the throat velocity u^* may be obtained by means of the maximum flow rate condition $\partial Q/\partial u = 0$. This condition allows one to find u^* for the "choked" flow regime as well as the dependence of u^* on Γ or w^*, the latter being the swirl velocity at the wall of a nozzle's throat [7.24]. Note that in a swirling flow the Mach number calculated by means of the axial velocity component at the throat is less than 1. The value of u^* at the throat allows one to obtain the gas flow rate Q:

$$Q = \rho_* a_* r_*^2 \int_{r_0}^{r_w} u\left[\frac{\gamma + 1}{2} - \frac{\gamma - 1}{2}\left(u^2 + w^2\right)\right]^{1/(\gamma-1)} r\,dr \quad , \tag{7.27}$$

if the nozzle geometry and stagnation parameters are known. Values of u in the remaining cross-sections may be obtained with the help of (7.27), which also allows one to obtain the dependence of the flow rate coefficient $\mu = Q/\rho_* a_* r_*^2$ (the ratio of the actual swirling flow rate through the same nozzle at the same critical parameters, but without swirl) on $\alpha_* = w_*/W_{max}$. A typical dependence of μ on α_* is presented in Fig.7.20. The flow rate coefficient decreases with an increase of swirl, since the circumferential velocity component does not contribute to the flow rate through a nozzle. As has been pointed out in [7.24], the flow rate coefficient depends on γ only slightly (if α_* is referred to a_* instead of W_{max}), though at intense swirls μ decreases more intensely with increasing γ.

In an "unchoked" flow regime when the flow parameters depend on the external pressure, the flow velocity u at the exit cross-section of a nozzle is obtained under the condition that the wall pressure is equal to the external pressure and the continuity equation (7.27) is used to get the gas velocity at any cross-section of a nozzle.

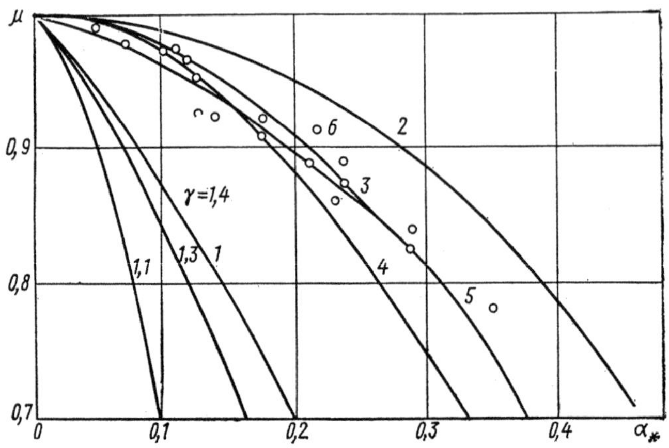

Fig.7.20. Dependence of the flow rate coefficient μ on the dimensionless swirl velocity α_* at the throat: (*1*) vortex swirl [7.24], (*2*) solid body swirl [7.23], (*3*) swirl according to $\Gamma(\psi) \simeq \psi$ [7.22], (*4*) vortexless swirl in an annular nozzle [7.22], (*5*) helical flow [7.21], (*6*) helical flow (experiment) [7.21]

The linearized theory of radially balanced flows has been developed in [7.20] and generalized for the case of flows with arbitrary thermodynamic properties in plug nozzles in [7.24]. Below, we consider an axisymmetric Laval nozzle without a central body. Let

$$S(\psi) = \bar{S} + \Delta S(\psi) \quad , \quad H_0(\psi) = \bar{H}_0 + \Delta H_0(\psi) \quad , \quad \Gamma(\psi) = wr \tag{7.28}$$

be known at the initial cross-section, where $\bar{S}$, $\bar{H}_0$ are known constants. If ΔS and ΔH_0 satisfy the conditions

$$\int_0^1 \Delta S \, d\psi = \int_0^1 \Delta H_0 d\psi = 0 \quad , \tag{7.29}$$

then integral characteristics do not depend on ΔS and ΔH_0 at a small swirl.

Assuming that swirl is small ($\Gamma \ll 1$) and $\Delta S/S \ll 1$, $\Delta H_0/H_0 \ll 1$, we may present the required parameters in the form

$$\begin{aligned} p(x,\psi) &= \bar{p}(x) + \Delta p(x,\psi) \quad , \\ \rho(x,\psi) &= \bar{\rho}(x) + \Delta \rho(x,\psi) \quad , \\ u(x,\psi) &= \bar{u}(x) + \Delta u(x,\psi) \quad , \end{aligned} \tag{7.30}$$

where $\bar{p}(x)$, $\bar{\rho}(x)$, $\bar{u}(x)$ are certain pressures, densities and velocities averaged over a cross-section and Δp, $\Delta \rho$, Δu are the respective increments. Below, the squares of $\Delta f(x,\psi)$ will be neglected. Since at a fixed nozzle geometry the values of pressure $p^0(x)$, density $\rho^0(x)$ and velocity $u^0(x)$ in the one-dimensional approximation are known, then the solution must give the dependence of the mean functions $\bar{f}(x)$ and increments on the one-dimensional parameters $f^0(x)$ and fixed values of $\Gamma(\psi)$, $\Delta S(\psi)$ and $\Delta H_0(\psi)$. The values of the flow parameters at the nozzle axis will be assumed equal to the mean values of the functions. From the first and the last equations of the system (7.25) one obtains, neglecting squares of increments,

$$\Delta p(x,\psi) = \frac{1}{2}\gamma\rho^{02}u^{o}\int_0^{\psi}\frac{\Gamma^2(\psi)}{\psi^2}\,d\psi \quad . \tag{7.31}$$

In accordance with [7.20], let us substitute $\Gamma(\varphi)$ by the integral swirl parameter

$$\varepsilon f(\psi) = \int_0^{\psi}\left[\int_{\psi}^{1}\frac{\Gamma^2(\psi)}{\psi^2}\,d\psi\right]d\psi \quad , \qquad \text{where} \tag{7.32}$$

$$\varepsilon = \int_0^{1}\left[\int_{\psi}^{1}\frac{\Gamma^2(\psi)}{\psi^2}\,d\psi\right]d\psi \quad , \qquad f(1) = 1 \quad ,$$

and ψ is normalized in such a way that it is zero at the axis and 1 at the nozzle contour. Then,

$$\Delta p(x,\psi) = \frac{1}{2}\gamma\rho^{02}u^{0}\varepsilon[f'(0) - f'(\psi)] \quad , \tag{7.33}$$

and hence, the pressure at the contour exceeds that at the axis by $0.5\gamma\rho^{02}u^{0}\varepsilon f'(0)$. The values of $\Delta\rho$ and Δu may be found at a known Δp, since $\Delta S(\psi)$ and $\Delta H_0(\psi)$ are given. We have

$$\Delta S(\psi) = S_p\Delta p + S_\rho\Delta\rho \quad ,$$

$$\Delta H_0(\psi) = \frac{1}{2}\Gamma^2(\psi)/r^2 + h_p\Delta p + h_\rho\Delta\rho + u\Delta u \quad ,$$

and hence,

$$\begin{aligned}\Delta\rho &= \frac{\Delta S(\psi)}{S_\rho} - \frac{S_p}{S_\rho}\Delta p \quad , \\ \Delta u &= \frac{1}{u}\left[\Delta H_0(\psi) - \frac{1}{2}\Gamma^2(\psi)r^{-2} - h_p\Delta p - h_\rho\Delta\rho\right] \quad .\end{aligned} \tag{7.34}$$

Here subscripts denote partial differentiation. Since the values of velocity increments are known, the flow rate coefficient may be easily obtained from the continuity equation. One gets

$$\frac{F}{\mu} = \frac{1}{\bar{\rho}\bar{u}}\left[1 - \int_0^1 \left(\frac{\Delta\rho}{\rho^0} + \frac{\Delta u}{u^0}\right)d\psi\right]$$

$$= \frac{1}{\bar{\rho}\bar{u}}\left\{1 + \frac{\varepsilon\rho}{2u}\left[M^2 + (1 - M^2)f(0)\right]\right\} . \tag{7.35}$$

Since the average values of the flow parameters practically coincide with the ideal ones, then in the case of a "choked" flow at the nozzle throat where

$$\frac{dq}{dM} = \frac{d}{dM}(\bar{\rho}\bar{u}) = \frac{d}{dM}(\rho^0 u^0) = 0 ,$$

$$\rho^0 = u^0 = \bar{F} = 1 , \quad (M^0 - 1)^2 = 0(\varepsilon)$$

we have

$$\mu = 1 - \frac{1}{2}\varepsilon . \tag{7.36}$$

Thus, the flow rate coefficient decreases with increasing swirl and does not depend on the adiabatic exponent in accordance with the above data for potential flows. The following relationships are used to find average values:

$$\bar{p} = p^0 + \frac{dp}{dq}\Delta q , \quad \bar{\rho} = \rho^0 + \frac{d\rho}{dq}\Delta q , \quad \bar{u} = u^0 + \frac{du}{dq}\Delta q , \tag{7.37}$$

where $\Delta q = \bar{q} - q^0$ and the gasdynamic function $q^0 = F^{-1}$. From (7.37) one gets, with the help of (7.33-35),

$$p(x,\psi) = p^0 + \frac{\gamma\rho^0 u^{02}}{M^{02} - 1}\left(\frac{\rho^0}{u^0} L - 1\right)\frac{\varepsilon}{2} ,$$

$$\rho(x,\psi) = \rho^0 + \frac{\Delta S(\psi)}{S_\rho} + \frac{\rho^0 M^{02}}{M^{02}-1}\left(\frac{\rho^0}{u^0} L - 1\right)\frac{\varepsilon}{2} , \tag{7.38}$$

$$u(x,\psi) = u^0 + \frac{1}{u}\left[\Delta H_0(\psi) - \frac{h_\rho \Delta S(\psi)}{S_\rho}\right] - \frac{u^0}{M^{02} - 1}\left(\frac{\rho^0}{u^0} N - 1\right)\frac{\varepsilon}{2} ,$$

where

$$L = M^{02} + (1 - M^{02})f'(\psi) , \quad N = L + (1 - M^{02})\psi f'(\psi) .$$

At the throat,

$$p^*(\psi) = 1 + \gamma L_* \frac{\varepsilon}{2} , \quad \rho^*(\psi) = 1 + \frac{\Delta S(\psi)}{S_\rho} + L_* \frac{\varepsilon}{2} ,$$

$$u^*(\psi) = 1 + \left[\Delta H_0(\psi) - \frac{h_\rho}{S_\rho}\Delta S(\psi)\right] - N_* \frac{\varepsilon}{2} , \tag{7.39}$$

while at the nozzle axis,

$$L_* = N_* = \left[1 + \left(\frac{1}{2\gamma}\frac{d^2p}{d\rho^2}\right)^{-1}\right]^{-1} - f'(0)$$

and at the contour

$$L_* = \left[1 + \left(\frac{1}{2\gamma}\frac{d^2p}{d\rho^2}\right)^{-1}\right]^{-1} \quad , \quad N_* = L_* - f''(1) \quad .$$

It follows from these formulas that for $\Gamma(\psi) = \text{const}$, $u^* < 1$ at the contour at the throat, i.e. the Mach number calculated by means of the circumferential velocity component is less than 1. From (7.39) it follows that at the throat wall

$$p^* = 1 + \gamma\frac{(\gamma - 1)}{\gamma+1}\frac{\varepsilon}{2}$$

for an ideal gas with $\gamma = \text{const}$. Then, one can easily obtain the critical pressure drop, equal to the ratio of the stagnation pressure to the throat wall pressure, at which the gas flow rate and the flow parameters cease to depend on the external pressure, i.e. nozzle "choking" takes place. We have

$$\pi^* = \frac{p_0}{p_H} = \frac{p_0}{p^*} = \pi(1)\left[1 - \frac{\gamma(\gamma - 1)}{\gamma+1}\frac{\varepsilon}{2}\right] \quad , \tag{7.40}$$

and, hence, swirl results in a nozzle "choking" at smaller pressure drops. Thus, at $\varepsilon = 0.5$ and $\gamma = 1.4$ the critical pressure drop is equal to 1.78 instead of 1.89 for a flow without swirl.

The above formulas allow expressions to be obtained for the nozzle impulse and specific impulse in a slightly swirling flow. We have

$$J = J^0 - \rho_* a_*^2(\rho + u)\frac{\varepsilon}{2} \quad , \tag{7.41}$$

$$J_s = J_s^0 - a_*\left(\rho - \frac{p}{\gamma\rho u}\right)\frac{\varepsilon}{2} \quad .$$

Calculations show that the nozzle impulse decreases considerably with increasing swirl, while the specific impulse stays practically unchanged. It follows from the above solution that the parameter ε is a rather universal similarity parameter usable for various types of swirl. It has been shown in [7.24] that the parameter allows integral and sometimes local characteristics to be correlated.

The above solution has been obtained for the case of a "choked" nozzle flow. If a flow is "unchoked", then the system of governing equations is closed by means of the condition that the wall pressure at the exit cross-section and the external pressure p_H are equal. Then, from (7.33) one gets for the exit cross-section $p_H = \bar{p} + 0.5\gamma\rho^{02}u_0 f'(0)\varepsilon$. From the adiabaticity condition and Bernoulli's integral, and using (7.33), one gets for the exit cross-section,

$$p(\psi) = p_H - [2\bar{H}_0 - h(p_H,\bar{S})]^{\frac{1}{2}}\rho^2(p_H,\bar{S})f'(\psi)\,\frac{\varepsilon}{2} \quad . \tag{7.42}$$

Using transformations similar to those above, the following expression for the flow rate coefficient may be easily obtained:

$$\mu = \frac{\rho^0 M^{02}}{u^0}\,\frac{\varepsilon}{2} \quad , \tag{7.43}$$

where ρ^0, u^0, M are the flow parameters at the exit cross-section and μ is the ratio of the true flow rate to that of an ideal flow without swirl at the same external pressure p_H. When $\rho^0 M^{02}/u^0 < 1$, it follows from the comparison of (7.36) and (7.43) that in an "unchoked" flow, the influence of swirl upon the flow rate coefficient is weaker than at the "choked" flow regime.

Let us consider another modification of a radially balanced flow, the so-called uniform helical flow [7.21] which takes place at $\lambda = \text{const}$ [see (7.23)]. From (7.22) and (7.23) we deduce that at a nozzle cross-section $x = \text{const}$,

$$\frac{du}{dr} = -\lambda w \quad , \quad \frac{d(rw)}{dr} = \lambda r u \quad , \qquad \text{or} \tag{7.44}$$

$$\frac{d^2u}{dr^2} + \frac{1}{r}\frac{du}{dr} + \lambda^2 u = 0 \quad . \tag{7.45}$$

The general solution of the equation has the form

$$u = C_0 I_0(\lambda r) + C_1 Y_0(\lambda r) \quad , \tag{7.46}$$

where $I_0(\lambda r)$ and $Y_0(\lambda r)$ are zeroth-order Bessel functions of the 1st and 2nd type. For the circumferential velocity component w we have

$$w = C_0 I_1(\lambda r) + C_1 Y_1(\lambda r) \quad . \tag{7.47}$$

The constants C_0 and C_1 will be obtained from the conditions $\Gamma(\psi) = wr$, $\Gamma = 0$ at $r = 0$ and $\Gamma = \Gamma_w$ at $r = r_w$, where r_w is the coordinate of the nozzle contour at a fixed cross-section. Hence,

$$u = \frac{\Gamma_w I_0(\lambda r)}{r_w I_1(\lambda r_w)} \quad , \quad w = \frac{\Gamma_w I_1(\lambda r)}{r_w I_1(\lambda r_w)} \quad . \tag{7.48}$$

Using (7.48), one obtains for the gas flow rate through a throat,

$$Q = 2\pi\sqrt{\frac{2}{\gamma - 1}}\,\rho_0 a_0 r_*^2 \alpha_* \int_0^1 \left[1 - \alpha_*^2 \frac{I_1^2(\bar\lambda\bar r) + I_0^2(\bar\lambda\bar r)}{I_1^2(\bar\lambda)}\right]^{1/\gamma-1} \frac{I_0(\bar\lambda\bar r)}{I_1(\bar\lambda)}\,\bar r\,d\bar r \quad , \tag{7.49}$$

where

$$\alpha_* = \sqrt{\frac{\gamma - 1}{2}}\,\frac{\Gamma_W}{r_* a_0} = \frac{W_*}{W_{max}} \quad , \quad \bar r = \frac{r}{r_*} \quad , \quad \bar\lambda = \lambda r_* \quad .$$

The dependence of α_* on $\bar\lambda$ is obtained from the condition that the flow rate through the throat is maximum, $(\partial Q/\partial\bar\lambda)\alpha_*=0$, which is valid in the "choked" flow regime.

Some results of numerical solutions are presented in [7.21,24]. The numerical scheme considered is applicable only up to a certain value of swirl intensity at which the velocity component u attains at the axis the maximum value corresponding to efflux into vacuum. Evidently, a further increase of swirl intensity results in the formation of a vacuum core similar to the one formed in potential flows. Figure 7.21 shows some radial profiles of the axial and circumferential velocity components built in isometric projection for various values of swirl. At small swirls the axial velocity component changes little along a radius while the circumferential velocity component changes almost linearly. However, at $\alpha_* = 0.2$ the dependence of the axial velocity component on the radial coordinate has already become significant; at the same time the circumferential velocity component ceases to be linear. At large swirls the profiles of both axial and circumferential velocity components are highly curved and the axial component changes from substantially subsonic values at the channel wall to rather high supersonic ones at the axis.

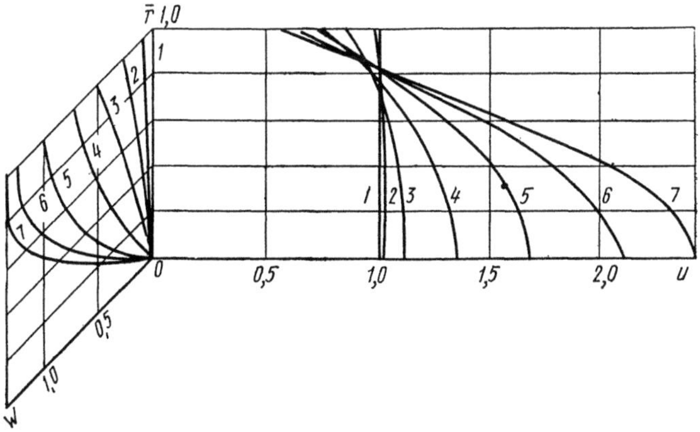

Fig.7.21. Profiles of the axial and circumferential velocity components in a uniform helical flow at various swirl intensities: (*1*) $\alpha_* = 0.01$, $\varepsilon = 0.0003$, (*2*) $\alpha_* = 0.1$, $\varepsilon = 0.03$, (*3*) $\alpha_* = 0.2$, $\varepsilon = 0.13$, (*4*) $\alpha_* = 0.3$, $\varepsilon = 0.33$, (*5*) $\alpha_* = 0.4$, $\varepsilon = 0.74$, (*6*) $\alpha_* = 0.5$, $\varepsilon = 1.63$, (*7*) $\alpha_* = 0.57$, $\varepsilon = 3$

An increase of the axial velocity component as the axis is approached is characteristic for uniform vortex flow too. and u attains at the axis larger values the more intense the flow swirl, and may become, as we see, supersonic even in the convergent part of a nozzle. Such behaviour is physically evident. In fact, according to the first of equations (7.25), the pressure gradient in the direction of the wall increases with increasing swirl and results in a substantial increase of the axial component compared to a one-dimensional flow without swirl, for which $\partial p/\partial r = 0$. On the contrary, the circumferential component becomes supersonic in the prewall region. Figure 7.20 shows the dependence of the flow rate coefficient on α_* for a uniform spiral flow.

Let us consider once again results obtained in [7.24] for the case when, at the initial cross-section, special swirling devices produced in the course of an experiment, a swirl angle $\varphi = w/u$ which changed linearly along the radius. Results of integrating the system (7.25) for a linearly changing φ specified at the initial cross-section agree qualitatively with the experimental data. They show that the number $M = u/a$ increases in the direction from the nozzle contour to its axis. At the nozzle wall at the throat, the Mach number is always subsonic, while at the axis it is supersonic. At large swirl angles the number M may also be supersonic at the initial cross-section. Pressure profiles at different cross-sections are similar if the pressure is referred to that at the wall. At the throat the swirl angle φ changes along the radius linearly just as at the initial cross-section. However, its value at the wall of the throat is smaller than at the initial cross-section, since φ decreases in a convergent nozzle and increases in a divergent one due to flow continuity and the condition $\Gamma(\psi) = \text{const}$. The experimental and calculated values of the flow rate coefficient are presented in Fig.7.20.

7.3.3 Axisymmetric Swirling Flows

In practice, nozzle flows are substantially two-dimensional; thus, the hypothesis of radially balanced flows often proves to be invalid. This stimulated studies of two-dimensional swirling nozzle flows in the framework of the direct and inverse problems [7.22-24]. Below some results of these investigations are presented. In [7.24] the method of stabilization has been used to solve the direct problem and to study a rather wide class of swirling flows. Various functions have been specified at the initial cross-section, among them vortex-like swirl, solid body rotation, uniform spiral flows, etc.

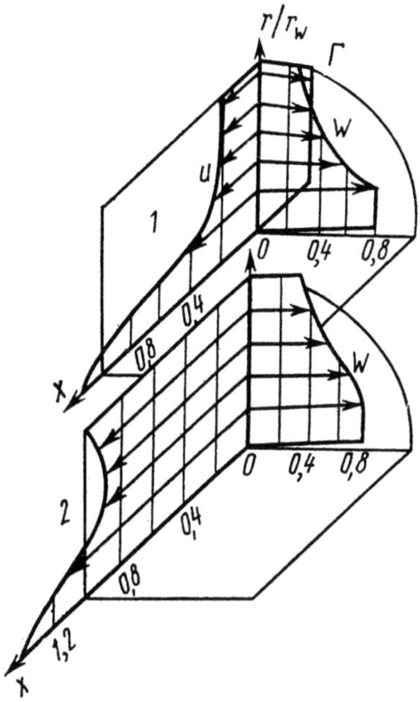

Fig.7.22. Profiles of the axial and circumferential velocity components at the initial (*1*) and throat (*2*) cross-sections of a nozzle with a swirling flow

In Fig.7.22 some characteristic profiles of circumferential and axial velocity components at the initial cross-section and at the throat are presented in isometric projection for a potential swirling flow ($\Gamma = \mathrm{const}$), which in the central part of flow transforms into a constant w flow (with the exception of the points at the axis, where $w = 0$).

In such flows the circumferential velocity component profile smoothes when approaching the throat, especially in intensely swirling flows. The axial velocity component is almost constant near the wall; then, approaching the axis, it increases rapidly and at large swirls may become supersonic at the axis both at the initial cross-section and at the throat, analogous to the cases of a helical flow and a flow with a linearly changing swirl angle considered above. Figure 7.23 shows a nozzle flow field transformation with increasing swirl. Here the lines of $M = (u^2 + w^2)^{\frac{1}{2}} a^{-1} = \mathrm{const}$ are presented for a flow without swirl (Fig.7.23a) and for a swirling flow at $\Gamma_W = 0.2$ and $\Gamma_W = 0.33$ at a nozzle wall (Fig.7.23b,c), for Γ and w profiles at the initial cross-section presented in Fig.7.22. Increasing swirl intensity results in a substantial change of the sonic line and the lines $M = \mathrm{const}$. At the central part of a flow the sonic line is drawn into the subsonic part of a nozzle and the larger the swirl intensity the stronger the effect. This transformation of the sonic line takes place at various modes of swirl,

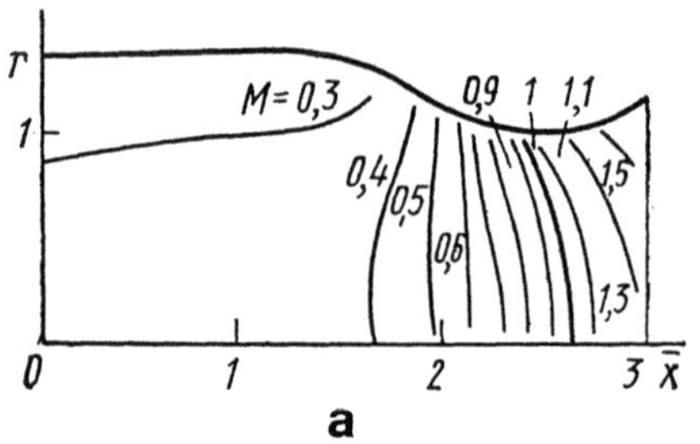

Fig.7.23a-c. Nozzle flow fields at various swirl intensities

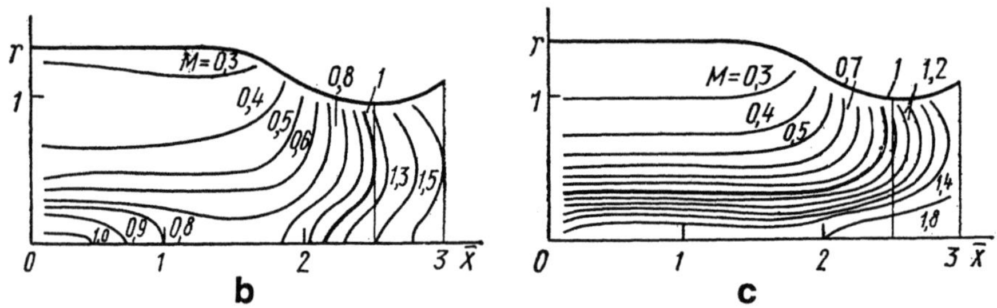

solid body rotation among them. An increase of swirl intensity results in the flow velocity at the axis becoming supersonic throughout a nozzle. Upon exceeding a certain value of intensity, a vacuum core may form as in the case of a potential flow.

If a flow is rotating as a solid body, then, depending on the initial conditions, closed or unclosed recirculation zones form near the nozzle walls whose formation is associated with the presence of a positive pressure gradient in the region of transition from the cylindrical part of a nozzle to the convergent one (Sect.5.1). The gradient is amplified by centrifugal forces appearing in a swirling flow. A further increase of swirl intensity results in its attaining the critical value (corresponding to $Fr = 1.92$), at which the flow separates and a recirculation zone forms in the prewall region of the subsonic part of a nozzle. A mighty toroidal flow separation has been observed in [7.18] near a nozzle wall where the initial swirl corresponded to solid body rotation. Other modes of flow swirl also produce recirculation zones at the axis and at a nozzle wall.

In [7.22] the effect of swirl in annular channels and channels of complicated geometry has been studied in the framework of the inverse problem. These results may be used for radial MGD generators in which swirl due to the Lorentz forces must be eliminated in the working part. Swirling flows in the general case of nonisentropic flows are calculated in the framework of the inverse problem with the help of (2.117-121) with independent variables ψ and σ, adopted for calculating flows in channels of complicated geometry. Some results obtained from this work are presented in Fig.7.20. The presence of swirl results in a displacement of the sonic line and an increase of the flow rate coefficient compared to the case of zero swirl.

The flow rate coefficient is an important characteristic of swirling flows. It characterizes both the flow rate variation and the variation of the nozzle impulse due to swirl, since the specific impulse practically does not depend on swirl intensity. In order to obtain a universal dependence of μ on the swirl intensity, various parameters are used, such as the integral swirl parameter ε introduced in the preceding section, the parameter α_*, the Frenkel number $Fr=(w/u)_W$ and the Rossby number $Ro=(0.5\ u/w)_W$. Each of the parameters has certain advantages and disadvantages compared to the rest and different authors prefer different criteria, though these produce practically the same correlations of the flow rate coefficient at various modes of swirl. Figure 7.20 shows the dependence of the flow rate coefficient μ on α_* for various modes of swirl obtained both theoretically and experimentally.

7.4 Three-Dimensional Nozzle Flows

The need to develop methods of studying three-dimensional nozzle gas flows is motivated by a number of reasons. For example, in axisymmetric nozzles, which are widely used for the solution of numerous technological and scientific problems, the flow may be asymmetric. This may be, in particular, associated with the presence of nonsymmetric flow disturbances at the initial cross-section of a nozzle.

A flow may also become asymmetric due to asymmetric distortion of the nozzle walls. Also, requirements for the geometrical forms of a nozzle may be dictated by the design parameters of jet engines or a flying vehicle as a whole. Sometimes, nozzles with noncircular cross-sections—rectangular, hexagonal, etc.—or nozzles with a "curvilinear axis" may be used.

In all these cases one has to be able to evaluate the effect of the three-dimensional character of a flow upon its local characteristics, the thrust characteristics of a nozzle and the aerodynamic stability of a vehicle's structure. Three-dimensional flows were studied by the methods of small perturbations [7.10,27,28], characteristics [7.29-31], stabilization

[7.32,33], shock-smearing [7.34-36] and by means of solving the inverse problem [7.37]. Some of the methods have been presented in Chaps.3 and 4.

Published results of experimental studies of three-dimensional nozzle flows are fragmentary. A sufficiently complete experimental investigation of the side force and the moment is presented in [7.35,36,38].

Below we present some results of theoretical and experimental studies of three-dimensional nozzle gas flows. Attention is mainly paid to studying the side forces and moments produced by asymmetric distortions of an axisymmetric conical or profiled nozzle or by asymmetric perturbations of the flow parameters at a nozzle cross-section.

7.4.1 Some Results of Analytical and Experimental Studies

We shall start by considering a transonic nozzle gas flow in a Laval nozzle. Near the centre of a nozzle, at a given axial velocity component $u = Ax$, the method of small perturbations may be used to obtain the following formulas for determining perturbations of the velocity components u, v, w with respect to a translational sonic flow in cylindrical coordinates x, r, φ [7.27]:

$$\begin{aligned} u &= Ax + (\gamma+1)r^2A^2\left(\frac{1}{4} - \ell\cos 2\varphi\right) \\ v &= A^2(\gamma+1)xr\left(\frac{1}{2} - 2\ell\cos 2\varphi\right) + A^3(\gamma+1)r^2 \\ &\quad\times\left(\frac{1}{16} - \frac{1}{3}\ell\cos 2\varphi + 4m\cos 4\varphi\right. \quad , \\ w &= 2(\gamma+1)A^2\ell xr\sin 2\varphi + (\gamma+1)A^3r\left(\frac{1}{6}\ell\sin 2\varphi - 4m\sin 4\varphi\right) \quad . \end{aligned} \tag{7.50}$$

In Cartesian coordinates we get

$$\begin{aligned} u &= Ax + (\gamma+1)A^2\left[\left(\frac{1}{4} - \ell\right)y^2 + \left(\frac{1}{4} + \ell\right)z^2\right] \quad , \\ v &= (\gamma+1)A^2\left(\frac{1}{2} - 2\ell\right)xy + (\gamma+1)^2A^3 \\ &\quad\times\left[\left(\frac{1}{16} - 12m\right)yz^2 + \left(\frac{1}{16} + \frac{1}{3}\ell + 4m\right)y^3\right] \quad , \\ w &= (\gamma+1)A^2\left(\frac{1}{2} + 2\ell\right)xz + (\gamma+1)^2A^3 \\ &\quad\times\left[\left(\frac{1}{16} - 12m\right)y^2z + \left(\frac{1}{16} + \frac{1}{3}\ell + 4m\right)z^3\right] \quad . \end{aligned} \tag{7.51}$$

Unlike plane and axisymmetric cases [see (4.37,38)], the expressions for u, v contain arbitrary constants ℓ and m, and specifying which, one can obtain nozzles of various configurations in the vicinity of the nozzle axis. The form of the sonic surface may be obtained under the condition $u=0$, which in Cartesian coordinates results in

$$x = -(\gamma + 1)A\left[\left(\frac{1}{4} - \ell\right)y^2 + \left(\frac{1}{4} + \ell\right)z^2\right] \quad . \tag{7.52}$$

It follows that at $|\ell| < 0.25$ the sonic surface is an elliptic paraboloid, at $\ell = 0.25$ it is a parabolic cylinder, and at $|\ell| > 0.25$, a hyperbolic paraboloid. Equations of the surface $v = w = 0$ (analogous to the lines $\theta = 0$ in plane and axisymmetric cases) have the following form in Cartesian coordinates:

$$\begin{aligned} y = 0 \quad , \quad x &= -\frac{2(\gamma + 1)A}{1 - 4\ell}\left[\left(\frac{1}{16} + \frac{1}{3}\ell + 4m\right)y^2 + \left(\frac{1}{16} - 12m\right)z^2\right] \quad , \\ z = 0 \quad , \quad x &= -\frac{2(\gamma + 1)A}{1 + 4\ell}\left[\left(\frac{1}{16} - 12m\right)y^2 + \left(\frac{1}{16} + \frac{1}{3}\ell + 4m\right)z^2\right] \end{aligned} \tag{7.53}$$

[see also (4.39)]. Generally these surfaces do not coincide and may be elliptical or hyperbolic paraboloids or parabolic cylinders.

The equations of the four characteristic surfaces contacting the sonic surface at the nozzle centre have the form

$$\begin{aligned} x &= \frac{\gamma + 1}{8} A[(1 - \Delta_1)y^2 + (1 - \Delta_2)z^2] \quad , \\ x &= \frac{\gamma + 1}{8} A[(1 + \Delta_1)y^2 + (1 - \Delta_2)z^2] \quad , \\ x &= \frac{\gamma + 1}{8} A[(1 - \Delta_1)y^2 + (1 + \Delta_2)z^2] \quad , \\ x &= \frac{\gamma + 1}{8} A[(1 + \Delta_1)y^2 + (1 + \Delta_2)z^2] \quad , \end{aligned} \tag{7.54}$$

where $\Delta_1 = \sqrt{5-16\ell}$ and $\Delta_2 = \sqrt{5+16\ell}$. At $|\ell| < 5/16$ the last of the formulas gives for the characteristic surface C_+^0 an equation of the elliptic paraboloid. Similarly to the case of plane and axisymmetric flows, C_+^0 and C_-^0 denote characteristic surfaces contacting the sonic line at the nozzle centre and stretching downstream and upstream respectively. At $|\ell| < 0.25$ the first of the formulas (7.36) gives for the surface C_-^0 the equation for an elliptic paraboloid, while the second and the third ones provide equations for hyperbolic paraboloids. Along the elliptic C_-^0 paraboloid disturbances propagate to the nozzle cnetre. Sources of disturbances inside the C_-^0 paraboloid reach the surface of transition and influence the subsonic part of a flow. It follows from (7.24) that C_-^0 surfaces exist at $|\ell| < 0.25$ only, and C_+^0

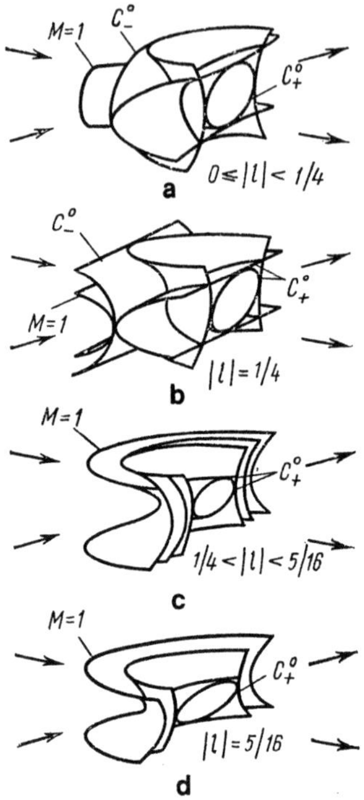

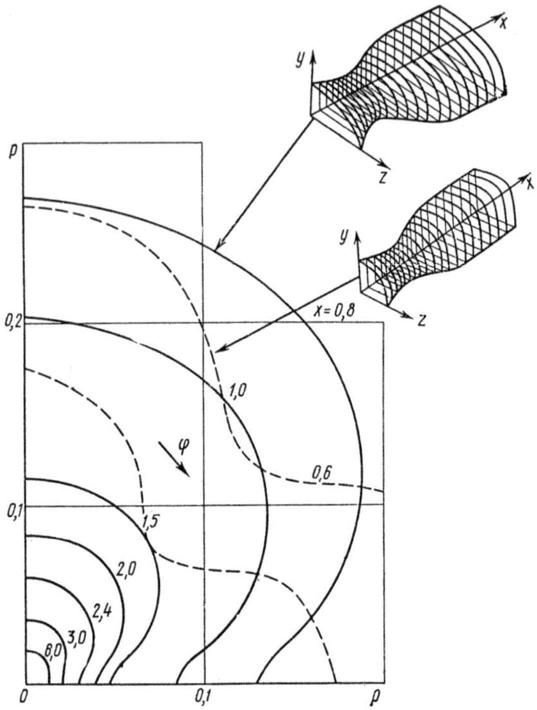

Fig.7.24a-d

Fig.7.25

Fig.7.24a-d. Configurations of the sonic surface and special characteristic surfaces in three-dimensional flows [7.27]

Fig.7.25. Pressure profiles in elliptic ($\omega = 2$, *solid curves*) and superelliptic ($\omega = 5$, *dashed curves*) nozzles

surfaces exist at $|\ell| \leq 5/16$ only. The absence of any special characteristic surfaces at such values of ℓ means that the corresponding mixed gas flows are unstable. Sonic surfaces, as well as characteristic surfaces for some values of ℓ, are presented in Fig.7.24.

Let us consider some results of numerical investigation of flows in three-dimensional nozzles. Figure 7.25 presents some data for gas flows in an elliptic nozzle whose surface is defined by the equation

$$(y/a)^{\omega} + (z/b)^{\omega} = 1 \qquad (7.55)$$

at $a/b = 2$ [6.34]. Axes y and z lie in the symmetry planes corresponding to $\varphi = 0$ and $\varphi = \pi/2$. The line of intersection of the nozzle surface with the plane $z = 0$, defined by the function $a(x)$, in the vicinity of the throat, is a circular arc ($\bar{R}_2^0 = 2$) smoothly conjugated with a straight line. At the throat of an elliptic nozzle the flow is uniform and has a constant velocity $\lambda = 1.1$. Figure 7.25 shows pressure distributions at the wall at various values of x. It is easily seen that the pressure is nonuniform with respect to φ, the minimum being attained at $\varphi = \pi/2$. The closer to the exit cross-section, the smoother the pressure distribution with respect to φ.

If $d = a/b$ stays constant at every cross-section of an elliptic nozzle, then at any value of x the ratio of the cross-section area to that of the throat is the same as for a Laval nozzle with the contour equation $y = b(x)$. In this case the one-dimensional theory gives the same longitudinal distribution of parameters for three-dimensional and axisymmetric nozzles. Figure 7.26 shows the influence of the parameter d upon the flow. Figure 7.26 a presents the lines M = const for the xy and xz planes of symmetry and demonstrates the function b(x). Here solid lines and $\zeta = y/a(o)$ correspond to the plane xy, and dashed lines and $\zeta = z/b(o)$, to the plane xz. The dash-dot lines show the results of calculations of a flow in an axisymmetric nozzle. It is easily seen that the flow is substantially three-dimensional in the trans- and supersonic regions. Increase of d results in an increasing nonuniformity in the distribution of parameters (Fig.7.26b).

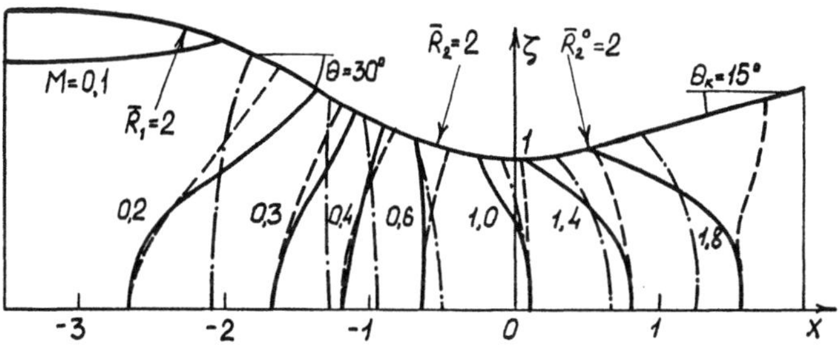

a

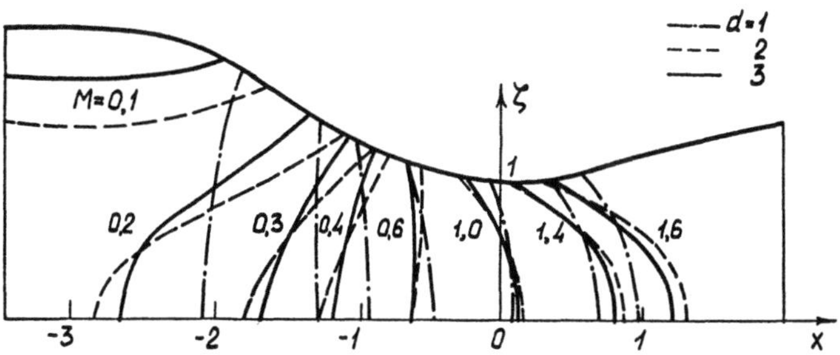

b

Fig.7.26a,b. The line M = const in an elliptic nozzle at the planes xy and xz (**a**) for d = 2 and (**b**) at the plane xy for various d

At $\omega \geq 2$, (7.55) defines the surface of the so-called "superelliptic" nozzles. At $a = b$ such nozzles have four symmetry planes ($\varphi = 0$, $\pi/4$, $\pi/2$, $3\pi/4$). Cross-sections of these nozzles are almost square. Equation (7.55) defines a wide class of nozzles including those with circular throats which have at the same time rather arbitrary configurations of the supersonic parts. In order to do this, ω must be considered a function of x. If $\omega(0) = 2$, and $a(0) = b(0) = 1$, then the throat is a unit radius circle. Pressure distributions for two cross-sections of a superelliptic nozzle are presented in Fig.7.25. We see that the minimum pressure occurs in the planes $\varphi = \pi/4$ and $\varphi = 3\pi/4$; however, increase of the expansion ratio is accompanied by a smoothing of the pressure distribution over a cross-section and a decreasing of the circumferential velocity component just as in the case of elliptic nozzles.

In the framework of the inverse problem three-dimensional nozzle flows and their geometry at a specified axial velocity distribution have been calculated by means of the asymptotic formulas (4.28). The results are presented in Fig.7.27. The figure shows nozzles which are axisymmetric up to the throat and have almost elliptic cross-sections downstream. In addition, loci of the points $\theta = \text{const}$, i.e. projections of the second stream surfaces upon the plane yz are shown and pressure distributions at various cross-sections are presented. Dashed lines show the axisymmetric nozzle geometry at the same initial velocity distribution. In the case of the nozzle presented in Fig.7.27a, the gas flows from the plane $\varphi = 0$ into the plane $\varphi = \pi/2$, as in the above cases, while for the nozzle of Fig.7.27b it flows from the plane $\varphi = 0$ into the plane $\varphi = \pi$. Note that for the latter nozzle the centres of gravity of the cross-sections (the black points in Fig.7.27b) are displaced to the left of the centre of gravity of the throat, and this results in the axis being curved and in the appearance of the side force.

A three-dimensional nozzle of a quite arbitrary configuration may be obtained by cutting off from an axisymmetric circular or annular nozzle surfaces formed by various streamlines of such nozzles. In this case, the geometry of various cross-sections of a three-dimensional nozzle is close to the geometry of the throat. Evidently, the circumferential velocity component in such nozzles is zero and no transverse gas flow takes place.

Usually, the impulse losses due to flow nonuniformity are equal to 1-3%. Hence, an axisymmetric nozzle and a three-dimensional nozzle with equal area ratios must have almost equal impulses. A rigorous proof of the statement may be found in [7.39] where the "equivalence" rule has been formulated.

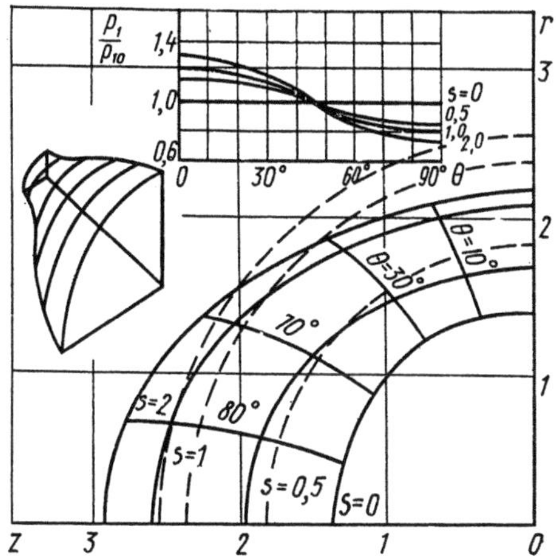

a

Fig.7.27a,b. Flows in three-dimensional nozzles with two (**a**) and one (**b**) symmetry planes

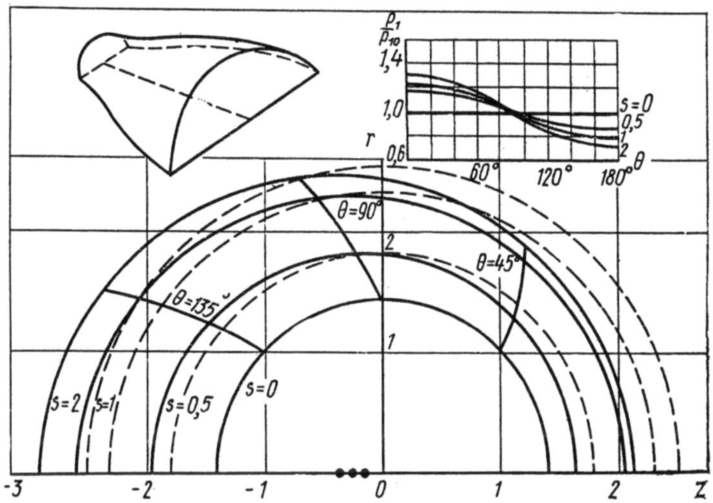

b

According to the rule, impulses of three-dimensional and axisymmetric nozzles are almost the same if the corresponding area ratios are equal. It has been shown for the examples of elliptic and superelliptic nozzles that even at a considerable nonuniformity of pressure with respect to φ and a substantial difference between the average pressure (i.e. the semisum of pressures at opposite generatrices) and the pressure in an axisymmetric nozzle with the same cross-section, the impulses at equal area ratios differ by no more than 0.5%.

The effect of an absence of symmetry in the subsonic part has been considered in [7.32] where it has been shown that when the angles of inclination of the upper and the lower generatrices lying in the symmetry plane of the subsonic part of a nozzle differ by $\Delta\alpha$, the difference of pressures at $x = \text{const}$ at the generatrices changes sign near the throat. At $\Delta\alpha = 15^{\circ}$, the side force in the vicinity of the throat is equal to 1% of the impulse. The side force depends linearly on $\Delta\alpha$.

7.4.2 Method of Small Perturbations and Determination of Side Forces and Moments

In Sect.4.5 we developed the method of small perturbations for studying flows close to radial. This allows one to determine both local and integral characteristics of flows.

For practical purposes it is most important to know the values of the side force and the moment generated by nonsymmetric deformations of the contour or by nonsymmetric perturbations at the nozzle's initial cross-section. Without loss of generality we shall assume that the nonsymmetric deformation of the contour takes place in the plane xy, i.e. the flow is symmetric about the plane. The vector of reactive force **R** will be characterized by its axial, J, and side, L, components and by the coordinate, y_c, of the point of intersection of the straight line DD', passing through the vector **R**, and the plane yz, passing through the centre of a conic nozzle (the point 0 in Fig.7.28a).

It follows from the laws of conservation of momentum and spin that the side force L and the moment M about the point 0 can be expressed as

$$L = \int_0^{\theta_k} \int_0^{2\pi} (p\mathbf{n} + \rho W_n \mathbf{W})\mathbf{e}_y \, dS \quad , \tag{7.56}$$

$$M = \int_0^{\theta_k} \int_0^{2\pi} \rho W_n (\mathbf{W} \times \mathbf{r})\mathbf{e}_z \, dS \quad , \tag{7.57}$$

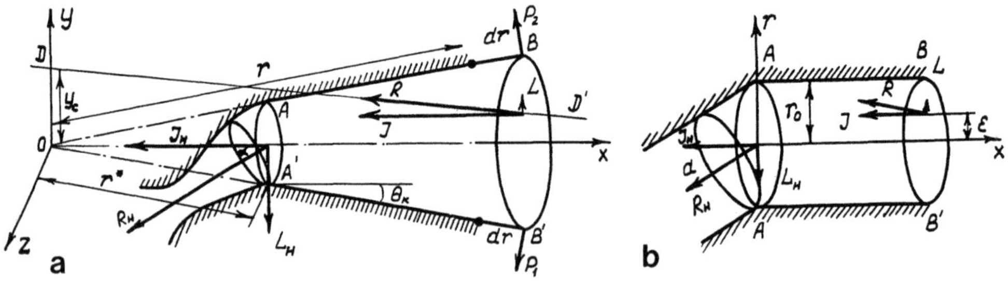

Fig.7.28a,b. Derivation of the side force and moment formulas

where $\mathbf{e}_y$, $\mathbf{e}_z$ are unit vectors and W_n is the normal component of the velocity vector. Integration in (7.56,57) is carried out over the spherical segment based upon the plane of the exit cross-section; therefore, in spherical coordinates $dS = r^2 \sin\theta \, d\theta \, d\varphi$. Assuming a small departure from the radial flow and omitting ε, we have

$$L = \rho_0 u_0 r^2 \int_0^{\theta_k} \int_0^{2\pi} \left(u \frac{1 - u_0^2}{1 - \kappa u_0^2} \sin\theta \cos\varphi \right.$$

$$\left. + v \cos\theta \cos\varphi - w \sin\varphi \right) \sin\theta \, d\theta \, d\varphi \quad , \tag{7.58}$$

$$M = y_c J_0 = \rho_0 u_0 r^3 \int_0^{\theta_k} \int_0^{2\pi} (w \cos\theta \sin\varphi - v \cos\varphi) \sin\theta \, d\theta \, d\varphi \quad . \tag{7.59}$$

Since the plane xy is a symmetry plane, the coefficients b_{1kn}, b_{2k} in (4.86-88) are equal to zero. Also, after integrating (7.58,59) with respect to φ, only the terms corresponding to $k = 1$ prove to be nonzero. In the case of nonsymmetric perturbations, to satisfy the impermeability condition at the wall ($v = 0$), the numbers n should be chosen in such a way that the condition

$$\frac{dP_n^{(1)}}{d\theta} = 0 \tag{7.60}$$

is fulfilled at $\theta = \theta_k$. The numbers of the corresponding associated Legendre polynomials can be found [with the help of (4.98)] from the approximate relationship

$$n \approx \frac{3}{4} \pi \theta_k^{-1} - 3 \quad . \tag{7.61}$$

This formula is valid for $\theta_k \leqslant 15^\circ$. For larger values of θ_k we have $n = 7, 6, 5, 4, 3, 2$ at $\theta_k = 14, 16, 19, 24, 31, 45^\circ$ respectively. The following relationship is valid for the associated Legendre function $P_n^{(1)}(\theta)$:

$$P_n^{(1)}(\theta) = \frac{dP_n}{d\theta} \quad . \tag{7.62}$$

From (7.58,59), using (4.86-88)

$$L = -\pi \sin\theta_k \cos\theta_k \rho_0 u_0 r^2 \left\{ \frac{1 - u_0^2}{r(1 - \kappa u_0^2)} \sum_{n=1}^{N} [a_{11n} Z_{1n}(r) \right.$$

$$+ a_{21n} Z_{2n}(r)] \frac{1}{2 - n(n+1)} \left. \frac{dP_n}{d\theta} \right|_{\theta=\theta_k} \tag{7.63}$$

$$+ \frac{1}{r} \sum_{n=1}^{N} [a_{11n}F_{1n}(r) + a_{21n}F_{2n}(r)] \frac{n(n+1)}{2 - n(n+1)} \left.\frac{dP_n}{d\theta}\right|_{\theta=\theta_k} \Big\} \quad ,$$

$$M = y_c J_0 = -\pi \sin\theta_k \rho_0 u_0 r^3 \sum_{n=1}^{N} [a_{11n}F_{1n}(r) + a_{21n}F_{2n}(r)] \left.\frac{dP_n}{d\theta}\right|_{\theta=\theta_k} \quad ,$$

$$\kappa = \frac{\gamma + 1}{\gamma - 1} \quad . \tag{7.64}$$

Considering the end element dr of the nozzle shown in Fig.7.28, we can obtain a relation between the increments in the side force and the moment. For a plane conic nozzle we have

$$dL = (p_2 - p_1)\cos\theta_k \, dr \quad , \tag{7.65}$$

where dL is the increment in the side force due to an increase in length of the nozzle's generatrix by dr and p_2 and p_1 are the values of pressure at the element dr at the upper and the lower halves respectively. The increment in the moment about the point 0 is equal to

$$dM = -(p_2 - p_1) r \, dr \quad . \tag{7.66}$$

A moment is considered positive if it is clockwise; a force is considered positive if it is directed along the x axis. Evidently,

$$r \, dL = -\cos\theta_k \, dM \quad . \tag{7.67}$$

This relation is valid for an axisymmetric conic nozzle too. It can be obtained by straightforward differentiation of L and M defined by (7.63) and (7.64). The relation (7.67) is exact and valid for any flow.

Since the numbers n defined by (7.61) are large, one can limit calculation of L and M from (7.63) and (7.64) by the first terms of the series. Using (4.91,92) and denoting u_0^2 by ψ, we get the following system of equations for L and M:

$$\frac{dL}{d\psi} = \frac{1}{4\psi} L[2 - n(n+1)] - \frac{1}{4\psi r} Mn(n+1)\cos\theta_k \quad , \tag{7.68}$$

$$\frac{dM}{d\psi} = -\frac{rL[2 - n(n+1)]}{4\psi \cos\theta_k} + \frac{Mn(n+1)}{4\psi} \quad . \tag{7.69}$$

For nondimensional functions,

$$L^0 = L/J_0 \quad , \quad M^0 = M\left(\frac{\gamma+1}{2\gamma} a_* r^* Q\right)^{-1} = z(u_0) y_c / r^* \quad .$$

We have

$$\frac{dL^0}{d\psi} = \frac{\cos\theta_k\sqrt{\psi}}{r^0(\psi + 1)}\frac{dM^0}{d\psi} - \frac{(\psi - 1)L^0}{2\psi(\psi + 1)} \quad , \tag{7.70}$$

$$\frac{dM^0}{d\psi} = -\frac{r^0(\psi + 1)[2 - n(n + 1)]L^0}{4\psi^{3/2}\cos\theta_k} + \frac{n(n + 1)M^0}{4\psi} \quad , \tag{7.71}$$

where r^* is the distance from the source centre to the nozzle throat, $r^0 = r/r^*$, $J_0 = [(\gamma + 1)/2\gamma]Qa_*z(u_0)$, $z(u_0) = u_0 + u_0^{-1}$ and Q is the total mass flow. The system (7.70,71) allows the side force and the moment to be found for various nonsymmetric perturbations both in the subsonic and the supersonic parts of a nozzle.

In the case of nonsymmetric flow in a cylindrical tube one can obtain, with the help of (4.102,7.56,57), finite relations for L^0 and ε^0, Fig.7.28:

$$L^0 = \frac{1.84\varepsilon_H^0}{\beta}\sin\left(\frac{1.84x^0}{\beta}\right) + L_H^0\cos\left(\frac{1.84x^0}{\beta}\right) \quad , \tag{7.72}$$

$$\varepsilon^0 = \varepsilon_H^0\cos\left(\frac{1.84x^0}{\beta}\right) - L_H^0\frac{\beta}{1.84}\sin\left(\frac{1.84x^0}{\beta}\right) \quad , \tag{7.73}$$

where $\varepsilon^0 = \varepsilon/r_0$, $x^0 = x/r_0$ and L_H^0 and ε_H^0 are the initial values of L^0 and ε^0 respectively [1.84 is the first root of (4.103) at $k = 1$].

The initial values of the side force L_H^0, the moment M_H^0 or ε_H^0 are necessary for integration of the system (7.70,71) and in order to be able to use (7.72,73). Generally, they are defined experimentally but in the case of relatively simple contour distortions, they can be found from geometrical considerations. For instance, in the case where some part of a nozzle is turned about the axis by an angle α, the linear theory gives for the cross-section AA' (Fig.7.28)

$$L_H^0 = -\left[\frac{2\gamma\alpha}{(\gamma + 1)z(u)}\right]_H \quad , \quad M_H^0 = \left[\frac{2\gamma\alpha r^0\cos\theta_k}{(\gamma + 1)}\right]_H \quad , \quad \varepsilon_H^0 = 0 \quad . \tag{7.74}$$

It follows from (7.72,73) that in a cylindrical tube L^0 and ε^0 oscillate with a constant amplitude defined by the value of the angle α. With the growth of the number M_0, zeros of the function L^0 move in the direction of larger x. Figures 7.29,30 show the dependence of the side force and the moment on x obtained by integration of (7.70,71) for conic nozzles with different θ_k and with the boundary conditions (7.74) when the subsonic part of a nozzle is turned at the throat by an angle α.

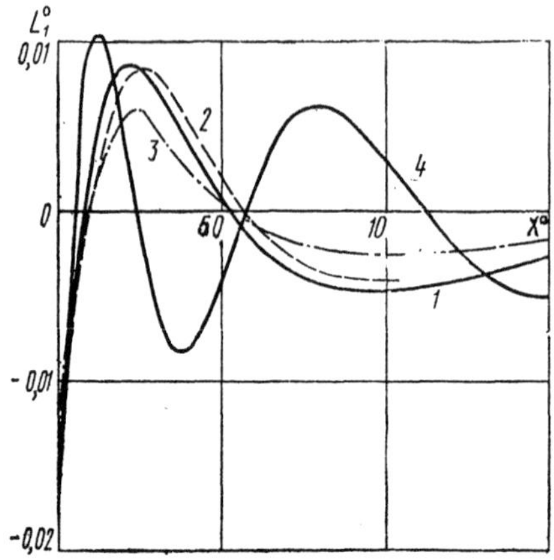

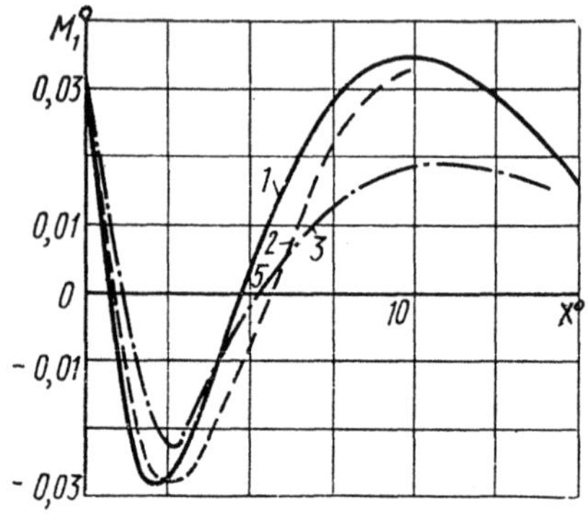

◄ Fig.7.29 ▲ Fig.7.30

Fig.7.29. Dependence of the side force on the length of a conic nozzle ($\gamma = 1.2$): (1) linear theory, $\theta_k = 10°$, (2) calculation [7.31], (3) calculated from (7.86), $\theta_k = 10°$, (4) linear theory, $\theta_k = 5°$ ($L_1^0 = [(\gamma+1)/2\gamma]z(u)(L^0/\alpha)$, α measured in degrees)

Fig.7.30. Dependence of the moment M_1^0 on the length of a conic nozzle ($\gamma = 1.2$): (1) linear theory, $\theta_k = 10°$, (2) calculation [7.31], (3) calculated from (7.87), $\theta_k = 10°$ ($M_1^0 = [(\gamma+1)/2\gamma]z(u)(M^0/\alpha)$, α measured in degrees)

The same figures allow comparison with the results of calculation using the three-dimensional method of characteristics [7.31]. To define the values of the side force L_a and the moment M_a developed by the stationary part of the nozzle AA'B'B in terms of L and M, one has to use the relations

$$L_a = L - L_H \quad , \quad M_a = M - M_H \quad . \tag{7.75}$$

Let us analyze the results of the calculations. At the cross-section AA' at $x = 0$ in the case of the turn shown in Fig.7.28, the side force L_H is negative. At the initial part of the contour AB, an elevated pressure develops under the impact of the jet, while at the initial part of the contour A'B' the pressure is evidently lower. The resulting positive side force L_a diminishes the initial negative side force L_H. If the length of the nozzle increases, there appears a point at which $|L_H| = |L_a|$. Since the forces have opposite signs, the resulting side force vanishes. In Fig.7.29 such points correspond to zeros of the function L, when the total reactive force vector is directed exactly along the x axis. With further increase in the nozzle's length the positive force L_a continues to grow, but only until the compression waves from the contour AB reach the contour A'B' and the rarefaction waves propagating from the contour A'B' reach the contour AB. From this moment on, the positive force L_a decreases and there may again appear points where $|L_a| = |L_H|$ and $L = 0$, while in the course of further increase in the length, the signs of L and L_H will coincide. Thus, the oscillatory character of the change in L is evident from physical considerations. It is also clear that the number of zeros of the function L is greater, the smaller the angle θ_k. Unlike the flow in a cylindrical tube, the amplitude of the oscillations decreases. Oscillatory character is typical for the moment M too; however, the increasing arm of the moment, i.e. the distance from the centre of the source, results in an increase of the amplitude of the oscillations as the nozzle length increases. The data presented in Figs.7.29,30 demonstrate the high accuracy provided by the linear theory. It is important that in the framework of the linear theory the amplitude of the perturbations is proportional to the angle α, while the zeros of the functions L and M do not depend on α and it is enough to calculate at only one of its values. At the same time determination of L and M in the framework of the linear theory is quite elementary and reduces to numerical integration of the ordinary differential equations (7.70,71). From this point of view, the linear theory has an advantage in comparison with a numerical solution. Note that the linear theory for the calculation of side forces and moments based upon linearization with respect to a one-dimensional flow has been developed in [7.41,42].

It would be interesting to get an approximate analytical solution for the side force and the moment in the sub- and supersonic parts of a nozzle. If we preserve the rule of signs adopted above, i.e. consider the side force positive if directed along the y axis, and the moment positive if it is clockwise about the sink centre, then the same system of equations can be used to calculate the side force and the moment upon reversing the signs of the left side of (7.68) and the first term on the right side of (7.69). Then, substituting the independent variable ψ by r and neglecting, at low subsonic velocities, $\psi = u^2$ compared to 1, one can easily obtain, with the help of (4.91,92,7.63,64), the following relationships for the determination of the side force and the moment about the sink centre:

$$L = L_H\left(\frac{r}{r_H}\right)^{n-1} \quad , \quad M = -\frac{r_H L_H[2 - n(n+1)]}{n^2 \cos\theta_k}\left(\frac{r}{r_H}\right)^n \quad , \tag{7.76}$$

where L_H is the initial value of the side force specified at $r = r_H$. It follows from (7.76) that in the subsonic part the side force and the moment decrease as a power law while approaching the throat. It is also evident that the relation

$$M = CrL \tag{7.77}$$

is approximately valid (here the value of the constant C depends on the initial conditions and the angle θ_k). To construct an approximate solution in the supersonic part we shall use the results of [7.43] where a method of constructing an asymptotic expansion of the solution of a second-order differential equation in terms of negative powers of n has been proposed. Let us write down the system (7.68,69) in two equivalent forms:

$$(\kappa - \psi)\frac{d^2L}{d\psi^2} + \left[\frac{1}{2(\gamma - 1)} + \frac{\kappa - \psi}{4\psi}\right]\frac{dL}{d\psi} + \frac{\kappa(\psi - 1)}{16\psi^2}[n(n + 1) - 2]L = 0 \quad , \tag{7.78}$$

$$(\kappa - \psi)\frac{d^2M}{d\psi^2} + \left[-\frac{1}{2(\gamma - 1)} + \frac{3(\kappa - \psi)}{4\psi}\right]\frac{dM}{d\psi} + \frac{\kappa(\psi - 1)}{16\psi^2}\, n(n + 1)M = 0 \quad , \tag{7.79}$$

and construct, in the manner of [7.43], an asymptotic expansion of the solution of the above equations in terms of negative powers of n near the singularity $\psi = \kappa$. It can be shown that to an accuracy of n^{-1} the approximate solution of (7.78,79) has the form

$$L = \frac{n_1\tilde{M}(\psi)}{r}\sqrt{\frac{\omega(\psi)}{\omega'(\psi)}}\left\{C_1 J_{+(p_0+1)}[2n_1\sqrt{\omega(\psi)}] + C_2 J_{-(p_0+1)}[2n_1\sqrt{\omega(\psi)}]\right\} \quad , \tag{7.80}$$

$$M = n_2\tilde{M}(\psi)\sqrt{\frac{\omega(\psi)}{\omega'(\psi)}}\left\{\bar{C}_1 J_{+(1-p_0)}[2n_2\sqrt{\omega(\psi)}] + \bar{C}_2 J_{-(1-p_0)}[2n_2\sqrt{\omega(\psi)}]\right\} \quad , \tag{7.81}$$

where $p_0 = 0.5(\gamma - 1)^{-1}$, C_1, C_2, $\bar{C}_1$, $\bar{C}_2$ are arbitrary constants, $J_{+(p_0+1)}$, $J_{-(p_0+1)}$, $J_{+(1-p_0)}$, $J_{-(1-p_0)}$ are the Bessel functions of orders $p_0 + 1$ and $1 - p_0$,

$$n_1 = \sqrt{n(n + 1) - 2} \quad , \quad n_2 = \sqrt{n(n + 1)} \quad ,$$

$$\tilde{M}(x) = (\kappa - \psi)^{-p_0/2}\psi^{-3/8} \quad , \tag{7.82}$$

$$\omega(\psi) = \left[\frac{1}{4}(f(\psi) - f(\kappa))\right]^2 \quad , \quad \omega' = \frac{d\omega}{d\psi}$$

and $f(\psi)$ is the Prandtl-Meyer function defined by (2.151). We remember that $\psi = u_0^2 = \lambda^2$. Using further asymptotic representations of the Bessel function at large values of the argument [7.44], we finally get from (7.80,81) analytic expressions for the side force and the moment:

$$L = \hat{L}[C_{1L}\ \sin\delta_1(\psi) + C_{2L}\ \cos\delta_1(\psi)] \quad , \tag{7.83}$$

$$M = \hat{M}[C_{1M}\ \sin\delta_2(\psi) + C_{2M}\ \cos\delta_2(\psi)] \quad , \qquad \text{where} \tag{7.84}$$

$$\hat{L} = \frac{\tilde{M}(\psi)}{r}\sqrt{\frac{2n_1}{\pi f'(\psi)}} = 2\sqrt{\frac{n_1}{\pi}}\ \frac{(\kappa-\psi)^{\gamma/4(\gamma-1)}}{\kappa^{\frac{1}{4}}(\psi-1)^{\frac{1}{4}}}\ \psi^{3/8} \quad ,$$

$$\hat{M} = r\hat{L} \quad , \quad \delta_1(\psi) = \frac{1}{2}\,n_1 f(\psi) \quad , \quad \delta_2(\psi) = \frac{1}{2}\,n_2 f(\psi) \quad , \tag{7.85}$$

$$f'(\psi) = \frac{1}{2\psi}\left(\frac{\psi-1}{1-\kappa^{-1}\psi}\right)^{\frac{1}{2}}$$

and C_{1L}, C_{2L}, C_{1M}, C_{2M} are arbitrary constants. It follows from (7.73,84) and (7.68,69) under the initial conditions (7.74) that at large n, $M \approx CrL$ [compare with (7.67)], i.e. the point of application of the side force lies in the plane of the exit cross-section of the nozzle, and knowing the value of the side force, one can define the moment about any point. It means also that zeros of the side force and the moment are close to each other (see Fig.7.29, 30) and hence in those planes of a nozzle where $L \approx M \approx 0$, the force vector is parallel to the nozzle axis and directed along it.

Let the initial values of the side force L_H and the moment M_H be known at a certain section of a nozzle and, due to (7.74), related by the equation $M_H = -r_H L_N \cos\theta_k$. Then, from (7.68,69,61) at large n we have

$$\left(\frac{dL}{d\psi}\right)_H \approx -\frac{9\pi^2}{64\psi_H}\ \frac{n(n+1)}{(n+3)^2}\,L_H \quad , \quad \left(\frac{dM}{d\psi}\right)_H = -r_H\left(\frac{dL}{d\psi}\right)_H \quad .$$

Knowing M_H, L_H, $(dL/d\psi)_H$, $(dM/d\psi)_H$, one can define the arbitrary constants in (7.83,84). Finally, for the calculation of L and M, we have to an accuracy of $n^{-3/2}$,

$$L = L_H\,\frac{\hat{L}(\Psi)}{\hat{L}(\psi_H)}\ \cos[\delta_1(\psi) - \delta_{1H}] \quad , \tag{7.86}$$

$$M = M_H\,\frac{\hat{M}(\psi)}{\hat{M}(\psi_H)}\ \cos[\delta_2(\psi) - \delta_{2H}] \quad , \tag{7.87}$$

where δ_{1H}, δ_{2H} are the initial values of functions $\delta_1(\psi)$, $\delta_2(\psi)$ defined by (7.85). It is evident from (7.86,87) that the side force and the moment have an oscillatory character with an attentuating amplitude of oscillations of the function L [due to the term $L(\psi)$]. The zeros ψ_k of the functions L and M are found from the relation

$$f(\psi_k) = f(\psi_H) + \frac{\pi}{n}(2k + 1) \quad , \quad (k = 0,1,\ldots n) \quad , \tag{7.88}$$

which shows that the number of zeros increases with the decrease of γ and θ_k (i.e. with the growth of n). Also, the first zero of the functions L and M, corresponding to $k = 0$, moves downstream with the growth of the angle θ_k.

In Figs.7.29,30 the above approximate solution is compared to the exact one. It is evident that the approximate solution predicts zeros of the functions L and M accurately enough. The maximum values of L and M are in error by 10-15%.

Concluding the section, let us compare the results of calculations for a gas flow in a conic nozzle using the method of small perturbations and the method of grids [7.9]. The initial data in the form of perturbations of the radial u and the circumferential v velocity components were found from the results of exact calculations. These perturbations were specified at the circular arc CD ($r = r_H$) placed downstream from the characteristic AB which limits the expansion flow generated in the flow of the "radius" part of the nozzle contour, Fig.7.31. Five terms of the series (4.99) proved to be sufficient to approximate the initial perturbations.

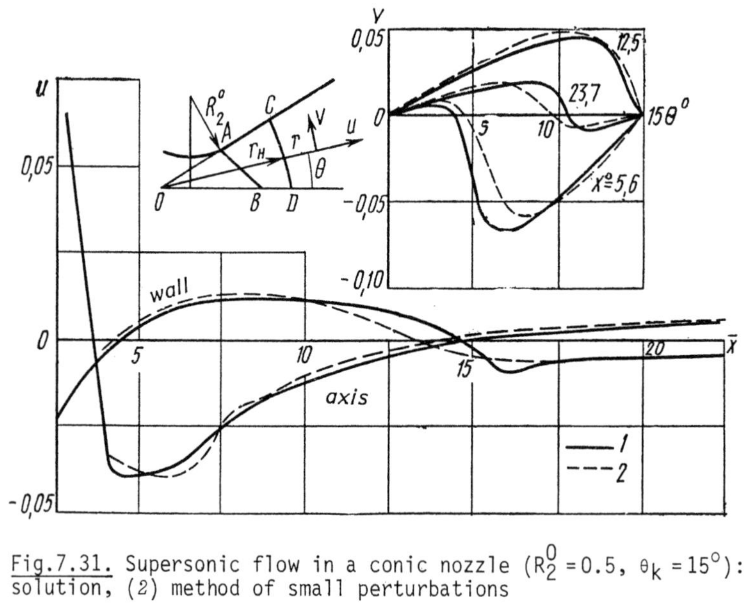

Fig.7.31. Supersonic flow in a conic nozzle ($R_2^0 = 0.5$, $\theta_k = 15°$): (1) numerical solution, (2) method of small perturbations

Calculations done by both methods demonstrate good agreement of the results. Discrepancy in the perturbations of the velocity components u and v do not exceed 10% in the flow field and 5% at the nozzle axis and its contour, Fig.7.31. There is some difference in the position of the zeros of the functions and in the character of the curves in the regions where shock waves terminate. Here the exact theory predicts discontinuities in the derivatives of the flow parameters which are poorly approximated by the theory of small perturbations, because the latter assumes analyticity of the solution and absence of swirl. However, generally these influences are weak, because the intensity of shock waves is small. Since perturbations of the flow parameters are calculated rather accurately, the total velocities will be defined with still higher precision. The results of the studies show that the use of the method of small perturbations is quite justified for calculations of flows in conic nozzles with $\theta_k < 20^\circ$ [7.9]. The conclusion is of principal importance and allows the method of small perturbations to be used for calculation of slightly perturbed three-dimensional flows in conic nozzles.

Let us consider selecting initial data for the method of perturbations, i.e. selecting the side force and the moment. At given distortions this is usually a simpler task compared to the determination of the parameter field at the initial cross-section, needed, for instance, for numerical integration of three-dimensional equations of gas dynamics.

Let us next consider the two most typical kinds of distortion of a nozzle: misalignment of the nozzle axis characterized by the angle α and its displacement h. We shall assume that at a cross-section AB (Fig.7.32a) which has radius r_0 and is situated in the supersonic part of a nozzle the vector $\mathbf{R}_0$ is displaced by ε_0 and turned through the angle $\delta = \arcsin(L_0/R_0)$. The turning of the nozzle axis through the angle α results in the formation of a transitory segment ABCD. This is needed to determine the side force L and the thrust displacement ε (or the moment about the centre of the source) at the cross-section AD, which may be used as initial data for calculating the propagation of disturbances downstream of the cross-section AD. The side force at the cross-section AD is equal to the sum of the projections of all the forces acting upon a nozzle up to the cross-section AD:

$$L = J_0 \sin\alpha + L_0 \cos\alpha + L_1 \cos\alpha + J_1 \sin\alpha + L_2 \quad .$$

We shall assume that at the region ABC the nozzle walls are subjected to the pressure p_0 which is constant through the segment and equal to the pressure at the cross-section AB. Then, in the spherical system of coordinates

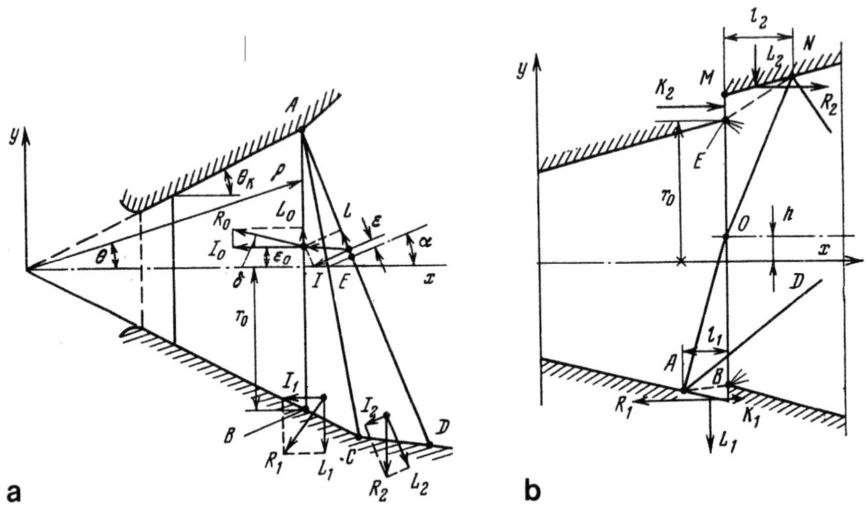

Fig.7.32a,b. Diagram of forces in the case of misalignment and displacement of nozzle parts

ρ, θ, φ the force L_1 is equal to

$$L_1 = -\pi r_0^2 p_0 \tan\frac{\alpha}{2}\left(1 - \theta\tan\frac{\alpha}{2}\right)$$

or (to the accuracy of $\alpha^2\theta$) $L_1 = -\pi r_0^2 p_0 \tan(\alpha/2)$. Under the same assumptions the longitudinal force acting upon the segment ABC is equal to $J_1 = -\pi r_0^2 p_0 \tan\theta\tan\alpha$.

At the crossing of the plane AC the pressure at the nozzle wall changes. It may be assumed that at small values of α, the pressure drop in the rarefaction wave issuing from the point A is equal to the pressure increase in the shock wave issuing from the point C while the flow stays attached. We shall assume that the pressure varies in the circumferential direction according to $p = p_0(1 - \Delta\cos\varphi)$, where $\Delta = (p_C - p_0)/p_0$ is the dimensionless pressure build-up in the shock wave issuing from the point C. It will be also assumed that the pressure stays constant along the nozzle generatrix in the region ACD. Upon integrating we conclude that the force L_2, to the accuracy of $\alpha^2\theta$, is equal to

$$L_2 = -\pi r_0^2 p_0 \tan\frac{\alpha}{2}(1 + \Delta) \quad .$$

Then, the side force at the cross-section AD is defined by the expression

$$L = J_0\sin\alpha + L_0\cos\alpha - \pi r_0^2 p_0\Big[\tan\frac{\alpha}{2}\cos\alpha + \tan\frac{\alpha}{2}(1+\Delta)$$

$$- \tan\theta\sin^2\alpha\Big] \approx J_0\alpha + L_0 - \pi r_0^2 p_0\left(\alpha + \frac{1}{2}\alpha\Delta\right) ,$$

or, in the dimensionless form,

$$L^0 = L_0^0 + \alpha\left(1 - \frac{\Delta}{2\gamma M_0^2}\right)\left(1 + \frac{1}{\gamma M_0^2}\right)^{-1} \tag{7.89}$$

where M_0 is the Mach number value at the cross-section AB. At small values of α the pressure increase at point C at the shock may be obtained from the compatibility equation at a characteristic. We have $\Delta = \alpha\gamma M_0^2(M_0^2 - 1)^{-\frac{1}{2}}$; then,

$$L_0 = L_0^0 + \alpha\,\frac{1 - 0.5\alpha(M_0^2 - 1)^{-\frac{1}{2}}}{1 + (\gamma M_0^2)^{-1}} . \tag{7.90}$$

Equations (7.89,90) may be used for an attached supersonic flow at the transition segment ABCD.

Let us determine the thrust eccentricity ε at the cross-section AD. Having done the necessary transformations, we get for small values of the angle α (to the accuracy of α^2),

$$\varepsilon^0 = \frac{\varepsilon}{r_*} = \varepsilon_0^0 - r_0^0\alpha\left[L_0^0 + \frac{\tan\theta_k(1 + 0.5\Delta)}{1 + \gamma M_0^2}\right] , \tag{7.91}$$

where $r_0^0 = r_0/r_*$ and $\varepsilon_0^0 = \varepsilon_0/r_*$ is the thrust eccentricity at the cross-section AB. The values of M^0 and ε^0 are referred to the moment about the centre of the throat section $M_* = -z(\psi)y_*/r_*$ (Fig.7.28) by the relations

$$M^0 = -\frac{1}{\sqrt{\mu}}\left[\frac{2\gamma M_*}{\pi(\gamma + 1)}\sin\theta - z(\psi)L^0\cos\theta\right] ,$$

$$\varepsilon^0 = -\frac{2\gamma M_*}{\pi(\gamma + 1)z(\psi)} + x^0 L^0 ,$$

where μ is the flow rate coefficient and $x^0 = x/x_*$.

Let us consider another typical distortion, a parallel displacement of the axis h (Fig.7.32b). In this case, in front of the shelf and behind it, separation zones form whose lengths in the plane of displacement are equal to ℓ_1 and ℓ_2 and the corresponding pressures are p_1 and p_2. The difference between these pressures and the pressure p_0 in the undisturbed flow results in the generation of additional forces $L_1, L_2, R_1, R_2, K_1, K_2$. Reducing all the forces to the point 0 lying at the axis of the subsequent segment in the plane of displacement, we get

$$L = L_0 - L_1 - L_2 \quad , \qquad \varepsilon = \varepsilon_0 + h_0 + \varepsilon_g \quad .$$

Here ε_0 and L_0 are the displacement and the side force of the preceding segment generated by the flow asymmetry independent of the axis displacement under consideration, ε_g is the thrust vector displacement equivalent to the torque produced by transferring the above forces to the point 0 and h_0 is the displacement of the thrust vector of the preceding segment with respect to the axis of the subsequent segment, which depends on the axes' displacement. The lower part of the flow in front of the displaced cross-section turns to the axis in the shock wave AD; therefore the value of h_0 is smaller than that of h. The distance between the axes of the cross-sections EB and MB is equal to h/2; therefore, one may assume in the first approximation (taking into account the rule of signs) that $h_0 = -h/2$. A more precise value of h_0 may be obtained experimentally.

Suppose that the boundary of the lower separation zone lies in the plane AO and that of the upper one lies in the plane NO. Assuming that the pressure is constant inside the separation zone, let us find the value of all additional forces and torques produced by transferring the forces to the point 0. As a result, we obtain the values of L^0 and ε^0 which may be used as initial data for calculating the asymmetric flow downstream of the cross-section MB:

$$L^0 = L_0^0 - \frac{\Delta_1 \ell_1^0 - \Delta_2 \ell_2^0}{2r_0^0(1 + \gamma M_0^2)} \frac{h}{|h|} f \quad ,$$

$$\varepsilon^0 = \varepsilon_0^0 - \frac{h^0}{2} + \frac{2}{\pi(1 + \gamma M_0^2)} \left[\frac{\Delta_1 (\ell_1^0)^2 - \Delta_2 (\ell_2^0)^2}{3r_0^0} \frac{h}{|h|} + h^0(\Delta_1 + \Delta_2) - \tan\theta(\Delta_1 \ell_1^0 - \Delta_2 \ell_2^0) \right] \quad . \tag{7.92}$$

Here $\Delta_1 = (p_1 - p_0)/p_0 = \text{const} > 0$, $\Delta_2 = (p_2 - p_0)/p_0 = \text{const} < 0$, $\ell_1^0 = \ell_1/r_*$, $\ell_2^0 = \ell_2/r_*$, $h^0 = h/r_*$ and f is the coefficient which takes into account the difference between the schematic and real flow patterns.

An analogous method of calculating the initial data, by means of reducing the excessive pressure forces to one cross-section, may be used for other kinds of distortions too, in particular for local asymmetric cavity-like and bulge-like distortions. The above scheme of calculating initial data may be used mainly in the supersonic part of a nozzle. In the subsonic part the use of a similar approach is complicated by the fact that here, as a rule, it is difficult to determine boundaries of the excessive pressure zones produced by the contour distortion and the pressure value itself cannot always be obtained reliably. However, requirements on the accuracy of initial data may be weaker here, since, as has been shown above, distortions in the subsonic part of a nozzle die out rapidly.

7.4.3 Numerical Investigation of Three-Dimensional Nozzle Gas Flows

Below we discuss the results of calculating nozzle flows by the method of small perturbations and the difference method analogous to the one described in Sect.3.2.2 [7.35,36]. Solution of the problem in the full nonlinear formulation is quite a difficult task; therefore, the difference method was not used for parametric studies. Due to its high accuracy, the method has been used for calculating a small number of variants in order to find the applicability boundaries of the method of small perturbations and to obtain solutions when the latter method is inapplicable.

The major features of the difference method are as follows. Let us consider the flow of a perfect gas in a nozzle, generally with a central body. We shall use the cylindrical system of coordinates x, r, φ and assume that the flow has at least one symmetry plane. The region of calculation is bounded by the walls of the nozzle and of a central body described by the equations $r=F(x,\varphi)$, $r=G(x,\varphi)$, by the cross-section $x=x^0$, the plane $\varphi=0$ and the nearest symmetry plane $\varphi=\varphi^*$. At the section $x=x^0$ all the values are considered known, at the surfaces $r=F(x,\varphi)$ and $r=G(x,\varphi)$ the impermeability condition,

$$v = uF_x + wF_\varphi/F \quad , \qquad v = uG_x + wG_\varphi/G \tag{7.93}$$

is put and at the planes $\varphi=0$, $\varphi=\varphi^*$ the symmetry conditions are used. Changing to the variables

$$x = x \quad , \qquad \xi = \frac{r - G(x,\varphi)}{F(x,\varphi) - G(x,\varphi)} \quad , \qquad \varphi = \varphi \quad ,$$

the region transforms into a rectangle limited by $x \geqslant x^0$, $0 \leqslant \xi \leqslant 1$, $0 \leqslant \varphi \leqslant \varphi^*$, while the system of gasdynamic equations in divergent form becomes

$$\frac{\partial \mathbf{A}}{\partial x} + \frac{\partial \mathbf{B}}{\partial \xi} + \frac{\partial \mathbf{C}}{\partial \varphi} + \mathbf{D} = 0 \quad , \tag{7.94}$$

where **A**, **B**, **C**, **D** are vectors with components A_k, B_k, C_k, D_k, $(k=1,\ldots,5)$:

$$A_k = P_k\delta r \quad , \quad B_k = Q_k r - P_k r(G_x + \xi\delta_x) - R_k(G_\varphi + \xi\delta_\varphi) \quad , \quad C_k = R_k\delta \; ,$$

$$P_1 = \rho u^2 + p \quad , \quad Q_1 = \rho uv \quad , \quad R_2 = \rho uw \quad , \quad D_1 = 0 \quad ,$$

$$P_2 = \rho uv \quad , \quad Q_2 = \rho v^2 + p \quad , \quad R_2 = \rho vw \quad , \quad D_2 = -(\rho w^2 + p)\delta \quad ,$$

$$P_3 = \rho uw \quad , \quad Q_3 = \rho vw \quad , \quad R_3 = \rho w^2 + p \quad , \quad D_3 = \rho vw\delta \quad ,$$

$$P_4 = \rho u \quad , \quad Q_4 = \rho v \quad , \quad R_4 = \rho w \quad , \quad D_4 = 0 \quad ,$$

$$P_5 = \rho uh \quad , \quad Q_5 = \rho vh \quad , \quad R_5 = \rho wh \quad , \quad D_5 = 0 \quad ,$$

$$h = \frac{\gamma}{\gamma - 1}\frac{p}{\rho} + \frac{1}{2}W^2 \quad , \quad \delta = F(x,\varphi) - G(x,\varphi) \quad ,$$

and u, v, w are the velocity components of W along x, r, φ.

The method is based on the application of an explicit difference scheme of second order of accuracy and a smoothing procedure; it belongs to the class of shock-smearing methods. The following net is introduced:

$$x^n = x^0 + n\Delta x \quad , \quad \xi_i = i\Delta\xi \quad , \quad \varphi_j = j\Delta\varphi \quad , \quad n = 0,1,\ldots,$$

$$i = 0,1,\ldots,M \quad , \quad j = 0,1,\ldots,N \quad , \quad \Delta\xi = 1/M \quad , \quad \Delta\varphi = \varphi^*/N$$

and semiinteger nodes $n + \sigma$, $i \pm 1/2$, $j \pm 1/2$. Calculation proceeds along x and is done in layers of $x = \text{const}$. Let us write down difference equations for (7.94). For a node $x^{n+\sigma}$, $\xi_{i+1/2}$, $\varphi_{j+1/2}$ we have

$$\begin{aligned} A^{n+\sigma}_{i+1/2,j+1/2} = A^n_{i+1/2,j+1/2} &- 0{,}5\sigma \frac{\Delta x}{\Delta\xi}(B^n_{i+1,j} - B^n_{i,j} + B^n_{i+1,j+1} \\ &- B^n_{i,j+1}) - 0{,}5\sigma \frac{\Delta x}{\Delta\varphi}(C^n_{i,j+1} - C^n_{i,j} + C^n_{i+1,j+1} - C^n_{i+1,j}) \\ &- \sigma\Delta x D^n_{i+1/2,j+1/2} \quad , \end{aligned} \tag{7.95}$$

where $\mathbf{A}^n_{i+1/2,j+1/2}$, $D^n_{i+1/2,j+1/2}$ are the values of $\mathbf{A}$ and $\mathbf{D}$ averaged over the four extreme nodes. Upon calculating $\mathbf{A}^{n+\sigma}_{i+1/2,j+1/2}$, one may obtain the values of u, v, w, p, ρ at a node $n+\sigma$, $i+1/2$, $j+1/2$. Similarly, the values at the nodes $i+1/2$, $j-1/2$; $i-1/2$, $j+1/2$; $i-1/2$, $j-1/2$ in the layer $n+\sigma$ are calculated. Using the "cross" scheme we obtain a system of equations for calculating $A_1,\ldots,A_5$ at the node $n+1$, i, j which allow u, v, w, p, ρ to be calculated.

In order to calculate the values of the parameters at the wall $r = F(x,\varphi)$, a one-sided scheme accurate to first order has been used. Let us consider (7.94) for the components $K = 1,3,4,5$. For them the difference equations have the form

$$(A_k)^{n+1}_{M,j} = (A_k)^n_{M,j} - \frac{\Delta x}{\Delta \xi}\left[(B_k)^n_{M,j} - (B_k)^n_{M-1,j}\right]$$

$$- 0{,}5\,\frac{\Delta x}{\Delta \varphi}\left[(C_k)^n_{M,j+1} - (C_k)^n_{M,j-1}\right] - \Delta x (D_k)^n_{M,j}\ , \quad k = 1,3,4,5\ . \tag{7.96}$$

Obtaining $(A_k)^{n+1}_{M,j}$, $(k=1,3,4,5)$, from (7.96) and using the first of the boundary conditions (7.93), $v^{n+1}_{M,j} = (uF_x + wF_\varphi/F)^{n+1}_{M,j}$, one can then find u, v, w, p, ρ at the at node $n+1$, M, j. The values of parameters at the surface of the central body, where the second of the boundary conditions is valid, are calculated analogously.

In order to calculate the flow parameters on the axis of the cylindrical system of coordinates, equations written in Cartesian coordinates are used.

Since shock waves may form in the flow, a smoothing procedure has been applied to the difference solution at each layer using five points in each direction. Smoothing has been applied not to the functions themselves, but to the complexes A_k; then, the final values of the gasdynamic functions at the layer $n+1$ were calculated. The size of Δx has been obtained from the Courant condition corresponding to variation of parameters along rays.

The main calculations of symmetric gas flows in conical nozzles have been carried out using the theory of slightly perturbed flows, using numerical solutions of (7.70,71). The distortion has been specified by rotating the nozzle axis at the throat through an angle of $\alpha = 1^\circ$. Note that in the framework of the linearized theory the amplitude of disturbances is proportional to α and zeros of the functions L and M do not depend on α. Therefore, the results of calculations may be used for other values of α too. The initial side force L^0_H is obtained from (7.89) where $L^0_0 = 0$ and $\Delta = 0$, since in the flow under consideration the shock wave degenerates into a characteristic. The initial value of the thrust eccentricity ε^0_H has been assumed to be zero. The calculations have been done in the interval of $\theta_k = 5\text{-}25^\circ$ at $\gamma = 1.2$ and 1.4.

Solid lines in Figs.7.33,34 show the values of L^0, M^0, ε^0 at $M_0 = 1$, $\gamma = 1.2$ for a nozzle contour at the initial segment of the supersonic part without any round-off ($R^0_2 = R_2/r_* = 0$). The variation of the side force is oscillatory and its amplitude dies out along the length of a nozzle, while the number of zeros increases with decreasing angle θ_k. The oscillatory character of the functions L^0, M^0 and ε^0 is associated with the subsequent reflections of the compression and rarefaction waves from the nozzle walls. It should be noted that at the cross-sections with $L^0 = 0$, maximum values of ε^0 are attained and, hence, the total moment is nonzero.

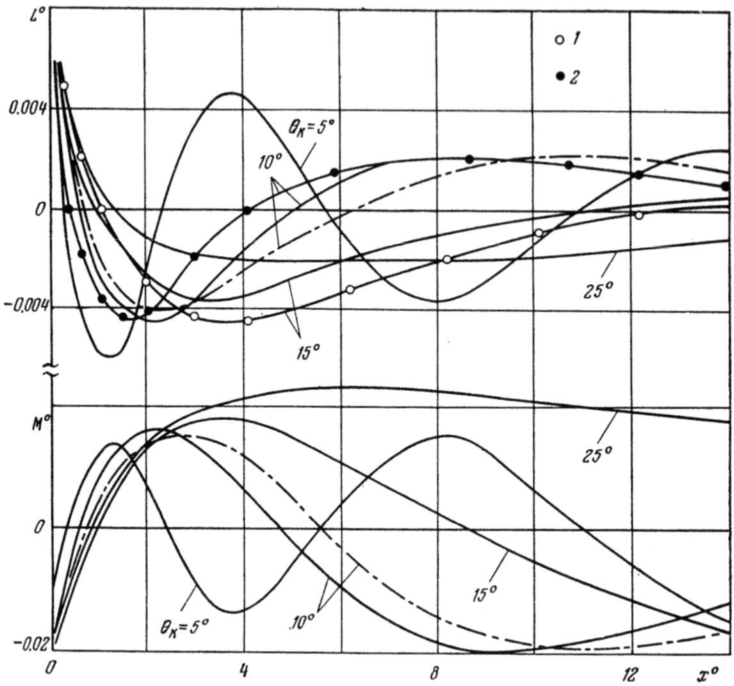

Fig.7.33

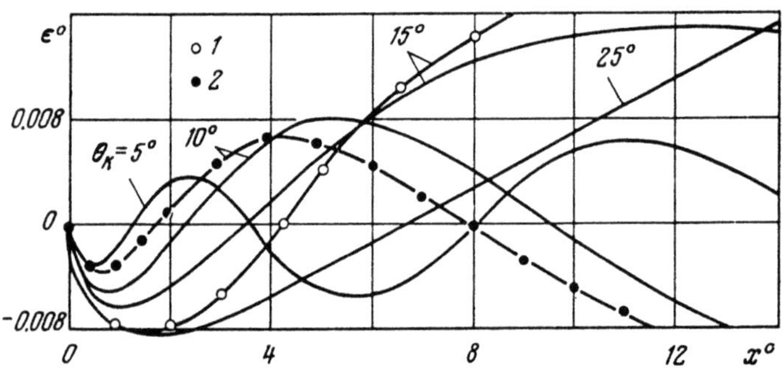

Fig.7.34

Fig.7.33. Longitudinal side force and moment variations for a conical nozzle: (1) M_0=1.2, γ=1.2, R_2^0=0, (2) M_0=1, γ=1.2, R_2^0=2, (-·-) M_0=1, γ=1.4, R_2^0=0

Fig.7.34. Longitudinal thrust displacement variation for a conical nozzle (see caption to Fig.7.33)

An increase in the nozzle length results in increasing amplitudes of oscillation of the functions M^0 and ε^0. Zeros of the functions L^0 and M^0 are slightly displaced in relation to each other, as one may see from (7.70,71). An increase in γ results in the displacement of the functions L^0 and M^0 to the right along the axis x, though the amplitude of oscillation changes insignificantly.

It is known that increasing the round-off radius of a nozzle contour in its supersonic part R_2^0 results in a displacement of the zeros of the functions L^0 and ε^0. This result is confirmed by calculations done with the help of the linearized theory when a round-off segment is replaced by a number of adjacent conical segments. With increasing R_2^0, the zeros of the functions L^0 and ε^0 move to the left and the amplitude of oscillations decreases. An increase of the angle θ_k results in the influence of R_2^0 becoming greater. Increasing M_0 results in the zeros of the functions L^0 and ε^0 being moved to the right.

Note that analogous data on the influence of the parameters γ, R_2^0 and M^0 obtained in [7.31] by the three-dimensional method of characteristics and by the above difference method agree well with the calculations using the method of small perturbations.

Now let us consider a flow in a cylindrical duct. Figure 7.35 presents L^0 (solid lines) and ε^0 (dashed lines). The numbers 1,2,4 correspond to $\gamma = 1.2$ and $M_0 = 1.05$, 1.2, 3.0, while 3 corresponds to $\gamma = 1.4$, $M_0 = 1.2$. The initial distortion is given as the axis misalignment, characterized by $\alpha = 1^\circ$. Just as in a conical nozzle, increasing the Mach number results in increasing the distance between the zeros of the functions L^0 and ε^0. Unlike the flow in a conical nozzle, the amplitudes with which the functions oscillate stay constant.

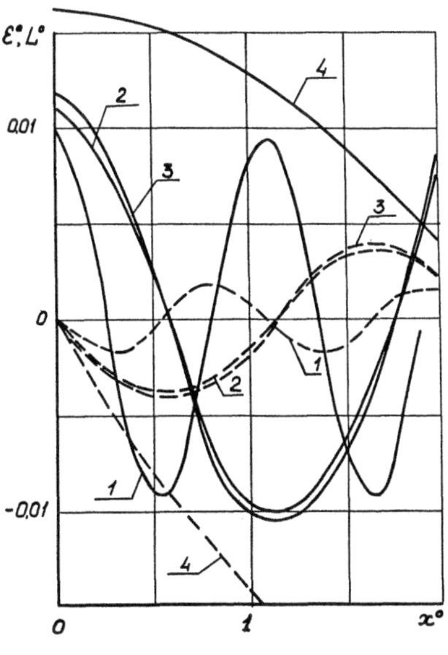

Fig.7.35. Longitudinal side force and thrust displacement variations for a duct: (*1*) $M_0 = 1.05$, $\gamma = 1.2$, (*2*) $M_0 = 1.2$, $\gamma = 1.2$, (*3*) $M_0 = 1.2$, $\gamma = 1.4$, (*4*) $M_0 = 3$, $\gamma = 1.2$

The data presented in [7.31] give practically the same dependence as that defined by (7.72,73).

The example of a cylindrical duct helps one to analyze the effect of the initial data upon the propagation of disturbances. The presence of the initial thrust eccentricity together with the side force results in increasing amplitudes of oscillation and in the zeros of the functions L^0 and ε^0 being moved to the right or to the left depending on whether the signs of L_H^0 and ε_H^0 coincide or not. The distance between zeros of the functions L^0 and ε^0 does not depend on the initial conditions. These conclusions are valid for nozzles too, which confirms the importance of specifying L_H and ε_H correctly.

The theory of slightly perturbed flows may also be used for studying asymmetric gas flows in profiled nozzles. This can be done by replacing the contour of a profiled nozzle by a broken line composed of segments whose inclinations are equal to the local inclinations of the contour. Results of calculations based on such an approach for conical nozzles with the initial part of the supersonic contour formed by a circular arc have been described above and have proved to be close to the results in [7.31]. However, an increase of the angle θ_k resulted in increasing differences in the results of calculations, while large contour inclinations are typical for profiled nozzles in the vicinity of the throat. The same approach has been used to calculate gas flows in profiled nozzles [7.31] using Rao's method [7.45] and the difference method described above. Figure 7.36 illustrates good agreement between these calculated results (solid lines) and the data of [7.31] (dashed lines). The substantial influence of $\bar{R}_2^0$ upon L^0 is connected with the presence of large gradients in the distributions of parameters in the transverse sections of a profiled nozzle, first of all in the region adjacent to the throat.

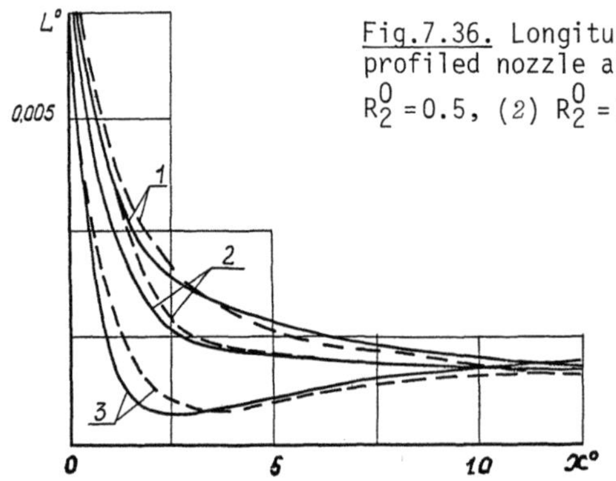

Fig.7.36. Longitudinal side force variation for a profiled nozzle at $M_0 = 1.1$, $\gamma = 1.2$, $\alpha = 1^0$: (1) $R_2^0 = 0.5$, (2) $R_2^0 = 1.0$, (3) $R_2^0 = 2.0$

Calculations of flows in profiled nozzles done with the help of the linearized theory give results which differ considerably from those presented in Fig. 7.36. Replacing the initial part of a nozzle by various equivalent cones or a cylinder gave no satisfactory results. A much better coincidence of results has been obtained in the case when calculations were started not from the critical cross-section, but from a certain cross-section in the supersonic region. The calculated values of L^0 and ε^0 presented in Fig.7.36 were used as the initial data. Propagation of disturbances in profiled nozzles may be evaluated approximately with the help of the linearized theory in the case where a contour is distorted downstream of the section $x^0 = x/r_* \approx 2\text{-}3$.

7.4.4 Experimental Studies of Side Forces and Moments

Below we present the results of experimental studies taken from [7.35,36]. Special attention has been paid to the accuracy of the experimental data obtained, to the qualitative characteristics of the initial distortions and to the preservation of identical conditions in tests for different kinds of distortions and different nozzles. The side force L may be conveniently studied. Coincidence of the calculated and experimental dependence of the side force on a nozzle's length also indicates the validity of the calculation for the remaining flow characteristics. Therefore, the present study has been dedicated mostly to measuring the side force for various kinds of contour distortions in conical and profiled nozzles.

The tests have been carried out at a differential installation described in [7.46]. In the process of the tests, the working medium (cold air) was fed in equal quantities into the reference and the tested nozzles installed along the same axis and looking in opposite directions. Thus, the measuring device of the installation is subject to the action of only a small difference of the nozzles' thrust which, when added to the known thrust of the reference nozzle, allows one to obtain the axial thrust of the tested nozzle. During the tests strain gauges measured the side force acting in the plane perpendicular to the axial thrust. The force is a sum/difference of the side forces developed by the tested nozzle and the reference one. In order to extract the side force of the tested nozzle from the sum, one has to do another experiment under the same conditions, but with the tested nozzle turned through 180° about the longitudinal axis and the reference nozzle in the same position. This allows one to measure the side force to an ac-

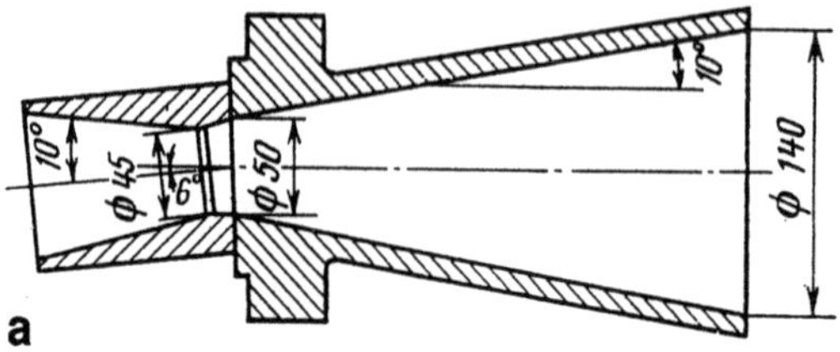

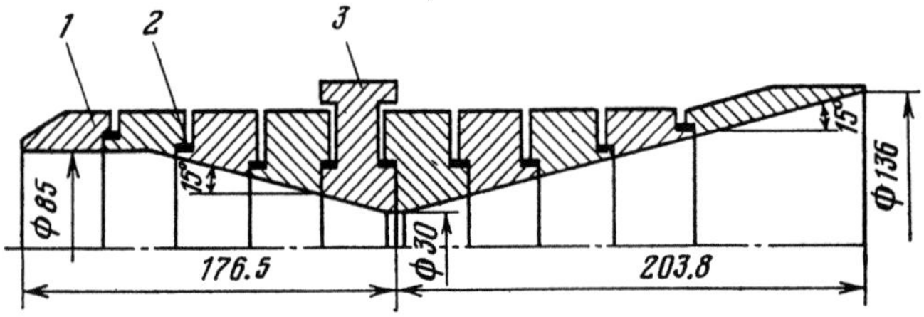

Fig.7.37a,b. Model diagrams: (**a**) conical nozzle with the axis misaligned in the supersonic part ($\theta_k = 10^\circ$, $\alpha = 6^\circ$), (**b**) conical nozzle for studying axis displacements ($\theta_k = 15^\circ$)

curacy of ~0.1% of the value of the axial thrust. Misalignment and displacement of the nozzle axis were the two major kinds of contour distortions selected for investigation. Mainly, the propagation of disturbances in the supersonic parts of nozzles was studied. Additionally, some kinds of local contour distortions which had the form of asymmetric transverse cavities were investigated.

Figure 7.37a shows schematically a conical nozzle with angle of axis misalignment $\alpha = 6^\circ$ in the supersonic part; the necessary dimensions are also presented. The calculated one-dimensional value of the Mach number M is equal to 1.58 at the misalignment cross-section (for $\gamma = 1.4$), while at the exit cross-section the Mach number $M_a = 3.9$. The subsonic part of the nozzle is also conical, rounded off at the throat with a radius $R_1^0 = R_1/r_* = 1$. Having measured the side force of the original nozzle, its supersonic part was subsequently truncated and the side force measured for other values of the exit cross-section diameter. All in all, the side force was measured at 15 different cross-sections of the nozzle with the same initial distortion.

Figure 7.37b shows a model of a conical nozzle used for studying distortions due to misalignment of the axis at various cross-sections. The model was composed of a number of adjacent sections. Parallel displacement of the axis $h = 0.5$, 1.0, 1.5, 2.0 mm was achieved by displacing the sections with respect to each other along the contact planes. This was done with the help of eccentric rings inserted instead of the symmetric ones 2. The nozzle contour was rounded off at the throat by circular arcs with $R_1^0 = R_2^0 = 1.0$; the cylindrical part of the throat was 3 mm long. In order to exclude additional components from the measured value of the side force, the experimental procedure was as follows. By means of an eccentric ring the axis was displaced at a certain cross-section and the model, devoid of subsequent sections, was tested twice, once in the original position and once after being turned through 180°. Then a subsequent section was installed and another double test was carried out. After testing the model with all the sections, the eccentric ring was exchanged for an axisymmetric one and another series of double tests was carried out with the sections being taken off one by one. The latter series of tests allowed the original level of the side force defined by technological errors of the assembly to be measured. Thus, the side force was obtained as a difference of the side forces measured in the first and the second series of tests.

Instead of the section 3 which has an axisymmetric contour of the channel, another section could be installed, in which the flow was turned through an angle $\alpha = 5^\circ$ at the throat. The mounting seats of section 3 allowed not only a conical supersonic part with $\theta_k = 15^\circ$ to be attached, but other models as well. Among these, two models with a profiled contour of the supersonic part ($M_0 = 4$, $r_a^0 = 3.16$, $\gamma = 1.4$ and $M_0 = 5$, $r_a^0 = 4.52$, $\gamma = 1.4$) were tested as well as a model with a conical supersonic part ($\theta_k = 5^\circ$).

Figure 7.38a shows the results of testing a conical nozzle with axis misalignment $\alpha = 6^\circ$ in the supersonic part. The side force of the nozzle varies over a wide interval, from the initial calculated value $L_H^0 = 0.08$ at the cross-section $r^0 = 1.11$ up to the maximum negative value $L^0 = -0.035$ at $r^0 = 1.9$. The same figure presents this dependence calculated with the help of the theory of slightly disturbed flows (at $L_H^0 = 0.08$ and $\varepsilon_H^0 = 0$) and with the help of the difference method; there is good agreement with the experimental data. This confirms the validity of the calculation schemes used to obtain the initial field of the flow parameters and the side force at the section of axis misalignment, and indicates a high accuracy of both methods of calculation used to analyze propagation of disturbances in conical nozzles. Figure 7.38b shows the side force for the case of axis misalignment at the throat. In a slender nozzle, L^0 reverses its sign sometimes while in the nozzle with $\theta_k = 15^\circ$ all the measured values of the side force, except the one at the cross-section

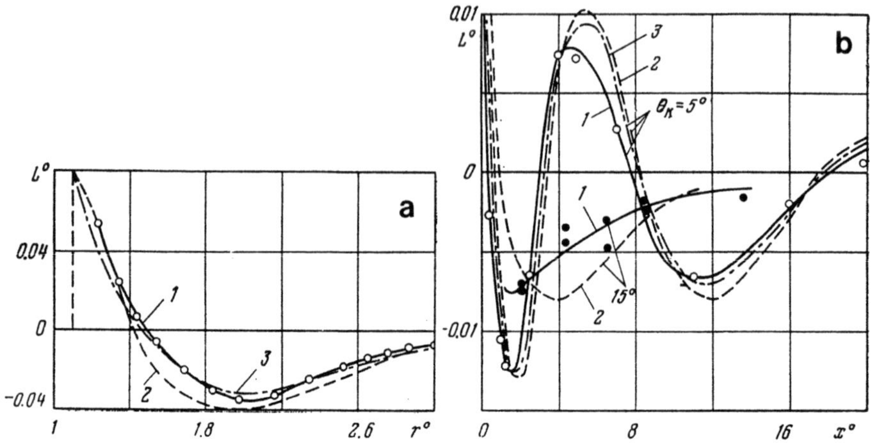

Fig.7.38a,b. Longitudinal side force variation for a conical nozzle at $\theta_k = 10^o$, $\alpha = 6^o$ (a) and $\theta_k = 5^o$, 15°, $\alpha = 5^o$ (b). (*1*) experimental, (*2*) calculated from linearized theory, (*3*) calculated by the difference method

$x^0 = 0$, are negative. The initial value $L_H^0 = 0.025$ obtained while testing the model without the supersonic part proved to be almost two times lower than the one calculated using (7.89). In order to find the reasons for the disagreement, the pressure was measured at the nozzle wall in the vicinity of the section of axis misalignment. The analysis of the data obtained shows that the region of elevated pressure exists not only behind the misalignment plane, but also in front of it, thus indicating the presence of a separation zone here. At the opposite wall the flow is accelerated up to $M \approx 1.3$. The side force calculated with the help of the experimental pressure distribution is close to the one measured with the strain gauge. The results obtained in the experiments show that calculation of the initial contour disturbance at the throat is a complicated task. This is equally true for the calculation of the initial field of the flow parameters used in various three-dimensional methods of calculation.

The calculated dependence of L^0 for the nozzle with $\theta_k = 5^o$ drawn at $L_H^0 = 0.025$, $\varepsilon_H^0 = 0$ (and at the corresponding flow parameter field in the difference method) agrees well with the experimental data. A certain disagreement means that at the initial cross-section ε^0 is evidently not zero. For the nozzle with $\theta_k = 15^o$ a larger disagreement between the calculated and experimental data is observed, associated probably with additional unaccounted for technological distortions of the contour at the locations where the nozzle sections are attached to each other.

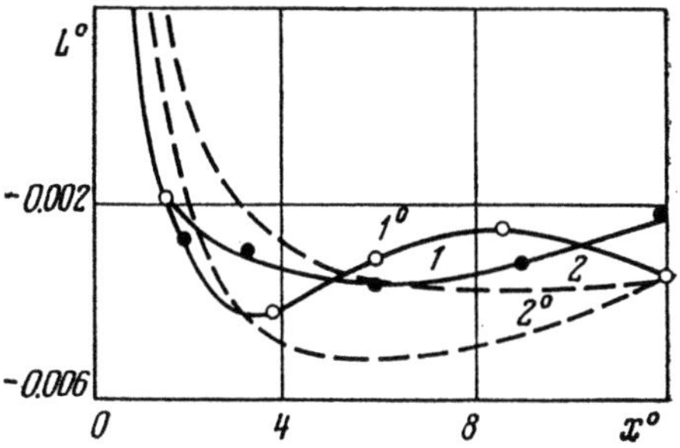

Fig.7.39. Longitudinal side force variation for a profiled nozzle: (1°, 2°) experimental and calculated by the difference method for a nozzle with $M_0 = 4$, (1,2) experimental and calculated by the difference method for a nozzle with $M_0 = 5$; $\alpha = 5^o$

Figure 7.39 presents longitudinal distributions of the side force for profiled nozzles obtained experimentally (curves 1) and by the difference method (curves 2) for the case when the nozzle axis was misaligned near the throat. In profiled nozzles the absolute value of the side force is smaller compared to that in conical nozzles of the same length. Also, it is negative at the larger part of the nozzles. For a nozzle with $M_0 = 4$ the calculated and experimental data agree favourably. Disagreement of some of the experimental points with the calculated ones is evidently explained by the same reason as in the case of the conical nozzle with $\theta_k = 15^o$. This conclusion is confirmed by the experimental dependence obtained for a model nozzle with $M_0 = 5$. In this case the supersonic part of the nozzle was manufactured as a single piece. Thus, the possibility of unaccounted for additional distortions was practically excluded and the obtained dependence $L^0(x^0)$ is quite smooth. The calculated dependence for the nozzle at $\varepsilon_H = 0$ is displaced along the axis x^0 to the right with respect to the experimental one, indicating the presence of a negative thrust vector displacement at the initial cross-section. This is true for other nozzles too (Figs.7.38,39); in all these models the axis misalignment was produced by the same component of the nozzle.

The dependence of the side force on the length of a conical nozzle with $\theta_k = 15^o$ was measured at various values of the axis displacement $h^0 = h/r_*$ (Fig.7.40). If the axis is displaced upwards, then the initial side forces are negative in all the cases considered as follows from the flow pattern discussed above. If the displacements take place in the supersonic part of a nozzle, then the side force stays negative downstream, while in the case when the displacement occurs at the throat, the sign of the side force is already reversed a short distance from the place where the disturbance originates. The value of the side force is approximately proportional to the displacement h.

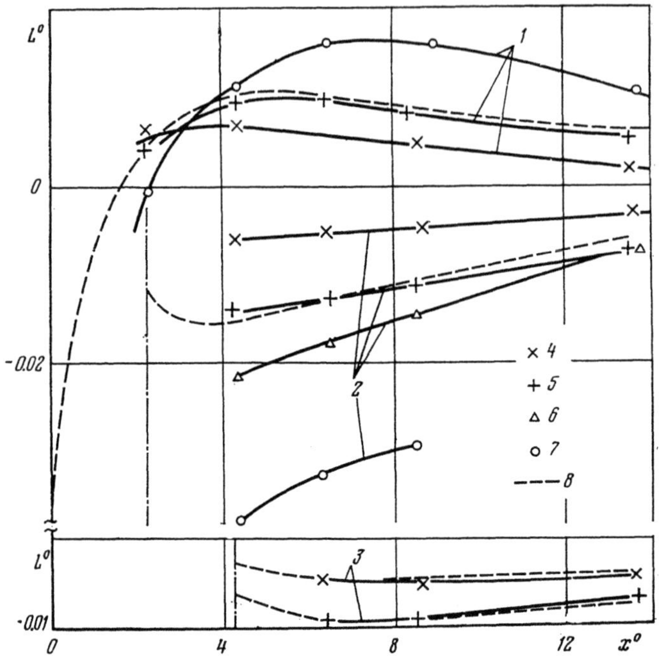

Fig.7.40. Longitudinal side force variation for a conical nozzle ($\theta_{k0} = 15^\circ$) whose axis is displaced at various cross-sections [(*1*) $x^0 = 0$, (*2*) $x^0 = 2.14$, (*3*) $x^0 = 4.25$] by h^0 [(*4*) $h^0 = 0.0033$, (*5*) $h^0 = 0.066$, (*6*) $h^0 = 0.1$, (*7*) $h^0 = 0.133$], (*8*) calculated by linearized theory

In order to compare experimental data with calculated results one has to know the value of the initial disturbances. However, as noted above, the formulas for the initial values of L_H^0 and ε_H^0 must be corrected using experimental data. Let us consider in this connection a method of obtaining the initial data by measuring the side force at any two cross-sections of a nozzle. This will be done with the use of the two facts stated above: firstly, the correctness of calculating the propagation of disturbances in conical nozzles with the help of the linearized theory and, secondly, the presence of a unique relation between the side force and the moment at a cross-section. Taking this into account, one may show that if the values of the side force are known at any two cross-sections of a nozzle, then there is only one combination of L_H^0 and M_H^0 at which these side forces may be realized.

The values of L_H^0 and M_H^0 may be readily calculated in the following way. Due to the linearity of the differential equations (7.70,71), their solution may be presented in the form

$$L^0 = L_H^0 a + M_H^0 b \quad , \quad M^0 = L_H^0 c + M_H^0 d \tag{7.97}$$

where a, c are calculated values of the side force and the moment at $L_H^0 = 1$, $M_H^0 = 0$ and b, d are calculated values of the side force and the moment at $M_H^0 = 1$, $L_H^0 = 0$. From (7.97) one gets

$$L_H^0 = \frac{L_2^0 b_1 - L_1^0 b_2}{a_2 b_1 - a_1 b_2} \quad , \quad M_H^0 = -\frac{L_2^0 a_1 - L_1^0 a_2}{a_2 b_1 - a_1 b_2} \quad . \tag{7.98}$$

Here L_1^0 and L_2^0 are the values of the side force at any two cross-sections of a nozzle measured in an experiment and a_1, a_2, b_1, b_2 are calculated values. Analogous formulas can be obtained for calculating L_H^0 and M_H^0 using known values of the moment at two cross-sections of a nozzle.

One can use another scheme of calculation too. Let the side forces L_1^0 and L_2^0 be known at nozzle cross-sections 1 and 2, produced by disturbances at a cross-section. Then, assuming that L_1^0 is the initial side force with respect to the force L_2^0, one may obtain the moment at the cross-section 1:

$$M_1^0 = (L_2^0 - L_1^0 a_2)/b_2 \quad . \tag{7.99}$$

Here the values of a_2 and b_2 are obtained from the calculation made assuming cross-section 1 to be the initial one. Knowing M_1^0 and L_1^0, one can determine the force and the moment at the initial cross-section:

$$L_H^0 = \frac{L_1^0 d_1 - M_1^0 b_1}{a_1 d_1 - b_1 c_1} \quad , \quad M_H^0 = \frac{M_1^0 a_1 - L_1^0 c_1}{a_1 d_1 - b_1 c_1} \quad . \tag{7.100}$$

The above procedure has been used to obtain the initial data for a displacement at the cross-section $x = 0$ with a subsequent construction of the calculated dependence. The calculated dependences 8 in Fig.7.40 agree well with experimental data not only near the cross-sections used to obtain the initial data, but throughout the nozzle. This confirms that the initial side force and the initial moment have been specified correctly. The initial data at other cross-sections where axis displacement took place were obtained in a similar way. It was found that if the axis is displaced in the supersonic part of a nozzle, then the thrust displacement $\varepsilon_H^0 \approx -h^0/2$. This confirms the assumption made while deriving (7.92), that $h_0 = -h/2$ (the remaining components of a displacement are small in the cases under consideration). However, these relationships do not hold if the displacement takes place at the throat. Here the thrust displacement proved to be so small that ε_H^0 could be assumed zero. Then, the axis displacement is equivalent to turning the nozzle axis through an angle α at the throat.

The values of L_H^0 obtained from (7.98,100) may be compared to those obtained from (7.92). The values on the right side of (7.92) are defined as follows: $p_1 = (0.31 + 0.69\, M_0)/p_0$ is the pressure in front of the shelf; p_2 is the pressure behind the shelf depending on M_0 and δ^{**}/h (δ^{**} is the momentum thickness) which can be calculated with the help of [7.47]; $\ell_1 = |h| \tan^{-1} \times (\delta_0 - \delta_1)$, $\ell_2 = |h| \tan^{-1}(\delta_2 - \delta_0)$ are the lengths of the separation zones in front of the shelf and behind it respectively; δ_0, δ_1, δ_2 are the Prandtl-Meyer angles corresponding to p_0, p_1, p_2. Calculations show that if the axis is displaced in the supersonic part of a nozzle, then (7.92) at $f = 0.5$ gives correct values of L_H^0. An analogous calculation for the throat is hardly possible, since reliable data for the values of pressure and the dimensions of separation zones are absent. Here one should use (7.98,100). The dependence of the side force on the length of a nozzle obtained with the help of the linearized theory for displacements at the cross-sections $x^0 = 2.14$, 4.25 at $\varepsilon_H^0 = -0.5\, h^0$ and L_H^0 calculated from (7.92), agrees well with the experimental data. This is also illustrated by Fig.7.41 where the experimental (1) and calculated (2) dependence of L^0 are presented for a conical nozzle with $\theta_k = 5^\circ$ whose axis is displaced at the throat by $h^0 = 0.066$. The calculated dependence was obtained for $L_H^0 = 0.033$ and $\varepsilon_H^0 = 0$.

The same figure presents experimental values of the side force for a profiled nozzle with $M_0 = 4$. The symbols 3-7 correspond to the cross-sections $x^0 = 0$, 1.55, 3.67, 6.04, 8.63 at which the nozzle axis was displaced by

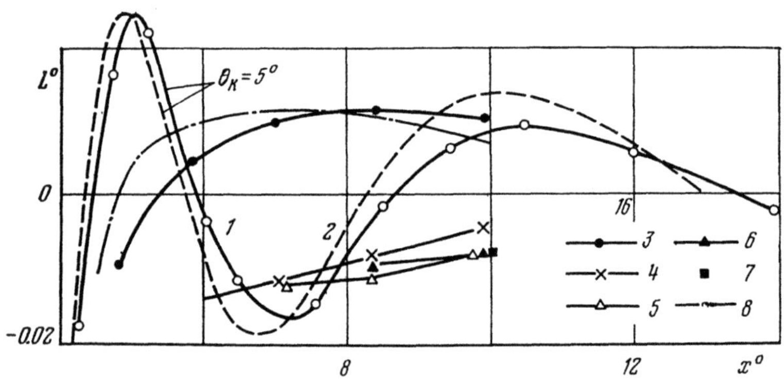

Fig.7.41. Longitudinal side force for a nozzle whose axis is displaced by $h^0 = 0.066$. Conical nozzle with $\theta_k = 5^\circ$, displacement at $x^0 = 0$, (*1*) experimental, (*2*) calculated by linearized theory. Profiled nozzle with $M_0 = 4$, displacement at the cross-sections: (*3*) $x^0 = 0$, (*4*) $x^0 = 1.55$, (*5*) $x^0 = 3.67$, (*6*) $x^0 = 6.04$, (*7*) $x^0 = 8.63$, (*8*) calculated by the difference method, $x^0 = 0$

$h^0 = 0.066$. Just as in the case of a conical nozzle with $\theta_k = 15°$, L^0 reverses its sign if the displacement occurs at the throat, and is negative throughout for displacements in the supersonic part. At the nozzle's exit cross-section (at $x_a^0 = 12$) L^0 stays practically constant over a broad interval of the coordinate of the displacement cross-section ($x_a^0 = 3.67$-8.63).

The quantitative comparison of experimental data with calculated results obtained by means of the difference method is made difficult by the initial conditions being indefinite. The only exception is the case of a displacement at the throat which, under the assumption $\varepsilon_H = 0$, is equivalent to turning the axis through an angle α. Let us find the value of the angle which matches the calculated value of the maximum positive side force and the experimental one ($L_{max} = 0.011$, Fig.7.41). Then the initial side force is equal to $L_H^0 = 0.051$ which corresponds to $\alpha = 4.6°$. The calculated dependence (curve 8) obtained by the difference method for an axis misalignment at the throat of $\alpha = 4.6°$ differs little from the experimental one; however, it is displaced along the axis x to the left, probably due to a slight displacement of the thrust vector at the initial cross-section. Similar differences in the positions of zeros of calculated and experimental dependences have been pointed out above while analyzing asymmetric flows and in other models. Note that the initial thrust displacement ε_H^0 may, in principle, be determined using the distance between the zeros of the experimental and calculated dependence. In order to use this method, one has to have an array of calculated curves $L^0 = f(x^0)$ at a single value of L_H^0 and different values of ε_H^0. Such dependences may be easily obtained for conical nozzles with the help of the linearized theory. For profiled nozzles the task is more complicated because in this case one has to solve numerically a three-dimensional problem and be able to specify at the initial cross-section the distributions of the flow parameters corresponding to a given thrust displacement.

Figure 7.42 presents the variation of the side force along a conical nozzle produced by asymmetric cavities whose dimensions are shown. Experiments have been done for two cases: $x_A^0 = 0.74$, $r_B^0 - r_{B_1}^0 = 0.12$, $\alpha_a = 10°30'$, $\alpha_2 = 10°$ (curve 1) and $x_A^0 = 2.65$, $r_B^0 - r_{B_1}^0 = 0.11$, $\alpha_1 = 8°36'$, $\alpha_2 = 7°30'$ (curve 2). We see that the direction of the side force coincides with the direction of the contour distortion. The side force produced by the cavity decreases with decreasing distance from the exit cross-section. A similar contour distortion in a profiled nozzle with $M_0 = 4$ produces a smaller side force which varies slightly along a nozzle's length ($L^0 = 0.0020$-0.0023 at $x_A^0 = 1.69$).

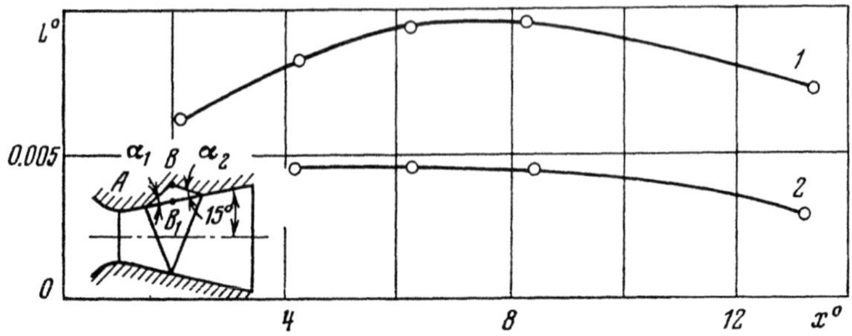

Fig.7.42. Longitudinal side force variation for a nozzle with $\theta_K = 15^\circ$ and local distortions of the contour (experimental)

The effect of a displacement of the contour axis in the subsonic part of a nozzle has been studied on the model presented in Fig.7.37 and assembled from the sections without the supersonic part. In order to obtain a maximum value of the side force, the displacement was made maximum and equal to $h = 2$ mm. If four sections 1 are displaced with respect to the section 3, then a side force appears equal to $L^0 = 0.0018$, parallel to the displacement of section 3. A similarly directed force $L^0 = 0.0017$ was registered when three sections 1 were displaced. Displacements of the sections situated further upstream give at the throat a side force which is less than 0.001. Thus, contour distortions in the subsonic part of a nozzle result in much smaller distortions than distortions in the supersonic part.

In conclusions, it should be noted that parametric experimental measurements of the side force produced by asymmetric geometrical forms of the subsonic or supersonic part of a nozzle have been made in [7.47]. Measurements were carried out at a differential installation (Sect.5.2.4). Reduction of the experimental data gave formulas for evaluating the side force for various asymmetric distortions of a nozzle contour.

7.5 Flows at Small Reynolds Numbers

The throat sizes and the total pressure in low-thrust rocket engines and hypersonic aerodynamic nozzles may be such that the Reynolds number $Re_0 = \rho_0 W_{max} r_* \eta_0^{-1}$ changes from 10 to 10^4. The number Re_0 is related to the Reynolds number Re_{W0} defined in Chap.5 by means of the relationship $Re_{W0} = Re_0 \bar{L}/\bar{T}_W$, where $\bar{L}$ is the length of a nozzle referred to the throat radius and $\bar{T}_W$ is the wall temperature referred to the stagnation temperature.

Thus, the number Re_{w_0} exceeds Re_0 by an order of magnitude. At the above values of Re_0 the viscosity must be considered not only in the thin prewall layer but over the whole of the cross-section too. When calculating a flow one can no longer just introduce corrections for the displacement thickness of the boundary layer, but one has also to solve the Navier-Stokes system of equations making some assumptions. Theoretical and experimental studies of nozzle flows at small Reynolds numbers have been reported in [7.48-50]. Below, some results of the studies are presented.

Theoretical investigations of viscous gas flows at small Reynolds numbers are based upon the solution of simplified Navier-Stokes equations under the assumptions that the ratio of the transverse velocity component to the longitudinal one and the ratio of the longitudinal gradient to the lateral one are much less than unity [7.51]. After appropriate simplifications, the terms of the Navier-Stokes equations containing second derivatives with respect to x vanish and, in order to calculate a nozzle flow, one can solve the initial-value problem with respect to x instead of the boundary problem as in the case of the full Navier-Stokes equations. The above approximation results in the following system of simplified Navier-Stokes equations:

$$
\begin{aligned}
&\frac{\partial}{\partial x}(\rho u r^{\nu}) + \frac{\partial}{\partial r}(\rho v r^{\nu}) = 0 \quad , \\
&u\frac{\partial u}{\partial x} + v\frac{\partial u}{\partial r} = -\frac{1}{\rho}\frac{\partial p}{\partial x} + \frac{1}{\rho r^{\nu}}\frac{\partial}{\partial r}\left(\eta r^{\nu}\frac{u}{\partial r}\right) \quad , \qquad (7.101)\\
&\frac{\partial p}{\partial r} = 0 \quad , \\
&\rho u\frac{\partial h}{\partial x} + \rho v\frac{\partial h}{\partial r} - u\frac{\partial p}{\partial x} = \frac{1}{r^{\nu}}\frac{\partial}{\partial r}\left(\frac{r^{\nu}}{Pr}\frac{\partial h}{\partial r}\right) + \eta\left(\frac{\partial u}{\partial r}\right)^2 \quad .
\end{aligned}
$$

This system of equations must be supplemented by thermal and caloric equations of state and relationships for determining the Prandtl number Pr and the viscosity coefficient η.

In order to construct a solution one has to specify initial conditions at the initial cross-section and boundary conditions at the nozzle wall and axis. The initial conditions are obtained while constructing a solution in the vicinity of the infinite point by expanding the required functions in a series in inverse powers of x ([7.48], see also Sect.4.4). At the nozzle axis the symmetry conditions are specified, according to which derivatives of u, p, ρ, T and h and the transverse velocity component are equal to zero. For a rarefied gas the slip condition and the condition of temperature jump at the walls of a nozzle are generally used [7.52], according to which

$$u_w = -\frac{(2-\alpha_u)\eta}{\alpha_u p}\left(\frac{\pi RT}{2\mu}\right)^{\frac{1}{2}}\left(\frac{\partial u}{\partial r}\right)_w \cos\theta_w \quad ,$$

$$T - T_w = -\frac{2\gamma(2-\alpha_T)\eta}{(\gamma+1)\alpha_T p Pr}\left(\frac{\pi RT}{2\mu}\right)^{\frac{1}{2}}\left(\frac{\partial T}{\partial r}\right)_w \cos\theta_w \quad , \qquad (7.102)$$

where α_u and α_T are accommodation coefficients for velocity and temperature, and the subscript w refers to parameters at the nozzle wall.

An iterative method of solving the system (7.101) has been proposed in [7.48]. It employs numerical integration starting from the subsonic part of a nozzle and allows one to determine, in the process of iteration, the gas flow rate and pressure distribution along the axis. In the case of solving the direct problem, just as in the case of inviscid flows, the system of equations has a saddle-type singular point where

$$\delta = \int_0^1 \frac{1-M^2}{M^2}\, dF = 0 \quad . \qquad (7.103)$$

This condition means choking of the flow. If the impermeability condition is put at the walls, then the integral in (7.103) diverges. In order to circumvent this difficulty, one can use the approach proposed in [7.53], according to which integration may be cut short near the nozzle wall.

In the process of solution a special value of the critical flow rate is chosen depending on Re_0, and when this is attained, downstream of the singular point an integrally supersonic flow is realized, i.e. $\delta < 0$ (Sect.2.2.1). At flow rates below critical the flow is integrally subsonic throughout, i.e. $\delta > 0$. Calculations show that with decreasing Re_0 the singular point moves downstream and may leave the nozzle. In this case no flow rate can ensure that the condition $\delta > 0$ is satisfied. At a certain value at the exit cross-section the pressure gradient becomes infinite and the flow becomes physically unrealizable. At large flow rates such a situation takes place inside a nozzle. At smaller flow rates, the pressure either attains a minimum and then increases, or decreases monotonically up to the exit cross-section if the pressure drop due to friction is large enough to overcome the effect of increasing the nozzle's cross-section area. At such flow rates the external pressure may influence the nozzle flow.

Some results of numerical and experimental studies are presented in Figs. 7.43-45. Figure 7.43 shows the dependence of the number M, temperature and pressure on the nozzle length obtained experimentally by the method of electronic beams [7.49] and by measuring pressure [7.50] and values calculated by the above method. We see that depending on the number Re_0, different flow

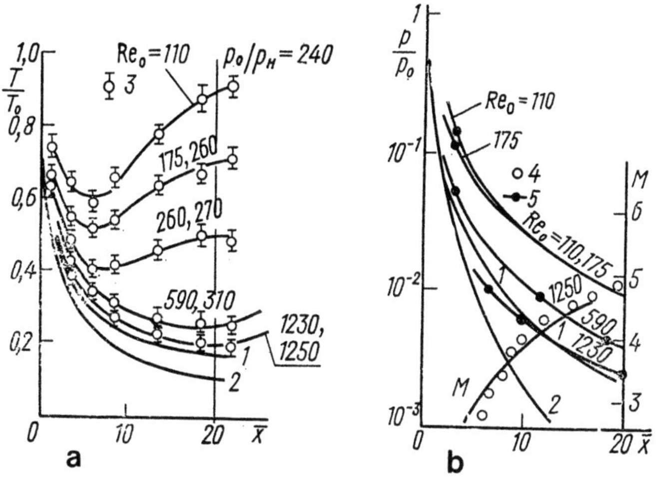

Fig.7.43. Temperature (**a**), pressure and Mach number distributions (**b**) along a conical nozzle axis ($\theta_k = 20^\circ$, $r_* = 25$ mm) at various values of the Reynolds number: (*1*) calculated at $Re_0 = 1230$ [7.48], (*2*) an ideal gas, (*3*) experimental [7.49], (*4*) experimental [7.50], (*5*) pressure calculated from the experimental data for temperature and density [7.49]

◄ Fig.7.44

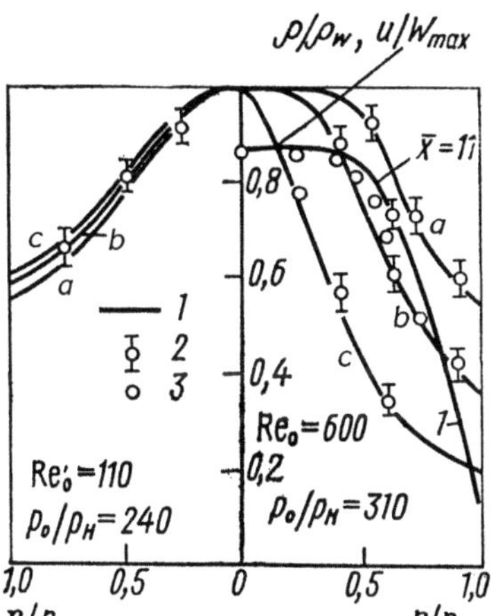

▼ Fig.7.45

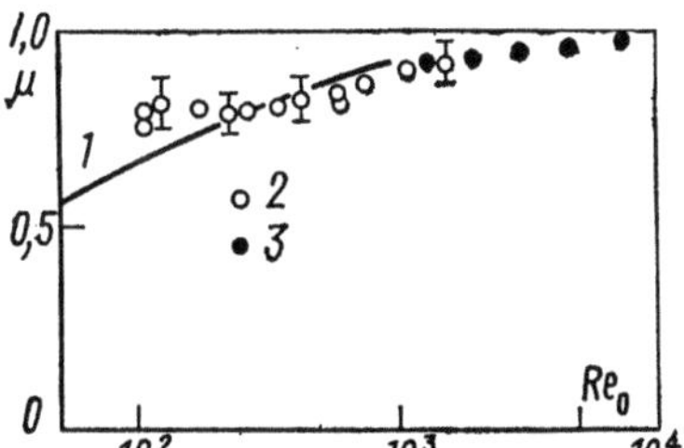

Fig.7.44. Density and velocity distributions at various cross-sections of a nozzle: (*1*) calculated at $Re = 1250$ [7.48], (*2*) experimental [7.49], (**a**) $\bar{x} = 3.7$, (**b**) $\bar{x} = 6.2$, (**c**) $\bar{x} = 18.7$, (*3*) experimental data at $Re_0 = 1250$ [7.50]

Fig.7.45. Dependence of the flow rate coefficient on the Reynolds number: (*1*) calculated [7.48], (*2*) experimental [7.49], (*3*) experimental [7.54]

regimes are realized. At $Re_0 > 500$ the temperature decreases monotonically, while at $Re_0 < 300$ it decreases down to $\bar{x} \approx 6$, and then starts growing. The increase of temperature is a result of viscous dissipation and must be accompanied by a decreasing velocity and Mach number. At $Re_0 \approx 100$ the value of $\bar{T} = T/T_0$ near the exit cross-section exceeds that at $M = 1$ for an isentropic flow of an ideal gas. At such values of the number Re_0 the flow expands near the axis up to supersonic speeds ($M \approx 1.8$ at $\bar{x} = 5$), and then a shock-free deceleration to subsonic speeds takes place. Such a flow character corresponds to the case when the singular point is outside a nozzle [7.48]. The temperature increase at small Reynolds number Re_0 cannot be explained by the formation of compression waves, since the experimental values of density and pressure decrease monotonically (Fig.7.43). The external pressure according to the experiments of [7.49] has a slight effect upon the shape of the profiles until the flow at the exit cross-section stays overexpanded ($p_0/p_H > 250$).

It follows from Fig.7.44 that at $Re_0 = 110$ the profiles of dimensionless density practically do not depend on the nozzle length and that the flow is fully viscous and does not have a definite core. At $Re_0 \approx 600$ a tiny inviscid core forms at $\bar{x} < 3.7$; however at the exit cross-section it disappears. At $Re_0 = 1250$ the inviscid core is preserved up to the exit cross-section. Experimental temperature and density profiles over cross-sections show that the derivatives $\partial T/\partial r$ and $\partial \rho/\partial r$ near the wall are nonzero. This indicates the presence of a slip and temperature jump in accordance with the boundary conditions (7.102). Near the walls the temperature is equal to about $0.8\ T_0$ agreeing with the calculated data for an adiabatic wall with a slip. With decreasing temperature the displacement boundary layer thickness decreases too. It has been shown that at $Re_0 > 1200$ the flow parameters at the axis may be calculated by means of the one-dimensional theory using an effective ratio of areas diminished by the displacement thickness [7.48].

Evidently, decrease of the number Re_0 results in the flow rate coefficient μ decreasing. The dependence of the flow rate coefficient on the number Re_0 is presented in Fig.7.45. Calculations of the impulse and the specific impulse show that at $800 < Re_0 < 1500$ the latter depends on the number Re_0 only slightly and decreases with decreasing $\bar{T}_w$ (the nozzle impulse decreases proportionally to the flow rate coefficient). Compared to inviscid flow, a less intense growth of the impulse and specific impulse takes place with an increasing ratio of areas. In practice, this means that for viscous flows, nozzles which realize maximum thrusts may be even more truncated.

Some special features of nozzle flows at small Reynolds numbers have been considered in [7.55], where experimental data for flow rate and specific impulse are presented, as well as results obtained with the help of the boundary layer theory and the method of narrow channels; the experimental data are compared with results of calculations.

Recently a number of numerical studies of flows in Laval nozzles have been published, based on solving complete Navier-Stokes equations [7.56-59]. In these studies stationary solutions are found with the help of the method of stabilization. Vibrationally nonequilibrium flows of a CO_2-N_2-O_2-H_2O mixture in plane Laval nozzles have been investigated in [7.58] at large and moderate Reynolds numbers. A number of peculiarities characteristic of such flows have been studied, for instance, processes of vibrational relaxation in an inviscid core and in the boundary flow, two-phase effects, influence of vibrational relaxation upon distribution of gasdynamic parameters and the reverse influence of the boundary layer upon the flow in the inviscid core.

The most comprehensive study of flows in Laval nozzles by means of solving complete Navier-Stokes equations has been reported in [7.59]. Below, some results of the work are presented. The complete system of nonstationary Navier-Stokes equations has the following form:

$$\frac{\partial\rho}{\partial t} + \frac{\partial\rho u}{\partial x} + \frac{\partial\rho v}{\partial r} + \nu\,\frac{v}{r} = 0 \tag{7.104}$$

$$\begin{aligned}\frac{\partial\rho u}{\partial t} = &- \frac{\partial\rho u^2}{\partial x} - \frac{\partial\rho uv}{\partial r} - \nu\,\frac{\rho uv}{r} - \frac{\gamma - 1}{2\gamma}\,\frac{\partial p}{\partial x} + \frac{1}{\mathrm{Re}}\left[2\eta\,\frac{\partial}{\partial x}\left(\frac{\partial u}{\partial x}\right)\right.\\ &\left.+ \frac{\partial}{\partial x}\,(\xi^k \mathrm{div}\,\mathbf{W}) + \frac{\nu}{r}\left(\frac{\partial v}{\partial x} + \frac{\partial u}{\partial r}\right) + \frac{\partial}{\partial r}\left(\frac{\partial v}{\partial x} + \frac{\partial u}{\partial r}\right)\right]\end{aligned} \tag{7.105}$$

$$\begin{aligned}\frac{\partial\rho v}{\partial t} = &- \frac{\partial\rho uv}{\partial x} - \frac{\partial\rho v^2}{\partial r} - \nu\,\frac{\rho v^2}{r} - \frac{\gamma - 1}{2\gamma}\,\frac{\partial p}{\partial r}\\ &+ \frac{1}{\mathrm{Re}}\left[\eta\left(\frac{\partial v}{\partial x} + \frac{\partial u}{\partial r}\right) + 2\eta\,\frac{\partial}{\partial z}\left(\frac{\partial v}{\partial r}\right) + \frac{\partial}{\partial r}\,(\xi^k \mathrm{div}\,\mathbf{W}) + \frac{\nu}{r}\,\eta\,\frac{\partial v}{\partial r}\right] \quad ,\end{aligned} \tag{7.106}$$

$$\begin{aligned}\frac{\partial(\rho c_v T)}{\partial t} = &- \frac{\partial(\rho u c_v T)}{\partial x} - \frac{\partial(\rho v c_v T)}{\partial r} - \frac{\partial}{\partial x}\left(\frac{\gamma\eta}{\mathrm{Pr}\;\mathrm{Re}}\;\frac{\partial T}{\partial x}\right)\\ &- (\gamma - 1)p\;\mathrm{div}\,\mathbf{W} - \nu\,\frac{\rho c_v T}{r} + \frac{\eta}{\mathrm{Re}}\left\{2\left[\left(\frac{\partial u}{\partial x}\right)^2 + \left(\frac{\partial v}{\partial r}\right)^2 + \nu\,\frac{v}{r^2}\right]\right.\\ &\left.+ \left(\frac{\partial u}{\partial r} + \frac{\partial v}{\partial x}\right)^2\right\} + \xi^k \mathrm{div}\,\mathbf{W} + \nu\,\frac{\gamma\eta}{r}\,\frac{1}{\mathrm{Pr}\;\mathrm{Re}}\,\frac{\partial T}{\partial r} \quad ,\end{aligned} \tag{7.107}$$

$$p = \rho T \quad , \tag{7.108}$$

where $\zeta^k = \eta_2 - 2\eta/3$ and η_2 is the second viscosity coefficient. Dimensionless variables have been constructed in the following way: velocities are referred to the maximum velocity of outflow, density, pressure and temperature have been divided by the corresponding stagnation parameters and viscosity is referred to its value at the stagnation temperature.

By introducing new independent variables $x = z$, $y = r/r_w(x)$, the domain under consideration has been transformed into the rectangle $\{a \leqslant x \leqslant b, 0 \leqslant y \leqslant 1\}$. Various methods of specifying boundary conditions at the entrance $x = a$ have been considered. After analyzing work by other authors, a combined approach was proposed [7.57]. Conditions at the boundary $x = a$ were sought with the help of asymptotic solutions of the "narrow channel" equations up to the moment $t = t_0$ when flow rate through the whole of the nozzle's section becomes constant; then, all the dependent variables were fixed and stayed further unchanged.

At the solid wall $(y = 1)$ boundary conditions may be specified either in the form of the attachment conditions and mixed conditions for temperature and $\partial T/\partial y$

$$u(t,x,1) = v(t,x,1) = 0 \tag{7.109}$$

$$\alpha_1 T + \alpha_2 \frac{\partial T}{\partial y} = f(t,x,1) \quad , \tag{7.110}$$

or in the form of slip conditions (7.39). The density at the wall $\rho(t,x,1)$ is obtained from the continuity condition.

At the right (exit) boundary, "soft" boundary conditions are used

$$\frac{\partial^2 f}{\partial x^2} = 0 \quad , \qquad f = \{u, v, \rho, T\} \quad .$$

Initial conditions are formulated on the basis of solution obtained in [7.55] by means of the "narrow channel" method in which truncated Navier-Stokes equations are used [7.56].

In [7.59] numerical investigation of the two-dimensional laminar flow of viscous heat-conductive gas in Laval nozzles in a wide range of Reynolds numbers and for various geometrical parameters of nozzles is discussed. A code for calculating flows on the basis of an explicit scheme of the "predictor-corrector" type has also been developed [7.56]. The method of stabilization was used for calculating stationary flows. The code developed enables swirling flows to be calculated. Both the attachment condition (at $Re > 10^4$) and the slip and temperature jump conditions (at $Re \leqslant 10^4$) were used as boundary conditions at the wall.

From a comparison of calculated and experimental data reported by other authors, it has been shown that the method provides trustworthy results for gas flows in the range of Reynolds numbers $Re = 10\text{-}10^5$, for nozzles with a rounding-off radius $R_2 = 0.15\text{-}15$, in profiled nozzles with a corner point for various heat exchange regimes at the wall, both in plane and axisymmetric nozzles.

The method includes an integral check of the laws of conservation of mass, momentum and energy, thus providing an additional control of the trustworthiness of the results. Integral checks of the laws of conservation have shown that for flows with $Re = 10^3$ the errors of closure for flow rate and momentum are equal to 0.1%.

It has been shown that calculation of viscous flows at $Re \leqslant 600$ by the "narrow channel" method results in an error which may reach 25% both for local parameters and for integral characteristics. These errors are especially large in the transonic part of a nozzle where both longitudinal and transverse gradients of pressure and velocity are considerable. Similar errors are characteristic for methods of integral relations.

For the first time flows of viscous heat-conductive gases in plane profiled nozzles with a corner point were investigated with the help of complete Navier-Stokes equations. The calculations have shown that transverse pressure drop in the transonic part of profiled nozzles may reach 60-70%. In addition, the presence of stagnation zones at a nozzle wall was discovered near the throat, and recirculation zones were registered for $Re \leqslant 500$.

For the first time, complete Navier-Stokes equations were used to study the influence of viscosity and thermal conductivity upon the intensity of shock waves in Laval nozzles. It was found that the finite difference scheme and the method used in the work enable physical peculiarities of the flow to be effectively traced, for instance, shock waves may be located, their intensity calculated, and so on. It has been shown that at $Re \geqslant 10^5$ results of calculations of flows in conic nozzles approach those obtained by alternative methods, in particular, by methods of calculating inviscid two-dimensional flows.

Some results obtained in [7.59] are presented in Figs.7.46,47. Figure 7.46 shows distribution of the Mach number M along a nozzle axis calculated in [7.59] (curve 1) by the "narrow channel" method [7.48,55] and obtained experimentally in [7.50]. The numerical method presented in [7.59] is seen to provide excellent agreement with the experiment. Figure 7.47 presents measured values of temperature and density at a nozzle axis for various values of Re

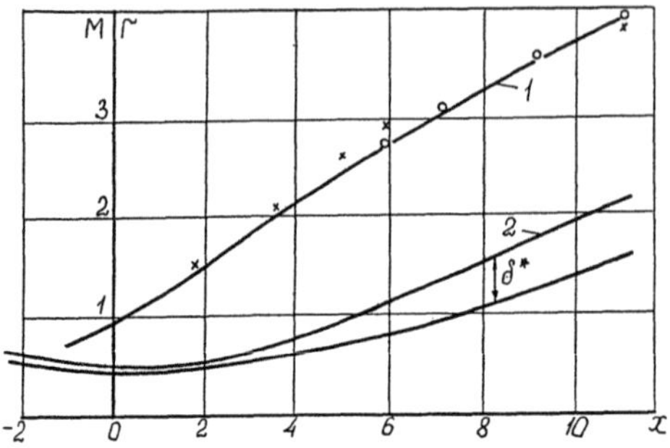

Fig.7.46. Distribution of Mach number M along the axis for viscous gas flow in a nozzle. (*1*) calculated by the method described in [7.59], (×) calculated by the "narrow channel" method [7.48], (o) experimental data [7.50], (*2*) the nozzle contour; δ^* is the displacement thickness

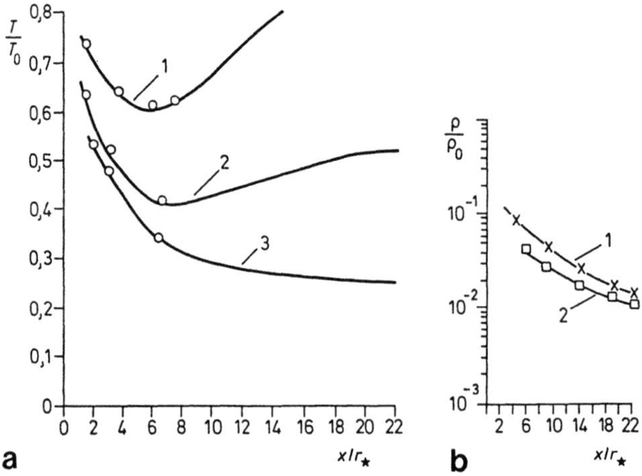

Fig.7.47a,b. Distributions of temperature (**a**) and density (**b**) along the axis for viscous gas flow in Laval nozzle at various Reynolds numbers. Comparison of calculated and experimental data. (*1*) Re = 110, (*2*) Re = 260, (*3*) Re = 590, (o, ×) experimental data [7.49]; solid lines present calculated results from [7.59]

[7.49] and the calculated results from [7.59] (solid lines). Good agreement between the calculated and experimental results is obvious, as is the temperature growth due to viscous dissipation mentioned above. Note that a turbulent boundary layer at a nozzle wall has been studied in [7.59] using the method of integral relations.

A method for solving the conjugated problem of calculating turbulent boundary layers and heat exchange in the case of regenerative cooling has been developed for the numerical study of flows with large Reynolds numbers. The method of integral relations for calculating boundary layers was modified in such a way that the necessity of using semiempirical relations for the form parameter δ^*/δ^{**} has been excluded. The method may be included into an automatic design system which solves multipurpose problems, such as choosing the optimum contour of a nozzle or profiles for channels of a cooling system, and a number of similar problems.

Nomenclature

a	Sonic speed 10 (2.25)
a_ℓ	Equilibrium sound velocity 23
a_f	Frozen sound velocity 23
b_k	Number of atoms of the kth type in a mole of the initial products 202 (6.6)
C_p	Pressure coefficient 151
c_p	Heat capacity at constant pressure 205
c_s	Heat capacities of particles 251 (6.75-81)
C_{xi}	Drag coefficient 252
C_0, C_+, C_-	Characteristics 20 (2.48)
D_i	Diffusion coefficient of a condensing component 279 (6.128)
d_{is}	Diameter of ith fraction particle 267
d_*	Throat diameter 263 (6.102)
e	Internal energy 8 (2.3)
F	Area of a nozzle's cross-section 41 (2.123)
$F_i(p,T,\alpha_i)$	Mass rate of substance formation 212 (6.31)
$\mathbf{f}$	Interaction force between gas and particles 12 (2.13)
$f(\alpha)$	Prandtl-Meyer angle 54 (2.151)
G_a	Nozzle's weight 196 (5.50)
H_0	Total enthalpy 15 (2.26)
h	Enthalpy 12 (2.11)
I	Nozzle's impulse 47 (2.132)
$I_s = I/G$	Specific impulse of a nozzle 49 (2.143)
I_+, I_-	Riemann's invariants 21 (2.49)
J	Rate of condensation nuclei formation 275 (6.125,126)
J	Nozzle's thrust 57 (2.154)
J	Axial force 366
J	Nozzle's impulse 174 (5.31)
$J_s = J/G$	Specific impulse of a nozzle 174

J^0	Impulse on an ideal nozzle 177 (5.34)
J_*	Impulse of the convergent part of a nozzle 179 (5.39)
K	Courant number 87
K_{ij}	Coagulation constant 266 (6.106)
K_j	Equilibrium constant 201 (6.3)
K_n, K_N	Coefficients of specific impulse and thrust 177 (5.36)
$K_+(T)$, $K_-(T)$	Rate constants of the direct and the reverse reaction 210 (6.26)
k	Boltzmann's constant 273 (6.116)
L	Condensation heat 271 (6.115)
L	Side force 366 (7.56)
L_a	Length of supersonic part of a nozzle 174 (5.26), Fig.5.1
L_c	Overall length of a nozzle 182 (5.43)
ℓ_0	Length of subsonic part 144, Fig.5.1
M	Mach number 16
M	Moment 366 (7.56)
Nu_{is}	Nusselt number 252 (6.82)
n_i, n_{si}	Numbers of moles of the element in gaseous and condensed states 202 (6.6)
n_{is}	Number of ith fraction of particles in unit volume 265 (6.105)
Pr	Prandtl number 252 (6.82)
p	Static pressure 10
p_H	External pressure 43
$p_{i\infty}$	Saturation pressure 271 (6.115)
p_k	Partial pressure 201 (6.3)
Q	Total mass flow 41 (2.123)
$\mathbf{Q}_i$	Heat flow vector 8 (2.3)
$q(\lambda)$	Dimensionless mass flow function 42 (2.126)
R	Universal gas constant 10
$R = J - p_H F_a$	Actual nozzle's thrust 174
R_s	Gas constant of the condensating component 271 (6.115)
R_1	Radius of matching of the conic and cylindrical parts 144, Fig.5.1
R_2, R_2^0	Radii of curvature of the contour in the vicinity of the throat 144, Fig.5.1
Re_0	Reynolds number 160
Re_{is}	Reynolds number 252 (6.82)
Re_s	Reynolds number 268 (6.113)

Re_{w0}	Reynolds number 182 (5.43)
r	Distance along the normal to curve 32, Fig.2.5
r	Radial cylindrical coordinate 144
r	Radial spherical coordinate 133
r,θ,φ	Spherical coordinate system 133, Fig.4.7
r_{cr}	Radius of critical drop (nuclei) 274 (6.121)
r_H	Radius of the initial line 135
r_i	Molar fraction of ith component of the mixture 205
r_{nk}	Average particle radius 250 (6.73)
r^0	Radius of an isobaric section of a jet 331 (7.16)
r_0, r^0	Radii of the initial and exit cross-sections 144, Fig.5.1
r_*	Radius of critical nozzle cross section 144, Fig.5.1
S	Entropy 11 (2.8)
S_a, S^0	Side surfaces of supersonic parts of the truncated non nontruncated nozzles with a uniform flow 174 (5.26)
St_{ij}	Stokes number 252 (6.82)
s	Supersaturation 274
s	Length of arc along $y = f_0(s)$ 32, Fig.2.5
s,r,φ	Curvilinear coordinate system 33, Fig.2.5
s,ψ,θ	Normal coordinate system 35
T	Temperature 8
T_k	Temperature at the point where condensation begins 284 (6.144)
T_v	Vibrational temperature 241, Fig.6.15
$\bar{T}_w = T_w/T_0$	Wall temperature referred to the stagnation temperature 182 (5.43)
t	Time 10
u	Velocity in one-dimensional flow 20
u,v	Projections of velocity vector in Cartesian coordinates 22
u,v,w	Projections of velocity vector in Cartesian coordinates 15; spherical coordinates 134; curvilinear coordinates 33
$\mathbf{W}$	Velocity vector 10
W	Velocity in one-dimensional flow 41
W	Absolute value of velocity 10
W_0	Velocity at the nozzle's centre 150 (5.4)
W_{max}	Maximum velocity 41
W_∞, $\bar{W}_\infty$	Asymptotic values of velocity 147 (5.3)
We	Weber number 267
X	Length of a nozzle in variational problem 57 (2.155)

x, y Cartesian coordinates 22

x, r Cartesian coordinates 111

x, y, z Cartesian coordinate system 14

x, r, φ Cylindrical coordinate system 97

$z(\lambda) = \lambda + 1/\lambda$ Gasdynamic function 50

$\alpha = \arcsin M^{-1}$ Mach angle 23

α Angle of rotation of the nozzle's axis 375, Fig.7.32a

α Oxidizer-to-fuel ratio 223

α_i Smoothing coefficients 87

α_i Mass concentration of ith component 10

α_{is} Weight fraction of the ith particle 252

α_{ik} Condensation coefficient 279

α_{is}^0, α_{is1}^0 Weight fractions of particles having fallen on a nozzle's wall in its subsonic and supersonic sections 294

α_s Total fraction of particles 252

α_u, α_T Accommodation coefficients for velocity and temperature 396 (7.102)

α_* Dimensionless swirl velocity 349

$\beta = \cot\alpha = \sqrt{M^2 - 1}$ 70

$\beta = p_0 F_* / Q$ Flow rate complex 204

β_i^0 Thermal accommodation coefficient 279

γ Adiabatic exponent 11

$\gamma_i = \alpha_i/\mu_i = r_i/\mu$ Mole-mass concentration of ith component 212

γ^0 Effective adiabatic exponent 254 (6.89)

ΔT Supercooling value 274

$\Delta\phi$ Formation energy of a nucleus 273 (6.116)

δ^* Displacement thickness 160 (5.19)

δ^{**} Impulse losses thickness 160 (5.19)

δ_a^{**} Impulse losses thickness refferred to a current radius 184 (5.44)

ε Thrust eccentricity 369 (7.73)

ε Swirl parameter 351 (7.32)

ε Small parameter 133 (4.64)

ε_i, ε_i^0 The ith-component's vibrational energy and its equilibrium value 235

$\varepsilon(\lambda)$ Dimensionless function of density 49 (2.139)

$\zeta = \tan\theta$ 70 (3.11)

ζ Coefficient of impulse losses 177 (5.36)

ζ_p	Divergence impulse losses 178 (5.39)
ζ_T	Coefficient of friction impulse losses 178 (5.44)
ζ_H	Coefficient of impulse losses due to nonequilibrium flow of chemical reactions 222
ζ_{KH}	Coefficient of impulse losses due to vibrational nonequilibrium 243
ζ_s	Coefficient of impulse losses associated with the velocity and temperature lags of particles 263 (6.102)
ζ_s^0	Coefficient of impulse losses due to absence of crystallization 288 (6.145)
ζ_w^0	Coefficient of impulse losses due to precipitation of particles 295 (6.158)
ζ_Σ	Total impulse losses in an annular nozzle 333 (7.17)
η	Viscosity coefficient 160
θ	Angle coordinate of spherical system r,θ,φ 133, Fig.4.7
θ	Angle between velocity and x direction 23
θ_v	Characteristic vibrational temperatures of molecules 232
κ	Effective adiabatic exponent 78
$\kappa=(\gamma-1)/(\gamma+1)$ 135	
$\lambda=W/a_*$	Velocity coefficient 41
λ_i	Slope of characteristic 19
λ_1, λ_2, λ_3	Lagrange multipliers 59
λ^0, λ^+, λ^-	Angle coefficients of C_0, C_+, C_- 70 (3.12,13)
μ	Molecular weight of the mixture 10
μ	Flow rate coefficient 159
$\nu=0;1$	for plane and axisymmetric flows, respectively 22
$\pi(\lambda)$	Dimensionless pressure function 49 (2.137)
ρ	Density 7
ρ^0, ρ_s^0	True densities of the gas and particles 252 (6.81)
σ	Total pressure ratio 259 (6.95)
σ	Surface tension 273 (6.117)
τ_i	Relaxation time 66 (3.1)
Φ	Velocity potential 37
Φ_i	Rate of ith relaxation processes 235
Φ_{ij}	Collision efficiency coefficient 266 (6.106)
$\varphi=J/J_0$	Coefficient of nozzle's impulse 176
φ	Angle coordinate in s,r,φ and other coordinate systems 33
φ	Velocity, potential 111

χ_k	Thermodynamic or chemical potential	11 (2.8)
Ψ	Scale parameter	241
ψ	Stream function	23 (2.68)
ψ,σ,φ	Orthogonal coordinate system	38

Superscripts

( )	Iterational index

Subscripts

a	Associated with nozzle's exit cross-section
is	Associated with i^{th} particle fraction
k	Parameters of source flow
w	Parameters on nozzle's wall
0	Stagnation parameters
*	Critical value

References

Chapter 2

2.1 S.R. de Groot, P. Mazur: *Non-Equilibrium Thermodynamics* (North-Holland, Amsterdam 1962)
2.2 A.N. Kraiko, R.I. Nigmatulin, V.K. Starkov, L.E. Sternin: "Mechanics of Multi-Phase Media", in *Hydrodynamics*, Vol.6 (VINITI, Moscow 1972) (in Russian)
2.3 L.I. Sedov: *A Course in Continuum Mechanics*, Vols.1-4 (Wolters-Noordhoff, Groningen 1971-1972)
2.4 R. Courant: *Partial Differential Equations* (Interscience Publishers, New York 1962)
2.5 R. Mises: *Mathematical Theory of Compressible Fluid Flow* (Academic, New York 1958)
2.6 V.V. Rusanov: J. Comput. Math. Math. Phys. **3**, 508-527 (1963) (in Russian)
2.7 V.H. Ransom, J.D. Hoffman, H.D. Thompson: AIAA J. **10**, 1573-1581 (1972)
2.8 D.S. Butler: Proc. Roy. Soc. (Ser. A) **267** (1962)
2.9 P.I. Chushkin: "Numerical Method of Characteristics for Three-Dimensional Supersonic Flows", in *Progress in Aeronautical Sciences*, Vol.9 (Pergamon, Oxford 1968) pp.41-122
2.10 L. Broer: J. Fluid Mech. **4**, 276-282 (1958)
2.11 L.E. Sternin: *Fundamentals of Gasdynamics of Two-Phase Nozzle Flows* (Mashinostroenie, Moscow 1974) (in Russian)
2.12 K.G. Guderley: *Theorie Schallnaher Strömungen* (Springer, Berlin, Göttingen, Heidelberg 1957) (in German)
K.G. Guderley: *The Theory of Transonic Flow* (Pergamon, Oxford 1962)
2.13 F.I. Frankl': Dokl. Akad. Nauk SSSR **58**, 381-384 (1947) (in Russian)
2.14 F.I. Frankl': Uchenye Zapiski Kabard.-Balk. Gos. Univ. **3**, 35-61 (1959) (in Russian)
2.15 O.S. Ryzhov: *Investigation of Transonic Flows in Laval Nozzles* (USSR Academy of Sciences, Computer Center, Moscow 1965) (in Russian)
2.16 U.G. Pirumov: Appl. Math. Mech. **36**, No. 2, 220-228 (1972)
2.17 J. Giese: J. Math. Phys. Cambridge, Mass. **30**, 31-35 (1951)
2.18 N.E. Kochin, I.A. Kibel', N.V. Rose: *Theoretical Hydrodynamics* (Fizmatgiz, Moscow 1963) (in Russian)
2.19 U.G. Pirumov: Fluid Dyn. **2**, 6-13 (1967)
2.20 A.M. Ovsyannikov: "Investigation of complex flows in radial nozzles for MGD-generators", in *Numerical Methods and Programming*, Vol.23 (Moscow University, Moscow 1974) (in Russian)
2.21 G.S. Roslyakov (ed.): *Tables of Gasdynamic Functions* (Moscow University, Moscow 1965) (in Russian)
2.22 A.M. Ovsyannikov: "Investigation of effect of swirl upon nozzle flows", in *Numerical Methods and Programming*, Vol.27 (Moscow University, Moscow 1977) (in Russian)

2.23 A. Bernstein, W.H. Heiser, C. Hevenor: J. Appl. Mech. **34**, 548-554 (1967)
2.24 K.G. Guderley, J.V. Armitage: Symp. on Extremal Problems in Aerodynamics (Boeing Scientific Research Laboratories, Flight Sciences Laboratory, Seattle, Washington 1962)
2.25 K.G. Guderley, E. Hantsch: Z. Flugwiss. **3**, 305-313 (1955)
2.26 A.A. Nikol'skii: "About ducted bodies of revolution with minimum wave drag in supersonic flow", in *Theoretical Works in Aerodynamics* (Oborongiz, Moscow 1957) (in Russian)
2.27 Yu.D. Shmyglevskii: *Some Variational Problems of Gas Dynamics* (USSR Academy of Science, Computer Center, Moscow 1963) (in Russian)
2.28 V.M. Borisov, A.V. Shipilin: Appl. Math. Mech. **28**, 220-222 (1964)
2.29 A.N. Kraiko: *Variational Problems of Supersonic Gas Flows with Arbitrary Thermodynamic Properties* (USSR Academy of Sciences, Computer Center, Moscow 1963) (in Russian)
2.30 A.N. Kraiko: Appl. Math. Mech. **28**, 348-360 (1964)
2.31 A.N. Kraiko, A.A. Osipov: Appl. Math. Mech. **32**, 617-626 (1968)
2.32 Yu.D. Shmyglevskii: Dokl. Acad. Nauk SSSR **122**, 782-784 (1958) (in Russian)
2.33 G.A. Bliss: *Lectures on the Calculus of Variations* (University of Chicago Press, Chicago 1946)
2.34 T.K. Sirazetdinov: Izv. Vuzov. Avia. Techn.,No.2, 11-21 (1963) (in Russian)
2.35 L.E. Sternin: Dokl. Akad. Nauk SSSR **139**, No.2, 335-336 (1961) (in Russian)

Chapter 3

3.1 P.I. Chushkin: "Numerical Method of Characteristics for Three-Dimensional Supersonic Flows", in *Progress in Aeronautical Sciences*, Vol.9 (Pergamon, Oxford 1968) pp.41-122
3.2 S.K. Godunov, V.S. Ryaben'kii: *Difference Schemes* (Nauka, Moscow 1973) (in Russian)
3.3 H.S. Kestenboim, G.S. Roslyakov, L.A. Chudov: *Point Explosion. Methods of Calculation. Tables* (Nauka, Moscow 1974) (in Russian)
3.4 R.D. Richtmyer, K.W. Morton: *Difference Methods for Initial-Value Problems*, 2nd ed. (Wiley, New York 1967)
3.5 B.L. Rozhdestvenskii, N.N. Yanenko: *Systems of Quasi-Linear Equations* (Nauka, Moscow 1968) (in Russian)
3.6 N.S. Galyun, A.N. Kraiko: Izv. Akad. Nauk SSSR. Mechanics and Machine-building **6**, 41-47 (1964) (in Russian)
3.7 V.N. Kamzolov, U.G. Pirumov: Fluid Dyn. **1**, 14-20 (1966)
3.8 V.F. Kolmogorov: Fluid Dyn. **9**, 275-280 (1974)
3.9 E.A. Ashratov, N.V. Dubinskaya: "Study of nozzle flows with vibrational relaxation", in *Numerical Methods and Programming*, Vol.27 (Moscow University, Moscow 1977) (in Russian)
3.10 V.N. Kamzolov, B.N. Maslov, U.G. Pirumov: Fluid Dyn. **6**, 849-856 (1971)
3.11 G.I. Averenkova, E.A. Ashratov, T.G. Volkonskaya, Yu.N. D'aykonov, N.I. Egorova, D.A. Mel'nikov, G.S. Roslyakov, V.I. Uskov: *Supersonic Ideal Gas Jets*, Part 1 (Moscow University, Moscow 1970) (in Russian)
3.12 E.A. Ashratov, T.G. Volkonskaya, G.S. Roslyakov, V.I. Uskov: "Investigation of supersonic gas flows in jets", in *Some Applications of Grid Methods in Gas Dynamic*, Vol.6 (Moscow University, Moscow 1974) (in Russian)

3.13 O.N. Katskova, I.N. Naumova, Yu.D. Shmyglevskii, N.P. Shulishnina: *Calculation of Plane and Axisymmetric Supersonic Gas Flows by Method of Characteristics* (USSR Academy of Sciences, Computer Center, Moscow 1961) (in Russian)
3.14 G.S. Roslyakov, L.A. Chudov: *Numerical Methods in Fluid Mechanics* (Moscow University, Computer Center, Moscow 1969) (in Russian)
3.15 G.S. Roslyakov, G.F. Telenin: "Review of numerical studies of external and internal problems of aerodynamics fulfilled in Moscow University", in *Numerical Methods and Programming*, Vol.11 (Moscow University, Moscow 1968) (in Russian)
3.16 O.M. Belotserkovskii, P.I. Chushkin: "The numerical solution of problems in gas dynamics", in *Basic Developments in Fluid Dynamics*, Vol.1 (Academic, New York 1965)
3.17 O.N. Katskova, Yu.D. Shmyglevskii: *Tables of Parameters of Axisymmetric Supersonic Freely Expanding Gas Flow with Plane Transition Surfaces* (USSR Academy of Sciences, Moscow 1962) (in Russian)
3.18 J.W. Chen: Commun. Pure Appl. Math. **6**, 179-229 (1953)
3.19 O.N. Katskova: *Calculation of Equilibrium Gas Flows in Supersonic Nozzles* (USSR Academy of Sciences, Computer Center, Moscow 1964) (in Russian)
3.20 I.N. Naumova: *Method of Characteristics for Equilibrium Flows of Non-Perfect Gas* (USSR Academy of Sciences, Computer Center, Moscow 1964)
3.21 U.G. Pirumov, V.A. Rubtsov, V.N. Suvorova: "Calculation of axisymmetric nozzles with equilibrium physico-chemical transformations", in *Numerical Methods in Gas Dynamics*, Vol.2 (Moscow University, Moscow 1963) (in Russian)
3.22 A.S. Predvoditelev, E.V. Stupochenko, E.V. Samuilov, I.P. Stakhanov, A.S. Pleshanov, I.B. Rozhdestvenskii: *Tables of Thermodynamic Functions for Air* (USSR Academy of Sciences, Moscow 1957-1961) (in Russian)
3.23 Yu.N. D'yakonov, L.V. Pchiolkina, I.D. Sandomirskaya: "Method of characteristics for calculating two-dimensional vortex flows of equilibrium and perfect gas", in *Numerical Methods and Programming*, Vol.7 (Moscow University, Moscow 1967) (in Russian)
3.24 O.N. Katskova, A.N. Kraiko: *Calculation of Plane and Axisymmetric Supersonic Flows with Irreversible Processes* (USSR Academy of Science, Computer Center, Moscow 1964) (in Russian)
3.25 L.E. Sternin: *Fundamentals of Gasdynamics of Two-Phase Nozzle Flows* (Mashinostroenie, Moscow 1974) (in Russian)
3.26 L.P. Vereshchaka, A.N. Kraiko, L.E. Sternin: *Method of Characteristics for Calculating Supersonic Dusted Gas Flows in Plane and Axisymmetric Nozzles* (USSR Academy of Sciences, Computer Center, Moscow 1969) (in Russian)
3.27 K.M. Magomedov, A.S. Kholodov: J. Comput. Math. Math. Phys. **9**, 373-386 (1969) (in Russian)
3.28 M.Ya. Ivanov, A.N. Kraiko, N.V. Mikhailov: J. Comput. Math. Math. Phys. **12**, 196-221, 309-320 (1972)
3.29 G.S. Roslyakov, V.P. Sukhorukov: "Difference method for calculating discontinuous gas flows", in *Numerical Methods and Programming*, Vol.19 (Moscow University, Moscow 1972) (in Russian)
3.30 N.V. Drozdova, U.G. Pirumov, G.S. Roslyakov, V.P. Sukhorukov: "Supersonic gas flows in conic nozzles", in *Some Applications of Grid Method in Gas Dynamics*, Vol.6 (Moscow University, Moscow 1974) (in Russian)
3.31 S.K. Godunov, A.V. Zabrodin, G.P. Prokopov: J. Comput. Math. Math. Phys. **1**, 1020-1050 (1961) (in Russian)
3.32 O.N. Katskova, P.I. Chushkin: J. Comput. Math. Math. Phys. **8**, 1049-1069 (1968) (in Russian)

3.33 V.H. Ransom, J.D. Hoffman, H.D. Thompson: AIAA J. **10**, 1573-1581 (1972)
3.34 D.S. Butler: Proc. Roy. Soc. (Ser. A) 267 (1962)
3.35 U.G. Pirumov: "Inverse problem of the theory of nozzles and numerical solution of internal problems of gas dynamics", in *Some Applications of Grid Method in Gas Dynamics*, Vol.6 (Moscow University, Moscow 1974) (in Russian)
3.36 A.N. Tikhonov: Dokl. Akad. Nauk SSSR **151**, 501-504 (1963) (in Russian)
3.37 A.M. Ovsyannikov: "Investigation of complex flows in radial nozzles for MGD-generators", in *Numerical Methods and Programming*, Vol.23 (Moscow University, Moscow 1974) (in Russian)
3.38 A.M. Ovsyannikov: "Investigation of effect of swirl upon nozzle flows", in *Numerical Methods and Programming*, Vol.27 (Moscow University, Moscow 1977) (in Russian)
3.39 A.M. Ovsyannikov: Fluid Dyn. **6**, 1028-1034 (1971)
3.40 D.J. Norton: J. Spacecr. Rockets **9**, 457-459 (1972)
3.41 Ya.I. Alikhashkin, A.P. Favorskii, P.I. Chushkin: J. Comput. Math. Math. Phys. **3**, 1130-1134 (1963) (in Russian)
3.42 Yu.D. D'yakonov, L.V. Pchiolkina: Dokl. Akad. Nauk SSSR **191**, 301-304 (1970) (in Russian)
3.43 M.Ya. Ivanov, A.N. Kraiko: Fluid Dyn. **4**, 49-53 (1969)
3.44 V.I. Kireev, Yu.B. Lifshitz, Yu.Ya. Mikhailov: Uchionye Zapiski ZAGI **1**, 8-13 (1970) (in Russian)
3.45 P. Laval: *Nonstationary Method of Calculating Transonic Nozzle Flows*, Proc. 2nd Intern. Conf. on Numerical Methods in Fluid Dynamics (Springer, Berlin, Heidelberg, New York 1971)
3.46 K.I. Babenko, G.P. Voskresenskii, A.N. Lyubimov, V.V. Rusanov: *Three-Dimensional Ideal Gas Flow Past Smooth Bodies* (Nauka, Moscow 1964) (in Russian)
3.47 U.G. Pirumov, V.N. Suvorova: "Direct problem of the theory of nozzles", in *Numerical Methods and Programming*, Vol.27 (Moscow University, Moscow 1977) (in Russian)

Chapter 4

4.1 N.E. Kochin, I.A. Kibel', N.V. Rose: *Theoretical Hydrodynamics* (Fizmatgiz, Moscow 1963) (in Russian)
4.2 R. Courant: *Partial Differential Equations* (New York 1962)
4.3 U.G. Pirumov: Fluid Dyn. **2**, 6-13 (1967)
4.4 Th. Meyer: Forschungsh. **62**, 31-67 (1908)
4.5 G.I. Taylor: Great Britain Aeronaut. Res. Committee Rep. Memor. No. 1381 (1930)
4.6 J.M. Hall: Quart. J. Mech. Appl. Math. **15**, Part 4, 487-508 (1962)
4.7 R. Sauer: Tech. Memor. NASA, 1147 (1947)
4.8 D.F. Hopkins, D.E. Hill: AIAA J. **4**, 1337-1343 (1966)
4.9 U.G. Pirumov: Appl. Math. Mech. **36**, No. 2, 220-228 (1972)
4.10 O.S. Ryzhov: *Investigation of Transonic Flows in Laval Nozzles* (USSR Academy of Sciences, Computer Center, Moscow 1965) (in Russian)
4.11 V.N. Kamzolov, U.G. Pirumov: J. Appl. Mach. Tech. Phys. **2**, 117-122 (1967) (in Russian)
4.12 K.G. Guderley: *Theorie Schallnaher Strömungen* (Springer, Berlin, Göttingen, Heidelberg 1957) (in German)
4.13 F.I. Frankl': Izv. Akad. Nauk SSSR. Mathematics **9**, 387-422 (1945) (in Russian)
4.14 G.H. Görtler: Z. Angew. Math. Mech. **19**, 325-337 (1939)
4.15 Th. von Kārmān: J. Math. Phys. **26**, 182-190 (1947)

4.16 L.V. Ovsyannikov: Trudy Leningrad Air Force Engineering Academy **33**, 3-24 (1950) (in Russian)
4.17 O.N. Katskova, Yu.D. Shmyglevskii: *Tables of Parameters of Axisymmetric Supersonic Freely Expanding Gas Flow with Plane Transition Surfaces* (USSR Academy of Sciences, Moscow 1962) (in Russian)
4.18 U.G. Pirumov: "Inverse problem of the theory of nozzles and numerical solution of internal problems of gas dynamics", in *Some Applications of Grid Method in Gas Dynamics*, Vol.6 (Moscow University, Moscow 1974) (in Russian)
4.19 W.J. Rae: AIAA J. **9**, 811-820 (1971)
4.20 R. Mises: *Mathematical Theory of Compressible Fluid Flow* (Academic, New York 1958)
4.21 I.S. Zhiguleva, U.G. Pirumov: *Investigation of Small Disturbances Propagation in Supersonic Conic Nozzles* (Oborongiz, Moscow 1959) (in Russian)
4.22 L. Crocco: Z. Angew. Math. Mech. **17**, 1-7 (1937)
4.23 G.A. Korn, T.M. Korn.: *Mathematical Handbook*, 2nd ed. (McGraw-Hill, New York 1968)

Chapter 5

5.1 V.P. Glushko: *Development of Rocketry and Space Technology in the USSR* (Novosti, Moscow 1973)
5.2 U.G. Pirumov: "Inverse problem of the theory of nozzles and numerical solution of internal problems of gas dynamics", in *Some Applications of Grid Method in Gas Dynamics*, Vol.6 (Moscow University, Moscow 1974) (in Russian)
5.3 C.W. Witoszynsky: "Über Strahlerweiterung und Strahlablenkung", in *Vorträge aus dem Gebiete der Hydro- und Aerodynamik* (Springer, Berlin 1924)
5.4 P. Laval: *Nonstationary Method of Calculating Transonic Nozzle Flows*, Proc. 2nd Intern. Conf. on Numerical Methods in Fluid Dynamics (Springer, Berlin, Heidelberg, New York 1971)
5.5 V.I. Kireev, Yu.B. Lifshitz: Fluid Dyn. **5**, 947-950 (1970)
5.6 D.A. Mel'nikov, U.G. Pirumov, A.A. Sergienko: "Nozzles of jet engines", in *Aerodynamics and Gas Dynamics* (Nauka, Moscow 1976) (in Russian)
5.7 L.H. Back, P.F. Massier, H.L. Gier: AIAA J. **3**, 1606-1614 (1965)
5.8 V.N. Kamzolov, U.G. Pirumov: J. Appl. Mach. Tech. Phys. 2, 117-122 (1967) (in Russian)
5.9 U.G. Pirumov, V.N. Suvorova: "Direct problem of the theory of nozzles", in *Numerical Methods and Programming*, Vol.27 (Moscow University, Moscow 1977) (in Russian)
5.10 F.I. Frankl': Dokl. Akad. Nauk SSSR **58**, 381-384 (1947) (in Russian)
5.11 D.F. Ferguson, M.J. Lighthill: Proc. Roy. Soc. **192**, 135-142 (1947)
5.12 D.E. Rothe: AIAA J. **9**, 804-811 (1971)
5.13 L.V. Ovsyannikov: Appl. Math. Mech. **13**, 537-542 (1949) (in Russian)
5.14 Yu.D. Shmyglevskii: Dokl. Acad. Nauk SSSR **122**, 782-784 (1958) (in Russian)
5.15 O.N. Katskova, Yu.D. Shmyglevskii: *Tables of Parameters of Axisymmetric Supersonic Freely Expanding Gas Flow with Plane Transition Surfaces* (USSR Academy of Sciences, Moscow 1962) (in Russian)
5.16 L.E. Sternin: Dokl. Akad. Nauk SSSR **139**, No. 2, 335-336 (1961) (in Russian)
5.17 Yu.D. Shmyglevskii: *Some Variational Problems of Gas Dynamics* (USSR Academy of Science, Computer Center, Moscow 1963) (in Russian)
5.18 K.G. Guderley, J.V. Armitage: Symp. on Extremal Problems in Aerodynamics (Boeing Scientific Research Laboratories, Flight Sciences Laboratory, Seattle, Washington 1962)

5.19 T.K. Sirazetdinov: Izv. Vuzov. Avia. Techn., No.2, 11-21 (1963) (in Russian)
5.20 A.N. Kraiko: Appl. Math. Mech. **28**, 348-360 (1964)
5.21 A.N. Kraiko, A.A. Osipov: Appl. Math. Mech. **32**, 617-626 (1968)
5.22 E.C. Guentert, H.E. Neumann: NASA Tr. R-53 (1958)
5.23 G. Rao: Jet Propul. **28**, 377-382 (1958)
5.24 L.V. Gogish: Fluid Dyn. **1**, 122-126 (1966)
5.25 U.G. Pirumov: "Particularities of one-phase nozzle flows", in *Thermodynamic and Thermophysical Properties of Combustion Products*, Vol.1 (Kerer Press, Jerusalem 1974)
5.26 L.I. Sedov, G.G. Chernyi: Theor. Hydromech. **4**, 17-30 (1954) (in Russian)
5.27 A.A. Sergienko, V.K. Gretsov: Dokl. Akad. Nauk SSSR **125**, No. 4, 746-747 (1959) (in Russian)
5.28 V.S. Avduevskii: Izv. Akad. Nauk SSSR. Mechanics and Machinebuilding No. 4, 3-12 (1962) (in Russian)
5.29 V.S. Avduevskii, R.M. Kopyatkevich: Izv. Akad. Nauk SSSR. Mechanics and Machinebuilding No. 1, 3-11 (1960) (in Russian)
5.30 W.J. Rae: AIAA J. **9**, 811-820 (1971)
5.31 E.N. Bondarev, G.I. Petrov: *Experimental Study of Interaction of Turbulent Boundary Layer with Shocks*, Proc. All-Union Congress on Theoretical and Applied Mechanics (USSR Academy of Sciences, Moscow 1960) (in Russian)
5.32 G.I. Petrov: "Determination of the position of the supersonic flow closing shock wave in channels of an air-breathing jet engine", in *2nd Intern. Conf. on Space Engineering* (Reidel, Amsterdam 1969)
5.33 A.N. Timoshin: J. Engineering Physics **18**, 167-169 (1972)

Chapter 6

6.1 V.E. Alemasov, A.F. Dregalin, A.P. Tishin, V.A. Khudyakov: *Thermodynamic and Thermophysical Properties of Combustion Products*, Vol.1, ed. by V.P. Glushko (Kerer Press, Jerusalem 1974)
6.2 Ya.B. Zel'dovich: J. Phys. Chem. **11**, 685-687 (1938) (in Russian)
6.3 A.P. Vanichev: *Thermodynamic Calculation of Combustion and Outflow at High Temperatures* (BNT, Moscow 1947) (in Russian)
6.4 M.P. Vukalovich, I.I. Novikov: *Equations of State of Real Gases* (Gasenergoizdat, Moscow 1948) (in Russian)
6.5 N.N. Bogolyubov: *Dynamic Theory Problems in Statistical Physics* (Gostekhizdat, Moscow 1946) (in Russian)
6.6 J.E. Mayer, M. Göppert-Mayer: *Statistical Mechanics*, 2nd ed. (Wiley, New York 1977)
6.7 L.M. Davydov, V.M. Zhiravov: Fluid Dyn. **9**, 827-831 (1974)
6.8 A.A. Vasserman, Ya.Z. Kazavchinskii, V.A. Rabinovich: *Thermophysical Properties of Air and Its Components* (Nauka, Moscow 1966) (in Russian)
6.9 V.P. Stulov, L.I. Turchak: Trans. NII Mech. MGU **5**, 35-49 (1970) (in Russian)
6.10 V.N. Kondrat'ev: *Rate Constants of Gas-Phase Reactions* (Nauka, Moscow 1970) (in Russian)
6.11 V.M. Glushko (ed.): *Thermodynamic Properties of Individual Substances* (USSR Academy of Sciences, Moscow 1962) (in Russian)
6.12 N.S. Galyun, A.N. Kraiko: Izv. Akad. Nauk SSSR. Mechanics and Machinebuilding **6**, 41-47 (1964) (in Russian)
6.13 V.N. Kamzolov, U.G. Pirumov: Fluid Dyn. **1**, 14-20 (1966)
6.14 U.G. Pirumov: "Inverse problem of the theory of nozzles and numerical solution of internal problems of gas dynamics", in *Some Applications of Grid Method in Gas Dynamics*, Vol.6 (Moscow University, Moscow 1974) (in Russian)

6.15 V.F. Kolmogorov: Fluid Dyn. **9**, 275-280 (1974)
6.16 V.M. Hailov: *Chemical Relaxation in Rocket Nozzles* (Mashinostroenie, Moscow 1975) (in Russian)
6.17 W.T. Olson: ARS J. **32**, 672-680 (1962)
6.18 F.K. Hill, H.J. Unger, W.P. Dickens: AIAA J. **5**, 873-881 (1967)
6.19 U.G. Pirumov: "Particularities of one-phase nozzle flows", in *Thermodynamic and Thermophysical Properties of Combustion Products*, Vol.1 (Kerer Press, Jerusalem 1974)
6.20 K.N.C. Bray: J. Fluid Mech. **6**, 1-32 (1959)
6.21 L.C. Franciscus, E.A. Legberg: AIAA J. **1**, 2077-2083 (1963)
6.22 E.A. Ashratov, N.V. Dubinskaya: "Study of nozzle flows with vibrational relaxation", in *Numerical Methods and Programming*, Vol.27 (Moscow University, Moscow 1977) (in Russian)
6.23 E.V. Stupochenko, S.A. Losev, A.I. Osipov: *Relaxation in Shock Waves*, Applied Physics and Engineering, Vol.1 (Springer, Berlin, Heidelberg, New York 1967)
6.24 L.D. Landau, E.K. Teller: "To the theory of sound dispersion", in *Collected Works*, ed. by L.D. Landau (Nauka, Moscow 1969) (in Russian)
6.25 A. Babiano, P. Chevalier: C.R. Acad. Sci. Paris **276**, 953-956 (1973)
6.26 V.V. Breev, S.N. Minin, U.G. Pirumov, V.P. Shevchenko: Fluid Dyn. **5**, 125-131 (1977)
6.27 V.I. Kireev, S.N. Minin, U.G. Pirumov: Fluid Dyn. **6**, 163-167 (1982)
6.28 V.A. Levin, Yu.V. Tunik: Fluid Dyn. **1**, 118-125 (1976)
6.29 S.A. Losev: *Gasdynamic Laser*, Springer Ser. Chem. Phys., Vol.12 (Springer, Berlin, Heidelberg 1981)
6.30 N.A. Fux: *Aerosol Mechanics* (Academy of Sciences, USSR, Moscow 1955) (in Russian)
6.31 S.R. de Groot, P. Mazur: *Non-Equilibrium Thermodynamics* (North-Holland, Amsterdam 1962)
6.32 A.P. Tishin: "Particularity of two-phase nozzle flows", in *Thermodynamic and Thermophysical Properties of Combustion Products*, Vol.1 (Kerer Press, Jerusalem 1974)
6.33 K.L. Goin, W.R. Lawrence: AIAA J. **6**, 961-962 (1968)
6.34 L.L. Kavanau: Trans. ASME **77**, 617-623 (1955)
6.35 N.B. Pashatskii, N.I. Syromyatnikov: Fluid Dyn. **5**, 682-683 (1970)
6.36 J.R. Kliegel: 9th Intern. Symp. Comb., Cornell Univ., pp.811-826 (Academic, New York 1963)
6.37 V.K. Starkov, L.E. Sternin, A.P. Tishin, V.A. Khudyakov: Fluid Dyn. **8**, 384-390 (1973)
6.38 L.E. Sternin, B.M. Maslov, A.A. Shraiber, A.M. Podvysotskii: *Two-Phase Mono- and Polydisperse Dusted Gas Flows* (Mashinostroenie, Moscow 1980)
6.39 A.P. Tishin, R.I. Khairutdinov: Fluid Dyn. **6**, 898-902 (1971)
6.40 M.S. Volynskii: Dokl. Akad. Nauk SSSR **18**, 237 (1949) (in Russian)
6.41 P. Wegener, L.M. Mack: "Condensation in supersonic wind tunnels", in *Advances in Applied Mechanics*, ed. by H.L. Dryden, Th. von Kārmān (Academic, New York 1958)
6.42 W. Willmarth, H. Nagamatsu: Appl. Phys. **23**, 1089-1095 (1952)
6.43 A.V. Chirikhin: Fluid Dyn. **1**, 137-145 (1977)
6.44 S. Kotake, I.I. Glass: "Flow with nucleation and condensation", in *Progress in Aerospace Science* (Pergamon, Oxford 1981)
6.45 Ya.B. Zel'dovich: J. Exp. Theor. Phys. **12**, 525-538 (1942) (in Russian)
6.46 Ya.I. Frenkel': *Kinetic Theory of Liquids* (USSR Academy of Sciences, Moscow 1945) (in Russian)
6.47 G. Gyarmathy, H. Meyer: VDJ-Forschungsh. N**508**, 5-30 (1965)
6.48 L.M. Davydov: Fluid Dyn. **6**, 419-424 (1971)
6.49 L.M. Davydov, U.G. Pirumov: Fluid Dyn. **1**, 173-175 (1979)

6.50 V.P. Bakhanov, M.V. Buikov: Trudy Ukr. NIGMI, 21-24 (Gidrometeoizdat, Moscow 1971) (in Russian)
6.51 V.N. Gorbunov, A.L. Itkin, U.G. Pirumov, Yu.A. Ryzhov: "Homogeneous nonequilibrium condensation of gases", in *12th Intern. Symp. on Rarefield Gas Dynamics* (University of Virginia, 1980)
6.52 A.L. Itkin, U.G. Pirumov, Yu.A. Ryzhov: "Studies of nonequilibrium homogeneous condensation", in *13th Intern. Symp. on Rarefield Gas Dynamics* (Novosibirsk 1982)
6.53 A.L. Itkin, U.G. Pirumov, Yu.A. Ryzhov: "Studies of nonequilibrium water condensation in rarefaction waves taking into account realistic properties", in *High-Temperature Gasdynamics, Shock Tubes and Shock Waves* (Institute of Heat-Mass Transfer, Minsk 1983)
6.54 K. Oswatitsch: Z. Angew. Math. Mech. **22**, 1-14 (1942)
6.55 L.M. Davydov, U.G. Pirumov: Fluid Dyn. **6**, 81-88 (1978)
6.56 V.N. Gorbunov, U.G. Pirumov, Yu.A. Ryzhov: *Nonequilibrium Condensation in Hypervelocity Gas Flows* (Mashinostroenie, Moscow 1984) (in Russian)
6.57 L.E. Sternin: *Fundamentals of Gasdynamics of Two-Phase Nozzle Flows* (Mashinostroenie, Moscow 1974) (in Russian)
6.58 L.P. Vereshchaka, A.N. Kraiko, L.E. Sternin: *Method of Characteristics for Calculating Supersonic Dusted Gas Flows in Plane and Axisymmetric Nozzles* (USSR Academy of Sciences, Computer Center, Moscow 1969) (in Russian)
6.59 G.V. Dritov, A.P. Tishin: Fluid Dyn. **4**, 21-24 (1969)
6.60 G.I. Averenkova, N.I. Egorova, U.G. Pirumov: "Motion of a polydisperse mixture of gas and particles in a supersonic nozzle", in *Numerical Methods and Programming*, Vol.30 (Moscow University, Moscow 1979) (in Russian)
6.61 U.G. Pirumov, V.N. Suvorova: "Calculation of two-phase gas flows in nozzles by means of the inverse method", in *Numerical Methods in Aerodynamics*, Vol.5 (Moscow University, Moscow 1980) (in Russian)
6.62 V.N. Kamzolov, B.N. Maslov, U.G. Pirumov: Fluid Dyn. **6**, 849-856 (1971)
6.63 W.S. Bailey, E.N. Nilson, R.A. Serra, T.F. Zupnik: ARS J. **31**, 793-799 (1961)
6.64 G.V. Dritov, A.P. Tishin: Fluid Dyn. **6**, 154-159 (1971)
6.65 R.A. Tkalenko: Fluid Dyn. **7**, 1009-1012 (1972)
6.66 A.V. Kurshakov, G.A. Saltanov, G.A. Tkalenko: J. Appl. Mach. Tech. Phys. **5**, 732-735 (1971) (in Russian)
6.67 A.G. Kulikovskii, G.A. Lyubimov: *Magnetic Hydrodynamics* (Addison-Wesley, MA 1965)
6.68 O.N. Katskova, P.I. Chushkin: Magnitnaya Hydrodynam. **4**, 11-18 (1966) (in Russian)
6.69 N.P. Isakova: Fluid Dyn. **6**, 730-735 (1971)
6.70 V.V. Breev, U.G. Pirumov, V.P. Shevchenko: "Complex axisymmetric nozzle conductive gas flow in the presence of a magnetic field", in *Numerical Methods and Programming*, Vol.27 (Moscow University, Moscow 1977) (in Russian)
6.71 A. Bernstein, W.H. Heiser, C. Hevenor: J. Appl. Mech. **34**, 548-554 (1967)
6.72 V.L. Zimont: Izv. Vuzov. Aviation Technique No. 2, 104-108 (1970)
6.73 U.G. Pirumov: Fluid Dyn. **5**, No. 4, 593 (1970)

Chapter 7

7.1 F.A. Vilenskii, T.G. Volkonskaya, V.P. Gryaznov, U.G. Pirumov: Fluid Dyn. **7**, 621-627 (1972)
7.2 E.G. Mezhibovskaya, U.G. Pirumov, V.A. Rubtsov, E.V. Sorokina: *Calculation of Annular Axisymmetric Nozzles* (Moscow University, Moscow 1961) (in Russian)

7.3 U.G. Pirumov, V.A. Rubtsov: Izv. Akad. Nauk SSSR. Mechanics and Machine-building **6**, 15-25 (1961) (in Russian)
7.4 A.M. Ovsyannikov: Fluid Dyn. **6**, 1028-1034 (1971)
7.5 W.P. Sule, T.J. Muller: AIAA Paper No. 73-137, AIAA 11th Aerospace Sciences Meeting (1973)
7.6 L.V. Pchiolkina, V.K. Solodkin: "Correction of boundary layer effect upon flows in nozzles with cusped generatrix", in *Numerical Methods in Gas Dynamics*, Vol.4 (Moscow University, Moscow 1965) (in Russian)
7.7 J.C. Sivells: J. Spacecr. Rockets **7**, 1292-1299 (1970)
7.8 A.M. Ovsyannikov, U.G. Pirumov: Fluid Dyn. **10**, 57-60 (1975)
7.9 N.V. Drozdova, U.G. Pirumov, J.S. Roslaykov, V.P. Sukhorukov: "Supersonic gas flow in conic nozzles", in *Some Applications of Grid Method in Gas Dynamics*, Vol.6 (Moscow University, Moscow 1974) (in Russian)
7.10 I.S. Zhiguleva, U.G. Pirumov: *Investigation of Small Disturbances Propagation in Supersonic Conic Nozzles* (Oborongiz, Moscow 1959) (in Russian)
7.11 L.H. Back, R.F. Cuffel: AIAA J. **4**, 2219-2221 (1966)
7.12 U.G. Pirumov, G.S. Roslyakov, V.P. Sukhorukov: Fluid Dyn. **9**, 415-420 (1974)
7.13 D. Migdal: Grumman Aircraft Eng. Corp. ADR-01-10-65 **1** (1965)
7.14 R. Mises: *Mathematical Theory of Compressible Fluid Flow* (Academic, New York 1958)
7.15 I.N. Naumova, Yu.D. Shmyglevskii: Fluid Dyn. **2**, 23-25 (1967)
7.16 A. Mager: ARS J. **31**, 1140-1148 (1961)
7.17 L.E. Fraenkel: Proc. Roy. Soc. (Ser. A) **233**, 506-526 (1956)
7.18 R. Dunlap: AIAA J. **7**, 2293-2300 (1969)
7.19 W.S. King: "A Theoretical Investigation of Swirling Flow Through a Nozzle"; Ph. D. Thesis, UCLA (1967)
7.20 G.G. Chernyi: Izv. Akad. Nauk Otd. Techn. Nauk **6**, 55-62 (1956) (in Russian)
7.21 Yu.A. Gostintsev: Fluid Dyn. **4**, 107-110 (1969)
7.22 A.M. Ovsyannikov: "Investigation of effect of swirl upon nozzle flows", in *Numerical Methods and Programming*, Vol.27 (Moscow University, Moscow 1977) (in Russian)
7.23 A.D. Rychkov: Fluid Dyn. **6**, 786-790 (1971)
7.24 N.N. Slavyanov: "Investigation of Swirling Compressible Gas Flows in Nozzles"; Ph. D. Thesis (Moscow 1974) (in Russian)
7.25 I.S. Gromeka: *Collected Works* (USSR Academy of Science, Moscow 1952) (in Russian)
7.26 U.G. Pirumov: Fluid Dyn. **2**, 6-13 (1967)
7.27 O.S. Ryzhov: *Investigation of Transonic Flows in Laval Nozzles* (USSR Academy of Sciences, Computer Center, Moscow 1965) (in Russian)
7.28 U.G. Pirumov: Fluid Dyn. **12**, 121-127 (1977)
7.29 V.H. Ransom, J.D. Hoffman, H.D. Thompson: AIAA J. **10**, 1573-1581 (1972)
7.30 O.N. Katskova, P.I. Chushkin: J. Comput. Math. Math. Phys. **8**, 1049-1069 (1968) (in Russian)
7.31 J.D. Hoffman, A.R. Maykut: J. Spacecr. Rockets **11**, 697-704 (1974)
7.32 V.M. Dvoretskii, M.Ya. Ivanov: Uchenye Zapiski ZAGI 5, 39-45 (1974) (in Russian)
7.33 M.Ya. Ivanov, O.A. Ryl'ko: Fluid Dyn. **7**, 507-509 (1972)
7.34 M.Ya. Ivanov, A.N. Kraiko, N.V. Mikhailov: J. Comput. Math. Math. Phys. **12**, 196-221, 309-320 (1972)
7.35 G.Z. Nikulin, U.G. Pirumov, L.V. Pchiolkina, G.S. Roslyakov: "Three-dimensional nozzle gas flows", in *Computation Methods and Programming*, Vol.30 (Moscow University, Moscow 1979) (in Russian)
7.36 G.Z. Nikulin, U.G. Pirumov, L.V. Pchiolkina, G.S. Roslyakov: Fluid Dyn. **15**, 231-244 (1980)

7.37 U.G. Pirumov: Appl. Math. Mech. **36**, No. 2, 220-228 (1972)
7.38 L.V. Bobovich, V.P. Kornilovich, B.N. Maslov, A.A. Shishkov: Fluid Dyn. **12**, 266-271 (1977)
7.39 V.M. Dvoretskii, M.Ya. Ivanov, B.A. Konyaev, A.N. Kraiko: Appl. Math. Mech. **38**, 951-961 (1974)
7.40 Th. von Kārmān: J. Math. Phys. **26**, 182-190 (1947)
7.41 H.M. Darwell, G.F.P. Trubridge: J. Spacecr. Rockets **5**, 30-41 (1968)
7.42 A.G. Walters: Phil. Trans. Roy. Soc. London A**273**, No. 1232, 185-235 (1972)
7.43 A.A. Dorodnitsyn: Usp. Math. Nauk **7**, 3-96 (1952) (in Russian)
7.44 G.A. Korn, T.M. Korn: *Mathematical Handbook*, 2nd ed. (McGraw-Hill, New York 1968)
7.45 G. Rao: Jet Propul. **28**, 377-382 (1958)
7.46 D.A. Mel'nikov, U.G. Pirumov, A.A. Sergienko: "Nozzles of jet engines", in *Aerodynamics and Gas Dynamics* (Nauka, Moscow 1976) (in Russian)
7.47 A.N. Antonov: Fluid Dyn. **6**, 391-397 (1971)
7.48 W.J. Rae: AIAA J. **9**, 811-820 (1971)
7.49 D.E. Rothe: AIAA J. **9**, 804-811 (1971)
7.50 G.A. Evseev: Izv. Akad. Nauk USSR. Mechanics No. 3, 165-172 (1965) (in Russian)
7.51 J.C. Williams III: AIAA J. **1**, 186-195 (1963)
7.52 G.N. Patterson: *Molecular Flow of Gases* (Wiley, New York 1956)
7.53 M.J. Lighthill: Proc. Roy. Soc. (Ser. A) **217**, 478-507 (1953)
7.54 P.F. Massier, L.H. Back, M.B. Noel, F. Sahel: AIAA J. **8**, 605-607 (1970)
7.55 V.Ya. Levin, U.G. Pirumov, O.I. Firsov, S.A. Shustov: Fluid Dyn. **3**, 90-97 (1980)
7.56 A.T. Fedorchenko: Dokl. Akad. Nauk SSSR **252**, 578-582 (1980) (in Russian)
7.57 L.V. Kuznetsova, B.M. Pavlov: "To the question of calculating viscous gas flows in Laval's nozzles", in *Numerical Methods and Programming*, Vol.27 (Moscow University, Moscow 1977) (in Russian)
7.58 A.A. Fedosov: "Investigation of Vibrationally Nonequilibrium Gas Flows in Laval's Nozzles", Ph. D. Thesis (Kazan 1982) (in Russian)
7.59 N.D. Kuvshinnikov: "Investigation and Calculation of Viscous Gas Flows in Laval's Nozzle", Ph. D. Thesis (Moscow 1983) (in Russian)

Subject Index

Topics in Applied Physics

Founded by **H.K.V. Lotsch**

Volume 45

Hydrodynamic Instabilities and the Transition to Turbulence

Editors: **H. L. Swinney, J. P. Gollub**
2nd edition. 1985. 82 figures. XIV, 306 pages
ISBN 3-540-13319-4

Contents: *H. L. Swinney, J. P. Gollub:* Introduction. – *O. E. Lanford:* Strange Attractors and Turbulence. – *D. D. Joseph:* Hydrodynamic Stability and Bifurcation. – *J. A. Yorke, E. D. Yorke:* Chaotic Behavior and Fluid Dynamics. – *F. H. Busse:* Transition to Turbulence in Rayleigh-Bénard Convection. – *R. C. DiPrima, H. L. Swinney:* Instabilities and Transition in Flow between Concentric Rotating Cylinders. – *S. A. Maslowe:* Shear Flow Instabilities and Transition. – *D. J. Tritton, P. A. Davies:* Instabilities in Geophysical Fluid Dynamics. – *J. M. Guckenheimer:* Instabilities and Chaos in Nonhydrodynamic Systems. – *F. H. Busse, J. P. Gollub, S. A. Maslowe, H. L. Swinney:* Recent Progress. – Subject Index.

Volume 44

Turbulent Reacting Flows

Editors: **P. A. Libby, F. A. Williams**
1980. 38 figures, 3 tables. XI, 243 pages. ISBN 3-540-10192-6

Contents: *P. A. Libby, F. A. Williams:* Fundamental Aspects. – *A. M. Mellor, C. R. Ferguson:* Practical Problems in Turbulent Reacting Flows. – *R. W. Bilger:* Turbulent Flows with Nonpremixed Reactants. – *K. N. C. Bray:* Turbulent Flows with Premixed Reactants. – *E. E. O'Brien:* The Probability Density Function (pdf) Approach to Reacting Turbulent Flows. – *P. A. Libby, F. A. Williams:* Perspective and Research Topics.

Volume 12

Turbulence

Editor: **P. Bradshaw**
2nd corrected and updated edition. 1978. 47 figures, 4 tables. XI, 339 pages. ISBN 3-540-08864-4

Contents: *P. Bradshaw:* Introduction. – *H.-H. Fernholz:* External Flows. – *J. P. Johnston:* Internal Flows. – *P. Bradshaw, J. D. Woods:* Geophysical Turbulence and Buoyant Flows. – *T. Cebeci, W. C. Reynolds:* Calculation of Turbulent Flows. – *B. E. Launder:* Heat and Mass Transport. – *J. L. Lumley:* Two-Phase and Non-Newtonian Flows.

Springer-Verlag
Berlin Heidelberg
New York Tokyo

Springer Series in Chemical Physics

Editors: V. I. Goldanskii R. Gomer F. P. Schäfer J. P. Toennies

1 **Atomic Spectra and Radiative Transitions** By I. I. Sobelman
2 **Surface Crystallography by LEED** Theory, Computation and Structural Results. By M. A. Van Hove, S. Y. Tong
3 **Advances in Laser Chemistry** Editor: A. H. Zewail
4 **Picosecond Phenomena** Editors: C. V. Shank, E. P. Ippen, S. L. Shapiro
5 **Laser Spectroscopy** Basic Concepts and Instrumentation By W. Demtröder 2nd Printing
6 **Laser-Induced Processes in Molecules** Physics and Chemistry Editors: K. L. Kompa, S. D. Smith
7 **Excitation of Atoms and Broadening of Spectral Lines** By I. I. Sobelman, L. A. Vainshtein, E. A. Yukov
8 **Spin Exchange** Principles and Applications in Chemistry and Biology By Yu. N. Molin, K. M. Salikhov, K. I. Zamaraev
9 **Secondary Ion Mass Spectrometry SIMS II** Editors: A. Benninghoven, C. A. Evans, Jr., R. A. Powell, R. Shimizu, H. A. Storms
10 **Lasers and Chemical Change** By A. Ben-Shaul, Y. Haas, K. L. Kompa, R. D. Levine
11 **Liquid Crystals of One- and Two-Dimensional Order** Editors: W. Helfrich, G. Heppke
12 **Gasdynamic Laser** By S. A. Losev
13 **Atomic Many-Body Theory** By I. Lindgren, J. Morrison
14 **Picosecond Phenomena II** Editors: R. M. Hochstrasser, W. Kaiser, C. V. Shank
15 **Vibrational Spectroscopy of Adsorbates** Editor: R. F. Willis
16 **Spectroscopy of Molecular Excitions** By V. L. Broude, E. I. Rashba, E. F. Sheka
17 **Inelastic Particle-Surface Collisions** Editors: E. Taglauer, W. Heiland
18 **Modelling of Chemical Reaction Systems** Editors: K. H. Ebert, P. Deuflhard, W. Jäger
19 **Secondary Ion Mass Spectrometry SIMS III** Editors: A. Benninghoven, J. Giber, J. László, M. Riedel, H. W. Werner
20 **Chemistry and Physics of Solid Surfaces IV** Editors: R. Vanselow, R. Howe
21 **Dynamics of Gas-Surface Interaction** Editors: G. Benedek, U. Valbusa
22 **Nonlinear Laser Chemistry** Multiple-Photon Excitation By V. S. Letokhov
23 **Picosecond Phenomena III** Editors: K. B. Eisenthal, R. M. Hochstrasser, W. Kaiser, A. Laubereau
24 **Desorption Induced by Electronic Transitions DIET I** Editors: N. H. Tolk, M. M. Traum, J. C. Tully, T. E. Madey
25 **Ion Formation from Organic Solids** Editor: A. Benninghoven
26 **Semiclassical Theories of Molecular Scattering** By B. C. Eu
27 **EXAFS and Near Edge Structures** Editors: A. Bianconi, L. Incoccia, S. Stipcich
28 **Atoms in Strong Light Fields** By N. B. Delone, V. P. Krainov
29 **Gas Flow in Nozzles** By U. Pirumov, G. Roslyakov
30 **Theory of Slow Atomic Collisions** By E. E. Nikitin, S. Ya. Umanskii
31 **Reference Data on Atoms, Molecules, and Ions** By A. A. Radzig, B. M. Smirnov
32 **Adsorption Processes on Semiconductor and Dielectric Surfaces I** By V. F. Kiselev, O. V. Krylov
33 **Surface Studies with Lasers** Editors: F.R. Aussenegg, A. Leitner, M.E. Lippitsch
34 **Inert Gases** Potentials, Dynamics, and Energy Transfer in Doped Crystals. Editor: M. L. Klein
35 **Chemistry and Physics of Solid Surfaces V** Editors: R. Vanselow, R. Howe
36 **Secondary Ion Mass Spectrometry, SIMS IV** Editors: A. Benninghoven, J. Okano, R. Shimizu, H. W. Werner
37 **X-Ray Spectra and Chemical Binding** By A. Meisel, G. Leonhardt, R. Szargan
38 **Ultrafast Phenomena IV** By D. H. Auston, K. B. Eisenthal
39 **Laser Processing and Diagnostics** Editor: D. Bäuerle

MIX
Papier aus verantwortungsvollen Quellen
Paper from responsible sources
FSC® C105338

If you have any concerns about our products,
you can contact us on
ProductSafety@springernature.com

In case Publisher is established outside the EU,
the EU authorized representative is:
Springer Nature Customer Service Center GmbH
Europaplatz 3, 69115 Heidelberg, Germany

Printed by Libri Plureos GmbH
in Hamburg, Germany